Advanced Strategies for Biodegradation of Plastic Polymers

Ravindra Soni • Prasenjit Debbarma
Deep Chandra Suyal • Reeta Goel
Editors

Advanced Strategies for Biodegradation of Plastic Polymers

Springer

Editors
Ravindra Soni
Department of Agricultural Microbiology
Indira Gandhi Krishi Vishwa Vidyalaya
Raipur, Chhattisgarh, India

Prasenjit Debbarma
Department of Botany
Iswar Chandra Vidyasagar College
Belonia, Tripura, India

Deep Chandra Suyal
Department of Science
Vidyadayini Institute of Science,
Management, and Technology
Bhopal, Madhya Pradesh, India

Reeta Goel
Department of Biotechnology
GLA University
Chaumuhan, Uttar Pradesh, India

ISBN 978-3-031-55660-9 ISBN 978-3-031-55661-6 (eBook)
https://doi.org/10.1007/978-3-031-55661-6

This Springer imprint is published by the registered company Springer Nature Switzerland AG
The registered company address is: Gewerbestrasse 11, 6330 Cham, Switzerland

Paper in this product is recyclable.

Contents

Chapter 1
Global Scenario of Plastic Production, Consumption, and Waste Generation and Their Impacts on Environment and Human Health

Bhanu Pratap Singh Gautam, Anjum Qureshi, Akanksha Gwasikoti, Vinod Kumar, and Manjul Gondwal

Abstract This review employs country-level statistics pertaining to trash management. The contemporary global landscape is currently grappling with the complexities surrounding the effective management and remediation of the escalating volumes of plastic trash. As the population grows, there is a corresponding growth in the need for plastic and plastic items. These products are accountable for the generation of plastic, the accumulation of plastic trash, and the consequent environmental degradation. The assortment of plastic materials utilized in the manufacturing process of diverse important commodities, including medical gadgets, food packaging, and water bottles, is known to harbor detrimental compounds such as bisphenol, phthalates, and phenanthrene. Despite the various benefits of plastic, the global plastic industry is widely recognized as the source of significant environmental issues. The plastic tool presents a potential source of ambiguity regarding human exposure to hazardous substances. This assessment examines the present state of plastic waste management, focusing on key aspects such as landfilling, recycling, and incineration. The objective of this review is to examine the environmental and human health implications of plastic. In addition, we put out a potential resolution for effectively tracking the escalation of this burgeoning catastrophe.

Keywords Plastic pollution · Environmental consequences · Human health · Waste management · Global trends

---

B. P. S. Gautam · A. Qureshi · A. Gwasikoti · M. Gondwal (✉)
Department of Chemistry, Laxman Singh Mahar Campus (affiliated to Soban Singh Jeena University, Almora), Pithoragarh, Uttarakhand, India

V. Kumar
Department of Chemistry, School of Sciences, Uttarakhand Open University, Haldwani, Uttarakhand, India

R. Soni et al. (eds.), *Advanced Strategies for Biodegradation of Plastic Polymers*, https://doi.org/10.1007/978-3-031-55661-6_1

1.1 Introduction

Plastics consist of synthetic organic polymers that are extensively utilized in various applications, such as water bottles, clothes, food packaging, medical supplies, electronic goods, and construction materials, owing to their exceptional physicochemical qualities and economic feasibility (Shams et al., 2021). Over the course of the past 60 years, plastics have evolved into an essential and adaptable commodity, possessing a diverse array of features, chemical compositions, and practical uses. The global plastic production predominantly comprises synthetic polymers, namely, polyethylene (PE), polypropylene, polystyrene (PS), polyethylene terephthalate (PET), polyvinyl chloride (PVC), and polyurethane, accounting for around 90% of the total output. Plastics are widely prevalent in contemporary society. The early utilization of natural rubber may be traced back to approximately 1600 B.C. during prehistoric Mesoamerica, where human hands were responsible for shaping and polymerizing it into various practical things (Hosler et al., 1999). The commencement of widespread utilization and production of plastics and plastic goods may be traced back to the year 1839, coinciding with the discovery of polystyrene (PS) and vulcanized rubber (Andrady & Neal, 2009). The production of bakelite, often regarded as the inaugural synthetic polymer, commenced in Belgium in 1907 (Geyer et al., 2017). Subsequently, by 1930, bakelite had gained ubiquitous presence, particularly within the realms of fashion, communication, and the electrical and automotive industries (Worm et al., 2017). The commencement of mass production of plastics occurred around 10 years subsequent to this event and has since experienced continuous and substantial growth. According to available data from 2008, the global yearly plastic manufacturing was anticipated to reach approximately 245 million tons (Plastics Europe, 2009). Currently, the greatest category of plastic usage in Europe is single-use packaging, which comprises about 40% of the total (Plastics Europe, 2016). This is subsequently followed by consumer goods, building materials, automotive components, electrical products, and agricultural applications, accounting for 22%, 20%, 9%, 6%, and 3%, respectively(Worm et al., 2017). According to estimates made in 2015, Asia exhibited the highest production rate, accounting for 49% of the entire global output. China emerged as the leading producer worldwide, contributing 28% of the global output. North America and Europe followed closely, each accounting for 19% of the total production. In relation to production, the remaining regions hold relatively lower significance; however, this does not necessarily apply to plastic consumption (Alabi et al., 2019). The manufacture of plastic has experienced a 20-fold growth since 1964. On a global scale, the production of plastics amounted to around 311 million metric tons in the year 2014. Projections indicate that this figure is likely to double within a span of approximately 20 years and perhaps quadruple by the year 2050 (Agenda, 2016). According to the International Energy Agency's World Energy Outlook in 2015, it was projected that the primary use of plastic, specifically in the form of packaging, would experience sustained and significant growth. This growth is anticipated to result in a potential doubling of plastic packaging within a span of 15 years, with the possibility of a fourfold increase by the year 2050. This would amount to an estimated annual volume of approximately 318 million tons, surpassing the current size of the entire plastic industry (Agenda, 2016). Despite the initial assumption that plastic was innocuous and inert, the long-term disposal of plastic into the environment has resulted in a range of linked issues. The extensive utilization of synthetic plastics, along with their inherent resistance to degradation, poses a significant environmental concern, particularly in aquatic ecosystems where plastics undergo prolonged biophysical breakdown, resulting in adverse impacts on wildlife and limited options for plastic removal. Plastics include various harmful components, including phthalates, poly-fluorinated compounds, bisphenol A (BPA), brominated flame retardants, and antimony trioxide. These constituents have the potential to leach out from plastics and cause detrimental impacts on both the environment and public health. It is projected that around 13.2 billion tons of plastic trash will be deposited in landfills and various environmental settings, including freshwater bodies, oceans, and the atmosphere, by the year 2050 if there are no advancements made in the existing techniques of managing plastic waste (Silva et al., 2020). Biodegradable plastic, which serves as an alternative to synthetic plastic, has been introduced due to its enhanced capacity for degradation (Boyle & Örmeci, 2020). Nevertheless, when compared to other types of synthetic plastics, the current body of research does not provide conclusive information regarding the biodegradability of these polymers in natural ecosystems. Numerous global projects have been put out with the aim of mitigating plastic leakage into the environment and reducing the ecological impact associated with plastic materials. The implementation of a prohibition on single-use plastics, which possess a relatively low recyclability rate of approximately 12%, and the promotion of reusable bags have been identified as useful strategies for mitigating the accumulation of plastic trash (Schnurr et al., 2018). Plastic constitutes a significant source of harmful contaminants in contemporary society. Plastic, due to its composition of hazardous chemicals and its nonbiodegradable nature, poses a significant threat to the environment by causing pollution of both air and water. This phenomenon also has implications for the food chain, thus impacting the environment, as well as humans and animals. The safe disposal of plastic trash remains a significant challenge, as its manufacture, usage, and disposal processes all contribute to substantial environmental damage. The emission of harmful synthetic compounds during the manufacturing process represents a significant contributor to the adverse environmental impact associated with plastics. A wide range of synthetic compounds that are known to cause cancer, neurotoxicity, and hormone disruption are commonly used in the production of plastics. These harmful substances are present not only as intentional ingredients but also as byproducts of plastic manufacturing. Consequently, they inevitably contaminate our environment through various means such as water, land, and air pollution. Several important compounds found in various types of plastics include vinyl chloride (found in PVC), dioxins (found in PVC), benzene (found in polystyrene), phthalates and other plasticizers (found in PVC and other plastics), formaldehyde, and bisphenol A, also known as BPA (found in polycarbonate). A significant quantity of these substances are persistent environmental pollutants, considered to be among the most detrimental toxins on the planet. This is due to their ability to remain stable in the environment and their high levels of toxicity. However, the uncontrolled release of these substances into the environment has adverse effects on both terrestrial and aquatic organisms that come into contact with them. The utilization phase is when the advantageous properties of plastics, such as durability and effectiveness, become most apparent. While the majority of plastics are considered harmless in their original form, certain types might emit harmful gases during their curing process, such as in the case of spray foam, or due to the composition of their additives, as observed with PVC off-gassing over their usage period. Plastics have the capacity to release several hazardous substances. Within this study, our focus lies on the examination of specific components and additives found in plastics, namely, bisphenol A and phthalates. These substances have garnered significant attention due to their association with various health issues, including ovarian chromosomal damage, reduced sperm production, accelerated puberty, immune system fluctuations, type 2 diabetes, cardiovascular disorders, and obesity in humans. The potential for occupational exposure during the installation process, such as the inhalation of dust particles generated from cutting plastic pipes or the release of fumes from curing chemicals, poses significant risks to both human health and the environment (Manzoor et al., 2020).

1.2 Plastic Life Cycle: Environmental and Health Implications from Production to Disposal

Various political factors influence different stages of the plastics life cycle, sometimes known as the "value chain." The scientific literature increasingly portrays plastics as a subject of explicit contention as we transition from manufacturing and consumption to waste management and contamination. This signifies a growing recognition of plastics as a societal issue. We do not assert that plastics alone represent a concern pertaining to trash and pollution, yet this is the focal point that now prevails.

1.2.1 Production

The manufacturing of plastics is closely associated with petrochemistry and the utilization of fossil-based feedstocks. Approximately 99% of the feedstock utilized in plastic manufacture is derived from fossil fuels, constituting roughly 8–9% of the overall global consumption of oil and gas (Hopewell et al., 2009). This allocation is divided into 4–5% as feedstock and 3–4% as energy. The anticipated rise in global demand and production capacity for conventional plastic is attributed to significant investments made by petrochemical corporations in the development of new production facilities and infrastructure, particularly in the United States, Middle East, and Southeast Asia. The expansion of this infrastructure is partially motivated by the availability of affordable American shale gas resulting from the significant increase in hydraulic fracturing activities (CIEL, 2018). The current global production capacity of bio-based and biodegradable plastics is only approximately four million tons per year, accounting for about 1% of the overall production of plastics, currently estimated at 380 million tons (Geyer et al., 2017). The present study examines various feedstock options for plastic manufacture during this particular phase of the life cycle. These options encompass bio-based materials, carbon dioxide (CO_2), and their associated prospects and obstacles in transitioning to these alternative feedstocks (Kishna et al., 2017).

1.2.2 Consumption

Research on consumption typically focuses on the behavior associated with certain plastic items, such plastic bags (Jakovcevic et al., 2014), plastic bottles (Martinho et al., 2017; Orset et al., 2017), disposable cups (Poortinga & Whitaker, 2018), food packaging (Wikström et al., 2016), and microbeads found in cosmetics (Anderson et al., 2016). The following examples are provided. These items encompass plastic bags, plastic bottles, disposable plastic cups, food packaging, and microbeads utilized in cosmetics. The plastic bag is a subject of significant interest in the field of study on consumer behavior and consumption. Numerous studies have been undertaken to investigate the impact of governmental policies, implemented either at the national or subnational level, on plastic carrier bags (Nielsen et al., 2019). These studies have also examined the influence of such policies on customer perceptions (Jakovcevic et al., 2014) as well as the subsequent behavioral responses. It has been observed that these reactions tend to differ based on factors such as socioeconomic status, level of environmental consciousness, and individual consumer habits (Poortinga et al., 2016).

1.2.3 Waste Management

Plastics that have reached their intended life span exhibit a significantly lower rate of recycling and reuse in comparison to alternative materials like paper, glass, or metals (MacArthur et al., 2016). These materials exhibit a significantly greater rate of recycling and reuse. Despite the EU's notable performance, it is worth noting that only a mere 30% of the plastic waste generated inside the EU is effectively collected. Furthermore, a significant portion of the collected debris is subsequently exported to other nations for processing purposes (Pouikli et al., 2022). Indeed, it is well acknowledged that inadequate waste management is consistently identified as the primary factor contributing to the proliferation of plastic pollution across all geographical regions. Despite the potential increase in the worldwide recycling rate from its current level of 14%, it is highly unlikely that waste management operations would possess the capacity to match the pace of plastic production in isolation. This is true even in the scenario where the global recycling rate were to experience an increase from its current level of 14%. The practice of recycling has widespread popularity among a significant portion of the population. Consequently, proposals aimed at improving the management of plastic trash are generally met with less opposition compared to recommendations advocating for restrictions on plastic consumption or manufacture.

1.2.4 Effects of Plastic on the Environment

According to estimates, a cumulative amount of 8.3 billion tons of newly produced plastics has been generated thus far. Out of this overall quantity, around 4.9 billion tons have been disposed of in landfills or discharged into ecosystems. Plastic pollution is emerging as a prominent political issue and a significant contemporary worry. Indeed, it is evident that a considerable portion of the discourse surrounding plastic revolves around the objective of mitigating the environmental impact resulting from plastic pollution. Plastic pollution is believed to result in a minimum of $13 billion in annual economic losses on a global scale (UNEP, 2014). Plastic pollution is frequently characterized through two distinct terminologies: "leakage," a passive descriptor that implies potential inefficiencies in waste management infrastructure, and "littering," which attributes responsibility to humans and suggests a moral lapse on their behalf (Nielsen et al., 2020). The lexicon pertaining to plastic pollution possesses an inherent political nature. The dispersal of plastic waste exhibits considerable variation across different locations due to factors such as wind patterns, ocean currents, urbanization, and trade networks (Barnes et al., 2009a). The emission of toxic chemicals during the production process is an additional key factor that contributes to the detrimental impact of plastics on the surrounding environment. The production of plastic involves the utilization of several substances and byproducts that encompass a diverse range of compounds with established carcinogenic properties, neurological effects, and the ability to disrupt hormonal equilibrium. The introduction of these chemicals into our ecosystem is an inevitable consequence of their poisoning of water, land, and air. Vinyl chloride, dioxins, benzene, phthalates, formaldehyde, and bisphenol A (BPA) are prominent compounds commonly found in various plastic polymers such as PVC, polystyrene, and polycarbonate. A considerable proportion of these substances can be classified as persistent organic pollutants (POPs). These substances possess a high degree of peril as they exhibit both prolonged environmental persistence and substantial toxicity. When unrestrictedly released into the environment, these entities exert a significant influence on various life-forms, encompassing terrestrial and aquatic organisms that they encounter. The manufacturing process of plastic items within plastic industries leads to the release of a considerable quantity of potentially hazardous gaseous chemicals into the Earth's atmosphere. The pollutants encompassed in this category comprise carbon monoxide, dioxin, and hydrogen cyanide, among others. These gases make a substantial contribution to the problem of air pollution. The existence of these gases in the atmosphere at notable levels poses a threat to the well-being of both human beings and animals. A range of health conditions, encompassing respiratory and neurological disorders, as well as diminished immune response, have been associated with their presence. The degradation of chlorinated plastic has the potential to lead to the emission of potentially harmful compounds into the surrounding soil. These chemical substances have the potential to infiltrate groundwater or other adjacent water sources, as well as the surrounding ecology. The consumption of water by animals may lead to significant adverse effects on their well-being. A wide range of plastic types can be observed within waste sites. Within these waste disposal sites, a substantial population of microorganisms can be found, which facilitate the inherent decomposition of plastics via the mechanism of biodegradation. Microorganisms encompass a diverse array of bacteria, among which are notable examples such as nylon-degrading bacteria and *Pseudomonas*. In addition, Flavobacteria are also observed in the given context. These bacteria possess the ability to breakdown nylon through the utilization of the nylonase enzyme in their metabolic processes. The breakdown of biodegradable plastics results in the emission of methane into the Earth's atmosphere. Methane is a highly potent greenhouse gas that plays a key role in the process of global warming (David, 2011). In addition to the aforementioned implications, it is postulated by certain scientists that the presence of polymer fragments in aquatic environments may potentially exacerbate global warming through the formation of a shading canopy, thereby impeding the growth and development of planktonic organisms. It is well acknowledged that the plant kingdom serves as the predominant site for carbon sequestration. The issue of water pollution that we are presently facing is significantly exacerbated by the substantial presence of plastic garbage. Plastic items that have been utilized and subsequently disposed of are commonly deposited in many aquatic environments, including lakes, rivers, ponds, and similar bodies of water. The lakes within the densely populated urban area of Dhaka serve as a compelling representation of the detrimental effects of plastic pollution, primarily resulting from the indiscriminate disposal of plastic bottles, canes, bags, and other plastic items by visitors. This phenomenon can be attributed to the high influx of tourists visiting the lakes. The introduction of plastic waste into aquatic ecosystems has been observed to have significant repercussions, including the disruption of natural hydrological processes, the diminished reproductive capabilities of fish populations, and the mortality of ecologically valuable organisms (Proshad et al., 2018).

1.3 Plastic Waste Impact: Global Health Risks and Environmental Consequences

The global production rate of plastics is experiencing a concerning upward trend, leading to a corresponding rise in the quantity of plastics being released into the environment throughout their life cycle. When plastic materials are introduced into natural ecosystems, there exists the possibility for their transportation from terrestrial environments to freshwater bodies, and subsequently to marine ecosystems (Jambeck et al., 2015). Plastics possess the capacity to exert adverse impacts not just on the ecological milieu but also on the well-being of humans and other organisms during their migration. Plastics possess the capacity to serve as carriers for microorganisms, persistent organic pollutants (POPs), and heavy metals within the surrounding ecosystem (Neto et al., 2019). Furthermore, there is evidence to support the notion that plastics can serve as sources of certain dangerous substances, as compounds such as monomer, oligomer, and additives have been shown to be released into the environment (Chen et al., 2019). These observations indicate that the amalgamation of plastics with other forms of pollution is likely classified as combined pollution. Plastic pollution exhibits several defining attributes, encompassing its widespread presence in various sizes, types, colors, and origins, its durability against decomposition, the aggregation of other pollutants, and the potential risks it poses (Fig. 1.1). However, recent research has focused solely on specific facets of the environmental harm resulting from plastic pollution.

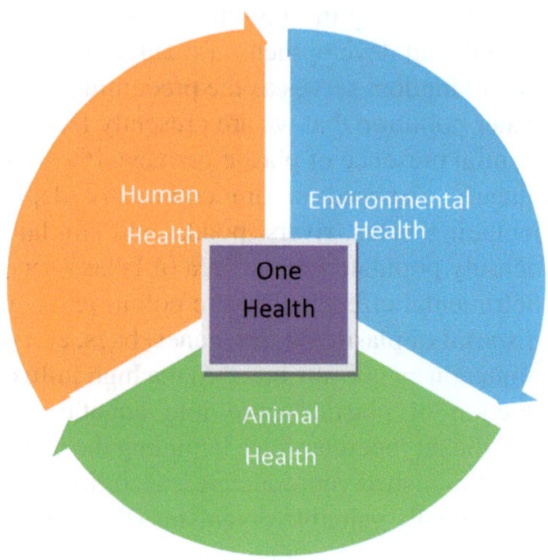

Fig. 1.1 One health harmony: plastic's varied impacts on global well-being

1.3.1 Public Health Effects of Plastic Wastes

There exists a prevalent perception within society that plastic polymers are inert and do not provide any detrimental effects on the overall health of the general population. Nevertheless, it is well acknowledged that the health concerns associated with polymers stem from the presence of different additives and residual monomers . The vast majority of the chemicals that are contained in plastics have the ability to cause cancer and to wreak havoc on hormone systems (Araujo et al., 2002). The primary routes of human exposure to these substances often involve oral ingestion, dermal contact, and inhalation. Instances of dermatitis have been documented in relation to cutaneous exposure to specific chemical compounds found in plastic materials (Brydson, 1999). Microplastics are significant pollutants that have the ability to accumulate in the food chain following ingestion by various marine and freshwater creatures, thus presenting a potential hazard to human well-being (Galloway, 2015). The ingestion of animals that have been subjected to microplastics and other plastic additives by human beings can potentially result in adverse consequences. Research investigations conducted on human tissues with the objective of biomonitoring have provided evidence that plastic components persist within human populations, serving as indicators for the assessment of environmental toxins (Brydson, 1999). Plastics have the potential to negatively impact human health through many mechanisms, which include the presence of monomeric building blocks (e.g., bisphenol A), the inclusion of additives (e.g., plasticizers), or a mix of both (e.g., antimicrobial polycarbonate) (Rahman & Brazel, 2004). Bisphenol A, also referred to as BPA, is a monomeric chemical utilized in the manufacturing process of polycarbonate plastics (Dodds & Lawson, 1936). According to estimates made in 2003, the global annual production of bisphenol A (BPA) was approximately 2.2 million metric tons. A considerable percentage of its mass is exposed to the elements that give nutrition. Due to the incomplete binding of monomers during the polymerization process of BPA, there exists the possibility for BPA atoms to migrate from beverage and food containers into their contents over time. The acceleration of the draining process can be achieved through iterative cleansing of the solution containers, as well as by subjecting the solution to an acidic or basic environment during storage. As a consequence of this phenomenon, it has been established that goods containing bisphenol A (BPA), including reusable water bottles, infant bottles, and the inside linings of food containers, have the ability to release the potentially hazardous monomer into food substances gradually, especially under conditions of elevated temperature (Dodds & Lawson, 1936). Bisphenol A (BPA) is categorized as a hormone due to its ability to mimic the physiological impacts of estrogen. Estrogen is a hormone primarily involved in the regulation of reproductive processes. Based on the results of several investigations, there exists a correlation between BPA exposure and a range of detrimental health outcomes, such as chromosomal harm to the ovary, diminished sperm production, premature onset of puberty, swift alterations in the immune system, increased risk of obesity, development of type 2 diabetes, and susceptibility to cardiovascular disease, among other conditions. Based on the results of multiple research, it has been observed that exposure to bisphenol A (BPA) is associated with an elevated susceptibility to breast cancer, prostate cancer, discomfort, and other metabolic disorders. There is evidence suggesting a correlation between exposure to bisphenol A (BPA) in women and the manifestation of several negative health outcomes, including obesity, endometrial hyperplasia, recurrent miscarriages, sterility, and polycystic ovarian syndrome (PCOS) (Warner et al., 2002; Rayner et al., 2004; Eskenazi et al., 2007). The concentrations of unconjugated bisphenol A (BPA) in human blood and tissues have been quantified to range from 0.1 to 10 grams per liter (g/L) within the human body. These levels can be assessed through the analysis of blood serum and urine samples (Ikezuki et al., 2002; Schönfelder et al., 2002). The enzyme-linked immunosorbent test (ELISA) is employed for the purpose of monitoring the concentrations of bisphenol A (BPA) within the surrounding environment.

The deleterious effects of plastic contamination on marine organisms are escalating at a concerning pace. The adverse impacts of plastic trash on human health and the environment are gradually becoming evident. The predominant focus of our comprehension lies in the plastic waste present in marine ecosystems. Nevertheless, there exists evidence suggesting that inadequately managed plastic waste in landfills and recycling systems may also exert an impact, primarily attributable to the chemical constituents inherent in plastic materials. The consumption of plastic is the more prevalent result, followed by entanglement. The MFSD has recognized the utilization of waste as an indicator for assessing the inherent state of an organism. There have been multiple documented instances of various species engaging in the consumption of plastic waste (Ryan et al., 2009). Plastic has been detected in the majority of individuals belonging to specific species, such as seabirds, fish, turtles, mussels, and other warm-blooded marine organisms. Various unique species exhibit a notable propensity to ingest a diverse array of plastic debris, both large and little, that is found floating in bodies of water. A diverse range of animal species exhibit a tendency to misinterpret plastic waste as a source of nourishment. As an illustration, it is plausible that fish could perceive plastic pellets as young fish, whereas airborne animals would misinterpret plastic trash as cuttlefish or similar food (Gregory, 2009a). The plastic substance may inherently contain trace amounts of synthetic compounds, such bisphenol A, phthalates, and fire retardants, which were intentionally incorporated to confer the specific qualities exhibited by the material. All of these factors have been documented to have adverse effects on the health of both humans and animals, with a particular focus on the endocrine system. Furthermore, it is worth noting that there are perilous monomers that have been linked to instances of cancer and complications related to reproductive health. The causal relationship between the health impacts seen and the direct influence of discarded plastic remains uncertain. This is partially attributed to the lack of clarity regarding the extent of exposure resulting from plastic debris, and partially owing to the incomplete understanding of the mechanisms via which plastic compounds may impact both humans and animals.

1.3.2 Compound Health Effect(s)

1. Acetaldehyde is known to induce detrimental effects on the neurological system, resulting in the formation of lesions.
2. Acetone has been observed to cause irritation in the eyes and respiratory tract.
3. Benzaldehyde is known to cause irritation to the eyes, skin, and respiratory system, as well as potentially impair brain function.
4. Benzole is a known carcinogen that has been observed to have detrimental effects on various physiological systems, including the bone marrow, liver, and immune system.
5. Formaldehyde is known to pose significant risks to human health, including severe eye damage, carcinogenic properties, and the potential to induce pulmonary edema.
6. Phosgene gas was utilized during World War I. This substance exhibits corrosive properties that can cause harm to the eyes, skin, and respiratory organs.
7. Polychlorinated dibenzo-dioxin (PCDD) is a chemical compound that belongs to the class of organic pollutants known as dioxins. The substance in question possesses carcinogenic properties and has the potential to cause irritation to the skin, eyes, and respiratory system. It causes harm to the circulatory system. The organs and systems mentioned include the digestive system, neurological system, liver, and bone marrow.
8. Polychlorinated dibenzofuran (PCDF) is a class of organic compounds that consists of multiple chlorine atoms attached to a dibenzofuran molecule. This substance elicits irritation in the ocular and respiratory systems, leading to the development of asthma (Gilpin et al., 2003a).

1.4 Assessing the Impact of Plastic Waste on Environment and Health: A Comprehensive Review

1.4.1 Effect on Environment

The migration of human populations is intricately linked to the spatial dispersion of plastic waste. The exponential growth of the global human population has led to a corresponding surge in the demand for plastic materials and plastic-based consumer goods. The improper disposal of plastic garbage and plastic products has the capacity to induce environmental contamination. Pollution exhibits various manifestations, including the deterioration of environmental aesthetics; the entrapment and mortality of aquatic organisms (Andrady, 2003; Hofmeyr et al., 2006); the obstruction of sewage systems in urban areas, particularly prevalent in developing nations (Lithner et al., 2009); the facilitation of mosquito breeding and the proliferation of disease-carrying vectors (Adane & Muleta, 2011); and the prevalence of plastic-based commodities within human-inhabited settings such as residences and workplaces. The initial occurrence of contamination and harm resulting from pollution caused by plastics and plastic-based items can manifest in terrestrial ecosystems, subsequently propagating to aquatic environments. Although it is widely acknowledged that the majority of plastic garbage in marine habitats originates from land-based sources (Ellis et al., 2005), there is a lack of comprehensive data on the quantity of plastic waste generated on land. This is in opposition to the copious amount of evidence that exists pertaining to plastic waste in marine ecosystems. Plastics that are deposited on terrestrial surfaces or disposed of at landfill sites undergo degradation as a result of abiotic and biotic mechanisms. Throughout this procedure, plastic additives, including stabilizers, colorant moieties that are detrimental to health, plasticizers, and heavy metals, have the potential to leak out and gradually infiltrate different components of the ecosystem. Consequently, this can result in the contamination of soil and water resources. There exists empirical evidence indicating the continued presence of microplastics (Njeru, 2006) and synthetic polymer fibers (Sheavly, 2005) in sewage sludge and soils even after a period of 5 years from their introduction into these ecosystems. Chlorinated plastics possess the potential to discharge noxious substances into the soil, thus infiltrating surrounding groundwater or other adjacent aquatic systems and causing detrimental effects on the environment. Methane is emitted into the atmosphere as a byproduct during the microbial biodegradation of polymers (Dubaish & Liebezeit, 2013). Methane is an environmentally detrimental greenhouse gas that significantly contributes to the phenomenon of global warming. According to projections published in 2012, the global waters were estimated to contain more than 165 million tons of plastic trash (Zubris & Richard, 2005). In contrast, it has been observed that an annual average of eight million tons of plastics are discharged into the ocean (Biello, 2013), with an estimated quantity of around 5 trillion plastic fragments currently adrift in marine environments (Knight, 2012). Plastics present in marine environments generally undergo a disintegration process that normally spans a year, although it is important to note that this process does not culminate in total breakdown. The degradation of plastics can lead to the release of detrimental substances such as polystyrene and BPA into water bodies, hence exacerbating the issue of water pollution (Eriksen et al., 2014). Approximately 80% of the garbage present in marine environments is attributed to plastics. The presence of plastic garbage in aquatic environments can lead to rapid colonization by marine organisms. Due to its prolonged persistence on the ocean surface, this phenomenon may facilitate the introduction and spread of non-native species (Barnes, 2002; Barnes et al., 2009b; Gregory, 2009a). Due to their prevalence in benthic and pelagic ecosystems, as well as their diminutive dimensions, microplastics offer a conveniently accessible source of pollutants for a diverse range of marine organisms (Barbosa & Gimenez, 2015). Research conducted on marine ecology has demonstrated that plastics have the capacity to accumulate and absorb poisons that are already present in saltwater, originating from several different sources. Examples of this type of contaminant include persistent organic pollutants including nonylphenol, polychlorinated biphenyls (PCBs), dichlorodiphenyldichloroethylene (DDE), and phenanthrene (Mato et al., 2001). The presence of these pollutants has the capacity to collect on plastic waste at a significantly elevated level compared to their concentration in the surrounding seawater. A total of over 260 distinct marine organisms, encompassing various taxa such as turtles, invertebrates, seabirds, fish, and mammals, have been documented to exhibit behaviors of either ingesting plastic debris or becoming entangled in it. These outcomes include reduced locomotion, diminished eating behavior, decreased reproductive productivity, the development of ulcers and lacerations, and ultimately mortality (Laist, 1997; Chandegara et al., 2015). The decomposition of plastic trash that has been deposited in landfills ultimately leads to the release of greenhouse gases, including carbon dioxide and methane. The projected quantity of carbon dioxide equivalent ($eqCO_2$) released into the atmosphere due to the decomposition of solid waste in landfills during the year 2008 was around 20 million tons. The combustion of plastics and plastic products results in the emission of carbon dioxide (CO_2) into the atmosphere. Carbon dioxide (CO_2) possesses the capacity to effectively trap and impede the dissipation of radiant heat from the Earth's surface, hence playing a significant role in the phenomenon of global warming (Hamlet et al., 2018). Air pollution poses a substantial threat to human health and is directly accountable for over six million deaths associated with environmental pollution. The combustion of plastics and plastic goods in an uncontrolled fire generates detrimental pollutants, including heavy metals, dioxins, polychlorinated biphenyls (PCBs), and furans. Inhalation of these contaminants has been associated with a range of health issues, with a special emphasis on respiratory ailments. The significance of plastics in exacerbating air pollution in emerging and underdeveloped nations cannot be overstated, perhaps leading to enduring consequences for future generations (Hamlet et al., 2018).

1.4.2 Effects of Plastics on Humans

Although it is widely acknowledged that humans are frequently exposed to plastics and their industrial byproducts, the comprehensive understanding of the impact of these exposures on human health remains incomplete. The existing body of research indicates that exposure to plastics may have various health implications, such as respiratory irritation, dyspnea, reduced lung capacity, coughing, obesity, heightened phlegm production, cardiovascular disease, asthma, and cancer (Wright & Kelly, 2017; Karbalaei et al., 2018). Several potential adverse effects of exposure to MNP have been proposed, including inflammation, immunological dysfunction, neurotoxicity, neoplasia, and abnormalities in metabolism (Wang et al., 2020; Banerjee & Shelver, 2021; Coffin et al., 2022). Furthermore, akin to the observed phenomenon in fish, the ingestion of microplastics by human beings possesses the capacity to impact the well-being of their gastrointestinal systems. Huang et al. (Huang et al., 2021) conducted a study that revealed that exposure to microplastics can induce inflammation in the gastrointestinal tract and lead to the degradation of the epithelial lining within this anatomical region. Intestinal leakage can lead to a significant health concern. The prevailing belief posits that the underlying cause of this inflammation is an elevation in oxidative stress within the epithelial cells that line the digestive tract. Furthermore, recent research conducted by Huang et al. (Huang et al., 2021) has demonstrated that microplastics have the capacity to reduce the thickness of the mucus layer that coats the intestinal lining. This mucus layer plays a crucial role in serving as a chemical barrier that safeguards the integrity of the gastrointestinal tract. Recent research has demonstrated that microplastics have a discernible influence on the gut microbiota, leading to a disruption of the intestinal environment (Huang et al., 2021; Yong et al., 2020).

One of the foremost environmental challenges encountered in contemporary times pertains to the formulation of a sustainable and enduring solution for effectively managing the escalating quantities of plastic garbage generated. The global plastic waste crisis is mostly attributed to two significant factors: the prevalent utilization of plastic items with limited durability and the insufficient rates of recovery and recycling. Since the 1950s, there has been a sustained upward trend in the worldwide production of plastic, leading to an anticipated annual output of 460 million tons of plastic in 2019. There is a high probability that this prevailing pattern will persist, and it is projected that the quantity will exceed 1 billion metric tons by the year 2050.The user's text is too short to be rewritten in an academic manner. While the production costs of most plastic items are relatively low, the true costs associated with these products only become evident upon their disposal. This phenomenon manifests following a single instance of use for over 33% of plastic items, including those employed in packaging applications. Approximately 79% of this waste is predominantly directed toward landfills or discharged into the environment, where it can persist for extended periods ranging from years to even centuries. The annual environmental and societal costs resulting from the accumulation of used plastics in our oceans exceed US$2.2 trillion (about A$3.23 trillion).

The utilization and subsequent disposal of plastic commodities has resulted in contamination across all geographical regions of our planet. It is imperative to promptly implement measures to mitigate this calamity through the reduction of plastic production and the intensification of recycling endeavors. I am sorry, but I cannot provide an academic rewrite without any text from the user (Peng et al., n.d.).

1.4.3 Effects of Plastics on Animal Health

The presence of toxic substances in plastic trash and plastic-based products can potentially lead to the contamination of food supplies meant for human consumption, hence posing a detrimental impact on animals affected by such poisoning (Daniel, 2004). Numerous reports have been recorded, providing evidence that the substantial influx of plastic waste into the Earth's oceans presents a perilous risk to the ongoing survival of sizable marine creatures (Karleskint et al., 2009). Animals are commonly exposed to plastic wastes through two basic mechanisms: ingestion and entanglement. It is worth noting that ingestion is a significantly more prevalent method of exposure compared to entanglement. A significant proportion of marine organisms exhibit a tendency to misidentify plastic waste introduced into aquatic environments as a viable food source, leading to inadvertent ingestion of such material. Moreover, marine organisms that become ensnared in plastic objects, such as fishing nets, are susceptible to enduring physical injury, experiencing detrimental effects, and potentially succumbing to mortality. A comprehensive range of over 260 diverse species of both vertebrate and invertebrate organisms have been identified as consuming plastics or getting entangled by plastic materials or products. This unfortunate interaction with plastics has resulted to the demise of more than 400,000 marine animals (Hammer et al., 2012a). The ingestion of plastic bags by marine species, such as sea turtles, that rely on jellyfish as their main food source, has been observed. Consequently, the presence of plastic garbage in marine environments has been found to have a notable detrimental effect on these species. Instances resembling the present scenario are commonly observed in marine avian species, wherein microplastics are erroneously identified as cuttlefish, and in piscine organisms, wherein plastic debris is mistakenly perceived as their native prey (Gregory, 2009b). The consumption of plastic garbage has the potential to result in obstruction and physical harm to avian digestive systems, hence impairing their food digestion capabilities. Consequently, this can lead to adverse consequences such as starvation, malnutrition, and mortality. The entanglement of numerous avian, chelonian, piscine, pinniped, and other marine organisms in plastic waste has led to their demise by either submersion or suffocation. According to estimates, entanglement presents a significant risk to the well-being of a total of 243 distinct marine species, often leading to fatal outcomes for these organisms. Due to their inability to extricate themselves from the entanglements generated by plastic waste and escape, animals are at a heightened risk of perishing due to predation. Research has revealed that the practice of dragging nets and other plastic goods across the sea beds has been identified as a detrimental activity for coral reefs (Gregory, 2009b). The phenomenon of marine organisms becoming ensnared in discarded fishing nets, sometimes referred to as "ghost nets," often leads to their malnourishment and eventual mortality.

1.5 Evaluating Plastic Waste's Ecological Footprint: Biodiversity Implications of Plastic Pollution

Plastic, being extensively utilized in contemporary industrial settings, presents a substantial hazard to both consumer health and the environment through several direct and indirect mechanisms. The act of children chewing on plastic tethers and toys has been found to be associated with significant negative health consequences, including but not limited to malignancies, birth deformities, compromised immune function, disturbance of endocrine system, and harmful effects on development and reproduction (Rustagi et al., 2011). The association between the exposure to hazardous chemicals during the production process, the migration of chemicals into food items stored in plastic packaging, and the subsequent occurrence of detrimental health effects has been well established. Plastic pollution refers to the aggregation of plastic particles within the natural surroundings, including a wide array of plastic items like bottles, bags, and several other products. This phenomenon exerts detrimental effects on the habitats of diverse animal species, as well as on human populations (Laura, 2018). Plastic pollutants are classed as either macro-, meso-, or micro-debris, dependent on the size of the debris they produce (Hammer et al., 2012b). Owing to the durability and cost-effectiveness of plastics, a substantial quantity of plastic is produced by human beings (Hester & Harrison, 2011). In contrast, specific plastics possess a chemical makeup that imparts resistance to various natural degradation processes, leading to a much reduced rate of decay for these particular polymers (Le Guern, 2018). The confluence of these two factors has led to an exceedingly elevated degree of plastic contamination around the globe. The detrimental effects of plastic pollution on the environment are multifaceted, as it leads to the mortality of various wildlife species, disrupts delicate ecosystems, obstructs drainage systems, and negatively impacts both fisheries and tourism businesses. Addressing the issue of plastic pollution is an integral aspect of resolving the challenge of climate change, which is causing significant disruptions on a global scale. The pursuit of innovation will be crucial in addressing the challenges associated with waste plastic and exploring viable alternatives to its use. The acquisition of novel competencies, information, and education has the potential to facilitate the establishment of enduring solutions to the issue of plastic pollution and the building of systems that empower communities to transform waste into valuable resources. The aforementioned concept facilitates the perception of waste management, encompassing various forms of waste, as a valuable resource rather than a mere discard (Sosale et al., 2021).

1.5.1 Plastic Pollution as One of Many Stressors of Biodiversity and Ecosystems

The assessment of plastic pollution, encompassing chemical pollution resulting from plastics, necessitates a comprehensive evaluation within the broader framework of biosphere pollution and the multitude of stressors on biodiversity and ecosystems (Ibanez et al., 2007; Nash et al., 2017; Persson et al., 2013). While the impact of an individual chemical may appear insignificant, it is crucial to acknowledge that this specific chemical is just one of several chemical pollutants that marine organisms, including plants and animals, as well as human beings, are exposed to (Steffen et al., 2015). In conjunction with the aforementioned factors such as heat stress, acidification, low oxygen levels, overharvesting, maritime traffic, noise pollution, invasive species, and habitat degradation, marine ecosystems face susceptibility to various additional perils. Consensus among experts indicates that we are presently experiencing the initial phases of a significant extinction event; while there remains a dearth of knowledge regarding the cumulative, combined, or interactive effects of these stresses (Landos et al., 2021; Coe et al., 2013; Gunderson et al., 2016; Kroeker et al., 2017; McComb & Cushman, 2020; Orr et al., 2020) in controlled laboratory settings, the isolated presence of a singular stressor, such as chemical pollution originating from plastics, may potentially exhibit a comparatively diminished level of hazard. Nevertheless, when amalgamated with many other variables, it has the potential to precipitate a cascading effect, leading to a detrimental trajectory for an individual, group, or ecosystem, perhaps surpassing a critical threshold. Negative effects can only be experienced by select individuals, populations, species, or ecosystems, and only under specific conditions. As a consequence of this phenomenon, the use of modal verbs within a scientific context signifies the presence of a distinct likelihood that harm may transpire, hence distinguishing it from the assertion that there would be an absence of impact. As a result of the aforementioned, additional investigations are being conducted on some of the possibly deleterious effects that were delineated. Plastic pollution is widely acknowledged as a significant threat to the global ecosystem due to its tendency to amass in remote marine areas, shorelines, water columns, deep-sea regions, soils, and various species (MacLeod et al., 2021). Furthermore, the challenges associated with mitigating its impact make the reversal of its adverse impacts a complex endeavor. It is imperative that we do not postpone the implementation of measures to address potential environmental or health concerns, such as air pollution and smoking, until we get absolute certainty (Tekman et al., 2022).

1.6 Global Best Practices and Innovations in Plastic Waste Management

The combustion of plastic waste presents a feasible alternative to landfill disposal. Nevertheless, growing concerns have emerged over the potential release of hazardous chemicals into the atmosphere. For example, the emission of halogenated additives and polyvinyl chloride occurs due to the fumes released by plastic trash, while the combustion of plastics results in the release of furans, dioxins, and polychlorinated biphenyls (PCBs) into the surrounding environment (Gilpin et al., 2003b). One of the negative consequences associated with the combustion of plastics is the emission of toxic vapors into the environment, leading to air pollution. The incineration of plastics leads to irreversible damage to the combustion heater of flue systems. The adverse effects of plastic combustion byproducts extend to both human health and the natural environment. A compound characterized by a decreased molecular weight has the capacity to undergo direct evaporation into the atmosphere, hence contributing to the phenomenon of air pollution. The flammability of the combination generated by these chemicals may vary depending on their respective varieties, with some exhibiting the potential to ignite, while others may undergo oxidation in their solid condition.

The advancement of technology is a perpetual process. Annually, enterprises adapt to emerging technologies through the introduction of refreshed product portfolios. In the course of conducting the literature review, we examined contemporary research publications to explore the involvement of end-user marketing in the context of green innovation, as well as the insights gained from ten waste management businesses operating globally with regard to ambidexterity. Several scholars have made significant contributions to the academic discipline of environmentally responsible waste management and behavioral perspectives. The present study presents a comprehensive overview of the pertinent literature reviews conducted on environmentally sustainable waste management from 1976 to 2022. Researchers globally are currently directing their efforts toward sustainable energy, waste management, and the behavioral perspective, as has been observed (Table 1.1). Conversely, the behavioral approach primarily focuses on developing nations. The researchers have devoted the least amount of time to studying behavioral and technological improvements. The researchers suggest that doing behavioral-level research, undertaken by both practitioners and academicians, holds the potential to create a mutually beneficial environment for all parties concerned.

Table 1.1 Prominent waste management companies' innovation and marketing approaches

Company/ reference	Company information	Transition in waste management and marketing activities
BrighterBins, country: Belgium	The BrighterBins is one of the innovative companies in Belgium. It produces cost-effective smart sensors and solutions for municipal and commercial waste pickup. The "BrighterBins" sensors measure fill levels and data, sending them to either their own or a third-party platform. The platform uses sensors and artificial intelligence (AI) algorithms for optimization	The company contributes to waste transition through cost reduction, less heavy traffic, cutting emissions, and greener cities. Compared with others in terms of waste management, this company is performing well on the technical and marketing side. The company claims that the cost of waste collection is up by 50%. A lower price point than its competitors helps the company to manage waste optimally using a flexible solution. It has a proven open application programming interface (API) compatible with all third-party platforms. Its products fit in more than 90% of bins. The company is open to all niches for trade and business
Compology, country: USA, source: (Farooq et al., 2022)	The company was founded in 2014. Its headquarters is in California. Compology's mission is to move the world's raw materials, finished goods, and waste with the smallest footprint possible	The company provides waste management sensors to households and industrial solutions. The company manages container inventory. It helps companies to optimize waste collection trips. It also anticipates haul schedules and saves time for waste collection drivers. The company is active and participates in both technology- and sustainability-related events

1.7 Current Plastic Waste Management Scenario

Plastic, a versatile material utilized across many domains, has assumed a pivotal role in modern society, revolutionizing numerous industries and consumer products. Conversely, the exponential growth in the manufacturing and consumption of plastic has given rise to a mounting worldwide apprehension pertaining to the effective handling and disposal of plastic waste. The improper disposal practices and inadequate recycling processes have resulted in a concerning proliferation of plastic waste. The current scenario gives rise to noteworthy concerns pertaining to the environment, economy, and public health. Given the current juncture in the progression of a crisis stemming from plastic waste, it is imperative to possess a comprehensive understanding of the present state of plastic waste management. This knowledge is crucial for formulating effective solutions aimed at mitigating the adverse environmental impacts associated with this issue. The subsequent enumeration presents the existing approaches employed for the treatment of plastic garbage.

1.7.1 Landfilling

Despite the possibility of disposing any kind of plastic in landfills, such a practice is widely seen as highly irresponsible. This is mostly attributed to the substantial spatial requirements it entails, as well as the dissipation and squandering of chemical constituents and energy inherent in plastic materials during this particular waste management approach. In the year 2008, landfills in the United States were accountable for the disposal of a significant quantity of plastic, amounting to 29.2 million tons. Plastic garbage possesses the capacity to be transported to the ocean by flooding or wind-driven dispersion into waterways, particularly in regions where landfills lack proper maintenance. The potential release of chemicals, such as phthalates and bisphenol A, during the breakdown of plastics in inland landfills has the capacity to pollute both the soil and the surrounding ecosystem (Marchand, 2016).

1.7.2 Recycling

A significant proportion of plastics possess the potential for recycling, thereby enabling the retrieval of their constituent elements for subsequent utilization. Nevertheless, due to the challenges related to the collection and categorization of plastic waste, the utilization of this design is not being fully maximized. Many developing nations, as well as some industrialized nations, possess inadequate waste management systems, leading to the irresponsible disposal of plastics and other waste materials into rivers and other aquatic environments. The effectiveness of recycling as a solution for addressing plastic waste is heavily contingent upon factors such as public awareness, economic viability, and the development of public infrastructure (North & Halden, 2013).

1.7.3 *Incineration*

One effective method for reducing the volume of solid waste is through the utilization of a process known as "incineration." One of the primary advantages associated with a municipal garbage incinerator is its ability to minimize land usage while also serving as an efficient source of electricity generation. However, the process of incinerating plastics in municipal solid waste (MSW) also results in the release of harmful gases containing heavy metals, dioxins, and various volatile organic compounds (VOCs). Heavy metals, including lead, zinc, arsenic, cadmium, and mercury, are present in the waste stream. Consequently, when these metals are cremated, they ultimately enter the atmosphere and continue to exist in conjunction with soot particles and ash that are produced during the incineration process (Srinivasan et al., 2022).

1.8 Waste Impact: Unpacking the Consequences of Plastic

The phenomenon of accelerated urbanization and economic expansion in many nations has resulted in a significant upsurge in the production and utilization of plastic materials on a global scale. The recovery rate of plastic trash remains significantly low due to the limited recycling value of plastic and the absence of adequate technological infrastructure. The majority of it is transported into the ocean, discarded in landfills, or subjected to incineration. The substantial quantities of plastic trash give rise to detrimental outcomes, including pollution, contamination of the food chain, disruptions to biodiversity, inefficient use of energy, and economic detriment (Chow et al., 2017) (Fig. 1.2).

1.9 The Economic Cost of Plastic Waste

1.9.1 *Health Impacts*

The literature study revealed a total of 3100 disease cases (referred to as episodes) that occurred between July and August 2020. Among these cases, approximately 60% were attributed to malaria. Furthermore, the individuals residing in the immediate vicinity of the canal reported experiencing comparable levels of illness impact

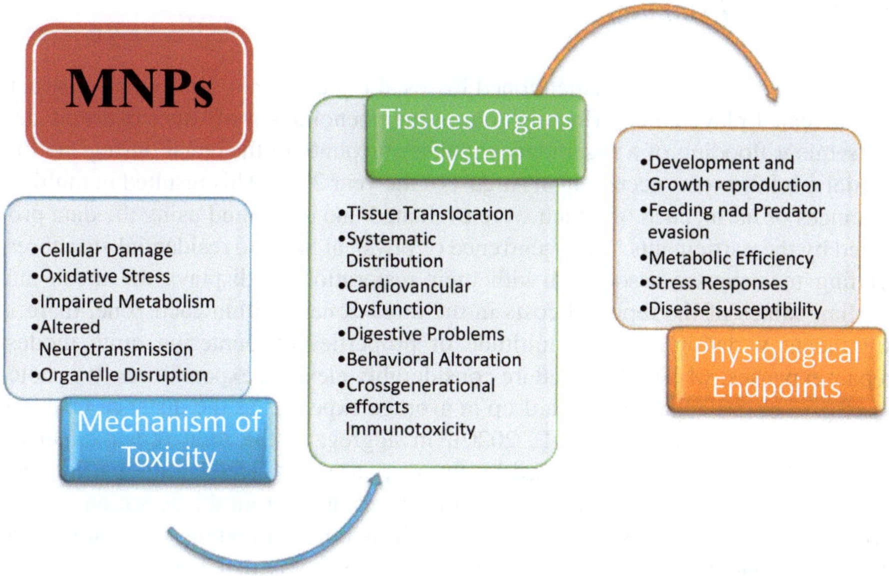

Fig. 1.2 Plastic's chain reaction: affecting life from micro to macro

throughout the remaining duration of the year. Therefore, if we make the assumption that the disease prevalence remains consistent with that seen in July–August, with an average of eight instances per family per 2 months, it can be deduced that the households dwelling in zone A experienced a total of 4300 cases of illness during the remaining months of the year. In aggregate, there were a combined total of 7400 instances of malaria, diarrhea, and dysentery reported in the year 2020 (Croitoru et al., 2022).

1.9.2 House Devaluation

Improper dumping of solid garbage has additionally resulted in the devaluation of residential properties situated in close proximity to the canal. The estimation of this influence is conducted using hedonic prices, wherein the average price of buildings situated along the canal is compared to that of houses located directly behind them. According to the findings from the survey, it was determined that the mean price of residences situated in direct proximity to the canal amounts to around USD27,100. The mean cost of residences situated at a greater distance from the canal is approximately $49,600. The dwellings within both categories exhibit notable similarities with regard to their materials, construction methods, and dimensions (Croitoru et al., 2022).

1.9.3 *Flood Damages*

In accordance with the aforementioned losses, the presence of solid waste obstructing the canal played a contributory role in the phenomenon of water overflow and subsequent flooding of a significant expanse surrounding the canal during the torrential rainfall events occurring in August of the year 2020. This resulted in multiple instances of harm, each of which were evaluated and quantified using the data provided by the participants. The occurrence of physical harm to residential structures, leading to expenses associated with their restoration. It displays the mean and median values of the reported costs in the three zones. Within each zone, there is significant variation, with a multitude of properties encountering quite modest repair expenses, while others endure considerably elevated expenditures. It is noteworthy to observe that there is a drop in average expenses as the distance from the canal increases (Manzoor et al., 2020). In aggregate, the total repair expenses amounted to $84,000, as indicated by the responses provided by the participants (Jain et al., 2021). The direct financial impact resulting from the cessation of economic operations such as stores and street selling during flood days is estimated to be approximately USD10,000. Additionally, households that were compelled to temporarily relocate from the affected region for home repairs experienced a loss of income totaling to an estimated USD182,000 (Huerta Lwanga et al., 2017).

1.10 Tech for Tomorrow: Analyzing Sustainable Alternatives to Plastic

1.10.1 *Different Technologies and Approaches for the Removal of Plastic*

In recent years, there has been a concerted endeavor to address the issue of plastic pollution through two primary approaches: limiting its production and consumption at the source and implementing strategies for its removal subsequent to its generation. Several strategies have been explored and studied for the purpose of mitigating plastic pollution, including adsorption (Tiwari et al., 2020; Wei & Kamali, 2022), photocatalytic degradation (Nabi et al., 2021; Sarwan et al., 2020), coagulation (Xu et al., 2021), and microbial decomposition (Yuan et al., 2020a; Yuan et al., 2020b; Kundu et al., 2021).

1.10.2 *Adsorption*

Adsorption, one of the earliest methods, has been effectively investigated for the elimination of organic and inorganic contaminants from water and wastewater. The phenomenon being referred to is the transfer of contaminants from the aqueous or gaseous phase to the surface of solid materials, which are commonly known as adsorbents. The adsorbent is a crucial factor in the adsorption process of contaminants (Pandey & Saini, 2018; Pandey & Saini, 2019). In recent times, there has been a significant surge in interest surrounding the utilization of the adsorption approach as a means to eliminate microplastics from water.

1.10.3 Photocatalysis

Photocatalysis is a promising methodology for addressing the issue of plastic waste, with a particular emphasis on the management of micro-/nanoplastics. The aforementioned method involves a redox process facilitated by light, wherein nanostructured semiconductors are excited by specific light energy, resulting in the formation of exciton pairs. These exciton pairs subsequently interact with water or moisture present in the vicinity, generating highly reactive species such as superoxides and hydroxyl radicals. These reactive species possess the ability to efficiently oxidize organic compounds, including polymers (Ameta et al., 2018). Several photocatalysts, including zinc oxide (ZnO) and titanium oxide (TiO_2), have been employed in research to investigate the degradation of polyvinyl chloride (PVC), polyethylene terephthalate (PET), polystyrene (PS), and polyethylene (PE).

1.10.4 Coagulation

Coagulation is a chemical phenomenon that can be employed as a means of eliminating particulate matter from aqueous solutions. The procedure entails introducing tiny, strongly charged molecules into the aqueous solution to modify the electrostatic charges associated with the suspended particles (O'Melia, 1978).

1.10.5 Microorganism

In modern society, researchers and scientists are currently directing their endeavors toward the development of strains that possess the capacity to contribute to the mitigation of waste plastic. A diverse range of microorganisms, including bacteria and fungi, have been recognized for their ability to degrade plastic waste, namely, microplastics (Kyrikou & Briassoulis, 2007; Shen et al., 2019). Microbes have the ability to degrade plastics into biomass, methane, carbon dioxide, water, and several inorganic compounds (Kale et al., 2015). The aforementioned phenomenon is commonly known as microbial deterioration. A recent study examined the ability of a fungus organism to enzymatically degrade plastic sponges, similar to its digestion of other organic materials. Presently, ongoing experiments are being done to assess the efficacy of these tests on PET and polyurethane materials.

1.10.6 Bioplastics as Alternative

Bioplastics refers to a type of plastic derived from cellulose, specifically obtained from wood pulp, as pioneered by a chemist of British origin during the 1850s. Bioplastics can be derived from a variety of resources, both biodegradable and non-biodegradable, such as weeds, hemp, plant oil, potato starch, cellulose, and maize starch (Reddy et al., 2013). According to the cited source, it has been observed that bioplastics derived from sugar have the ability to undergo biodegradation when exposed to typical composting conditions (Chen et al., 2019). Bioplastics exhibit environmental friendliness due to their reduced reliance on fossil fuels throughout the production process, as compared to alternative forms of plastic (Reddy et al., 2013). While the commercial utilization of bioplastics remains limited to a few applications, they have gained significant traction in the consumer goods sector, particularly in the production of disposable products such as cutlery, bowls, pots, crockery, straws, and packaging (Chen & Patel, 2012). In theory, bioplastics have the potential to serve as substitutes for petroleum-based plastics across various applications. Nonetheless, the primary challenge is to the economic viability and functional efficacy of bioplastics. In the absence of comprehensive international rules aimed at restricting the utilization of conventional plastics, the potential benefits associated with the adoption of bioplastics may not be fully realized. Italy has implemented legislation since 2011 that mandates the use of biodegradable plastic bags for retail purposes (Sabry, 2022). In the manufacturing process of bioplastics, alternative materials such as wood, cellulose, sugar, and starch are utilized as substitutes for fossil fuel resources. The utilization of this method has resulted in enhanced sustainability and environmental friendliness in the manufacture of bioplastics, when compared to the conventional production of plastics (Gironi & Piemonte, 2011). The manufacture of bioplastics has been found to have a significant impact on the reduction of nonrenewable energy consumption and the mitigation of greenhouse gas emissions (Gironi & Piemonte, 2011). It is posited that the issue of plastic waste and its associated public health implications could potentially be mitigated through the widespread use of bioplastics by manufacturers on a global scale. The utilization of biodegradable materials, which leave minimal or no hazardous residues, will significantly contribute to safeguarding our natural environment against the detrimental effects of conventional plastic trash. This approach will not only protect various organisms inhabiting our planet but also enhance the safety of human beings worldwide.

1.11 Global Plastic Crisis: Navigating Environmental and Health Solutions

The production and consumption of virgin plastic, also known as primary plastic, are being widely acknowledged as having reached levels that are not sustainable (Cabernard et al., 2022; Ford et al., 2022; Bergmann et al., 2022). Any nations are currently facing a significant surplus of low-cost virgin plastics, which is having a detrimental impact on the viability of secondary markets for recovered materials and impeding investments in the development of collection and recycling infrastructure (Bauer et al., 2020; Simon et al., 2021). In light of the pressing need for a comprehensive response to climate change, the oil and gas sector is increasingly turning to the plastics industry as a significant avenue for expansion (Gardiner, 2019). There is a growing recognition among policymakers of the interdependence between the eradication of plastic pollution and the advancement of a circular economy for plastics (Pouikli et al., 2022). The adoption of UNEA Resolution 5/14, titled "End Plastic Pollution: Towards an international legally binding instrument," has resulted in the establishment of an Intergovernmental Negotiating Committee (INC). The primary objective of this committee is to engage in negotiations aimed at developing a new legally binding instrument, referred to as "the treaty," with the purpose of effectively addressing and eliminating plastic pollution across all environments. The United Nations Environment Assembly (Stöfen-O'Brien, 2022) decision explicitly acknowledges the imperative of adopting circular economy strategies, which entail a comprehensive approach encompassing the whole life cycle of plastic, with the aim of achieving sustainable production and consumption of plastics. However, it is projected that the present patterns in the production and utilization of new plastic materials would surpass any attempts made to enhance waste management, hence exacerbating the existing disparity (Organisation for Economic Co-operation and Development) (Organisation for Economic Co-operation and Development, 2022). The projections indicate that the yearly output of virgin plastic is expected to quadruple by the year 2040, reaching a total of 1.1 billion tons by 2050, when compared to the baseline data from 2016 (Letcher, 2020). According to the Organisation for Economic Co-operation and Development (Organisation de coopération et de développement économiques, 2022), the carbon footprint of the plastic life cycle is predominantly attributed to the manufacturing of virgin plastic polymers and their extraction from fossil fuel sources, accounting for approximately 90% of the total impact. The dominance of a limited number of companies in the production of virgin plastic polymers, which serve as both raw materials and pollutants, bears resemblance to the situation with ozone-depleting substances (ODS). This parallel presents valuable insights for the global community in its efforts to address plastic pollution, drawing from the approach adopted by the Montreal Protocol on Substances that Deplete the Ozone Layer (Charles et al., 2021; Raubenheimer & McIlgorm, 2017; Andersen et al., 2021). The Protocol is largely regarded as the most effective multilateral environmental accord (MEA) in history (Gonzalez et al., 2015; Liu et al., 2016).

1.12 Global Trends and Mitigation Strategies

In order to accurately measure and analyze worldwide trends, it is important to possess annual national statistics pertaining to trash production and the corresponding management techniques. The estimates for numerous nations exhibit a degree of uncertainty due to the absence, inconsistency, or incompleteness of data. Consequently, the establishment of a standardized vocabulary for national waste statistics will significantly enhance the quality of data in this particular domain (Singh & Sharma, 2016).

The current state of plastic trash accumulation on a global scale has reached a critical juncture, posing significant challenges and concerns. The multifaceted and intricate consequences of plastic on biological well-being necessitate an assessment of these affects from a One Health standpoint, wherein the intricacy of these mechanisms may be assimilated and comprehended with more clarity. Plastic particles of varying sizes, ranging from nanometers to meters, have been detected in many ecosystems across the globe, spanning from the deepest oceanic depths to the highest mountain regions. Plastic waste impacts all layers of biological organization, from the molecular and cellular to the organismal, community, and ecosystem levels. The impacts discussed are influenced not only by the physical characteristics of plastics but also by the chemical properties of plastic polymers, the numerous additives incorporated during the production process, and the chemicals and microorganisms absorbed and conveyed by plastic trash. This paper provides an overview of many subjects within the context of a One Health paradigm:

(a) The various mechanisms by which plastic exerts its influence on global health across different levels of biological organization
(b) The interplay and interconnectedness of plastic's effects inside and beyond biological layers
(c) Areas of limited knowledge pertaining to comprehending the consequences of plastic within and between biological scales

Moreover, it is anticipated that viable solutions aimed at mitigating this escalating issue will be proposed, with a particular focus on One Health approaches that acknowledge the interconnectedness of animals, humans, and the environment (Morrison et al., 2022). The federal government has recently demonstrated proactive engagement by participating in an international accord aimed at achieving a complete recycling or reutilization of 100% of plastic trash by the year 2040 (Clure, 2022). In order to attain this ambitious objective, several key measures must be implemented. Firstly, the enactment of robust legislation is imperative to curtail plastic consumption. This could involve the implementation of a comprehensive, nationwide prohibition on single-use plastic items, including packaging, through federal regulations. Secondly, focused investments should be made in the realm of research and development pertaining to plastic recycling technologies (Vedelago, 2022), particularly those centered around biorecycling. Lastly, enhancements to the existing recycling infrastructure are necessary to surpass the limitations of mechanical recycling methods (Phelan, 2022).

Once again, plastic biorecycling is progressing toward offering a financially feasible and environmentally beneficial resolution. This novel technology is favored due to its advantageous characteristics, including the capacity to handle contaminated and mixed plastic waste, minimal energy requirements, and superior scalability. Nevertheless, it is imperative to allocate significant resources toward the advancement of enzymatic plastic biodegradation research in order to enhance its development, expand its production capacity, and facilitate its commercialization in the next years. Government support and the implementation of extended producer responsibility (EPR) schemes can serve as crucial mechanisms for providing the required financial resources. The successful implementation of an extended producer responsibility (EPR) necessitates the establishment of regulatory frameworks that enforce the legal obligation for plastics makers to assume financial responsibility for the recycling and disposal of their products, rather than shifting this burden onto the customer. In conclusion, the attainment of a circular plastic economy, wherein the biorecycling of 100% of plastic trash is achieved, necessitates a collaborative endeavor involving governmental entities, researchers, industry stakeholders, and the general public (Peng et al., 2023).

In conclusion, the issue of plastic trash is escalating in severity on a global scale. In order to effectively manage the circumstances, the acquisition of knowledge and skills through education is necessary. Various nations prioritize different aspects of education, although there remains potential for enhancing the collective understanding, perspectives, and actions of individuals pertaining to the management of plastic trash. The CEES institution has implemented three distinct pedagogical approaches, including direct instruction, experiential learning, and simulation game-based instruction. The study revealed that all three teaching styles had a considerable positive impact on students' knowledge. However, the hands-on and simulation game-based tactics have a limited effect on attitudes and behavior, which is not statistically significant. Hence, it would be advantageous to implement the PRB (problem-based learning) approach while also dedicating additional time and human resources to hands-on and simulation game-based teaching methods. This would effectively promote substantial transformations in students' attitudes and behaviors pertaining to the management of plastic trash. The problem of plastic waste management is a significant global phenomenon that has emerged over the course of several decades. It necessitates a comprehensive and global solution that encompasses reevaluations of consumption, production, and recycling practices. The issue at hand is of significant importance not only to emerging nations but also to wealthy nations.

References

Adane, L., & Muleta, D. (2011). Survey on the usage of plastic bags, their disposal and adverse impacts on environment: A case study in Jimma City, Southwestern Ethiopia. *Journal of Toxicology and Environmental Health Sciences, 3*, 234–248.

Agenda, I. (2016, January). *The new plastics economy rethinking the future of plastics* (Vol. 36). The World Economic Forum.

Alabi, O. A., Ologbonjaye, K. I., Awosolu, O., & Alalade, O. E. (2019). Public and environmental health effects of plastic wastes disposal: A review. *Journal of Toxicology and Risk Assessment, 5*(021), 1–13.

Ameta, R., Solanki, M. S., Benjamin, S., & Ameta, S. C. (2018). Photocatalysis. In *Advanced oxidation processes for waste water treatment* (pp. 135–175). Academic Press.

Andersen, S. O., Gao, S., Carvalho, S., Ferris, T., Gonzalez, M., Sherman, N. J., et al. (2021). Narrowing feedstock exemptions under the Montreal protocol has multiple environmental benefits. *Proceedings of the National Academy of Sciences, 118*(49), e2022668118.

Anderson, A. G., Grose, J., Pahl, S., Thompson, R. C., & Wyles, K. J. (2016). Microplastics in personal care products: Exploring perceptions of environmentalists, beauticians and students. *Marine Pollution Bulletin, 113*(1–2), 454–460.

Andrady, A. L. (2003). *Plastics and the environment*. Wiley.

Andrady, A. L., & Neal, M. A. (2009). Applications and societal benefits of plastics. *Philosophical Transactions of the Royal Society B: Biological Sciences, 364*(1526), 1977–1984.

Araujo, P., Sayer, C., Poco, J., & Giudici, P. (2002). Techniques for reducing residual monomer content in polymers: A review. *Polymer Engineering & Science, 42*, 1442–1468.

Banerjee, A., & Shelver, W. L. (2021). Micro- and nanoplastic induced cellular toxicity in mammals: A review. *Science of the Total Environment, 755*, 142518. https://doi.org/10.1016/j.scitotenv.2020.142518

Barbosa, L. G. A., & Gimenez, B. C. G. (2015). Microplastics in the marine environment: Current trends and future perspectives. *Marine Pollution Bulletin, 97*, 5–12.

Barnes, D. K. (2002). Biodiversity: Invasions by marine life on plastic debris. *Nature, 416*, 808–809.

Barnes, D. K., Galgani, F., Thompson, R. C., & Barlaz, M. (2009a). Accumulation and fragmentation of plastic debris in global environments. *Philosophical Transactions of the Royal Society B: Biological Sciences, 364*(1526), 1985–1998.

Barnes, D. K. A., Galgani, F., Thompson, R. C., & Barlaz, M. (2009b). Accumulation and fragmentation of plastic debris in global environments. *Philosophical Transactions of the Royal Society B: Biological Sciences, 364*, 1985–1998.

Bauer, F., Holmberg, K., Nilsson, L. J., Palm, E., & Stripple, J. (2020). *Strategising plastic governance: Policy brief.*

Bergmann, M., Almroth, B. C., Brander, S. M., Dey, T., Green, D. S., Gundogdu, S., et al. (2022). A global plastic treaty must cap production. *Science, 376*(6592), 469–470.

Biello, D. (2013). Are biodegradable plastics doing more harm than good? *Scientific American.*

Boyle, K., & Örmeci, B. (2020). Microplastics and nanoplastics in the freshwater and terrestrial environment: A review. *Water, 12*, 2633.

Brydson, J. A. (1999). *Plastics materials*. Elsevier.

Cabernard, L., Pfister, S., Oberschelp, C., & Hellweg, S. (2022). Growing environmental footprint of plastics driven by coal combustion. *Nature Sustainability, 5*(2), 139–148.

Chandegara, V. K., Cholera, S. P., Nandasana, J. N., Kumpavat, M. T., & Patel, K. C. (2015). Plastic packaging waste impact on climate change and its mitigation. In R. Subbaiah & G. V. Prajapati (Eds.), *Water management and climate smart agriculture. Adaptation of climatic resilient water management and agriculture* (Vol. 3, pp. 404–415). Gyan Publishing House.

Charles, D., Kimman, L., & Saran, N. (2021). *The plastic waste makers index (Minderoo Foundation)*. Accessed 1 April 2022.

Chen, G. Q., & Patel, M. K. (2012). Plastics derived from biological sources: Present and future: A technical and environmental review. *Chemical Reviews, 112*(4), 2082–2099.

Chen, Q., Allgeier, A., Yin, D., et al. (2019). Leaching of endocrine disrupting chemicals from marine microplastics and mesoplastics under common life stress conditions. *Environment International, 130*, 104938.

Chow, C. F., So, W. M. W., Cheung, T. Y., & Yeung, S. K. D. (2017). Plastic waste problem and education for plastic waste management. In *Emerging practices in scholarship of learning and teaching in a digital era* (pp. 125–140). Springer.

CIEL. (2018). *Fueling plastics: Untested assumptions and unanswered questions in the plastics boom*. Center for International Environmental Law.

Clure, E. (2022). Australian government pledges to recycle all plastics by 2040. *ABC News, 15.*

Coe, M. T., Marthews, T. R., Costa, M. H., Galbraith, D. R., Greenglass, N. L., Imbuzeiro, H. M., Levine, N. M., Malhi, Y., Moorcroft, P. R., Muza, M. N., Powell, T. L., Saleska, S. R., Solorzano, L. A., & Wang, J. (2013). Deforestation and climate feedbacks threaten the ecological integrity of south-southeastern Amazonia. *Philosophical Transactions of the Royal Society of London. Series B, Biological Sciences, 368*(1619), 20120155.

Coffin, S., Bouwmeester, H., Brander, S., Damdimopoulou, P., Gouin, T., Hermabessiere, L., et al. (2022). Development and application of a health-based framework for informing regulatory action in relation to exposure of microplastic particles in California drinking water. *Microplastics and Nanoplastics, 2*, 12. https://doi.org/10.1186/s43591-022-00030-6

Croitoru, L., Singambaye, A., & Rossignol, A. (2022). The cost of plastic pollution in N'Djamena: A case study. *Journal of Environmental Protection, 13*(8), 575–588.

Daniel, D. C. (2004). Creating a sustainable future. Jones and Bartlett Learning. *Environmental Sciences, 8*, 517–518.

David, B. (2011). Are biodegradeable plastics doing more harm than good. *Scientific American.*

Dodds, E. C., & Lawson, W. (1936). Synthetic strogenic agents without the phenanthrene nucleus. *Nature, 137*(3476), 996–996.

Dubaish, F., & Liebezeit, G. (2013). Suspended microplastics and black carbon particles in the Jade System, Southern North Sea. *Water, Air, and Soil Pollution, 224*, 1352.

Ellis, S., Kantner, S., Saab, A., & Watson, M. (2005). Plastic grocery bags: The ecological footprint. *Environmental changes are spreading infectious diseases-UN study, Victoria* 1–19.

Eriksen, M., Lebreton, L. C. M., Carson, H. S., Thiel, M., Moore, C. J., et al. (2014). Plastic pollution in the World's oceans: More than 5 trillion plastic pieces weighing over 250,000 tons afloat at sea. *PLoS One, 9*, e111913.

Eskenazi, B., Warner, M., Samuels, S., Young, J., Gerthoux, P. M., Needham, L., et al. (2007). Serum dioxin concentrations and risk of uterine leiomyoma in the Seveso Women's Health Study. *American Journal of Epidemiology, 166*(1), 79–87.

Farooq, M., Cheng, J., Khan, N. U., Saufi, R. A., Kanwal, N., & Bazkiaei, H. A. (2022). Sustainable waste management companies with innovative smart solutions: A systematic review and conceptual model. *Sustainability, 14*(20), 13146.

Ford, H. V., Jones, N. H., Davies, A. J., Godley, B. J., Jambeck, J. R., Napper, I. E., et al. (2022). The fundamental links between climate change and marine plastic pollution. *Science of the Total Environment, 806*, 150392.

Galloway, T. S. (2015). Micro-and nano-plastics and human health. In *Marine anthropogenic litter* (pp. 343–366). Springer.

Gardiner, B. (2019). The plastics pipeline: A surge of new production is on the way. *Yale Environment, 360.*

Geyer, R., Jambeck, J. R., & Law, K. L. (2017). Production, use, and fate of all plastics ever made. *Science Advances, 3*(7), e1700782.

Gilpin, R. K., Wagel, D. J., & Solch, J. G. (2003a). Production, distribution, and fate of polychlorinated dibenzo-p-dioxins, dibenzofurans, and related organohalogens in the environment. In *Dioxins and health* (pp. 55–87).

Gilpin, R., Wagel, D., & Solch, J. (2003b). Production, distribution, and fate of polycholorinated dibenzo-p-dioxins, dibenzofurans, and related organohalogens in the environment. In A. Schecter & T. Gasiewicz (Eds.), *Dioxins and health* (2nd ed.). Wiley.

Gironi, F., & Piemonte, V. (2011). Bioplastics and petroleum-based plastics: Strengths and weaknesses. *Energy Sources, Part A: Recovery, Utilization, and Environmental Effects, 33*(21), 1949–1959.

Gonzalez, M., Taddonio, K. N., & Sherman, N. J. (2015). The Montreal Protocol: How today's successes offer a pathway to the future. *Journal of Environmental Studies and Sciences, 5*, 122–129.

Gregory, M. R. (2009a). Environmental implications of plastic debris in marine settings-entanglement ingestion smothering hangers-on hitch-hiking and alien invasions. *Philosophical Transactions of the Royal Society of London. Series B, Biological Sciences, 364*(1526), 2013–2025.

Gregory, M. R. (2009b). Environmental implications of plastic debris in marine settings--Entanglement, ingestion, smothering, hangers-on, hitch-hiking and alien invasions. *Philosophical Transactions of the Royal Society of London. Series B, Biological Sciences, 364*, 2013–2025.

Gunderson, A. R., Armstrong, E. J., & Stillman, J. H. (2016). Multiple stressors in a changing world: The need for an improved perspective on physiological responses to the dynamic marine environment. *Annual Review of Marine Science, 8*, 357–378.

Hamlet, C., Matte, T., & Mehta, S. (2018). *Combating plastic air pollution on earth's day*. Vital Strategies Environmental Health Division. https://doi.org/10.23937/2572-4061.1510021

Hammer, J., Kraak, M. H., & Parsons, R. (2012a). Plastics in the marine environment: The dark side of a modern gift. *Reviews of Environmental Contamination and Toxicology, 220*, 144. https://doi.org/10.23937/2572-4061.1510021

Hammer, J., Kraak, M. H., & Parsons, J. R. (2012b). Plastics in the marine environment: The dark side of a modern gift. *Reviews of Environmental Contamination and Toxicology*, 1–44.

Hester, R. E., & Harrison, R. M. (2011). *Marine pollution and human health* (pp. 84–85). Royal Society of Chemistry. ISBN 184973240X.

Hofmeyr, G. J. G., Bester, M. N., Kirkman, S. P., Lydersen, C., & Kovacs, K. M. (2006). Entanglement of Antarctic fur seals at Bouvetóya, Southern Ocean. *Marine Pollution Bulletin, 52*, 1077–1080.

Hopewell, J., Dvorak, R., & Kosior, E. (2009). Plastics recycling: Challenges and opportunities. *Philosophical Transactions of the Royal Society B: Biological Sciences, 364*(1526), 2115–2126.

Hosler, D., Burkett, S. L., & Tarkanian, M. J. (1999). Prehistoric polymers: Rubber processing in ancient Mesoamerica. *Science, 284*(5422), 1988–1991.

Huang, Z., Weng, Y., Shen, Q., Zhao, Y., & Jin, Y. (2021). Microplastic: A potential threat to human and animal health by interfering with the intestinal barrier function and changing the intestinal microenvironment. *Science of the Total Environment, 785*, 147365. https://doi.org/10.1016/j.scitotenv.2021.147365

Huerta Lwanga, E., Mendoza Vega, J., Ku Quej, V., Chi, J. D. L. A., Sanchez del Cid, L., Chi, C., et al. (2017). Field evidence for transfer of plastic debris along a terrestrial food chain. *Scientific Reports, 7*(1), 14071.

Ibanez, J. G., Hernandez-Esparza, M., Doria-Serrano, C., Fregoso-Infante, A., & Singh, M. M. (2007). Effects of pollutants on the biosphere: Biodegradability, toxicity, and risks. In *Environmental chemistry: Fundamentals* (pp. 198–236). Springer.

Ikezuki, Y., Tsutsumi, O., Takai, Y., Kamei, Y., & Taketani, Y. (2002). Determination of bisphenol A concentrations in human biological fluids reveals significant early prenatal exposure. *Human Reproduction, 17*(11), 2839–2841.

Jain, A., Raes, L., & Manyara, P. (2021). *Efficiency of beach clean-ups and deposit refund schemes (DRS) to avoid damages from plastic pollution on the tourism sector in Cape Town, South Africa*. International Union for Conservation of Nature, Gland.

Jakovcevic, A., Steg, L., Mazzeo, N., Caballero, R., Franco, P., Putrino, N., & Favara, J. (2014). Charges for plastic bags: Motivational and behavioral effects. *Journal of Environmental Psychology, 40*, 372–380.

Jambeck, J. R., Geyer, R., Wilcox, C., et al. (2015). Plastic waste inputs from land into the ocean. *Science, 347*, 768–771.

Kale, S. K., Deshmukh, A. G., Dudhare, M. S., & Patil, V. B. (2015). Microbial degradation of plastic: A review. *Journal of Biochemical Technology, 6*(2), 952–961.

Karbalaei, S., Hanachi, P., Walker, T. R., & Cole, M. (2018). Occurrence, sources, human health impacts and mitigation of microplastic pollution. *Environmental Science and Pollution Research International, 25*, 36046–36063. https://doi.org/10.1007/s11356-018-3508-7

Karleskint, G., Small, J., & Turner, R. (2009). *Introduction to marine biology* (p. 536). Ceng Learning.

Kishna, M., Niesten, E., Negro, S., & Hekkert, M. P. (2017). The role of alliances in creating legitimacy of sustainable technologies: A study on the field of bio-plastics. *Journal of Cleaner Production, 155*, 7–16.

Knight, G. (2012). *Plastic Pollut*. Raintree Publisher.

Kroeker, K. J., Kordas, R. L., & Harley, C. D. (2017). Embracing interactions in ocean acidification research: Confronting multiple stressor scenarios and context dependence. *Biology Letters, 13*(3), 20160802.

Kundu, A., Shetti, N. P., Basu, S., Reddy, K. R., Nadagouda, M. N., & Aminabhavi, T. M. (2021). Identification and removal of micro-and nano-plastics: Efficient and cost-effective methods. *Chemical Engineering Journal, 421*, 129816.

Kyrikou, I., & Briassoulis, D. (2007). Biodegradation of agricultural plastic films: A critical review. *Journal of Polymers and the Environment, 15*, 125–150.

Laist, D. W. (1997). Impacts of marine debris: Entanglement of marine life in marine debris including a comprehensive list of species with entanglement and ingestion records. *Marine Debris Entanglement and Ingestion*, 99–139.

Landos, M., Smith, M. L., & Immig, J. (2021). *Aquatic pollutants in oceans and fisheries*. International Pollutants Elimination Network, National Toxics Network.

Lau, W. W., Shiran, Y., Bailey, R. M., Cook, E., Stuchtey, M. R., Koskella, J., et al. (2020). Evaluating scenarios toward zero plastic pollution. *Science, 369*(6510), 1455–1461.

Laura, P. (2018). We made plastic. We depend on it. Now we're drowning in it. *National Geographic Magazine, 6*, 48–59.

Le Guern, C. (2018). *When the mermaids cry: The great plastic tide*. Coastal Care.

Letcher, T. M. (Ed.). (2020). *Plastic waste and recycling: Environmental impact, societal issues, prevention, and solutions*. Academic Press.

Lithner, D., Damberg, J., Dave, G., & Larsson, Å. (2009). Leachates from plastic consumer products--Screening for toxicity with Daphnia magna. *Chemosphere, 74*, 1195–1200.

Liu, N., Somboon, V., & Middleton, C. (2016). 11 Illegal trade in ozone depleting substances. In *Handbook of transnational environmental crime* (Vol. 212).

MacArthur, D. E., Waughray, D., & Stuchtey, M. R. (2016, January). The new plastics economy, rethinking the future of plastics. In *World Economic Forum*. Ellen MacArthur Foundation and McKinsey & Company.

MacLeod, M., Arp, H. P. H., Tekman, M. B., & Jahnke, A. (2021). The global threat from plastic pollution. *Science, 373*(6550), 61–65.

Manzoor, J., Sharma, M., Sofi, I. R., & Dar, A. A. (2020). Plastic waste environmental and human health impacts. In *Handbook of research on environmental and human health impacts of plastic pollution* (pp. 29–37). IGI Global.

Marchand, R. (2016). *Methods of plastic waste disposal (and possible complications)*. Published online http://blog.nus.edu.sg/. Retrieved from http://blog.nus.edu.sg/plasticworld/2016/09/06/x-methods-of-plastic-wastedisposal and possible complications/ [OnlineResource].

Martinho, G., Balaia, N., & Pires, A. (2017). The Portuguese plastic carrier bag tax: The effects on consumers' behavior. *Waste Management, 61*, 3–12.

Mato, Y., Isobe, T., Takada, H., Kanehiro, H., Ohtake, C., et al. (2001). Plastic resin pellets as a transport medium for toxic chemicals in the marine environment. *Environmental Science & Technology, 35*, 318–324.

McComb, B. C., & Cushman, S. A. (2020). Synergistic effects of pervasive stressors on ecosystems and biodiversity. *Frontiers in Ecology and Evolution, 8*, 398.

Morrison, M., Trevisan, R., Ranasinghe, P., Merrill, G. B., Santos, J., Hong, A., et al. (2022). A growing crisis for One Health: Impacts of plastic pollution across layers of biological function. *Frontiers in Marine Science, 9*, 980705.

Nabi, I., Ahmad, F., & Zhang, L. (2021). Application of titanium dioxide for the photocatalytic degradation of macro-and micro-plastics: A review. *Journal of Environmental Chemical Engineering, 9*(5), 105964.

Nash, K. L., Cvitanovic, C., Fulton, E. A., Halpern, B. S., Milner-Gulland, E. J., Watson, R. A., & Blanchard, J. L. (2017). Planetary boundaries for a blue planet. *Nature Ecology & Evolution, 1*(11), 1625–1634.

Neto, J. A. B., Gaylarde, C., Beech, I., et al. (2019). Microplastics and attached microorganisms in sediments of the Vitória bay estuarine system in SE Brazil. *Ocean Coast Manage, 169*, 247–253.

Nielsen, T. D., Holmberg, K., & Stripple, J. (2019). Need a bag? A review of public policies on plastic carrier bags – Where, how and to what effect? *Waste Management, 87*, 428–440.

Nielsen, T. D., Hasselbalch, J., Holmberg, K., & Stripple, J. (2020). Politics and the plastic crisis: A review throughout the plastic life cycle. *Wiley Interdisciplinary Reviews: Energy and Environment, 9*(1), e360.

Njeru, J. (2006). The urban political economy of plastic bag waste problem in Nairobi, Kenya. *Geoforum, 37*, 1046–1058.

North, E. J., & Halden, R. U. (2013). Plastics and environmental health: The road ahead. *Reviews on Environmental Health, 28*(1), 1–8.

O'Melia, C. R. (1978). Coagulation in wastewater treatment. In *The scientific basis of flocculation* (pp. 219–268). Springer Netherlands.

Organisation de coopération et de développement économiques. (2022). *Global plastics outlook: Economic drivers, environmental impacts and policy options*. OECD Publishing.

Organisation for Economic Co-operation and Development. (2022). *Global plastics outlook: Policy scenarios to 2060*. OECD Publishing.

Orr, J. A., Vinebrooke, R. D., Jackson, M. C., Kroeker, K. J., Kordas, R. L., Mantyka-Pringle, C., Van den Brink, P. J., De Laender, F., Stoks, R., Holmstrup, M., Matthaei, C. D., Monk, W. A., Penk, M. R., Leuzinger, S., Schafer, R. B., & Piggott, J. J. (2020). Towards a unified study of multiple stressors: Divisions and common goals across research disciplines. *Proceedings of the Biological Sciences, 287*(1926), 20200421.

Orset, C., Barret, N., & Lemaire, A. (2017). How consumers of plastic water bottles are responding to environmental policies? *Waste Management, 61*, 13–27.

Pandey, P., & Saini, V. K. (2018). *Pillared interlayered clays for pollution remediation* (pp. 353–376). Innovative Materials.

Pandey, P., & Saini, V. K. (2019). Pillared interlayered clays: Sustainable materials for pollution abatement. *Environmental Chemistry Letters, 17*, 721–727.

Peng, Y., Prabhu, A., & Rinke, C. (2023). Facing our plastic waste crisis: Biorecycling as a promising solution. *Microbiology Australia, 44*(1), 52–56.

Peng, Y., Prabhu, A., & Rinke, C.. *Facing our plastic waste crisis: Biorecycling as a promising solution.* https://www.oecd.org/environment/plastics/plastics-lifecycle-is-far-from-circular.htm

Persson, L. M., Breitholtz, M., Cousins, I. T., de Wit, C. A., MacLeod, M., & McLachlan, M. S. (2013). Confronting unknown planetary boundary threats from chemical pollution.

Phelan, A. (2022). REDcycle's collapse is more proof that plastic recycling is a broken system. *The Conversation, 17*.

Plastics Europe. (2009). *The compelling facts about plastics. An analysis of European plastics production, demand and recovery for 2008*. Plastics Europe, European Plastics Converters, European Association of Plastics Recycling and Recovery Organisations and European Plastics Recyclers.

Plastics Europe. (2016). *Plastics – The facts. An analysis of European plastics production, Demand and waste data*. Plastics Europe.

Poortinga, W., & Whitaker, L. (2018). Promoting the use of reusable coffee cups through environmental messaging, the provision of alternatives and financial incentives. *Sustainability, 10*(3), 873.

Poortinga, W., Sautkina, E., Thomas, G. O., & Wolstenholme, E. (2016). *The English plastic bag charge: Changes in attitudes and behaviour.*

Pouikli, K., Delegkou, A., & Tsakalogianni, I. (2022). Circular economy for a Climate-Neutral Europe: Mapping the latest policy and legislative developments. *Carbon & Climate Law Review, 16*, 281.

Proshad, R., Kormoker, T., Islam, M. S., Haque, M. A., Rahman, M. M., & Mithu, M. M. R. (2018). Toxic effects of plastic on human health and environment: A consequences of health risk assessment in Bangladesh. *International Journal of Health, 6*(1), 1–5.

Rahman, M., & Brazel, C. S. (2004). The plasticizer market: An assessment of traditional plasticizers and research trends to meet new challenges. *Progress in Polymer Science, 29*(12), 1223–1248.

Raubenheimer, K., & McIlgorm, A. (2017). Is the montreal protocol a model that can help solve the global marine plastic debris problem? *Marine Policy, 81*, 322–329.

Rayner, J. L., Wood, C., & Fenton, S. E. (2004). Exposure parameters necessary for delayed puberty and mammary gland development in Long–Evans rats exposed in utero to atrazine. *Toxicology and Applied Pharmacology, 195*(1), 23–34.

Reddy, R. L., Reddy, V. S., & Gupta, G. A. (2013). Study of bio-plastics as green and sustainable alternative to plastics. *International Journal of Emerging Technology and Advanced Engineering, 3*(5), 76–81.

Rustagi, N., Pradhan, S. K., & Singh, R. (2011). Public health impact of plastics: An overview. *Indian Journal of Occupational And Environmental Medicine, 15*(3), 100.

Ryan, P. G., Moore, C. J., Franeker, J. A., & Moloney, C. L. (2009). Monitoring the abundance of plastic debris in the marine environment. *Philosophical Transactions of the Royal Society of London. Series B, Biological Sciences, 364*, 1999–2012.

Sabry, F. (2022). *Bioplastica: Lífið í lífplasti er frábærra. Er það lífrænt eða niðurbrjótanlegt plast? Er það sigur eða hreinn skáldskapur?* (Vol. 3). One Billion Knowledgeable.

Sarwan, B., Acharya, A. D., Kaur, S., & Pare, B. (2020). Visible light photocatalytic deterioration of polystyrene plastic using supported BiOCl nanoflower and nanodisk. *European Polymer Journal, 134*, 109793.

Schnurr, R. E. J., Alboiu, V., Chaudhary, M., Corbett, R. A., Quanz, M. E., Sankar, K., Srain, H. S., Thavarajah, V., Xanthos, D., & T.R. (2018). Walker reducing marine pollution from single-use plastics (SUPs): A review. *Marine Pollution Bulletin, 137*, 157–171.

Schönfelder, G., Wittfoht, W., Hopp, H., Talsness, C. E., Paul, M., & Chahoud, I. (2002). Parent bisphenol A accumulation in the human maternal-fetal-placental unit. *Environmental Health Perspectives, 110*(11), A703–A707.

Shams, M., Alam, I., & Mahbub, M. S. (2021). Plastic pollution during COVID-19: Plastic waste directives and its long-term impact on the environment. *Environmental Advances, 5*, 100119.

Sheavly, S. B. (2005). *Marine debris – An overview of a critical issue for our oceans*. p. 7.

Shen, M., Zeng, G., Zhang, Y., Wen, X., Song, B., & Tang, W. (2019). Can biotechnology strategies effectively manage environmental (micro) plastics? *Science of the Total Environment, 697*, 134200.

Silva, A. L. P., Prata, J. C., Walker, T. R., Campos, D., Duarte, A. C., Soares, A. M. V. M., Barcelò, D., & Rocha-Santos, T. (2020). Rethinking and optimising plastic waste management under COVID-19 pandemic: Policy solutions based on redesign and reduction of single-use plastics and personal protective equipment. *Science of the Total Environment, 742*, Article 140565.

Simon, N., Raubenheimer, K., Urho, N., Unger, S., Azoulay, D., Farrelly, T., et al. (2021). A binding global agreement to address the life cycle of plastics. *Science, 373*(6550), 43–47.

Singh, P., & Sharma, V. P. (2016). Integrated plastic waste management: Environmental and improved health approaches. *Procedia Environmental Sciences, 35*, 692–700.

Sosale, S., Shepardson, K., Aedo, C., & Jha, A. (2021). Tackling plastic pollution and climate change through education. *Worldbank Blogs*. https://blogs.worldbank.org

Srinivasan, S. K., Deivasigamani, U. M., Chidambaram, B., Raju, G. P., Kamaraj, Y., Manikantan, C. S., ... & Rengaraj, J. (2022, October). Converting plastics to fuel using solar energy. In AIP conference proceedings (Vol. 2527, No. 1). AIP Publishing.

Steffen, W., Broadgate, W., Deutsch, L., Gaffney, O., & Ludwig, C. (2015). The trajectory of the Anthropocene: The great acceleration. *Anthropocene Review, 2*(1), 81–98.

Stöfen-O'Brien, A. (2022). The prospects of an international treaty on plastic pollution. *The International Journal of Marine and Coastal Law, 37*(4), 727–740.

Tekman, M. B., Walther, B., Peter, C., Gutow, L., & Bergmann, M. (2022). *Impacts of plastic pollution in the oceans on marine species, biodiversity and ecosystems*. WWW Germany.

Tiwari, E., Singh, N., Khandelwal, N., Monikh, F. A., & Darbha, G. K. (2020). Application of Zn/Al layered double hydroxides for the removal of nano-scale plastic debris from aqueous systems. *Journal of Hazardous Materials, 397*, 122769.

UNEP. (2014). *Valuing plastics: The business case for measuring, managing and disclosing plastic use in the consumer goods industry*. United Nations Environment Programme.

Vedelago, C. (2022). The fire that sparked the end of Coles and woolies plastic recycling program. *The Sydney Morning Herald, 9*.

Wang, Q., Bai, J., Ning, B., Fan, L., Sun, T., Fang, Y., et al. (2020). Effects of bisphenol a and nanoscale and microscale polystyrene plastic exposure on particle uptake and toxicity in human caco-2 cells. *Chemosphere, 254*, 126788. https://doi.org/10.1016/j.chemosphere.2020.126788

Warner, M., Eskenazi, B., Mocarelli, P., Gerthoux, P. M., Samuels, S., Needham, L., et al. (2002). Serum dioxin concentrations and breast cancer risk in the Seveso Women's health study. *Environmental Health Perspectives, 110*(7), 625–628.

Wei, S., & Kamali, A. R. (2022). Trifunctional mesoporous magnetic adsorbent-photocatalyst nanocomposite for efficient removal of potassium ethyl xanthate from mining wastewater. *Journal of Water Process Engineering, 49*, 103067.

Wikström, F., Williams, H., & Venkatesh, G. (2016). The influence of packaging attributes on recycling and food waste behaviour—An environmental comparison of two packaging alternatives. *Journal of Cleaner Production, 137*, 895–902.

Worm, B., Lotze, H. K., Jubinville, I., Wilcox, C., & Jambeck, J. (2017). Plastic as a persistent marine pollutant. *Annual Review of Environment and Resources, 42*, 1–26.

Wright, S. L., & Kelly, F. J. (2017). Plastic and human health: A micro issue? *Environmental Science & Technology, 51*, 6634–6647. https://doi.org/10.1021/acs.est.7b00423

Xu, Q., Huang, Q. S., Luo, T. Y., Wu, R. L., Wei, W., & Ni, B. J. (2021). Coagulation removal and photocatalytic degradation of microplastics in urban waters. *Chemical Engineering Journal, 416*, 129123.

Yong, C. Q. Y., Valiyaveettil, S., & Tang, B. L. (2020). Toxicity of microplastics and nanoplastics in mammalian systems. *International Journal of Environmental Research and Public Health, 17*, 1509. https://doi.org/10.3390/ijerph17051509

Yuan, F., Yue, L., Zhao, H., & Wu, H. (2020a). Study on the adsorption of polystyrene microplastics by three-dimensional reduced graphene oxide. *Water Science and Technology, 81*(10), 2163–2175.

Yuan, J., Ma, J., Sun, Y., Zhou, T., Zhao, Y., & Yu, F. (2020b). Microbial degradation and other environmental aspects of microplastics/plastics. *Science of the Total Environment, 715*, 136968.

Zubris, K. A. V., & Richard, B. K. (2005). Synthetic fibers as an indicator of land application of sludge. *Environmental Pollution, 138*, 201–211.

Chapter 2
Constraints of Conventional Strategies in Managing Plastic Waste and Future Challenges

Adyasha Sahu, Sanjay Chandravanshi, Rishikesh Venkatrao Kadam, Narsale Swapnil Ananda, Abhinanda Jena, and Sudhan Chandran

Abstract In recent years, managing plastic garbage has become one of the most important environmental challenges. Conventional approaches that have been used to solve this issue have run across a number of obstacles that have reduced their efficacy in reducing plastic pollution. Inefficient recycling practices, the prevalence of single-use plastics, inadequate waste management infrastructure, difficulties with collection and sorting and the ongoing plastic waste discharged into the environment are now the biggest issues of conventional plastic waste management strategies that are highlighted in this paper. Looking ahead, future challenges in management of waste plastic are emerging as the problem continues to escalate. These difficulties include addressing the growing levels of waste plastic the fragmentation of plastic into hazardous microplastics and nanoplastics, the problem of pollution from plastic waste in the ocean, the creation and application of efficient policies and regulations and the promotion of international cooperation. In order to prevent plastic trash, communities, businesses, researchers, government agencies and individuals must work together. Preserving our planet for future generations requires concerted action to reduce plastic pollution and foster a cleaner and healthier environment for all living beings.

A. Sahu (✉) · S. Chandravanshi · S. Chandran
Department of Fisheries Biology and Resource Management, Fisheries College and Research Institute, TNJFU, Thoothukudi, Tamil Nadu, India

R. V. Kadam
Department of Aquatic Environment Management, Fisheries College and Research Institute, TNJFU, Thoothukudi, Tamil Nadu, India

N. S. Ananda
Department of Fish Pathology and Health Management, Fisheries College and Research Institute, TNJFU, Thoothukudi, Tamil Nadu, India

A. Jena
Department of Fish Processing Technology, Fisheries College and Research Institute, TNJFU, Thoothukudi, Tamil Nadu, India

R. Soni et al. (eds.), *Advanced Strategies for Biodegradation of Plastic Polymers*, https://doi.org/10.1007/978-3-031-55661-6_2

Keywords Waste management · Conventional techniques · Plastic waste · Microplastic · Environmental health

2.1 Introduction

Plastics have globally expanded commodity due to their exceptional qualities and vast range of applications at low cost (Liang et al., 2021). It is primarily a non-biodegradable material made from petroleum that has a distinctive polymeric structure and low specific weight and thermal conductivity, and excellent durability and strong mechanical characteristics, and affordable prices, among other benefits (Eagle et al., 2016; Wang et al., 2016; Landon-Lane, 2018; Mazhandu et al., 2020). The steadily rising population and quick urbanisation and industrial growth are some of the major human-made problems that ultimately put life on this planet in jeopardy (Kibria et al., 2023). This offers significant challenges to existing waste management systems (Hopewell et al., 2009; Horodytska et al., 2018). Polybags, eating utensils, packaging materials and other single-use plastic items are thought to account for around 50% of the total amount of waste of plastics finally meet in the ocean (Duru et al., 2019). Single-use plastic (SUP) is increasingly commonly utilised in packaging due to its exceptional preservation, protection and cost-effectiveness qualities (Gahleitner & Paulik, 2017). In India, 43% of manufactured plastics are used for packaging purpose, in which most of them are one-time use. Multilayered plastics are recyclable and energy recoverable or have another alternative application; however, recycling them is an expensive procedure. The effects of plastic waste on the environment are evident in the rising levels of plastic pollution both on land and in the world ocean. Similar to inland plastic waste, marine pollution from living things is getting increasingly concerned about plastic garbage, which is thought to be present in between 60% and 80% of marine competitions (Paletta et al., 2019; Peng et al., 2020). Almost 80% of the marine waste are found in the ocean's surface water and the deep sea, and 14 million tonnes of plastic waste are released into the oceans annually. Certainly, the world's polluted oceans are mostly found in Mumbai, Kerala and the Andaman and Nicobar Islands (fact sheet of plastic, 2018). Investigating existing and new legally enforceable agreements is urgently required to prevent marine plastic pollution. The coastal ecosystems are negatively impacted by significant quantities of heavy metals such as Pb, Zn, Cu and Cd which are collected from plastic debris at seashores. Lead and cadmium pigments, which are frequently employed as additives in the majority of plastics, are dangerous to the environment and have a history of leaching out (fact sheet of plastic, 2018). Despite the low cost of plastic production, 90% of products are only used once and then discarded, contributing to the environmental disaster worldwide. 6.3 billion tonnes of plastic garbage have been produced globally since the advent of plastic. The current review also intended to list the limitations of traditional plastic waste management approaches and upcoming difficulties.

2.2 Status of Plastic Generation

Since plastics are utilised practically in every industry, particularly in food packaging and other products, a lot of plastic waste is produced, which has an impact on the ecosystem. According to a survey by the World Bank Group, plastics make up between 5% and 12% (20–30% by weight) of the entire amount of waste generated worldwide, and 60% of plastic waste end up in the environment (Zhang et al., 2020). Globally, the use of plastic has reached 385 Mt. in 2020 alone, with 250 Mt. of garbage produced overall (Global Plastics Flow, 2018). According to estimates, 70% of plastic packaging items are quickly converted into plastic waste. The country produces 26,000 TPD2 worth of thermoplastic polyamide elastomer (TPA) plastic waste, or around 9.4 million TPAs. About 60% of the plastic waste is recycled, which is mostly done by the unorganised sector. Some plastics do not disintegrate at all, while others may take up to 450 years. The majority of plastic waste, approximately 8% of the total solid waste, is produced in Delhi and then followed by Kolkata and Ahmedabad. The current recycling rate for plastic waste is only 60% (Fig. 2.1).

Figure 2.2 displays the per-person waste creation trend over the previous 5 years (2016–2020). It has been noted that the output of plastic garbage per capita has increased by about double in the last 5 years.

The anticipated volume of plastic waste for the 2019–2020 fiscal year is 3,469,780 TPA, according to data from the 35 states/UTs. The distribution of plastic garbage generation by state and UT is shown in Table 2.1 as a percentage. Punjab, Tamil Nadu and Maharashtra are known to create the highest plastic waste, in that order (Fig. 2.3).

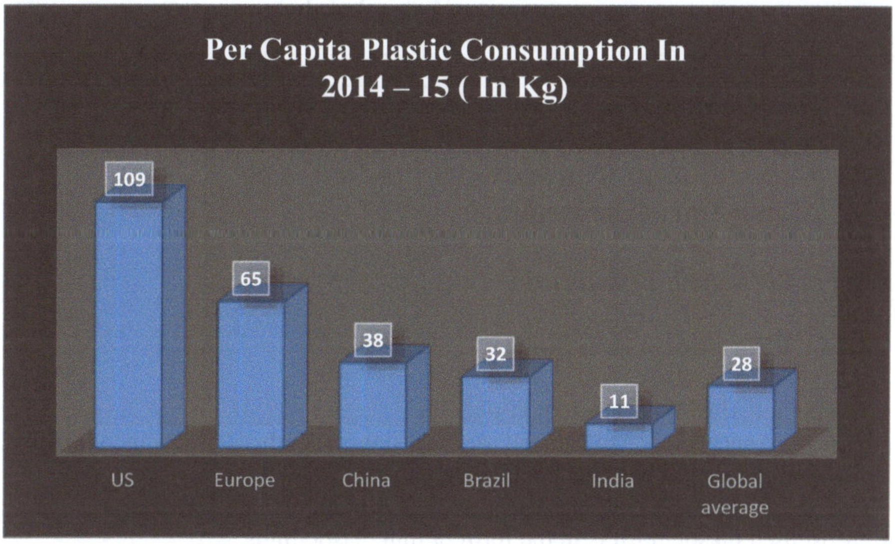

Fig. 2.1 Per capita plastic consumption in 2014–2015 (in kg). (Source: FICCI)

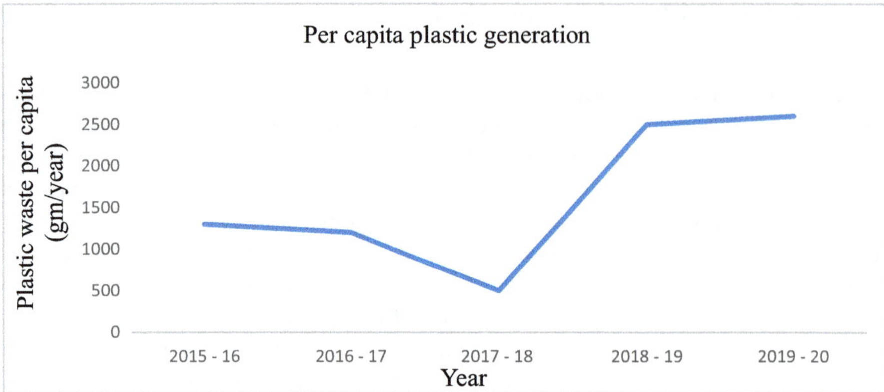

Fig. 2.2 Per capita plastic generation. (Source: FICCI)

Table 2.1 Plastic waste generation in India

States/UTs	Plastic waste generation (%)
Maharashtra	13
Tamil Nadu	12
Gujarat	12
West Bengal	9
Karnataka	9
Telangana	7
Delhi	7
Uttar Pradesh	5
Haryana	4
Kerala	4
Madhya Pradesh	3
Punjab	3
Others	11

Source: CPCB, Delhi

2.3 Constraints of Conventional Plastic Waste Management Strategies

2.3.1 Recycling Limitations

2.3.1.1 Difficulties in Recycling

- Difficult to distinguish from non-plastics.
- The different chemical makeup of plastic resins makes them largely incompatible.
- Recycling leads to polymer chain degradation.
- Recycled polymer is of poorer quality than virgin polymer.
- Most used plastic trash is in the form of thin plastic films.
- Because films have a limited commercial value, further effort is required.

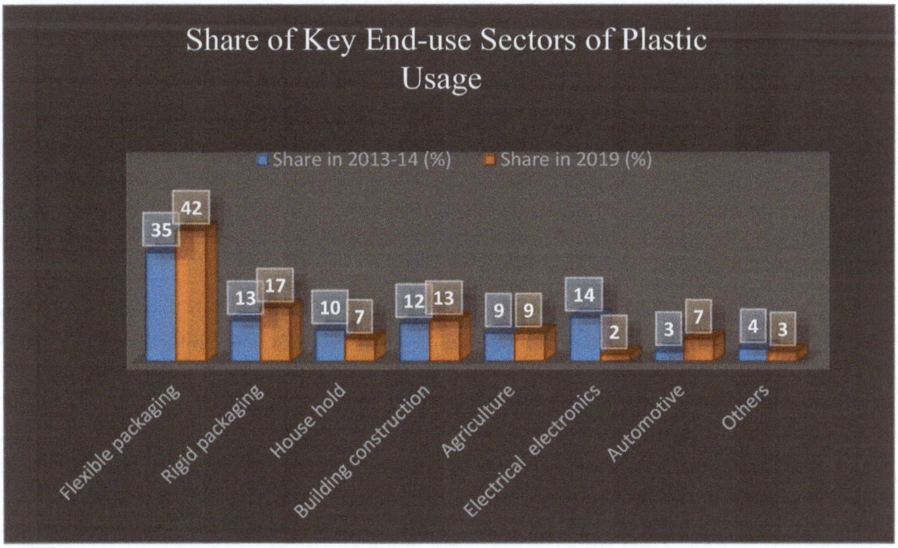

Fig. 2.3 Share of key end-use sectors of plastic usage. (Source: Plastindia Foundation, 2014 and 2019)

- Not invested in gathering them.
- Recognising opportunities for reusing and recycling.
- Lack of infrastructure and plastics markets.
- Low value of recycled plastics.
- Recycling programme subsidies.

The practice of recycling plastics can be complicated by a number of issues, including the gathering of plastic waste, sorting different types of plastics, cleaning the garbage and potential plastic pollution. The low value of the majority of the products that can be made from recycled plastics adds to the complexity. Not all plastics can be recycled readily, and doing so can be expensive and energy-intensive. Recycling plastic is better than reusing it since reusing it requires fewer resources and less energy, yet recycling plastic still requires less energy than producing plastic from the ground up.

2.3.1.2 Collection and Sorting of the Plastic Waste

Waste collection, storage spaces and transport are necessary components of any kind of waste management system, but these tasks can be tough in urban areas. The collection and sorting of plastic trash are key steps in the mechanical recycling method since they occur at the start of the chain and establish the quantity and complexity of the materials supplied for recycling (Feil et al., 2017). Reaching European recycling goals, improving resource efficiency and completing a circular economy cycle all depend on the collection of waste using source separation waste collection systems (Tallentire & Steubing, 2020). In order to collect plastic waste, bring-schemes can be used, such as the deposit-refund system, also known as the deposit-return system (DRS) (for polyethylene terephthalate), or collection of kerbside (pertaining to plastic packaging materials not protected by deposit systems) (Hopewell et al., 2009). Along with aluminium cans and glass bottles, many nations have DRS programmes in place that target PET bottles specifically, and they have demonstrated strong collection rates (Watkins et al., 2019). In 2017, Germany used 97.3% of PET bottles for DRS, while 96.2% were recycled and used to make new fibres, films or food packaging.

Garbage collection in India is the responsibility of the municipal government, and garbage cans that are both biodegradable and inert are frequently offered (Guria & Tiwari, 2010; Das & Bhattarcharya, 2014; Bhoyar et al., 1999). Mixed biodegradable and inert wastes are frequently discarded, with open burning being a regular practice. Local authorities invest between Rs. 500 and Rs. 1000 per tonne in solid waste management (SWM), with 70% of the expenditure going towards collecting and 20% towards transportation. Some cities have established segregation centres across the city to permit secondary segregation of plastic trash into 25–27 categories and aid in plastic recycling, in addition to segregating waste at the source.

To preserve high-value materials, collection and sorting are essential tasks, yet the current infrastructures are not standard between cities and nations. To enable high-quality recycling, the infrastructure generally needs to be reorganised (European Commission, 2011). Plastic is a complex material. It is critical to split plastic into distinct groups so that it can be kept as pure as possible.

2.3.1.3 Contaminants

According to Pivnenko et al. (2015), the main contaminations that can be found in plastics along the entire value chain of plastics were enumerated. They consist of additives, polymer contamination caused by other polymers, non-polymer contaminants and by-products of polymer degradation. Every stage of the value chain, with the exception of the extraction procedure, is contaminated. Production-stage contamination is caused by additives. The majority of contamination happens during the manufacturing process (as a result of the design requirements for several layers, components, labels and inks). In the application stage, there may be additional contamination (often made up of breakdown products and non-polymer contaminants), segregation, collection, sorting and reprocessing (Brouwer et al., 2020). Chemical additives used to give preset attributes for application and marketing demands are just one of the many pollutants found in plastics (Sanchez et al., 2020; Bartsch, 2019; Singh et al., 2012). Additionally, contamination and the mixing of different kinds of plastic hamper recycling efficiency and lower the quality of the recycled material.

2.3.1.4 Additives in Plastics

The resins made of polymers virtually are not ever utilised as pure polymers; instead, they are typically blended with other chemicals (also known as additives) (Alassali et al., 2020; Thompson et al., 2009). In order to improve the qualities of plastics, additives are inserted into the polymeric matrix. This is related to little to no change in the polymer's chemical composition. According to Andrady and Neal (2009), additives are occasionally the most valuable components of a formulation. They can either protect a substance from damage (e.g. heat- and UV-stabilising agents, flame retardant antioxidants) or enhance its physical properties (e.g. fillers, colourants, lubricants and antistatic agents). Based on their function, additives for the production of plastics could be categorised (Hansen et al., 2014):

- Functional additives (e.g. cross-linkers, hardeners, plasticisers, flame retardants, stabilisers, blowing agents, and UV stabilisers).
- Colourants (e.g. carbon black (black), titanium dioxide (white) and various inorganic oxides).
- Fillers (e.g. silica, calcium carbonate, talc, carbon black, zinc oxide, and titanium dioxide).
- Reinforcements.

2.3.1.5 Limitations to Mechanical Recycling Technologies

Technologies for mechanical recycling have limitations since any type of plastics reacts to the procedure differently based on its chemical composition, mechanical behaviour and thermal characteristics. Temperature-sensitive polymers, composite materials and polymers (such thermosets) that do flow at high temperature cannot be treated mechanically. Due to this, only two types of plastic, polyethylene and polypropylene, collectively constitute 9% and 37% of the world's annual plastic production, are mechanically recovered and recycled. The recovery rate for other solid waste of plastic is either 0% or less than 1% of production (MacArthur, 2017). The recycling rate of plastic waste is a little higher in Europe, at 30% (Geyer et al., 2017).

Chemical recycling refers to modern technology that extends beyond mechanical recycling like thermolysis which uses catalysts to selectively produce fuels, gases or waxes (Garcia & Robertson, 2017). Because of the high cost of energy, chemical recycling is a practice that is not often used. The burning of materials while gathering energy in the form of heat is an additional alternative. Because it eliminates the necessity for sifting, incinerating mixed garbage is more convenient, but once the initial materials are burned, they cannot be recovered or used again. Additionally, it uses up more energy than recycling does (Geyer et al., 2017).

2.3.2 Single-Use Plastic (SUP)

SUP is a great alternative for packaging materials because of its outstanding potency to weight ratio, strong barrier properties and relatively inexpensive cost of production (Lewis et al., 2010). According to data released in 2018, the polyethylene (mainly LDPE and HDPE) industry is currently an industry with a value of close to "100 million tonnes" and a volume of "close to 100 million tonnes" (Hutley & Ouederni, 2016). Polyethylene accounts for more than 31% of the world market for plastics, while polypropylene accounts for 27% of the market (Hutley & Ouederni, 2016).

The growth of microalgae, primary producers at the base of the food chain, was affected by the significant rise in pollution caused by plastic debris. When exposed to uncharged polystyrene particles at higher concentrations, it has been found that the expansion of microalgae, which is also an important oxygen sources, was negatively impacted by up to 45%, but its photosynthesis was unaffected (Sjollema et al., 2016).

Despite having so many benefits, the substance is not biodegradable; thus, it can persist in nature without significantly losing weight. There is additional evidence to suggest that UV rays and weathering cause large plastic particles to slowly biodegrade into smaller, less dense particles. The slow rates of degradation endanger the environment over time. Numerous academic publications have discussed the potential for harmful pollutants such as heavy metals, organic compounds and reaction or breakdown products to migrate and contaminate packaged food products. Many health risks are now caused by the hazardous chemicals that SUPs release into the environment. Consequently, it is one of the main issues with the usage of SUPs as food contact materials nowadays.

These non-recyclable products have a double-edged effect on the environment since consumers throw them away in the wild despite the fact that they are not recyclable or reusable. These might promote more severe environmental pollution. In addition, using multicomponent composites as food contact materials increases the likelihood of food contamination. Posing practical solutions to collect, reuse and reprocess the SUPs after usage is a significant task in and of itself.

2.3.3 Infrastructure and Technology Gaps

Infrastructure for waste management is crucial for achieving sustainable development. Waste materials have potential resources, and solid waste management requires efficient waste management combined with resource extraction. Many individuals may be able to make a living by extracting value from trash in the form of materials, energy or nutrition. The only option to turn wastes into resources is to invest in SWM since this involves a planned series of actions to create markets and maximise the recovery of recyclables and reusable materials (Sridevi et al., 2012).

In developing nations, the issue of improperly and inadequately managing municipal solid waste, including plastic garbage, is particularly significant. According to Jambeck et al. (2018), this is caused by a lack of infrastructure, an unfair distribution of monetary resources, systematic poverty and environmental injustice caused by the removal of waste plastic generated by developed nations, as well as numerous instances of inadequate capacity for responsible, equitable and efficient governance of politics. It is possible to restore resources from debris using modern technology and India's long tradition of recycling.

Governments in underdeveloped nations are frequently unable to provide the infrastructure necessary for the subsequent generations, upkeep and management of solid waste, including roads that are secure and large enough to accommodate trash trucks, a workforce that is paid and landfills that are secure and expansive. Furthermore, political unrest and natural disasters frequently make it difficult for typical municipal solid waste management infrastructure to function (Ferronato & Torretta, 2019; Ferronato et al., 2019). Therefore, developing a national strategy that is effective and efficient for managing municipal solid waste necessitates a substantial investment of time, money and resources – all of which impoverished nations lack (Yukalang et al., 2017).

2.3.4 Leakage into the Environment

According to the UNEP, ten rivers in Asia and Africa are to blame for over a quarter of the marine plastic litter because of improper waste management of plastics, which has caused plastics to leak into seas. The estimated global leakage (terrestrial and marine) in 2019 was 22 Mt. By 2060, it is expected that this value will be doubled, i.e. 44 Mt. (OECD).

They are primarily macroplastics, which are easily recognisable objects like beverage bottles, and the bulk of them entered the environment due to insufficient waste disposal and collection. Fly-tipping, trash and marine activities are some other leakage pathways. A sizable portion of the total leakage is also made up of microplastics that are small fragments of plastic less than 5 mm (0.2 in.) long. These particles primarily enter the environment through tyre wear and road markings, but they can also accidentally fall out of plastic pellets and wash off of synthetic textile fibres. These figures highlight how urgent it is to address waste management practices, taking into account littering, leakage from maritime activities and the continuous rising leakage of microplastics globally.

According to a study, the anticipated amount of plastic waste in aquatic habitats is currently significant and will increase. Plastic leakage is affecting marine and terrestrial ecosystems while posing significant risks to people's livelihoods which depend on the integrity of such places, like fishing and tourism. In 2019 alone, 6.1 Mt. of plastic waste leaked into rivers, lakes and oceans. With 1.7 Mt. of plastics pouring into the ocean in 2019, it is likely that 109 Mt. of plastic will have accumulated in rivers around the world by 2019. The majority of plastics reach the ocean through rivers through a sluggish process that can take years or even decades. The quantity of leaky plastics in rivers is still concerning, even though inflow estimates are lower than those from earlier studies that did not take that into account. The quantity of plastics accumulating in aquatic habitats is anticipated to increase to more than triple from 140 Mt. in 2019 to 493 Mt. in 2060 due to the increased plastic use and waste.

2.3.5 Consumer Behaviour

An established state of mind, feeling or perspective is referred to as an approach when it is in response to an outside occurrence (Nagaraju & Thejaswini, 2016; Yusliza et al., 2020). A consumer's attempts to evaluate a good, service or similar item might be hampered by an attitude that may develop towards the thing being measured, whether it be temporary or long-lasting (Nagaraju & Thejaswini, 2016). External cues have an impact on how someone feels about something, which can facilitate the rise of specific habits (Yusliza et al., 2020). To put it in another way, according to Arifani and Haryanto (2018), the practical explanation of attitude is how a customer evaluates their likes and dislikes in light of any outside influences. Yadav and Pathak (2017) contend that attitude is the consequence of cognitive ideas on which an individual believes and the resulting judgements of a good or bad choice on the repercussions of an event (Arifani & Haryanto, 2018).

It is still difficult to persuade customers to adopt more environmentally friendly habits like cutting back on their use of plastic and properly discarding their plastic garbage (Hohmann et al., 2016). According to studies by Madigele et al. (2017) and Afroz et al. (2017), those with lower levels of education are more likely to use disposable plastic packaging, such as bags, than those with higher levels of education (Jeewska-Zychowicz & Jeznach, 2015). On the basis of income analysis, wealthier consumers use their bags more frequently when shopping and less frequently choose to purchase eco-plastic bags.

In the issue of plastics, many businesses are seeking for solutions in the realm of the "nudge theory" and the developing science of behaviour economics. Brands have a crucial role in influencing consumer behaviour. The nudge theory contends that by making minor changes to the environment, one can influence people's decisions without limiting their freedom of choice.

The growing popularity of online shopping and e-commerce reflects how customer's preferences are always changing. Packaging waste has increased as a result of the use of boxes and packaging supplies. As a result, managing the changing customer behaviour would be challenging.

2.3.6 Lack of Awareness

The most effective and practical methods for collecting and disposing of waste are not currently employed, which has led to India's current poor state of waste management. Techniques for managing trash are not adequately trained, and there are few certified waste management consultants (Khajuria et al., 2010). Local administrations in India are responsible for managing waste, although their finances are too tight to cover the costs of building effective systems for collecting, storing, treating and disposing of trash. Key obstacles to effective waste management in India include an absence of a systematic waste management plan, trash collection/segregation and a monetary regulatory structure from the government. Implementation of cutting-edge technology that might improve India's waste handling has been hampered by a lack of creativity and low enthusiasm. Public attitudes about rubbish are a big barrier to better trash disposal in India.

2.4 Future Challenges

2.4.1 Emerging Plastic Substitutes

2.4.1.1 Bioplastic

Bioplastics are produced from biomass, primarily sugars, and are subsequently transformed into plastics. This sugar is made from corn, sugar cane, sugar, beetroot or potatoes. Although not all bioplastics are biodegradable, some are (Bakar & Othman, 2019). Bioplastics are broken down into three groups, viz. those that are both biobased and biodegradable, those that are entirely bio-based and those that are only biodegradable. Examples of bioplastics that are both bio-based and biodegradable include poly(lactic acid) (Garlotta, 2001; Nampoothiri et al., 2010), polyhydroxy-alkanoates (PHA) (Chanprateep, 2010) and bio-based polybutylene succinate (Xu & Guo, 2010; Siracusa & Blanco, 2020). Last but not least, biodegradable bioplastics made from fossil resources include polycaprolactone (PCL) (Labet & Thielemans, 2009), polyvinyl alcohol (PVA) (Aslam et al., 2018) and polybutylene adipate terephthalate (PBAT) (Ferreira et al., 2019). Large-scale production of bioplastics is accomplished using poly-hydroxyl alkanoates (PHA) and polylactide acid (PLA). Biodegradable plastics, on the other hand, are those that can be broken down by microbes.

The recycling of bioplastics is possible, but it requires division into various streams of the recycled material (Narancic et al., 2018). Assuming a biodegradable material is added to the regular plastic stream and completely breaks down during recycling, in that instance, the characteristics and specifications of the conventional material may be changed (Shamsuddin et al., 2017). Energy recovery is widely used because of the significant amount of heat produced by plastics (Gioia et al., 2021).

PLA and other biodegradable plastics made from renewable resources typically include carbon, oxygen and hydrogen atoms, with no chlorine atoms. They do not degrade in landfills (Chong et al., 2021). They may therefore typically be burned safely without emitting dioxins or heavy metals (Dilkes-Hoffman et al., 2019). Due to the significant potential for recuperating energy, it should be preferred over the terminal phase methods including recycling, composting and anaerobic digestion (Havstad, 2020; Shen et al., 2020).

2.4.2 Microplastics and Nanoplastics

The upper limits for microplastics had previously been established at 5 mm by numerous investigations (Frere et al., 2018; Cai et al., 2020; Stock et al., 2021), whereas the upper limits for nanoplastics had previously been established at 100 nm (Lehner et al., 2019; Gonçalves & Bebianno, 2021). In addition, according to Boverhof et al. (2015), nanoparticles are classified by the International Organization for Standardisation as objects having their exterior dimensions lying in the nanoscale scale. In particular, microplastics are defined as plastic particles with a size between 100 and 5 mm, while nanoplastics are defined as plastic particles with a size of less than 100 nm (Amobonye et al., 2021). According to their sources, microplastics and nanoplastics have been divided into primary and secondary categories. Primary MNPs are small particles that are released into the environment as a result of numerous industrial processes and are associated with their specific applications and products for sale (Gonçalves & Bebianno, 2021). The formation of secondary MNPs results from the decomposition of plastic trash as a result of its exposure to physical, animal and microbiological forces (Laskar & Kumar, 2019).

Cosmetics containing microbeads and plastic pellets are the key sources of primary MNPs (Smyth et al., 2021). Secondary MNPs are created when macroplastic materials are broken down by a variety of environmental deterioration processes, including chemical (corrosion, photooxidation, temperature), mechanical (abrasion erosion, wave action) and chemical (photooxidation) biodegradation (Karbalaei et al., 2018). Municipal wastes, such as farming films, plastic bags and bottles, fishing gear, shipping, tyre wear from cars and other major plastic wastes, have been observed to be the principal sources of secondary MNPs in this regard. According to recent estimations, other sources are the main sources of MNPs in ecosystems of both land and water (An et al., 2020).

Microplastics and nanoplastics have primarily been recognised as marine pollutants due to estimations that they are present in hundreds of thousands of metric tonnes at the summit of various marine ecological communities (Wright et al., 2020). However, new research has also revealed that MNPs are contaminating inland water bodies (Li et al., 2020; Dong et al., 2021) and landmass (Mohajerani & Karabatak, 2020; Tan et al. ,2021). Moreover, as noted in numerous research (Jakubowicz et al., 2021; Sharma et al., 2021), due to their diversity and other inorganic plastic fragments and organic components, MNPs can form both homo- and hetero-aggregates. These unwanted aggregates harm the biotic elements in various ecosystems through the bioaccumulation and bioamplification phenomena that are caused by their creation (da Costa, 2018).

Aquatic, terrestrial and atmospheric ecological chambers are all impacted by MNP pollution, which is connected by different kinds of source-pathway-sink links that could influence MNP flow and retention (Bergmann et al., 2019; Zhang et al., 2020). Numerous studies have concentrated on the negative impact of MNPs on aquatic life, including microorganisms, invertebrates like zooplankton and vertebrates like fish, seabirds and amphibians (Frydkjaer et al., 2017; Sana et al., 2020). According to reports, MNPs prevent some yeast, bacteria and algae from growing, which has an impact on the important roles they play in various ecosystems (Nomura et al., 2016; Sun et al., 2018; Mammo et al., 2020). Additionally, it has been noted that MNPs impede the digestive tracts of zooplankton and marine benthic creatures like mussels and oysters, which frequently results in decreased appetite, malnutrition and fatalities (Lee et al., 2013; Van Cauwenberghe et al., 2013). MNPs can present health concerns to human beings because they are consumed by a range of freshwater and marine animals and can build up through the food web (Sana et al., 2020). It has been recommended that specialised bioindicator species should be chosen as sentinel creatures to biomonitor the effects of MNPs in various topographical habitats in order to ensure the integrity of aqueous foodstuffs. For an instance, lugworm (*Arenicola marina*), a powerful benthic food web deposit feeder, is frequently used in experiments to determine the toxicity of marine sediment (Besseling et al., 2017), and mussel (*Mytilus galloprovincialis*) is a widely recognised sentinel species used to monitor oceanic pollution (Al-Thawadi, 2020).

Most plastic wastes that wind up in aquatic environments were initially produced, used and carelessly abandoned on ground (de Souza Machado et al., 2018). The biota in terrene systems may be exposed through various paths, mainly affecting the geochemistry and leading to environmental toxicity. Consequently, terrestrial environments are viewed as massive MNP stores (Allouzi et al., 2021). MNPs have been known to persist in soil for hundreds of years (Wahl et al., 2021). They also interfere with their ability to grow, reproduce and survive and their longevity through a number of harmful pathways, such as bioaccumulation, damage to the genome, imbalance of the gut microbiota, histopathological destruction and endocrine disruption (Wang et al., 2021).

MNPs are thus one of the key issues that must be handled in the future to preserve both aquatic and terrestrial ecosystems.

2.4.3 *Plastic Waste in the Oceans*

On an annual basis, approximately 14 million tonnes of plastic enter into the sea. Plastic currently makes up 80% of all aquatic trash, which includes garbage from surface waters to underwater sediments, making it the most common type of garbage in the sea (IUCN 2021). Microplastics (particles less than 5 mm) and nanoplastics (particles smaller than 100 nm) are the tiny particles that are formed when plastic breaks down in the presence of ultraviolet (UV) rays from sunlight, airborne, waves and other naturally occurring events. Due to their tiny size, aquatic organisms can easily mistakenly devour them. Ecosystems may be impacted by the world's traffic in discarded plastics, both officially and unofficially, in countries where waste management practice are inadequate to regulate plastic waste.

A serious issue that impacts the aquatic environment is the presence of plastic. It poses a threat to the well-being of the sea, aquatic life, the quality and safety of food, the well-being of humans, coastal recreation and climate change. The most obvious repercussions of plastic waste are the eating, asphyxiation and entanglement of hundreds of aquatic species. Seabirds, whales, fish and other aquatic animals assume plastic waste as food since their stomachs are filled with it, and as a result they starve to death. Additionally, they experience internal wounds, scrapes, illnesses and a decline in swimming capability. Meanwhile, floating debris serves as a breeding ground for invasive marine species, which threatens aquatic biodiversity and the food web. Plastic waste reduces the aesthetic appeal of tourist destinations, which affects tourism revenue. Additionally, it results in large financial outlays for site maintenance and cleanup.

2.4.4 Circular Economy Implementation

One of the most pressing concerns confronting society currently is developing a business model that incorporates circular economy (CE) concepts. CE presupposes sensible utilisation of raw commodities by enabling them to move across the market, provided that doing so is ethically and environmentally sound (Katarzyna Joachimiak-Lechman et al., 2020). The circular economy is implemented at the community level through better waste collection and management procedures, consumer behaviour change, plastic recycling, reduction and reuse of new products. By incorporating waste back into the supply network in order to create on-demand commodities for a variety of business sectors, this technique improves economic sustainability (Nascimento et al., 2019; Sandvik & Stubbs, 2019; Kocí et al., 2016).

The circular economy aims to preserve commodities in use for as many years as feasible, get maximum benefit from them while they are utilised and recover and regenerate items and supplies at the end of their productive lifespans as a substitute to the existing linear, make, use, discard economy model (WRAP). The circular economy boosts a restorative and regenerative production and consumption paradigm by design. It aims to reduce waste creation by planning out waste and hazardous materials (EC, 2011) while maintaining the worth of products, resources and materials in the economy as high as is practical for as long. Both biological and technical materials can be used in the circular economy. Through a "value circle," in which manufacturers, consumers, businesses and the government all play significant roles, it encourages systems approach and invention to maintain the continuous flow of materials.

Though CE plays an important role in managing plastic wastes these days, it faces some challenges in implementing the circular economy:

- Being restricted into a linear plastic production method renders altering costly.
- Hazards and significant upfront expenses associated with switching to a circular model.
- Complicated worldwide delivery systems for manufacturing and consumption.
- A lack of expertise for expanding circular models, particularly among small to medium businesses.
- Challenges in company-to-company collaboration, such as transaction costs.
- Uncompetitive circular products because subsidies support the linear production and usage model.
- Insufficient implementation ability and skill.
- Plastics are not sufficiently considered in important legislation.
- Inadequate standards and unfriendly regulations.
- Insufficient data collection and reporting, especially in developing countries.
- A deficit of customer attention or passion, as well as a reluctance to accept recycled products.

To overcome these challenges, extensive legislative and regulatory assistance will be required to stimulate innovation, increase the circular model's competitive edge and generate consumer appetite for circular plastic items. Collaboration with businesses will also be necessary to spur change, as will cooperation with the general public to promote societal norms changes and customer demand for circular products.

2.4.5 Policy and Regulations

It is vital to establish effective laws and regulations to promote responsible plastic manufacture, use and disposal. Striking a balance between environmental concerns, economic interests and industry needs, on the other hand, can be difficult.

The MoEF published the MSW (Management and Handling) Rules 2000 to promote efficient waste management in India, and more recently, revised drafted rules have been published (MoEF, 2015). Municipal governments are in charge of developing the infrastructure for MSW gathering, keeping, segregation, distribution and dumping as well as of enforcing these requirements. Chandigarh has improved waste management and is the first city in India to utilise SWM in a systematic way (Rana et al., 2015).

2.4.6 Global Collaboration

Plastic garbage is a global issue, and effectively addressing it necessitates international collaboration and cooperation. It is critical to bridge the waste management infrastructure and technological gap between industrialised and developing countries.

The worldwide task of eliminating plastic waste requires unique regional solutions. Supporting the implementation of waste management systems in communities without basic waste collection, particularly in areas with substantial plastic leakage, is necessary. We need to develop the waste hierarchy in other areas and demonstrate the effectiveness and originality of fresh approaches, including advanced recycling. Our work is supported by regional and national task groups acting in particular geographies in order to adopt a genuinely global perspective. The current task groups are concentrated in the Middle East and Africa, North America, South America, India, China and Japan. Each task group directs the choice and implementation of initiatives utilising crucial context that harmonises our approach with local and national priorities (David Taylor, 2020).

The 2030 Sustainable Development Goals are becoming more difficult to achieve. There are currently three billion people without access to basic waste management services in the earth, and as the global population increases, so does the yearning for higher living standards.

2.5 Conclusion

In conclusion, traditional approaches to managing plastic waste have run into a number of issues that have limited their ability to effectively address the problem of global plastic pollution. Some of the major obstacles are the difficulty in recycling, the prevalence of single-use plastics, the inefficient infrastructure for waste management, the difficulties in collecting and sorting waste and the ongoing removal of plastic waste into the environment.

In the area of managing plastic garbage, there are new difficulties appearing as we look to the future. These difficulties include managing the expanding plastic waste stream, addressing the decentralisation of plastic into micro- and nanoplastics, combating ocean plastic pollution, establishing efficient policies and regulations, converting to a circular economy and encouraging international cooperation.

A thorough and integrated strategy is required to get around existing limitations and future obstacles. This entails encouraging the creation and use of environmentally friendly plastic substitutes and biodegradable substitutes, investing in cutting-edge recycling technologies, improving waste collection and sorting systems, putting in place efficient policies and regulations and promoting environmentally friendly consumer behaviour. Additionally, to address the global aspect of the plastic trash problem, international engagement and cooperation are essential.

In the end, combating plastic waste involves a team effort from governmental entities, businesses, researchers, communities and people. We can take steps towards a more sustainable and responsible approach to managing plastic trash and reducing its environmental impact by cooperating and accepting creative ideas. Prioritising the protection of our world for the coming generations by eliminating plastic pollution and fostering a cleaner, healthier environment for all living things is essential.

References

Afroz, R., Rahman, A., Masud, M. M., & Akhtar, R. (2017). The knowledge, awareness, attitude and motivational analysis of plastic waste and household perspective in Malaysia. *Environmental Science and Pollution Research, 24*(3), 2304–2315. https://doi.org/10.1007/s11356-016-7942-0. [2019 10 01].

Alassali, A., Barouta, D., Tirion, H., Moldt, Y., & Kuchta, K. (2020). Towards a high quality recycling of plastics from waste electrical and electronic equipment through separation of contaminated fractions. *Journal of Hazardous Materials, 387*, 121741.

Allouzi, M. M. A., Tang, D. Y. Y., Chew, K. W., Rinklebe, J., Bolan, N., Allouzi, S. M. A., & Show, P. L. (2021). Micro (nano) plastic pollution: The ecological influence on soil-plant system and human health. *Science of the Total Environment, 788*, 147815. https://doi.org/10.1016/j.scitotenv.2021.147815

Al-Thawadi, S. (2020). Microplastics and nanoplastics in aquatic environments: challenges and threats to aquatic organisms. *Arabian Journal for Science and Engineering, 45*, 4419–4440. https://doi.org/10.1007/s13369-020-04402

Amobonye, A., Bhagwat, P., Raveendran, S., Singh, S., & Pillai, S. (2021). Environmental impacts of microplastics and nanoplastics: A current overview. *Frontiers in Microbiology, 12*, 3728.

An, L., Liu, Q., Deng, Y., Wu, W., Gao, Y., & Ling, W. (2020). Sources of microplastic in the environment. In *Microplastics in terrestrial environments: Emerging contaminants and major challenges* (pp. 143–159). Springer.

Andrady, A. L., & Neal, M. A. (2009). Applications and societal benefits of plastics. *Philosophical Transactions of the Royal Society B: Biological Sciences, 364*(1526), 1977–1984.

Arifani, V. M., & Haryanto, H. (2018, November). Purchase intention: implementation theory of planned behavior (Study on reusable shopping bags in Solo City, Indonesia). In *IOP conference series: Earth and environmental science* (Vol. 200, p. 012019). IOP Publishing.

Aslam, M., Kalyar, M. A., & Raza, Z. A. (2018). Polyvinyl alcohol: A review of research status and use of polyvinyl alcohol based nanocomposites. *Polymer Engineering & Science, 58*(12), 2119–2132.

Bakar, N. F. A., & Othman, S. A. (2019). Corn bio-plastics for packaging application. *Journal of Design for Sustainable and Environment, 1*(1).

Bartsch, N. (2019). *Polymer additives, contaminants and non-intentionally added substances in consumer products: Combined migration, permeation and toxicity analyses in skin.* Doctoral dissertation.

Bergmann, M., Mützel, S., Primpke, S., Tekman, M. B., Trachsel, J., & Gerdts, G. (2019). White and wonderful? Microplastics prevail in snow from the Alps to the Arctic. *Science Advances, 5*(8), eaax1157. https://doi.org/10.1126/sciadv.aax1157

Besseling, E., Foekema, E. M., van den Heuvel-Greve, M. J., & Koelmans, A. A. (2017). The effect of microplastic on the uptake of chemicals by the lugworm Arenicola marina (L.) under environmentally relevant exposure conditions. *Environmental Science & Technology, 51*(15), 8795–8804. https://doi.org/10.1021/acs.est.7b02286

Bhoyar, R. V., Titus, S. K., Bhide, A. D., & Khanna, P. (1999). Municipal and industrial solid waste management in India. *Journal of Indian Association for Environment Management, 23*, 53–64.

Boverhof, D. R., Bramante, C. M., Butala, J. H., Clancy, S. F., Lafranconi, M., West, J., et al. (2015). Comparative assessment of nanomaterial definitions and safety evaluation considerations. *Regulatory Toxicology and Pharmacology, 73*, 137–150. https://doi.org/10.1016/j.yrtph.2015.06.001

Brouwer, M. T., Thoden van Velzen, E. U., Ragaert, K., & ten Klooster, R. (2020). Technical limits in circularity for plastic packages. *Sustainability, 12*(23), 10021.

Cai, H., Chen, M., Chen, Q., Du, F., Liu, J., & Shi, H. (2020). Microplastic quantification affected by structure and pore size of filters. *Chemosphere, 257*, 127198.

Chanprateep, S. (2010). Current trends in biodegradable polyhydroxyalkanoates. *Journal of Bioscience and Bioengineering, 110*(6), 621–632.

Chong, J. W. R., Yew, G. Y., Khoo, K. S., Ho, S. H., & Show, P. L. (2021). Recent advances on food waste pretreatment technology via microalgae for source of polyhydroxyalkanoates. *Journal of Environmental Management, 293*, 112782.

Da Costa, J. P. (2018). Micro-and nanoplastics in the environment: research and policymaking. *Current Opinion in Environmental Science & Health, 1*, 12–16. https://doi.org/10.1016/j.coesh.2017.11.002

Das, S., & Bhattarcharya, B. K. (2014). Estimation of municipal solid waste generation and future trends in greater metropolitan regions of Kolkata, India. *Journal of Industrial Engineering and Management Innovation, 1*, 31–38. https://doi.org/10.2991/jiemi.2014.1.1.3

David S. Taylor. (2020). *Progress-report-2020.*

De Souza Machado, A. A., Kloas, W., Zarfl, C., Hempel, S., & Rillig, M. C. (2018). Microplastics as an emerging threat to terrestrial ecosystems. *Global Change Biology, 24*, 1405–1416. https://doi.org/10.1111/gcb.14020

Dilkes-Hoffman, L., Ashworth, P., Laycock, B., Pratt, S., & Lant, P. (2019). Public attitudes towards bioplastics–knowledge, perception and end-of-life management. *Resources, Conservation and Recycling, 151*, 104479.

Dong, Y., Gao, M., Qiu, W., & Song, Z. (2021). Effects of microplastic on arsenic accumulation in Chlamydomonas reinhardtii in a freshwater environment. *Journal of Hazardous Materials, 405*, 124232. https://doi.org/10.1016/j.jhazmat.2020.124232

Duru, R. U., Ikpeama, E. E., & Ibekwe, J. A. (2019). Challenges and prospects of plastic waste management in Nigeria. *Waste Disposal and Sustainable Energy, 1*(2), 117–126.

Eagle, L., Hamann, M., & Low, D. R. (2016). The role of social marketing, marine turtles and sustainable tourism in reducing plastic pollution. *Marine Pollution Bulletin, 107*(1), 324–332.

European Commission. (2011). *Communication from the Commission to the European Parliament, the Council, the European Economic and Social Committee and the Committee of the Regions Youth Opportunities Initiative*. European Commission.

Feil, A., Pretz, T., Jansen, M., & Thoden van Velzen, E. U. (2017). Separate collection of plastic waste, better than technical sorting from municipal solid waste? *Waste Management & Research, 35*(2), 172–180.

Ferreira, F. V., Cividanes, L. S., Gouveia, R. F., & Lona, L. M. (2019). An overview on properties and applications of poly (butylene adipate-co-terephthalate)–PBAT based composites. *Polymer Engineering & Science, 59*(s2), E7–E15.

Ferronato, N., & Torretta, V. (2019). Waste mismanagement in developing countries: A review of global issues. *International Journal of Environmental Research and Public Health, 16*(6), 1060. https://doi.org/10.3390/ijerph16061060

Ferronato, N., Rada, E. C., Portillo, M. A. G., Cioca, L. I., Ragazzi, M., & Torretta, V. (2019). Introduction of the circular economy within developing regions: A comparative analysis of advantages and opportunities for waste valorization. *Journal of Environmental Management, 230*, 366–378. https://doi.org/10.1016/j.jenvman.2018.09.095

Frère, L., Maignien, L., Chalopin, M., Huvet, A., Rinnert, E., Morrison, H., et al. (2018). Microplastic bacterial communities in the bay of Brest: influence of polymer type and size. *Environmental Pollution, 242*, 614–625. https://doi.org/10.1016/j.envpol.2018.07.023

Frydkjaer, C. K., Iversen, N., & Roslev, P. (2017). Ingestion and egestion of microplastics by the cladoceran Daphnia magna: effects of regular and irregular shaped plastic and sorbed phenanthrene. *Bulletin of Environmental Contamination and Toxicology, 99*, 655–661. https://doi.org/10.1007/s00128-017-2186-3

Gahleitner, M., & Paulik, C. (2017). Polypropylene and other polyolefins. In M. Gilbert (Ed.), *Brydson's plastics materials* (8th ed., pp. 279–309). Butterworth-Heinemann.

Garcia, J. M., & Robertson, M. L. (2017). The future of plastics recycling. *Science, 358*(6365), 870–872.

Garlotta, D. (2001). A literature review of poly (lactic acid). *Journal of Polymers and the Environment, 9*, 63–84.

Geyer, R., Jambeck, J. R., & Law, K. L. (2017). Production, use, and fate of all plastics ever made. *Science Advances, 3*(7), e1700782.

Gioia, C., Giacobazzi, G., Vannini, M., Totaro, G., Sisti, L., Colonna, M., et al. (2021). End of life of biodegradable plastics: composting versus Re/upcycling. *ChemSusChem, 14*(19), 4167–4175. https://doi.org/10.1002/cssc.202101226

Gonçalves, J. M., & Bebianno, M. J. (2021). Nanoplastics impact on marine biota: a review. *Environmental Pollution, 273*, 116426. https://doi.org/10.1016/j.envpol.2021.116426

Guria, N., & Tiwari, V. K. (2010). Municipal solid waste management in Bilaspur city (C.G.) India. *National Geographer, Allahabad, 1*, 1–16.

Hansen, E., Nilsson, N. H., & Vium, K. S. R. (Eds.). (2014). *Hazardous substances in plastics*. Danish Ministry of the Environment, Environmental Protection Agency.

Havstad, M. R. (2020). Biodegradable plastics. In *Plastic waste and recycling* (pp. 97–129). Academic Press.

Hohmann, R., Wattana, C., Sracheam, P., Siriapornsakul, S., Ruckthum, V., & Clapp, R. (2016). An exploration of the factors concerned with reducing the use of plastic carrier bags in Bangkok, Thailand. *ABAC ODI Journal: Vision. Action. Outcome, 3*(2), 162–181.

Hopewell, J., Dvorak, R., & Kosior, E. (2009). Plastics recycling: Challenges and opportunities. *Philosophical Transactions of the Royal Society B: Biological Sciences, 364*(1526), 2115–2126.

Horodytska, O., Valdés, F. J., & Fullana, A. (2018). Plastic flexible films waste management–A state of art review. *Waste Management, 77*, 413–425.

Hutley, T. J., & Ouederni, M. (2016). Polyolefins – The history and economic impact. In *Polyolefin compounds and materials: Fundamentals and industrial applications* (pp. 13–50). Springer.

Jakubowicz, I., Enebro, J., & Yarahmadi, N. (2021). Challenges in the search for nanoplastics in the environment – A critical review from the polymer science perspective. *Polymer Testing, 93*, 106953.

Jambeck, J., Hardesty, B. D., Brooks, A. L., Friend, T., Teleki, K., Fabres, J., et al. (2018). Challenges and emerging solutions to the land-based plastic waste issue in Africa. *Marine Policy, 96*, 256–263.

Jeżewska-Zychowicz, M., & Jeznach, M. (2015). Consumers' behaviours related to packaging and their attitudes towards environment. *Journal of Agribusiness and Rural Development, 37*(3).

Joachimiak-Lechman, K., Garstecki, D., Konopczyński, M., & Lewandowska, A. (2020). Implementation of life cycle based tools in the circular economy context –Case study of plastic waste. *Sustainability, 12*(23), 9938.

Karbalaei, S., Hanachi, P., Walker, T. R., & Cole, M. (2018). Occurrence, sources, human health impacts and mitigation of microplastic pollution. *Environmental Science and Pollution Research, 25*, 36046–36063.

Khajuria, A., Matsui, T., Machimura, T., & Morioka, T. (2010). Assessment of the challenge of sustainable recycling of municipal solid waste management in India. *International Journal of Environmental Technology and Management, 13*(2), 171–187. https://doi.org/10.1504/IJETM.2010.034300

Kibria, M. G., Masuk, N. I., Safayet, R., Nguyen, H. Q., & Mourshed, M. (2023). Plastic waste: challenges and opportunities to mitigate pollution and effective management. *International Journal of Environmental Research, 17*(1), 20.

Kočí, V., Rocha, J. L., & Zakuciová, K. (2016, February) *The concept of circular economy applied to CCS, waste and wastewater treatment technologies*. In Proceedings of the 5th Annual International Conference on Sustainable Energy and Environmental Sciences (SEES 2016), Global Science & Technology Forum (GSTF), Singapore (pp. 22–23).

Labet, M., & Thielemans, W. (2009). Synthesis of polycaprolactone: A review. *Chemical Society Reviews, 38*(12), 3484–3504.

Landon-Lane, M. (2018). Corporate social responsibility in marine plastic debris governance. *Marine Pollution Bulletin, 127*, 310–319.

Laskar, N., & Kumar, U. (2019). Plastics and microplastics: A threat to environment. *Environmental Technology and Innovation, 14*, 100352. https://doi.org/10.1016/j.eti.2019.100352

Lee, K.-W., Shim, W. J., Kwon, O. Y., & Kang, J.-H. (2013). Size-dependent effects of micro polystyrene particles in the marine copepod Tigriopus japonicus. *Environmental Science & Technology, 47*, 11278–11283. https://doi.org/10.1021/es401932b

Lehner, R., Weder, C., Petri-Fink, A., & Rothen-Rutishauser, B. (2019). Emergence of nanoplastic in the environment and possible impact on human health. *Environmental Science & Technology, 53*, 1748–1765. https://doi.org/10.1021/acs.est.8b05512

Lewis, H., Verghese, K., & Fitzpatrick, L. (2010). Evaluating the sustain ability impacts of packaging: The plastic carry bag dilemma. *Packaging Technology and Science International Journal, 23*, 145–160. https://doi.org/10.1002/pts.886

Li, C., Busquets, R., & Campos, L. C. (2020). Assessment of microplastics in freshwater systems: A review. *Science of the Total Environment, 707*, 135578. https://doi.org/10.1016/j.scitotenv.2019.135578

Liang, Y., Tan, Q., Song, Q., & Li, J. (2021). An analysis of the plastic waste trade and management in Asia. *Waste Management, 119*, 242–253.

MacArthur, E. (2017). Beyond plastic waste. *Science, 358*(6365), 843–843.

Madigele, P. K., Mogomotsi, G. E. J., & Kolobe, M. (2017). Consumer willingness to pay for plastic bags levy and willingness to accept eco-friendly alternatives in Botswana. *Chinese Journal of Population Resources and Environment, 15*(3), 255–261. https://doi.org/10.1080/10042857.2017.1369243. [2019 10 01].

Mammo, F., Amoah, I., Gani, K., Pillay, L., Ratha, S., Bux, F., et al. (2020). Microplastics in the environment: interactions with microbes and chemical contaminants. *Science of the Total Environment, 743*, 140518. https://doi.org/10.1016/j.scitotenv.2020.140518

Mazhandu, Z. S., Muzenda, E., Mamvura, T. A., & Belaid, M. (2020). Integrated and consolidated review of plastic waste management and bio-based biodegradable plastics: challenges and opportunities. *Sustainability, 12*(20), 8360.

Ministry of Environment and Forests (MoEF). (2015). *The Gazette of India*. Municipal solid waste (Management and Handling) rules, New Delhi, India.

Mohajerani, A., & Karabatak, B. (2020). Microplastics and pollutants in biosolids have contaminated agricultural soils: An analytical study and a proposal to cease the use of biosolids in farmlands and utilise them in sustainable bricks. *Waste Management, 107*, 252–265. https://doi.org/10.1016/j.wasman.2020.04.021

Nagaraju, B., & Thejaswini, H. D. (2016). A study on consumer attitude towards eco-friendly FMCG products with reference to Hubli city in Karnataka. *Journal of Business and Management, 18*(11), 58.

Nampoothiri, K. M., Nair, N. R., & John, R. P. (2010). An overview of the recent developments in polylactide (PLA) research. *Bioresource Technology, 101*(22), 8493–8501.

Narancic, T., Verstichel, S., Reddy Chaganti, S., Morales-Gamez, L., Kenny, S. T., De Wilde, B., & O'Connor, K. E. (2018). Biodegradable plastic blends create new possibilities for end-of-life management of plastics but they are not a panacea for plastic pollution. *Environmental Science & Technology, 52*(18), 10441–10452.

Nascimento, D. L. M., Alencastro, V., Quelhas, O. L. G., Caiado, R. G. G., Garza-Reyes, J. A., Rocha-Lona, L., & Tortorella, G. (2019). Exploring Industry 4.0 technologies to enable circular economy practices in a manufacturing context: A business model proposal. *Journal of Manufacturing Technology Management, 30*(3), 607–627.

Nomura, T., Tani, S., Yamamoto, M., Nakagawa, T., Toyoda, S., Fujisawa, E., & Konishi, Y. (2016). Cytotoxicity and colloidal behavior of polystyrene latex nanoparticles toward filamentous fungi in isotonic solutions. *Chemosphere, 149*, 84–90.

Paletta, A., Leal Filho, W., Balogun, A. L., Foschi, E., & Bonoli, A. (2019). Barriers and challenges to plastics valorisation in the context of a circular economy: Case studies from Italy. *Journal of Cleaner Production, 241*, 118149.

Peng, L., Fu, D., Qi, H., Lan, C. Q., Yu, H., & Ge, C. (2020). Micro- and nanoplastics in marine environment: source, distribution and threats – A review. *Science of the Total Environment, 698*, 134254.

Pivnenko, K., Jakobsen, L. G., Eriksen, M. K., Damgaard, A., & Astrup, T. F. (2015, October). *Challenges in plastics recycling*. In Proceedings of the Sardinia.

Rana, R., Ganguly, R., & Gupta, A. K. (2015). An assessment of solid waste management system in Chandigarh City, India. *Electronic Journal of Geotechnical Engineering, 20*(6), 1547–1572.

Sana, S. S., Dogiparthi, L. K., Gangadhar, L., Chakravorty, A., & Abhishek, N. (2020). Effects of microplastics and nanoplastics on marine environment and human health. *Environmental Science and Pollution Research, 27*, 44743–44756. https://doi.org/10.1007/s11356-020-10573-x

Sanchez, F. A. C., Boudaoud, H., Camargo, M., & Pearce, J. M. (2020). Plastic recycling in additive manufacturing: A systematic literature review and opportunities for the circular economy. *Journal of Cleaner Production, 264*, 121602.

Sandvik, I. M., & Stubbs, W. (2019). Circular fashion supply chain through textile-to-textile recycling. *Journal of Fashion Marketing and Management, 23*(3), 366–381.

Shamsuddin, I. M., Jafar, J. A., Shawai, A. S. A., Yusuf, S., Lateefah, M., & Aminu, I. (2017). Bioplastics as better alternative to petroplastics and their role in national sustainability: A review. *Advances in Bioscience and Bioengineering, 5*(4), 63.

Sharma, V. K., Ma, X., Guo, B., & Zhang, K. (2021). Environmental factors mediated behavior of microplastics and nanoplastics in water: A review. *Chemosphere, 271*, 129597. https://doi.org/10.1016/j.chemosphere.2021.129597

Shen, M., Song, B., Zeng, G., Zhang, Y., Huang, W., Wen, X., & Tang, W. (2020). Are biodegradable plastics a promising solution to solve the global plastic pollution? *Environmental Pollution, 263*, 114469.

Singh, P., Saengerlaub, S., Abas Wani, A., & Langowski, H. C. (2012). Role of plastics additives for food packaging. *Pigment & Resin Technology, 41*(6), 368–379.

Siracusa, V., & Blanco, I. (2020). Bio-polyethylene (Bio-PE), Bio-polypropylene (Bio-PP) and Bio-poly (ethylene terephthalate)(Bio-PET): Recent developments in bio-based polymers analogous to petroleum-derived ones for packaging and engineering applications. *Polymers, 12*(8), 1641.

Sjollema, S. B., Redondo-Hasselerharm, P., Leslie, H. A., Kraak, M. H., & Vethaak, A. D. (2016). Do plastic particles affect microalgal photosynthesis and growth? *Aquatic Toxicology, 170*, 259–261. https://doi.org/10.1016/j.aquatox.2015.12.002

Smyth, K., Drake, J., Li, Y., Rochman, C., Van Seters, T., & Passeport, E. (2021). Bioretention cells remove microplastics from urban stormwater. *Water Research, 191*, 116785. https://doi.org/10.1016/j.watres.2020.116785

Sridevi, V., Modi, M., Lakshmi, M. V. V. C., & Kesavarao, L. (2012). A review on integrated solid waste management. *International Journal of Engineering Science and Advance Technology, 2*, 1491–1499.

Stock, V., Laurisch, C., Franke, J., Dönmez, M. H., Voss, L., Böhmert, L., et al. (2021). Uptake and cellular effects of PE, PP, PET and PVC microplastic particles. *Toxicology In Vitro, 70*, 105021. https://doi.org/10.1016/j.tiv.2020.105021

Sun, X., Chen, B., Li, Q., Liu, N., Xia, B., Zhu, L., et al. (2018). Toxicities of polystyrene nano- and microplastics toward marine bacterium Halomonas alkaliphila. *Science of the Total Environment, 642*, 1378–1385. https://doi.org/10.1016/j.scitotenv.2018.06.141

Tallentire, C. W., & Steubing, B. (2020). The environmental benefits of improving packaging waste collection in Europe. *Waste Management, 103*, 426–436.

Tan, M., Liu, L., Zhang, M., Liu, Y., & Li, C. (2021). Effects of solution chemistry and humic acid on the transport of polystyrene microplastics in manganese oxides coated sand. *Journal of Hazardous Materials, 413*, 125410. https://doi.org/10.1016/j.jhazmat.2021.125410

Thompson, R. C., Moore, C. J., Vom Saal, F. S., & Swan, S. H. (2009). Plastics, the environment and human health: current consensus and future trends. *Philosophical Transactions of the Royal Society B: Biological Sciences, 364*(1526), 2153–2166.

Van Cauwenberghe, L., Vanreusel, A., Mees, J., & Janssen, C. R. (2013). Microplastic pollution in deep-sea sediments. *Environmental Pollution, 182*, 495–499. https://doi.org/10.1016/j.envpol.2013.08.013

Wahl, A., Le Juge, C., Davranche, M., El Hadri, H., Grassl, B., Reynaud, S., et al. (2021). Nanoplastic occurrence in a soil amended with plastic debris. *Chemosphere, 262*, 127784. https://doi.org/10.1016/j.chemosphere.2020.127784

Wang, J., Tan, Z., Peng, J., Qiu, Q., & Li, M. (2016). The behaviors of micro plastics in the marine environment. *Marine Environmental Research, 113*, 7.

Wang, Q., Adams, C. A., Wang, F., Sun, Y., & Zhang, S. (2021). Interactions between microplastics and soil fauna: A critical review. *Critical Reviews in Environmental Science and Technology*, 1–33. https://doi.org/10.1080/10643389.2021.1915035

Watkins, E., Schweitzer, J. P., Leinala, E., & Börkey, P. (2019). *Policy approaches to incentivise sustainable plastic design.*

Wright, S., Ulke, J., Font, A., Chan, K., & Kelly, F. (2020). Atmospheric microplastic deposition in an urban environment and an evaluation of transport. *Environment International, 136*, 105411. https://doi.org/10.1016/j.envint.2019.105411

Xu, J., & Guo, B. H. (2010). Poly (butylene succinate) and its copolymers: Research, development and industrialization. *Biotechnology Journal, 5*(11), 1149–1163.

Yadav, R., & Pathak, G. S. (2017). Determinants of consumers' green purchase behavior in a developing nation: Applying and extending the theory of planned behavior. *Ecological Economics, 134*, 114–122.

Yukalang, N., Clarke, B., & Ross, K. (2017). Barriers to effective municipal solid waste management in a rapidly urbanizing area in Thailand. *International Journal of Environmental Research and Public Health, 14*(9), 1013.

Yusliza, M. Y., Amirudin, A., Rahadi, R. A., Nik Sarah Athirah, N. A., Ramayah, T., Muhammad, Z., & Mokhlis, S. (2020). An investigation of pro-environmental behaviour and sustainable development in Malaysia. *Sustainability, 12*(17), 7083.

Zhang, Y., Kang, S., Allen, S., Allen, D., Gao, T., & Sillanpää, M. (2020). Atmospheric microplastics: A review on current status and perspectives. *Earth Science Reviews, 203*, 103118. https://doi.org/10.1016/j.earscirev.2020.103118

Chapter 3
Current Progress and Potential Microbial Cornucopia for Plastic Degradation

Shriniketan Puranik, Amanda Shylla, M. Manoj, and D. Vijaysri

Abstract Plastic pollution has emerged to be the biggest global concern in the modern day. The production of large volumes and different varieties of plastics has led to their huge accumulation and contamination in environmental matrices. Along with plastics, there are additional problems posed by microplastics and related chemicals like phthalates that leach into surrounding resources, thereby polluting them and causing health hazards to living organisms. Thus, their degradation becomes an immediate concern. Although some strategies like photodegradation have been suggested to tackle plastic menace, microbial degradation has been found to be the most effective one. Several microbes have been isolated, characterized, and screened for plastic degradation. Various screening procedures provide more potent microorganisms that break down a variety of plastics by producing several specific and non-specific enzymes. These microbes constitute bacteria, fungi, and other classes like diatoms and microalgae. In-depth studies linked to potent microbes like *Ideonella sakaiensis* and establishment of databases like PlasticDB make the process more convenient. Further, advancements in molecular techniques like metagenomics and genetic engineering have opened doors for prospecting newer tactics against this global issue. This chapter deals with the microbial cornucopia that are capable of metabolizing various types of plastics and their allied aspects.

Keywords Plastic pollution · Biodegradation · Microbial wealth · Plastic remediation · Genetic engineering

S. Puranik (✉)
Division of Crop Protection, Indian Institute of Horticultural Research, Bengaluru, India

A. Shylla
Meghalaya State Biological Control Laboratory, Shillong, India

M. Manoj
Department of Agricultural Microbiology, University of Agricultural Sciences, GKVK, Bengaluru, India

D. Vijaysri
Division of Microbiology, Indian Agricultural Research Institute, New Delhi, India

R. Soni et al. (eds.), *Advanced Strategies for Biodegradation of Plastic Polymers*, https://doi.org/10.1007/978-3-031-55661-6_3

3.1 Introduction

With continuous evolution in science, newer materials have surfaced, which have revolutionized human existence. One such invention is a class of materials called plastics, which are known for their versatile applications across a number of industries. Due to their different chemical and chemical properties like inertness, durability, long shelf life, and tensile strength, they have been extensively used worldwide. By 2020, it has been estimated that the world plastic production was 367 megatons (Plastics Europe, 2021). However, their excessive production, overuse, and resistance to degradation have led to their accumulation in the environment, threatening terrestrial, aerial, and aquatic lives including plants and soil microorganisms (Huerta Lwanga et al., 2016; Blettler & Mitchell, 2021; Hartmann et al., 2022). Their large quantity, light weight, and ease of movement via winds, tides, and dumping make them ubiquitous. They are found in almost every corner of the globe, even in remotest habitats like deserts and both poles (Zylstra, 2013; Vermeiren et al., 2016; Kurniawan & Imron, 2019; Bergmann et al., 2022; Sakti et al., 2023). Plastics are also known to break down into smaller pieces (<0.5 cm) due to wear, tear, shredding, etc. These are called "microplastics" and are known to have an adverse impact on living systems (Wong et al., 2020). Additionally, chemicals used in plastic production also leach into surrounding matrices, thereby contaminating and threatening the health of inhabitant biota (Wang et al., 2018).

Although alternatives like biodegradable plastics of plant and microbial origin have been introduced in the market, they have failed to compete with synthetic hydrocarbon-based nonbiodegradable ones, due to their relatively lesser cost and more convenience of production (Goel et al., 2021). Further, different types of synthetic plastics with varying chemical and physical properties like polyethylene (PE), polypropylene (PP), polyethylene terephthalate (PET), polystyrene (PS), polyester-polyurethane (PE-PU), polybutylene adipate-co-terephthalate (PBAT), polycaprolactone (PCL), poly(butylene succinate) (PBS), high-density polyethylene (HDPE), low-density polyethylene (LDPE), and polyvinyl chloride (PVC) make them more recalcitrant to degradation (Chamas et al., 2020). Considering the rate of production and disposal into the environment, Geyer et al. (2017) projected an aggregation of 12,000 Mt. of plastics on earth by the year 2050. Also, Jambeck et al. (2015) estimated that plastic wastes in oceans would increase by threefold (150 Mt) by 2025, in comparison to 2015 (50 Mt). An accumulation of a variety of plastics makes them even more resistant to degradation through natural processes. While some plastics take days to decompose naturally, others take centuries (Oluwoye et al., 2023). Sunlight, UV, and thermal treatments are certain factors that aid in plastic structural aberration (Yu et al., 2016; Cai et al., 2018; Wang et al., 2022). However, it has been reported that microbe-mediated processes increase the rate of plastic decomposition in environmental matrices. Microorganisms like bacteria, fungi, and microalgae have been studied in relation to the metabolism of plastics (Kale et al., 2015; Ru et al., 2020; Wei & Wierckx, 2021). They possess a variety of enzymes that aid in cleaving large polymers under congenial conditions (Loredo-Treviño et al., 2012;

Bollinger et al., 2020; Wright et al., 2020). It is estimated that only less than 1% of the total microbial life on the globe is culturable. Of this, a small portion of microbes is studied at the lab level. Considering this, the huge fraction of unexplored culturable microorganisms gives a wide scope of deciphering potent strains for combating plastic pollution. This chapter deals with the methods to screen potent plastic-degrading microorganisms, their enzymatic and genetic advancements, various class of capable microbes, and challenges.

3.2 Recent Advancements Pertaining to Plastic Biodeterioration

3.2.1 Screening Procedures for Potent Microorganisms in Plastic Breakdown

Screening for potent microorganisms is a crucial step for any biological process, especially plastic decomposition, given the highly diverse physicochemical characteristics of different plastic types/polymers, which get depolymerized in a variable fashion (Orhan & Büyükgüngör, 2000). At the outset, enrichment technique with plastic of interest as C and energy source is the first move for successful screening, since it is useful to restrict the growth of non-useful microbes and enhance the load of potent degraders (Patil & Bagde, 2012). Once a set of microbes are isolated on the same solid medium containing a substrate of interest after serial dilution, they can be subjected to further screening, based on biochemical and microbiological reactions. Usually, the organisms are provided the same substrate as powder or liquid on solid agar plates and observed for growth and colonization. Usha et al. (2011) isolated PU-utilizing organism by providing a PU powder in the solid medium as a C-source. Charnock (2021) supplemented a mineral salt medium with eight different types of plastic pellets/powder as the C-source, separately, viz., poly[(R)-3-hydroxybutyrate, poly(3-hydroxybutyrate-co-3-hydroxyvalerate), poly(ethylene succinate), PBS, polycaprolactone, poly(L-lactide), poly(D,L-lactide-co-glycolide), and PET. This method yielded 11 potent microbial species belonging to genera *Acidovorax*, *Rugamonas*, *Rhodococcus*, *Nakamurella*, and *Streptomyces*. Using the same technique, Kim et al. (2023) also isolated the potent PU-degrading strains *Acinetobacter guillouiae* and *Pseudomonas aeruginosa* from larvae of superworms (*Zophobas atratus*), by providing PU (8 g/l) as the C-source in the mineral medium. Once the potent microorganisms are isolated, they are screened qualitatively for a clear zone that indicates polymer utility. Penkhrue et al. (2015) inoculated 7 bacteria, 67 actinomycetes, and 5 fungi on the PLA (1 g/l) agar medium (30 °C for 30 days) and found that all the isolates showed a clear zone, thereby proving substrate degradation. Similarly, Rüthi et al. (2023) subjected their microbial strains (19 bacterial and 15 fungal) from alpines and arctic areas to qualitatively screen for substrate utilization by inoculating on the mineral medium containing PU

(Impranil: 10 ml/l). They found clear zones around the colonies that indicated positive for substrate breakdown.

Apart from clear zones, biochemical reactions (redox) might also indicate substrate utilization, usually by enzyme production. Charnock (2021) screened the isolates for esterase/lipase activity by using tributyrin and coconut/olive oil (supplemented with Rhodamine B) agar, which are linked to plastic degradation. Further, quantitative methods like enzyme production assay, weight reduction, scanning electron microscopy of damaged polymer (SEM), Fourier transform infrared (FTIR) spectroscopy, and fluorescence-based assay also depict the polymer decomposing ability of microorganisms. Skariyachan et al. (2015) used weight reduction as a parameter for determining the degradability of low-density polyethylene powder (LDPE, 0.1% w/v) by using *Pseudomonas* sp. and reported 35–40% mass loss by the end of 4 months. Zhang et al. (2022a) evaluated the PE consuming ability of *A. baumannii* Rd-H2 by assessment of weight loss and FTIR and SEM images. They reported a significant reduction in weight (0.62%) in inoculated PE, as compared to control. Also, there were ununiform rough cracks and crevices in bacterial-treated plastic as against smooth control. FTIR showed that a characteristic peak representing carbonyl moiety (1600/cm–1850/cm) appeared in bacterial-inoculated PE, which confirmed PE decomposition. They also reported a laccase-like multi-Cu oxidase enzyme catalyzing the degradation. Rüthi et al. (2023) used a fluorescence-based assay for evaluating PBAT degradation by incorporating the polymer with fluorogenic probe called 4-methylumbelliferyl laurate (4-MUL). The co-hydrolysis of MUL along with PBAT was the basis of the assay. In addition to these assays, quantifying the evolving CO_2 or CH_4 and visual observations can also be indirect screening methods for efficient plastic-degrading microorganisms. Thus, screening for these characteristics will lead to a better selection of potent microorganisms. Further, "omics" approaches like metagenomics, metatranscriptomics, and metaproteomics also aid in screening for efficient enzymes at the molecular/genomic level (Viljakainen & Hug, 2021).

3.2.2 **Ideonella sakaiensis*: The Prime "Plastic Eater"***

Previously, it was believed that only a few fungal species could enzymatically degrade PET, making biodegradation impractical for remediation or recycling. However, an important microorganism that is evidently reported for plastic degradation, like PET, is *Ideonella sakaiensis*, which was first isolated from PET-contaminated sediment (Sakai, Japan) and identified by Yoshida et al. (2016). It was isolated while studying natural microbial communities exposed to PET. This bacterium can utilize PET as its primary energy and carbon source. It is a Gram-negative, rod-shaped bacterium that does not form spores. It is aerobic and has a polar flagellum for movement. It is capable of producing cytochrome oxidase and catalase enzymes. It can grow at pH 5.5–9.0 and temperatures of 15–42 °C. Some major lipids that it possesses are phosphatidylethanolamine, lyso-phosphatidylethanolamine, phosphatidylglycerol, and diphosphatidylglycerol (Tanasupawat et al., 2016). Taxonomically, it belongs to family *Comamonadaceae* and class *Betaproteobacteria* and is nutritionally a chemoorganotroph. It was found to completely degrade powdered PET under in vitro conditions. Also, another study shows that the genome of *I. sakaiensis* contains a polyhydroxyalkanoate (PHA) (biodegradable plastic) synthesizing gene cluster, thus converting PET into PHA in vivo (Fujiwara et al., 2021).

It has been observed that, when grown on amorphous PET, it produced two enzymes that efficiently break down PET and its intermediate, mono(2-hydroxyethyl) terephthalic acid. Both enzymes were required for optimal enzymatic conversion of PET into nontoxic monomers, terephthalic acid, and ethylene glycol (Yoshida et al., 2016). The first enzyme, *Is*PETase, converts PET into a substance called mono-(2-hydroxyethyl) terephthalate (MHET). The second enzyme, MHETase, further breaks down MHET into the original components of PET, terephthalate (TPA), and ethylene glycol. Palm et al. (2019) examined the crystalline morphology of MHETase and discovered a specific domain that enables it to target MHET. By understanding the active site of MHETase and conducting experiments to modify its substrate specificity, scientists believe that this enzyme has the potential to greatly contribute to the degradation of plastic through enzymatic processes. Janatunaim and Fibriani (2020) also conducted in silico designing of MHETase gene, inserted in *Escherichia coli*, and found good levels of expression. Shi et al. (2021) stated that the *Is*PETase enzyme has very high specificity for PET, which can break down it into MHET, TPA, and bis-2(hydroxyethyl) TPA (BHET). However, it has low structural stability and solubility, making it challenging to apply in industrial settings using standard laboratory procedures. To address this issue, the researchers improved the expression of the enzyme by using a secretion system (Seo et al., 2019). They successfully produced an extracellular *Is*PETase by incorporating *Sec*-dependent translocation signal peptides (*pel*B) from *Escherichia coli* into the enzyme. The study confirmed that the secreted *Is*PETase retained its degradation activity. Further, two high-yielding variants of the enzyme were designed (*Is*PETaseS242T and *Is*PETaseN246D) by Son et al. (2019). They had better substrate attachment and tolerance to higher temperature. The succeeding variants had long-lasting activity (20 days) as against the wild type (<1 day) at 37 °C. Also, they found a multifold relative increase in enzymatic activity (58-fold). Sevilla et al. (2023) introduced single-point mutations in *Is*PETase and observed that percent reduction in thermal crystallinity of the PET film was two times higher in the mutated enzyme than in the wild type. This was attributed to the enhanced activity of the enzyme due to mutations. Such modifications will assist in the degradation of certain highly crystalline PET formed due to the expansion process that does not support the growth of *I. sakaiensis* (Wallace et al., 2020), in contrast to amorphous PET that promotes the establishment of the bacterium.

Thus, to fully exploit the capabilities of *I. sakaiensis* and its enzymes, further research is required. Future work should focus on enzyme engineering to optimize catalytic activity and stability, as well as understanding the metabolic pathways involved in PET degradation to improve the efficiency of PET assimilation.

Bioprocess optimization is essential for scaling up PET degradation processes, and the environmental applications of *I. sakaiensis*-based systems should be assessed for effective PET waste management and recycling. In conclusion, *I. sakaiensis* and its PET-degrading enzymes offer promising solutions for tackling PET waste. Continued research and development efforts are needed to unlock the full potential of these microorganisms and enzymes, paving the way for sustainable and environmentally friendly strategies to address the global plastic waste problem.

3.2.3 Databases Aiding in Plastic Degradation

There is a huge gap in comprehensive data concerning plastic degradation. The wide spectra of plastic polymers with varying physicochemical properties due to addition of different types of additives make it tedious to corroborate the mediating biological entities like microorganisms and related enzyme systems. To tackle this environmental problem, deep knowledge of the same is very crucial. Well-curated data often helps in proper diagnosis, partitioning, troubleshooting, and building effective strategies against the problems. It is very important to have databases at global levels to collaborate research and dissemination of real-time information, in order to cut down time in excess research and to understand multilocational attempts to mitigate certain global issues. There are three global databases pertaining to plastics: PMBD (Plastics Microbial Biodegradation Database), PAZy (Plastics-Active Enzymes Database), and PlasticDB (Plastic DataBase) (Gan & Zhang, 2019; Buchholz et al., 2022; Gambarini et al., 2021, 2022).

The first database associated with plastic decomposition, i.e., PMBD, was developed by Gan and Zhang et al. (2019) by manually sequestering information relevant to plastic breakdown, from the National Center for Biotechnology Information (NCBI) database. Further, the designated sequences of related enzymes were also taken from the UniProt database (UniProt Consortium, 2019). It was developed by using several keywords along with plastics including enzymes, depolymerization, bioremediation, and biodegradation. The search was made more effective by using both full names and abbreviations of plastics in abstract and title sections. Both the microbes and genes involved (with GenBank ID) were included. It has more than 390 microbial and 79 protein sources in it (Gambarini et al., 2022). The enzymes were searched and selected from the TrEMBL section of the UniProt database. Any identical or redundant data were manually trimmed, so as to reduce the repetition and complexity of the database. The website comprises sequence alignment tools like Hmmer and enzymatic function predicting model like CNN. At the initial training levels, sequences pertaining to several depolymerases (1221), dehydrogenases (732), polyurethanases (205), and other related phthalate breaking enzymes (2301) were added. Similarly, PAZy is yet another database comprising plastic hydrolyzing enzymes (Buchholz et al., 2022). It was derived by studying around 60 research articles that involved characterization of plastic-degrading enzymes at the biochemical level. The database mainly concentrated on enzymes related to depolymerization of PET and ester-based PUR, since scantly characterization of enzymes linked to other polymers was found. The database depended on PMBD and NCBI for protein sequences; based on structural and biochemical information, a total of 44 (34 PET and 10 PUR) hydrolyzing enzymatic sequences were selected. PETase homologs were curated and Hmmer-based hidden Markov models (HMM) were predicted by multiple sequence alignment. The PURase homologs were searched using four protein sequences as queries at NCBI-BLASTp (Basic Local Alignment Search Tool protein). PAZy also constitutes the Lipase Engineering Database (LED), which earlier contained sequences of α/β hydrolase superfamilies. The PAZy database has an additional 3034 sequence and 33 protein database (PDB) entries (currently 283,672 and 1590, respectively) as compared to LED entries in June 2019.

The drawbacks of these databases are that they are not updated until recently. Either of them does not involve structural data of proteins. Additionally, genomic, metagenomic, and taxa table data cannot be analyzed in these databases. This is overcome by the PlasticDB, initiated by Gambarini et al. (2021), which provides tools for annotating genes, genomes, taxa table, genomic comparison, analysis of pathways, and interactive phylogeny (Gambarini et al., 2022). While DIAMOND is used for multiple sequence alignment, Annotate taxa table tool used Quantitative Insights Into Microbial Ecology (QIIME) and Divisive Amplicon Denoising Algorithm (DADA2) for outputs (Callahan et al., 2016; Bolyen et al., 2019). PathVisio 3 was the pathway analysis tool used for deciphering the metabolic process involved (Kutmon et al., 2015). The protein structure prediction and visualization of their 3D structures were done by AlphaFold2 and iCn3D, respectively (Wang et al., 2020; Jumper et al., 2021). Further, as per the recent update (September 13, 2023), there are 633 microorganisms and over 212 proteins in PlasticDB which are potent of plastic degradation. According to the database, Japan and India have the highest contributions to the total number of potent microorganisms reported (14.1% and 13.8%, respectively) (Gambarini et al., 2021).

Although all these databases provide a bird eye's view on capable and authentic biological entities related to plastic degradation, they have different focal points and data compositions. PAZy comprises curated and functionally confirmed enzymes. But PMBD and PlasticDB are much bigger datasets containing predictive enzymes and microbes. Elucidation of results in these databases becomes challenging, as some enzymes cannot be found in PAZy, while there can be unverified hits in the other two databases. However, in PlasticDB, clear-cut labeling of predicted enzymes can be found. Further, wrong annotation of datasets, especially genomics/metagenomics, leads to aggravation of false-positive enzymes in PlasticDB and PMBD. However, this is not the case in PAZy, which biochemically verifies the enzymes using manual curation and shows only authentically functional enzymes. Thus, selection of proper database according to the purpose is a crucial step toward interpretation of results.

3.2.4 *Molecular Advancements Involving Enzymes, Genes, and Engineering*

Microorganisms employ a variety of enzymes for degradation of plastic polymers (Fig. 3.1). Enzymes are the most important biological entities that catalyze plastic decomposition by binding to the substrates and cleaving them into monomers. It is an effective technique to depolymerize long-chain plastics which can later get metabolized to CO_2, H_2O, or cellular material along with byproducts of high value (Montazer et al., 2019). While some microbes produce enzymes and are excreted outside the cell for cleaving the polymers (called extracellular enzymes), others metabolize polymers in vivo by producing intracellular enzymes. Extracellular enzymes attach to the plastic surface, thereby hydrolyzing polymers to smaller units which are taken up by microbes as the C-source. Morever, membrane-bound extracellular enzymes are more efficient than secreted ones, since the former are reported to tackle the hydrophobic property of plastics in a better way (Barcoto & Rodrigues, 2022). These enzymes are coded by genes that are expressed usually in the presence of suitable substrates. In case of less expressing genes, the process of genetic engineering is carried out that encompasses the deliberate alteration of the genetic material within microbial cells in order to enhance their ability to degrade plastic pollutants found in the natural environment (Dvořák et al., 2017). Notably, Chakrabarty's work in 1981 revolutionized this area by identifying multi-plasmid *P. putida* strains derived from four distinct *Pseudomonas* species. The plasmid was designed to enhance the catabolic efficiency of a *Pseudomonas* strain, enabling it to

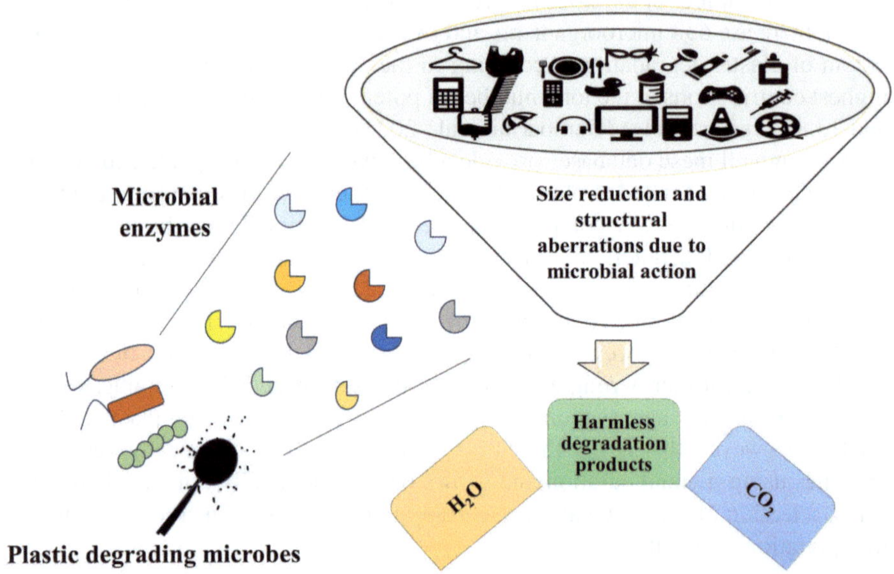

Fig. 3.1 Action of plastic-degrading microorganisms on various plastics

degrade various hydrocarbons, including plastics (Kumari & Chaudhary, 2020). The field of synthetic biology and metabolic engineering has played a crucial role in enabling the modified microbial strains and related microbiome that can effectively and safely degrade and recycle plastic waste in an environmentally friendly manner (Priya et al., 2022). The genetic manipulation of the microbiome can also be accomplished by two main approaches: ex situ design of customized microbes and in situ modification of the metagenome. The genetically modified microorganisms have the potential to produce multiple enzymes, regulate the quorum sensing mechanism, and facilitate biofilm formation. Genetic engineering enables the expression of *PET*ase, a crucial enzyme involved in the degradation of PET, within various bacterial cell systems such as *B. subtilis* and *E. coli*, thereby providing supporting evidence. The *Pseudomonas* spp. exhibit a wide range of metabolic capabilities and have been observed to exhibit differential activity in the degradation of various types of plastics (Wilkes & Aristilde, 2017). Several genes have been found in a limited number of bacteria, actinomycetes, yeast, and fungi. In some cases, these genes have been isolated and replicated using conventional molecular biology techniques.

With a number of microbes assimilating plastics, a plethora of enzymes produced during the process are reported. Alkane hydroxylases (AHs) are a big class of enzymes responsible for alkane degradation mediated by bacteria (Watanabe et al., 2003). These include enzymes like monooxygenases that can de-chain straight alkanes like PE via β-oxidation pathway (Matthews et al., 2017). The genes responsible for encoding enzymes derived from various microbial species, such as *Thermobifida* sp., *Pseudomonas* sp., *Bacillus* sp., *Fusarium* sp., and *Saccharomonospora* sp., belong to the class of hydrolases. The *alkB* gene, derived from *Rhodococcus* sp. TMP_2, coded for five AHs: $alkB_{1-5}$ (Takei et al., 2008). The *P. aeruginosa* strain E7 also possessed *alkB* that metabolized low-molecular-weight polyethylene (LMWPE) into CO_2 (Yoon et al., 2012). Further, *alkB* gene from the *Pseudomonas* sp. E4 was successfully inserted into the *E. coli* BL21 strain, and the engineered strain was able to degrade LMWPE at the rate of 19.3%. The rubredoxin and rubredoxin reductase proteins play a crucial role in facilitating electron transfer from alkane monooxygenase. The coexpression of rubredoxin and rubredoxin reductase with alkane monooxygenase has the potential to enhance the rate of conversion. The *alkB* gene was fused and expressed in conjunction with its coenzyme genes rubA1, rubA2, and rub in the *E. coli* BL21 strain. The findings revealed that 30.5% of the carbon content in LMWPE-1 underwent degradation over a period of 78 days (Jeon & Kim, 2015). Another investigation demonstrated that the *A. johnsonii* JNU01 possesses an alkane-1-monooxygenase (*AlkB*) enzyme capable of degrading PS (Ndahebwa Muhonja et al., 2018). Some of these enzymes have also been named based on their respective substrates, including PETase, MHETase, and BHETase (Kumari & Chaudhary, 2020). Yoshida et al. (2016) reported a low-crystallinity PET film-degrading MHETase and *Is*PETase from *I. sakaiensis. Is*PETase was capable of degrading PET film (~1.9%) under a wide range of temperature (20–45 °C) and outperformed the other mesophilic PET hydrolases (~30 °C). Joo et al. (2018) further identified the crystalline structure of the *Is*PETase enzyme. They substituted Arg 280 amino acid with alanine in wild type (calling it *Is*PETase^{R280A}), which increased substrate (PET)-enzyme interaction time, altering the substrate binding site, thereby increasing the enzymatic activity by ~22% and ~ 32% at the end of 0.75 and 1.5 days, respectively, as against the wild type.

Santo et al. (2013) found another class of copper-encoded enzymes called laccases (phenol oxidases) which have been reported to catalyze PE breakdown in *Rhodococcus ruber*, an actinomycete. They reported 2.5% weight loss in the LDPE film after 30 days of incubation. Laccases were also associated with PP degradation mediated by *Phanerochaete chrysosporium* and *Engyodontium album.* Ihssen et al. (2015) identified and characterized five previously unreported bacterial laccase-like multicopper oxidases (LMCOs) derived from various sources. The utilization of laccase in the biodegradation of PE has been extensively documented, and the heterologous expression of laccase in *E. coli* presents a promising approach for the biodegradation of PE (Johnnie et al., 2021). Gao et al. (2022) expressed a laccase gene derived from a marine fungus in *E. coli*, and the expressed laccase gene demonstrated the capability to degrade PE. Many recent studies have demonstrated that engineered *E. coli* can be effectively used for the purpose of laccase heterologous expression and secretion (Guo et al., 2017). Mo et al. (2022) investigated the expression of three distinct laccases derived from various organisms, specifically Lac1326 from marine sediment samples, fungal Tvel5 laccase obtained from *Trametes versicolor*, and bacterial BPUL laccases derived from *Bacillus pumilus*.

Similar kinds of enzymes like hydroquinone peroxidase are also reported in the degradation of PS films (Nakamiya et al., 1997; Jeon & Kim, 2015). Additionally, lignin peroxidases are important class of nonspecific polymer degraders and have been reported to act upon PVC films (Khatoon et al., 2019). These are generally produced by basidiomycete fungi like *P. chrysosporium.* Apart from these, lipases are noteworthy enzymes that aid in plastic degradation. Numerous lipases have been reported from a variety of microorganisms, especially in PU degradation. Takamoto et al. (2001) and Liu et al. (2016) independently reported lipase from *Candida antarctica* for the degradation of thermoplastic polyester and polyester coating PU, respectively. Lipases also aided in thermoplastic poly(ester ether) PU (Feng et al., 2017). The enzyme was found to be present in Asian mantis (*Hierodula patellifera*)-dwelling *Serratia* sp. HY-72 that catalyzed the degradation of PU wastes with a specific activity of 2.9 U/g. They used 100 g/l Impranil as a substrate, which is a colloidal counterpart of PU for dispersion, and found 85.24% reduction after 3 days of inoculation. They also reported around 10.95–23.95% weight decrease in PU foams after incubation of 14 days (Kim et al., 2022).

Further, esterases are important class of enzymes that are involved in the cleavage of plastic-containing ester bonds. These enzymes are reported to bind to hydrophobic surfaces of plastics like PU at the substrate-binding moiety and cleavage of ester bonds at the catalytic moiety (Crabbe et al., 1994; Akutsu et al., 1998; Magnin et al., 2020). Barth et al. (2016) developed a cocktail of two enzymes, viz., PE-hydrolase and immobilized carboxylesterase (sourced from *Thermobifida fusca* TfCa), which enhanced PET film hydrolysis at 60 °C. Puranik et al. (2023) observed that degradation of dibutyl phthalate was mediated by esterases that were produced by *A. flavus*. Further, *A. fumigatus* also produces esterase for cleaving diethyl hexyl phthalate (unpublished data). Cutinases are also well-established plastic-degrading enzymatic systems (subclass of esterases) that exhibit significant activity, especially against PET and its analogous polymers, thereby enhancing the hydrophilic properties of the material's surface. Sulaiman et al. (2012) observed that a thermophilic cutinase found in compost metagenomic library caused 50% weight loss in low-crystallinity PET film at 7 days of incubation, under high temperature of 50 °C. Another thermophilic cutinase (63 °C), *Cut 190* isolated from *Saccharomonospora viridis*, resulted in 27% weight reduction in the same substrate in 3 days. Likewise, Schmidt et al. (2017) found an LC cutinase enzyme sourced from compost metagenomic library that reduced the weight of polyester PUR cubes by 0.3–3.2% within 100 h of incubation. Tseng et al. (2023) reported an increased expression of cutinase gene in the presence of PBAT films in the fungus *Purpureocillium lilacinum* BA1S. They also observed that the rate of degradation of the substrate was found to be 15.27% by the eukaryote, at the end of 2 months. Cloning has also been found effective in case of cutinase genes. In a study conducted by Herrero Acero et al. (2013), two *cutinases (Thc_Cut1 and Thc_Cut2)* were identified, and genes from *Thermobifida cellulosilytica* were cloned in order to facilitate the hydrolysis of PET. Kawai et al. (2014) also reported the cutinase gene *(cut190)* isolated from the thermophilic bacterium *S. viridis* and introduced in *E. coli* for the purpose of enzymatic degradation of polyester films. Additionally, the active site modeling of the enzyme was performed through mutational substitution of specific amino acids, resulting in improved thermostability and enhanced catalytic activity. Apart from genetic manipulation, some microorganisms have also shown multiple enzymatic expressions in the presence of plastic. Transcriptome analysis of a marine fungus, *Cladosporium halotolerans*, exhibited expression of enzyme-coding genes like lipase, cutinase, peroxidase, and other proteins involved in binding to hydrophobic surfaces (*HsbA*) in the presence of Impranil PU (Zhang et al., 2022b).

Thus, it is evident that various organisms have different enzymes for breaking down the plastic polymers. Further, enzymes and genes are variable as far as different substrates are concerned. Additionally, genetic engineering for efficient degradation is also documented. It has been observed that *E. coli* possesses numerous advantages as a model microorganism, which includes a simple genetic background, ease of genetic modification, and simple growth conditions. Zhang et al. (2022c) employed genetically modified *E. coli* to facilitate the expression of alkane hydroxylase for the purpose of degrading LMWPE. Further, the exploration of gene editing tools and techniques, such as zinc-finger proteins, transcription activator-like effector nucleases (TALENs), and the clustered regularly interspaced palindromic repeats/CRISPR-associated protein 9 (CRISPR/Cas9) system, can also be undertaken to assess potentiality of microbes in the field of plastic pollution mitigation. The incorporation of gene-encoding enzymes, such as esterase, PETase, depolymerase, and laccase, can be implemented as strategies to enable plastic degradation in a non-degrading microorganism (Gaj et al., 2013). Techniques like metagenomics, metatranscriptomics, and metaproteomics further provide a bird eyes' view for exploring potent microbes, genes, and enzymes (Viljakainen & Hug, 2021). Identification and exploitation of such efficient genes and enzymes gives us an upper hand over the plastic problem in years to come.

3.3 Potential Microbial Wealth for Plastic Biodegradation

3.3.1 Bacteria

Plastics are synthetic materials made from polymers, which are long chains of repeating units. Traditional plastics are typically resistant to natural degradation processes, leading to their persistence in the environment and causing serious pollution issues. However, certain species of bacteria have evolved the capability to utilize plastics as a carbon source, utilizing them as an energy and nutrient source for their growth. Plastic-degrading bacteria have the ability to break down and degrade various types of plastic. These bacteria possess specific enzymes that enable them to metabolize or digest the polymer chains found in plastic materials, converting them into simpler compounds. Some plastic-degrading bacteria have been discovered in diverse environments, including soil, water bodies, and even the guts of certain organisms. These bacteria can break down different types of plastics, including PET, PE, PP, and PS. They secrete enzymes that target and break the chemical bonds within the plastic polymers, resulting in the breakdown of the plastic into smaller fragments. The discovery of plastic-degrading bacteria has sparked significant interest in their potential applications for addressing plastic pollution. Researchers and scientists are actively studying and isolating these bacteria to better understand their mechanisms of plastic degradation and to develop biotechnological solutions for plastic waste management. By harnessing the capabilities of these bacteria, there is hope for developing more sustainable and environmentally friendly approaches to dealing with plastic waste.

The first microbe reported for synthetic plastic degradation was *Aureobasidium pullulans*, a fungus, isolated by Fields et al. (1974), which consumed PCL as the C-source. Further, another PCL-degrading fungus, *Penicillium* sp., was reported by Tokiwa and Suzuki (1977). Shimao et al. (1983) discovered a PVA-degrading bacterium belonging to *Pseudomonas* genus. Kay et al. (1991) investigated the ability of 16 bacteria to degrade PE-PU. They found that when supplementary organic nutrients were added in the form of low-level aliquots of bacterial broth cultures, significant degradation was observed by *Corynebacterium* sp. and *P. aeruginosa*. Organic nutrients, specifically yeast extract, significantly enhanced the degradation over 12 weeks. Nakajima-Kambe et al. (1995) isolated *Comamonas acidovorans* with significant PU-degrading activity (Table 3.1). The degradation products identified during the process were adipic acid and diethylene glycol. PE degradation has also been documented. Three *Pseudomonas* strains isolated from sewage sludge sump, household garbage dump, and textile effluent drainage were capable of

Table 3.1 List of various plastic-related polymer biodeterioration agents and their impact

Microorganism	Isolation source	Plastic type	Significance of the study	References
Bacteria				
Corynebacterium sp. and *P. aeruginosa*	PE-PU samples buried in soil	PE-PU	Degraded significant amounts of PU in the presence of yeast extract (1% w/v) over 3 months	Kay et al. (1991)
C. acidovorans	Soil	PUR	Complete degradation (initial 50 mg) in a week	Nakajima-Kambe et al. (1995)
K. oxytoca	Mangrove sediment (Hong Kong's Mai Po Nature Reserve)	Dimethyl isophthalate (DMI)	Complete degradation at 36 h (initial: 108 mg/l)	Gu et al. (2005)
B. cereus strain NBAII B7	Soil	PE, PVC	Time: 30 days Mass reduction of PE: 17.39% Mass reduction of PVC: 22.22%	Sharma et al. (2014)
I. sakaiensis	Sediment, soil, wastewater, and activated sludge surrounding a PET bottle recycling site	PET	PET film surface degradation rate: 0.13 mg/cm^2/day at 30 °C Degradation: 75% at 28 °C	Yoshida et al. (2016)
Vibrio sp. and *Aspergillus* sp.	Marine environment	Synthetic polymers from plastic bottle waste	*Vibrio* sp.: 35% degradation *Aspergillus* sp.: 22% degradation Time of incubation: 6 weeks	Sarkhel et al. (2020)
Pseudomonas sp.	Beach soil polluted with crude oil	LDPE	Degradation: 64%	Lomwongsopon and Varrone (2022)
Peribacillus frigoritolerans (JZ1, JZ2, JZ3, JZ4, and JZ5)	Soil	PBAT	The degradation rate order: JZ1 (3.98%) > JZ4 (3.51%) > JZ2 (3.33%) > JZ5 (3.14%) > JZ3 (2.27%) Uninoculated control: 0.57%	Wufuer et al. (2022)
Fungi				
Aspergillus spp.	Seawater	LDPE	Degradation rate: 83% Time: 7 days of incubation Structural changes: Cracks, pits, and minute holes were formed	Pramila and Ramesh (2011)

(continued)

Table 3.1 (continued)

Microorganism	Isolation source	Plastic type	Significance of the study	References
P. chrysosporium and *Engyodontium album*	Soil from plastic dumping zone	PP	Size of film: 8 × 2.5 cm and 0.05 mm thick, Initial weight: 0.5 g Weight loss of ~4–18% in 365 days of incubation coupled with laccase production of 1.8 nanokatals/ml Intermediates like propylene and 1,2-epoxypropane were found	Jeyakumar et al. (2013)
Cladosporium sp.	Soil samples collected from municipal solid waste	LDPE	In vitro assessment of LDPE degradation assay was done using 1 cm × 1 cm small pieces of LDPE for 30 days using fungal isolates Degradation of 0.3%, coupled with laccase production Microscopic observations: Adherence and proliferation on the hydrophobic surface of (LDPE) the film	Kumar et al. (2013)
P. chrysosporium	Soil	PVC	Weight reduction of PVC: 178,292/Da as against control: 200,000/Da Structural changes: Hexagonal rings appeared, significant change in color, and surface deterioration	Ali et al. (2014)
Lasiodiplodia theobromae	Plant samples	Irradiated LDPE and PP	Weight loss: 0.3 mg after 90 days of incubation. This was coupled with laccase production and biofilm formation	Sheik et al. (2015)

A. oryzae	Soils from garbage dumping sites	Plastic strips collected from dumping sites	Weight reduction to 35 mg (30%) after 2 months of incubation (initial: 50 mg) Degradation pattern: Microcracks and embrittlement on the plastic strips	Indumathi and Gayathri et al. (2016)
C. cladosporioides, Xepiculopsis graminea, and *P. griseofulvum*	Floating plastic debris	PE and PU	The fungi were inoculated on sole C-source media with 10 g/l powder PE and 10 ml/l PU. Halo zones were formed by fungi on petri plates	Brunner et al. (2018)
T. viridae RH03 and *A. nomius* RH06	Local landfill soil	LDPE	Weight reduction (and tensile strength reduction): 5.13% (58%) and 6.63% (40%), respectively Time: 45 days	Munir et al. (2018)
Aspergillus sp. S45	Soil	PU	Weight reduction: 250 mg PU films were used for 28 days and found 15–20% of weight loss Observations: Halos and pits were observed on films. In the presence of PU film, esterase activity of isolate increased to 2.687 μM/min/mg at the 15th day	Osman et al. (2018)
P. chrysosporium	Plastic dump site	PVC film	Maximum PVC weight reduction (31% of initial 0.1 g) and lignin peroxidase production occurred at 25 °C, pH 5 in the fourth week of incubation (~0.9 IU/ml)	Khatoon et al. (2019)
A. flavus	Activated sludge	HDPE	Weight reduction: 3.9% Time: 28 days	Taghavi et al. (2021)

(continued)

Table 3.1 (continued)

Microorganism	Isolation source	Plastic type	Significance of the study	References
Embarria clematidis	Not available	PE-PU (Impranil)	Highest optical density (~0.9) at 14 days of incubation in the presence of 10 ml/l Impranil Highest esterase activity (~2.45 mM/min/mg) at 8 days after incubation	Khruengsai et al. (2022)
P. citrinum	Municipal landfill soils of Bhopal	LDPE	*P. Citrinum* showed 38.82 ± 1.08% weight loss of untreated LDPE pieces; biodegradation was significantly stimulated by 47.22 ± 2.04% after pre-treatment with nitric acid	Khan et al. (2023)
Cladosporium sp.	Not available	PE-PU (Impranil), poly(1,4-butylene adipate)-based PU (PBA-PU)	Impranil: Degradation: 94.5% Time: 4 days Degradation: 94.5% Time: 4 days PBA-PU film: Degradation: 32.42%–43.91% Time: 28 days Metabolites found: Adipic acid, 1,4-butanediol, and 4,4′-methylenedianiline	Liu et al. (2023)
Purpureocillium lilacinum strain BA1S	Arable soil	PBAT	Weight loss: ~15% (size, 2.5 cm × 5 cm; thickness, 30 μm) films coupled with increased expression of cutinase gene and cytochrome P450-related genes Days of inoculation: 30 Observations: Cracks and rough patches on the films, analyzed in SEM	Tseng et al. (2023)

Bacteria (*Burkholderia cepacia*, *B. diffusa*, *B. aenigmatica*, *Chryseobacterium nepalense*) **Fungi** (*Geotrichum candidum*, *F oxysporum*, and *Trichoderma* sp.)	Mulch-covered decade-old vegetable soil	PE	Weight loss by individual strains: 1.58–2.44% Weight loss by mixed strains: 1.34–1.62% Time: 3 months	Lin et al. (2024)
Other microbes				
Anabaena spiroides	PE bags	LDPE	Initial weight: 6.07 mg Inoculum size: 1% Period: 30 days Degradation: 8.18% Observation: Cracks and rough patches on the films, analyzed in SEM	Kumar et al. (2017)
Spirulina sp.	Not available	PP and PET	Initial weight: 150 mg Period: 112 days Reduction in tensile strength (and carbon): PET: 0.9939 MPa/day (48.61%) PP: 0.1977 MPa/day (36.7%)	Khoironi and Anggoro (2019)
Cylindrotheca closterium	Marine rock	DBP	Degradation: 6 1.5%	Zhang et al. (2019)
Uronema africanum	LDPE sheet	LDPE	Initiation of degradation at 30 days after inoculation as confirmed by aberrations in SEM images	Sanniyasi et al. (2021)
Picochlorum maculatum	Marine water	LDPE	Weight reduction: 20.16%	Gowthami et al. (2023)

Note: polyethylene (PE), polypropylene (PP), polyethylene terephthalate (PET), polystyrene (PS), polyester-polyurethane (PE-PU), polybutylene adipate-co-terephthalate (PBAT), polycaprolactone (PCL), poly(butylene succinate) (PBS), high-density polyethylene (HDPE), low-density polyethylene (LDPE), and polyvinyl chloride (PVC) degrading up to 46.2% natural PE and 29.1% synthetic PE at 40 °C by day 21 (Nanda et al., 2010). Likewise, two strains isolated from municipal solid waste belonging to *Bacillus* genera, viz., *B. amyloliquefaciens* BSM-1 and BSM-2, degraded 30 mg LDPE in 60 days coupled with weight reduction of 11% and 16% films, respectively (Das & Kumar, 2015). Hou et al. (2022) in their study have identified PE-degrading bacteria and determined the PE degradation efficiency of two *Pseudomonas* sp. identified by 16S rDNA analysis and elucidated their potential mechanisms through whole genome sequencing. During an 8-week period, PE mulch lost $5.95 \pm 0.03\%$ and $3.62 \pm 0.32\%$ of its mass after incubation with *P. knackmussii* N1–2 and *P. aeruginosa* RD1–3 strains, respectively. Moreover, considerable pits and wrinkles were observed on PE. The hydrophobicity of PE films also decreased, and new oxygenic functional groups were detected on PE mulch by Fourier transform infrared (FTIR) spectrometry. Complete genome sequencing analysis indicated that two *Pseudomonas* strains encode genes for enzymes and metabolism pathways involved in PE degradation. Wu et al. (2023) isolated a novel bacterium (*B. paramycoides*) from a waste mulch recycling plant and showed an extraordinary ability to customize PE film. It was observed by scanning electron microscopy that a large number of pits and wrinkle cracks existed on the PE, indicating that the strain used PE film as the sole carbon source. Meanwhile, the loss of weight of the film was tested continuously, and approximately 12% of the initial weight of the film was found to be lost within 45 days. X-ray diffraction (XRD) analysis supported the results by exhibiting relative crystallinity reduction (33–11.51%).

Sharma et al. (2014) investigated the in vitro biodegradation of PE and PVC strips using microorganisms isolated from soil. The *B. cereus* strain NBAII B7 exhibited plastic-degrading capacity. A study published in the journal science by Yoshida et al. (2016) reported the discovery of a bacterium called *I. sakaiensis* that can degrade PET, a common type of plastic used in bottles. The researchers identified the enzymes involved in the breakdown of PET and demonstrated the bacterium's ability to utilize it as a carbon source. It was also found that *I. sakaiensis* degraded another type of plastic called polyethylene furanoate (PEF). Espinosa et al. (2020) isolated a bacterium, *Pseudomonas* sp., from a site contaminated with brittle plastic. It was found to grow on PU-diol solution, an oligomer of PU. The organism produced substrate-specific enzymes like dioxygenases, which aid in the degradation of the oligomers. Likewise, another bacterium belonging to same genera, *P. aeruginosa* and *B. subtilis*, were tested for PU degradation (Shah et al., 2016). They noted that the weight of the PU film (250 mg) reduced to ~195 mg after 1 month of incubation, in case of individual isolates. However, their consortium significantly reduced more weight (~150 mg) during the same time interval. Further, esterase production by the consortium was approximately twofold higher than the single strains, at 25 days of incubation. Gui et al. (2023) reported that *B. velezensis*, inhabitant of deep sea, is a potent of PU degradation. Transcriptomic analysis showed upregulation of lipase genes in the presence of PU as the substrate.

Singh et al. (2016) reported that *Staphylococcus* sp. (P1A), *Pseudomonas* sp. (P1B), and *Bacillus* sp. (P1C) bacteria were recovered from different sites like petrol pump, hospital, and local area. They were assessed for their degradation capacity of 40-μ-thick PE. The results indicated that degradation was in the following order: PIC (42.5%) > P1A (20%) > P1B (7.5%). Their consortium (PID) showed weight loss of 5% in 40 days in 40-μ-thick PE. Montazer et al. (2018) further reported 11 LDPE film-metabolizing potent bacteria, isolated from landfill. They were identified as *P. aeruginosa, P. putida, Sphingobacterium multivorum, Delftia tsuruhatensis, Stenotrophomonas humi, S. maltophilia, Ochrobactrum oryzae, O. humi, Micrococcus luteus, A. pitti*, and *Citrobacter amalonaticus*. The highest degradation was found in treatment containing *A. pitti* (26.8% gravimetric weight over 4 weeks). The highest cell mass production of 6.29 log cfu/cm^2 on LDPE was by *S. multivorum.* Lomwongsopon and Varrone (2022) also reported that a *Pseudomonas* sp. was potent of LDPE degradation.

Soud (2019) reported two actinobacteria, i.e., *Arthrobacter* sp. and *Streptomyces* sp., that are capable of using a PE film as a sole carbon source. *Arthrobacter* sp. mainly grew in the suspension phase of the culture, and *Streptomyces* sp. formed substantial biofilms on the surface of the PE film, indicating that these strains were of different metabolic types and occupied different microenvironments with contrasting nutritional access. Individual strains were able to degrade the PE film to some extent in a 90-day inoculation experiment, as indicated by decreased hydrophobicity, increased carbonyl index and CO_2 evolution, and the formation of biofilms on the film surface. However, a consortium of both strains had a much greater effect on these degradation properties. Han et al. (2020) also prospected LDPE-degrading actinobacteria and found different strains pertaining to *Streptomyces* sp. (SSP2, SSP4, SSP14). These isolates caused significant weight reduction (6–19%), depending on the type of LDPE strips.

Phthalate esters (PAEs) are types of microplastics that contaminate soils. Their degradation by bacteria has also been reported by several researchers. Chang et al. (2004) isolated bacteria capable of degrading aerobic PAEs, from river sediment and petrochemical sludge. The strains, *Sphingomonas* sp. (DK4) and *Corynebacterium* sp. (O18), exhibited rapid degradation of PAEs with shorter alkyl-chains, while PAEs with longer alkyl-chains showed poor degradation. Strain DK4 demonstrated higher degradation rates compared to strain O18, and their simultaneous presence enhanced the degradation of PAEs. They also found that the addition of nonylphenol or selected polycyclic aromatic hydrocarbons (PAHs) at a concentration of 1 μg/g in the sediment delayed PAE degradation. A study was carried out by Gu et al. (2005) who explored mangrove sediment in Hong Kong for dimethyl isophthalate (DMI)- and dimethyl phthalate ester (DMPE)-metabolizing bacteria. They observed that co-culture of two bacteria, *K. oxytoca* and *Methylobacterium mesophilicum*, demonstrated degradation through intermediate compounds before aromatic ring opening. The consortium consisting of the two species degraded 450 mg/l DMPE within 3 days, in contrast to incapability of individual species to transform DMPE individually. Sharma et al. (2021) demonstrated that *Acinetobacter* sp. 33F could utilize diethyl phthalate (DEP), mono-butyl phthalate (MBP), benzyl butyl phthalate (BBP), di-octyl phthalate (DIOP), and di-decyl phthalate (DIDP) as sole carbon sources. The above properties make *Acinetobacter* sp. 33F a promising candidate for phthalate ester removal from the contaminated agricultural sites.

Sarkhel et al. (2020) inoculated mangrove-dwelling *Vibrio* sp. and *Aspergillus* sp. in media containing PET bottle films. They observed that, after 6 weeks of incubation, bacteria degraded higher substrate weight reduction (35%) as compared to fungi (22%). The study was also supported by SEM analysis and XRD, wherein they found cracks on the surface of the plastic and reduction in crystallinity of the film. The degradation capacity also varied with pH. They found that as pH increased (pH 1–11), the degradation (%) also increased. This study also shows that fungi can also be explored for plastic degradation. This is discussed in the next section.

3.3.2 Fungi

Fungi are a diverse group of eukaryotic microorganisms that include yeasts and molds (Schmit & Mueller, 2007). They exhibit heterotrophic nutrition, obtaining their sustenance from diverse sources through the formation of specialized structures called mycelium, which facilitates the absorption of nutrients from host cells (Pathak, 2017). They possess the ability to proliferate and establish themselves on diverse substrates, exhibiting resilience in both natural and human-altered environments. Fungi are involved in various biogeochemical processes and inhabit a wide range of ecological niches from a metabolic standpoint. They are of significant importance in the process of decomposing recalcitrant substrates, thereby serving as an important component within the food web. Fungi possess the capability to break down synthetic compounds, such as persistent organic pollutants (POPs), polycyclic aromatic hydrocarbons (PAHs), benzene, toluene, ethylbenzene, and xylenes (BTEX compounds), as well as pesticides (Olicón-Hernández et al., 2017; Vaksmaa et al., 2021). The metabolic adaptability exhibited by fungi, along with their capacity to break down complex compounds, suggests that certain fungal species may possess the metabolic capability to biodegrade plastics within the environment (Zeghal et al., 2021).

Fungi are known to produce hydrophobins for better attachment. Hydrophobins are a class of proteins characterized by their water-repelling nature, typically produced by fungi. These proteins play a crucial role in the development of aerial structures in filamentous fungi, including hyphae, fruiting bodies, and spores. Additionally, they are responsible for facilitating the attachment of hyphae to hydrophobic surfaces. On hydrophobic-hydrophilic interfaces, fungal hydrophobins self-assemble into amphipathic films (monolayers) (Kulkarni et al., 2017). Numerous hydrophobins have been extracted from various fungal sources. Typically, individual fungi encompass a diverse array of hydrophobins. The remarkable characteristics exhibited by hydrophobins include their robust adhesion, their elevated surface activity, and their ability to form diverse self-assembled structures with a tendency toward high insolubility, thereby facilitating their application for growth on hydrophobic surfaces. The process of biodegradation involves the catalytic activity of microbial enzymes acting upon the surface of plastic materials. The fungi possess a notable characteristic of exocytosis, wherein they secrete a diverse range of digestive enzymes including cutinase, lipase, proteases, carboxylesterases, esterases, lignocellulolytic enzymes, and certain prooxidant ions. This secretion occurs outside their hyphae and aids in the degradation of plastic materials (Srikanth et al., 2022). The fungi adhere to the plastic film and secrete enzymes, subsequently utilizing it as a substrate for growth. Through enzymatic activity, they break down macro and organic molecules into smaller organic compounds, which are then utilized as a nutritional source. Consequently, the polymers undergo gradual depolymerization, leading to degradation that is subsequently facilitated by a mineralization process. This process results in the production of water (H_2O), carbon dioxide (CO_2), and methane (CH_4) as end products (Montazer et al., 2019). Environmental factors, including moisture, pH, temperature, and other variables, can influence the impact of fungi on the surface of plastics (Srikanth et al., 2022).

There are mainly two types of enzymatic systems involved in fungal degradation, namely, intracellular and extracellular enzymatic systems. The intracellular enzymatic system acts as an internal mechanism for detoxification and plays a major role in fungal adaptation. The enzymatic system under consideration is facilitated by the cytochrome P450 family (CYP) epoxidases, which are classified as Phase I enzymes, as well as transferases, which are categorized as Phase II enzymes. The enzymes that perform oxidation and conjugation reactions are classified as phase I and phase II enzymes, respectively (Schwartz et al., 2018). However, the extracellular enzymatic system comprises a hydrolytic system that synthesizes hydrolases, which are responsible for the degradation of polysaccharides. Additionally, there is an exceptional and nonspecific oxidative system that participates in the breakdown of complex structures, such as lignin degradation (Sánchez, 2009). They primarily consist of nonspecific oxidoreductases, which encompass various enzymes such as class II peroxidases (specifically manganese peroxidase, lignin peroxidase, and versatile peroxidase), laccases, and unspecific peroxygenases, like bacteria (Karich et al., 2017). Maeda et al., (2005) purified the plastic-degrading enzyme from the culture supernatant of *Aspergillus oryzae*. By using proteolysis and gel electrophoresis analysis, the enzyme was identified as cutinase, based on the internal amino acid sequences. Similarly, Sowmya et al., (2015) isolated *Penicillium simplicissimum* from a local dumpsite and screened for the enzymes responsible for PE degradation. Laccase and manganese peroxidase were observed in significant quantities. Thus, the aforementioned strategies encompass a robust enzymatic system, the capacity for adsorption, and the synthesis of natural biosurfactants, specifically hydrophobins. These mechanisms facilitate the utilization of polymers, such as plastics, as both a carbon source and an electron donor, thereby supplying the microorganisms with cellular constituents and a source of energy, respectively.

In addition to xenobiotic substances, several fungi are known to break down plastics. Some of which are *A. flavus*, *P. adametzii*, *Paecilomyces variotii*, *Pullularia pullulans*, *Trichoderma viride*, *Mucor* sp., *Rhizopus arrhizus*, *Plectosphaerella* sp., *Fusarium* sp., *Coriolus versicolor*, *Aureobasidium pullulans*, *Chaetomium globosum*, *Cephalosporium* spp., *Alternaria* sp., *Phoma* sp., etc. (Renstad et al., 1999;

Maeda et al., 2005; Russell et al., 2011; Sakhalkar & Mishra, 2013; Batista-García et al., 2017; Ojha et al., 2017; Rani & Singh, 2017; Ndahebwa Muhonja et al., 2018; Brunner et al., 2018). Zahra et al., (2010) isolated fungi from landfills with the purpose of investigating their potential for degrading LDPE and were identified as *A. fumigatus*, *A. terreus*, and *F. solani*. In the same way, Kumar et al., (2013) were able to effectively isolate eight fungal strains from municipal solid waste that exhibited the ability to degrade plastic. Of these, two were recognized as *Fusarium* sp., while the remaining six belonged to *Aspergillus* sp. Ojha et al., (2017) isolated fungi capable of decomposing plastic from a soil sample obtained from a site of plastic waste disposal. The fungal strains were identified as *P. oxalicum* NS4 and *P. chrysogenum* NS10. A notable decrease in the overall weight was found in the treatment with NS10 with significant total weight loss in HDPE (58.6%) and LDPE (34.35%) during the 3-month incubation period. Findings also revealed that the LDPE and HDPE films, upon treatment with the fungal isolates, exhibited surface roughness along with the formation of fractures and grooves. On the other hand, the control group exhibited a consistently smooth surface even after a 90-day incubation period under similar conditions. In an insightful study conducted by Munir et al., (2018), the researchers explored the identification of fungi with the potential to degrade plastic, with a specific focus on LDPE. Fungi were isolated from local landfill soil and cultured in a medium containing LDPE powder. Two out of nine isolates, RH03 (identified as *T. viride*) and RH06 (identified as *A. nomius*), showed the best growth response and were selected for degradation tests. After 45 days of cultivation, RH03 and RH06 reduced the weight of LDPE film by 5.13% and 6.63%, respectively (Table 3.1). The treated films also exhibited significant reductions in tensile strength compared to the untreated film. Electron microscopy analysis revealed the formation of grooves and roughness on the surface of the LDPE film in the treated samples. Sangale et al. (2019) have found efficient fungi from rhizosphere soil of *Avicennia marina* and were identified as *A. terreus* and *A. sydowii* capable of degrading PE. The significant weight loss was found with the use of *A. terreus* (58.51%) at pH 7. Conversely, a considerable decrease in tensile strength (94.44) was reported with *A. sydowii* at pH 3.5. This was complemented by SEM images of PE strip that demonstrated cracks, holes, and scions. Taghavi et al. (2021) investigated the biodegradation capabilities of different microbial strains derived from various sources. The study specifically focused on the biodegradation of HDPE, PS, PP, and PET using *P. raperi*, *A. flavus*, *P. glaucoroseum*, and *Pseudomonas* sp. Among them, *A. flavus* showed the highest weight loss (5.5%) when incubated with PE in the unstimulated mixed condition. Also, Maroof et al. (2022) recovered several beneficial fungal species from soil. These fungal isolates had the potential to degrade plastic and were identified as *A. niger*, *A. flavus*, white-rot fungi, and brown-rot fungi. Further, Khruengsai et al. (2022) prospected fungal strains for PE-PU (Impranil) degradation. They obtained cultures from Thailand Institute of Scientific and Technological Research, Bangkok, Thailand, and culture collection of the Institute of Excellence in Fungal Research, Mae Fah Luang University, Thailand. They found that incubation of cultures with the substrate had an effect on optical density during 14 days of incubation. The highest optical density was found to be that of the fungus *Embarria clematidis* (~0.9). This proves that the fungus was able to grow in the presence of a single C-source, PE-PU. They also observed that increased esterase production during the course of incubation was highest at 8 days after incubation (2.45 mM/min/mg). It was followed by *A. niger* that produced the highest amount of esterase, i.e., 0.88 mM/min/mg, on day 12. Gong et al., (2023) studied biodegradation of LDPE by the fungus *Cladosporium* sp. CPEF-6. They found that a 30-day incubation reduced the weight of LDPE by 0.3%. This was also in association with the production of laccase enzyme. Tseng et al., (2023) reported the cultivable soil-dwelling *Purpureocillium lilacinum* strain BA1S for efficient degradation of PBAT, which could cause a weight reduction of 15% in 1 month of inoculation and was coupled with overexpression of *cutinase* gene. Apart from degradation of plastics, microplastics like DBP are also reported to be metabolized by fungi. Puranik et al., (2023) reported degradation of DBP by *A. flavus.* They also found that esterases were involved in the cleavage of DBP.

3.3.3 Other Class of Microbes

Microorganisms other than bacteria and fungi have been documented to degrade plastics by several researchers. Some of these belong to microalgae that are capable of degrading a variety of plastic polymers. Carson et al., (2013) observed diatom abundance on plastic items with rough surfaces and at sites with high plastic concentrations in the North Pacific Gyre. Oberbeckmann et al., (2016) studied microbial biofilm communities on single-use PET drinking bottles in the North Sea and found an abundance of diatoms (like *Coscinodiscophytina* and *Bacillariophytina*). Kumar et al., (2017) explored the potential of microalgae, including *Scenedesmus dimorphus*, *Anabaena spiroides*, and *Navicula pupula*, to biologically degrade LDPE and HDPE sheets. They isolated microalgae from PE bags in suburban water bodies and observed that *A. spiroides* was particularly effective, leading to an 8.18% degradation of LDPE (Table 3.1). Mishra et al., (2021) prospected a filamentous blue-green alga, *Jacksonvillea* sp. ISTCYN1, that exhibited growth and biofilm-forming potential on HDPE, LDPE, and PP surfaces. The genomic analysis identified genes related to biofilm formation and key enzymes like laccase, esterase, lipase, thioesterase, and peroxidase, shedding light on its bioremediation potential.

Zhang et al., (2019) investigated the effect of *Cylindrotheca closterium*, a marine benthic diatom, on di-n-butyl phthalate (DBP) fate in a water-sediment system and observed a 38.5% lower DBP residue in the presence of the diatom compared to uninoculated treatment. The presence of *C. closterium* increased the net flux of DBP from water to sediment by 7.3 times, leading to a 25.6% increase in total biodegradation flux of DBP. Vingiani et al., (2022) identified two putative DBP hydrolases (*DBPH1* and *DBPH2*) in *C. closterium* through in silico analysis. They also deciphered the structure of these enzymes and their interactions with DBP and further found that the expression increased with concentration of the substrate (5 and 10 mg/l) at 24 and 48 h. Su et al., (2022) explored the microplastic-degrading ability of *Chaetoceros muelleri* and expressed that the microalga was potent of breaking microplastic PS (mPS) and mPET.

Gowthami et al. (2023) investigated the degradation of LDPE by five marine microalgae strains. They found that *Picochlorum maculatum* exhibited the most efficient degradation, with a significant weight loss (20.16%), a higher reduction rate (0.005/day), and a shorter half-life (138.4 days) for LDPE. Laroche et al., (2023) explored the factors influencing diatom communities on plastic substrates in marine environments. They found that geographic location and submersion time had the most significant impact on diatom colonization, with noticeable changes occurring within 2 weeks. Several taxa were identified as early colonizers (such as *Cylindrotheca*, *Navicula*, and *Nitzschia* spp.) with known strong adhesion capabilities.

Apart from harboring bacteria, fungi, and certain microalgae, plastics also inhabit toxin- producing microbial lives. Casabianca et al. (2019) conducted a study to decipher the role of marine debris in carrying harmful microalgae. They quantified the presence of harmful microalgal taxa, like *Pseudo-nitzschia* and *Ostreopsis*, on plastic surfaces and found a direct relationship between their abundance and toxin production and analyzed samples of marine plastics floating at the sea surface using a molecular qPCR assay to quantify the attached microalgal taxa, in particular harmful species. The most abundant harmful microalgal taxon was the diatom *Pseudo-nitzschia* spp., including at least 12 toxic species, and the dinoflagellate *O. ovata* with 6606 and 259 cells/cm^2, respectively. The abundance of other harmful microalgal species including the toxic allochthonous dinoflagellate *Alexandrium pacificum* ranged from 1 to 73 cells/cm^2. However, such microbes should be screened well for potent harm to other living forms.

3.4 Challenges and Future Direction in Microbe-Mediated Plastic Deterioration

Although there is a plethora of microorganisms for degradation of plastics, there are many challenges pertaining to the issue. The first hurdle is their bulk volume. A multiple fold increase in plastic manufacturing from 1964 (15 Mt) to 2017 (350 Mt) and years to come, the amount of plastic piling up is humongous (Plastics Europe, 2018). These end up in environmental matrices, especially soil, water, and sediments, thereby threatening the resources. The second important criterion is their recalcitrant nature. PP, PE, PVC, PU, and PET are the most widely produced plastics and are nonbiodegradable on their own. Their huge half-lives add up to the environmental issues due to hyperaccumulation. Not only do these plastics pose a threat, but the chemicals that they leach into the surrounding matrices are equally dangerous as far as health is concerned. Many chemicals like bisphenol A, phthalate esters, and polyaromatic hydrocarbons are released into water, soil, etc. and end up in the food chain, ultimately harming biotic entities (Chakraborty et al., 2019). These problems make it important for researchers to optimize their quest for alternate tactics.

Finding green alternatives to plastics is a crucial step toward tackling plastic problem. Going for nonbiodegradable, chemical-based, and microbe-based plastics and bioplastics of plant origin is a good option because they are completely biodegradable and do not leach harmful molecules into surrounding matrices (Nandakumar et al. 2021). This reduces the load of nonrenewable-based plastic products in the environment. Further, reduction of existing plastic polymers on earth needs to be taken up. This can be done by strategies that involve recycling, reusing, and biodegradation. The decomposition of plastics using potent microorganisms is the best way to lessen plastic load, since they are highly efficient in acclimatizing to diverse habitats and do not release harmful byproducts into the environment in most cases (Wei & Wierckx, 2021). Additionally, they are known to metabolize the toxic compounds in leachate of plastics, viz., phthalates, bisphenol A, and PAHs (Ghosal et al. 2016; Chouhan et al. 2014; Boll et al. 2020). Prospecting for microorganisms that can metabolize diverse range of plastics saves resources and time, a strategy as good as employing mixed cultures (consortium).

Also, genetically engineering the strains for better enzyme production and expression of genes adds up to the advantages of microbial degradation of polymers. In recent years, significant strides have been made in the realm of plastic degradation through genetic advancements. This groundbreaking field has revolutionized our understanding of how we can harness nature's intrinsic capabilities to break down and recycle persistent plastics. One pioneering technique that has emerged is metagenomics, which has opened doors to uncover novel plastic-degrading genes within diverse microbial communities. This method has enabled scientists to delve into the genetic composition of entire microbial populations, offering a more comprehensive grasp of their potential to degrade plastic waste. The advent of high-throughput screening and next-generation sequencing technology has also played a pivotal role in this journey. These cutting-edge tools have facilitated the pinpointing of specific gene structures and enzymes responsible for plastic degradation (Yadav et al., 2022). Yet, the deluge of genetic data generated by metagenomic studies necessitates innovative computational biology methods to analyze and interpret this wealth of information effectively. Such genetic advancements have profoundly deepened our insights into the potential of plastic biodegradation. They have unveiled the intricate mechanisms through which microbes disintegrate plastic polymers, shedding light on a formerly enigmatic process. Moreover, global metagenomic analyses have confirmed predictions posited by earlier studies, revealing the widespread presence of plastic-degrading microbes and their correlation with pollution trends (Vidiella & Solé, 2022). These findings underscore the viability and scalability of using genetically engineered microorganisms to combat the dire issue of plastic and microplastic contamination in our oceans (Caruso et al., 2022).

By ingeniously combining genetic engineering approaches, researchers have showcased the remarkable potential of manipulating microorganisms to augment their plastic-degrading abilities. Notable among these is the transformation of native marine bacteria into hydrocarbon degraders through the introduction of genes from hydrocarbon-degrading bacteria like *P. putida*. This innovative genetic modification has led to a heightened capacity to break down plastic polymers, offering a blueprint for addressing plastic waste challenges across various types of plastics. The shift toward genetic engineering is not confined to the laboratory; it offers a promising prospect for tackling the burgeoning problem of plastic and microplastic pollution in our oceans. Our expanding comprehension of pollutant degradation metabolism, coupled with advancements in genetic engineering techniques, empowers us to envision new strains of microorganisms with unparalleled plastic-degrading capabilities. Through these methods, researchers have successfully engineered bacteria that exhibit remarkable efficiency in degrading pollutants (Viljakainen & Hug, 2021; Pinnell et al., 2022; Sun et al., 2023). These engineered microorganisms hold tremendous promise as a pragmatic solution to the escalating issue of plastic pollution in marine environments.

In essence, the realm of microbial role in plastic degradation has illuminated a hopeful pathway toward a cleaner, more sustainable future. By tapping into the intricacies of microbial genetics and applying cutting-edge technologies, we are unraveling nature's secrets and crafting innovative solutions to one of the most pressing challenges of our time. As these studies continue to evolve, they hold the potential to reshape the landscape of plastic waste management and safeguard the health of our oceans for generations to come.

3.5 Conclusion

It is evident that plastic menace is global and has very limited and difficult solutions. However, further delay in solving the issue would greatly affect the generations to come. Firstly, reducing the use of plastic is the first step, followed by recycling and reusing. Although steps are being taken toward these solutions, production of plastic polymers at humongous scale has led to minimization of their contribution. Microbial degradation has been a sustainable and most effective strategy for reducing plastic impact. These microscopic lives have great potential in plastic degradation, by employing a number of enzymes. Besides their wide adaptability, newer techniques in science like metagenomics and genetic engineering have strengthened their potential in breaking down plastic. However, further studies are required which provide more effective microbial alternatives for decreasing the impacts of this global menace.

References

Akutsu, Y., Nakajima-Kambe, T., Nomura, N., & Nakahara, T. (1998). Purification and properties of a polyester polyurethane-degrading enzyme from *Comamonas acidovorans* TB-35. *Applied and Environmental Microbiology, 64*(1), 62–67.

Ali, M. I., Ahmed, S., Robson, G., Javed, I., Ali, N., Atiq, N., & Hameed, A. (2014). Isolation and molecular characterization of polyvinyl chloride (PVC) plastic degrading fungal isolates. *Journal of Basic Microbiology, 54*(1), 18–27.

Barcoto, M. O., & Rodrigues, A. (2022). Lessons from insect fungiculture: From microbial ecology to plastics degradation. *Frontiers in Microbiology, 13*, 812143.

Barth, M., Honak, A., Oeser, T., et al. (2016). A dual enzyme system composed of a polyester hydrolase and a carboxylesterase enhances the biocatalytic degradation of polyethylene terephthalate films. *Biotechnology Journal, 11*(8), 1082–1087.

Batista-García, R. A., Sutton, T., Jackson, S. A., et al. (2017). Characterization of lignocellulolytic activities from fungi isolated from the deep-sea sponge Stelletta normani. *PLoS One, 12*(3), e0173750.

Bergmann, M., Collard, F., Fabres, J., et al. (2022). Plastic pollution in the Arctic. *Nature Reviews Earth & Environment, 3*(5), 323–337.

Blettler, M. C., & Mitchell, C. (2021). Dangerous traps: Macroplastic encounters affecting freshwater and terrestrial wildlife. *Science of the Total Environment, 798*, 149317.

Boll, M., Geiger, R., Junghare, M., & Schink, B. (2020). Microbial degradation of phthalates: Biochemistry and environmental implications. *Environmental Microbiology Reports, 12*(1), 3–15.

Bollinger, A., Thies, S., Knieps-Grünhagen, E., et al. (2020). A novel polyester hydrolase from the marine bacterium *Pseudomonas aestusnigri*–structural and functional insights. *Frontiers in Microbiology, 11*, 114.

Bolyen, E., Rideout, J. R., Dillon, M. R., et al. (2019). Reproducible, interactive, scalable and extensible microbiome data science using QIIME 2. *Nature Biotechnology, 37*(8), 852–857.

Brunner, I., Fischer, M., Rüthi, J., Stierli, B., & Frey, B. (2018). Ability of fungi isolated from plastic debris floating in the shoreline of a lake to degrade plastics. *PLoS One, 13*(8), e0202047.

Buchholz, P. C., Feuerriegel, G., Zhang, H., et al. (2022). Plastics degradation by hydrolytic enzymes: The plastics-active enzymes database—PAZy. *Proteins: Structure, Function, and Bioinformatics, 90*(7), 1443–1456.

Cai, L., Wang, J., Peng, J., et al. (2018). Observation of the degradation of three types of plastic pellets exposed to UV irradiation in three different environments. *Science of the Total Environment, 628*, 740–747.

Callahan, B. J., McMurdie, P. J., Rosen, M. J., et al. (2016). DADA2: High-resolution sample inference from Illumina amplicon data. *Nature Methods, 13*(7), 581–583.

Carson, H. S., Nerheim, M. S., Carroll, K. A., & Eriksen, M. (2013). The plastic-associated microorganisms of the North Pacific Gyre. *Marine Pollution Bulletin, 75*(1–2), 126–132.

Caruso, F., Tedesco, P., Della Sala, G., Palma Esposito, F., Signore, M., Canese, S., et al. (2022). Science and dissemination for the UN Ocean decade outcomes: Current trends and future perspectives. *Frontiers in Marine Science, 9*, 863647.

Casabianca, S., Capellacci, S., Giacobbe, M. G., et al. (2019). Plastic-associated harmful microalgal assemblages in marine environment. *Environmental Pollution, 244*, 617–626.

Chakraborty, P., Sampath, S., Mukhopadhyay, M., Selvaraj, S., Bharat, G. K., & Nizzetto, L. (2019). Baseline investigation on plasticizers, bisphenol a, polycyclic aromatic hydrocarbons and heavy metals in the surface soil of the informal electronic waste recycling workshops and nearby open dumpsites in Indian metropolitan cities. *Environmental Pollution, 248*, 1036–1045.

Chamas, A., Moon, H., Zheng, J., et al. (2020). Degradation rates of plastics in the environment. *ACS Sustainable Chemistry & Engineering, 8*(9), 3494–3511.

Chang, B. V., Yang, C. M., Cheng, C. H., & Yuan, S. Y. (2004). Biodegradation of phthalate esters by two bacteria strains. *Chemosphere, 55*(4), 533–538.

Charnock, C. (2021). Norwegian soils and waters contain mesophilic, plastic-degrading bacteria. *Microorganisms, 9*(1), 94.

Chouhan, S., Yadav, S. K., Prakash, J., & Swati, & Singh, S. P. (2014). Effect of Bisphenol A on human health and its degradation by microorganisms: A review. *Annals of Microbiology, 64*, 13–21.

Crabbe, J. R., Campbell, J. R., Thompson, L., et al. (1994). Biodegradation of a colloidal ester-based polyurethane by soil fungi. *International Biodeterioration & Biodegradation, 33*(2), 103–113.

Das, M. P., & Kumar, S. (2015). An approach to low-density polyethylene biodegradation by bacillus amyloliquefaciens. *3 Biotech, 5*(1), 81–86.

Dvořák, P., Nikel, P. I., Damborský, J., & de Lorenzo, V. (2017). Bioremediation 3.0: Engineering pollutant-removing bacteria in the times of systemic biology. *Biotechnology Advances, 35*(7), 845–866.

Espinosa, M. J. C., Blanco, A. C., Schmidgall, T., et al. (2020). Toward biorecycling: Isolation of a soil bacterium that grows on a polyurethane oligomer and monomer. *Frontiers in Microbiology, 11*, 404.

Feng, X., Wang, G., Neumann, K., et al. (2017). Synthesis and characterization of biodegradable poly (ether-ester) urethane acrylates for controlled drug release. *Materials Science and Engineering: C, 74*, 270–278.

Fields, R. D., Rodriguez, F., & Finn, R. K. (1974). Microbial degradation of polyesters: Polycaprolactone degraded by *P. pullulans*. *Journal of Applied Polymer Science, 18*(12), 3571–3579.

Fujiwara, R., Sanuki, R., Ajiro, H., et al. (2021). Direct fermentative conversion of poly (ethylene terephthalate) into poly (hydroxyalkanoate) by *Ideonella sakaiensis*. *Scientific Reports, 11*(1), 19991.

Gaj, T., Gersbach, C. A., & Barbas, C. F. (2013). ZFN, TALEN, and CRISPR/Cas-based methods for genome engineering. *Trends in Biotechnology, 31*(7), 397–405.

Gambarini, V., Pantos, O., Kingsbury, J. M., et al. (2021). Phylogenetic distribution of plastic degrading microorganisms. *mSystems, 6*, e01112–e01120.

Gambarini, V., Pantos, O., Kingsbury, J. M., et al. (2022). PlasticDB: A database of microorganisms and proteins linked to plastic biodegradation. *Database*, baac008.

Gan, Z., & Zhang, H. (2019). PMBD: A comprehensive plastics microbial biodegradation database. *Database*, baz119.

Gao, R., Liu, R., & Sun, C. (2022). A marine fungus Alternaria alternata FB1 efficiently degrades polyethylene. *Journal of Hazardous Materials, 431*, 128617.

Geyer, R., Jambeck, J. R., & Law, K. L. (2017). Production, use, and fate of all plastics ever made. *Science Advances, 3*(7), e1700782.

Ghosal, D., Ghosh, S., Dutta, T. K., & Ahn, Y. (2016). Current state of knowledge in microbial degradation of polycyclic aromatic hydrocarbons (PAHs): A review. *Frontiers in Microbiology, 1369*.

Goel, V., Luthra, P., Kapur, G. S., & Ramakumar, S. S. V. (2021). Biodegradable/bio-plastics: Myths and realities. *Journal of Polymers and the Environment, 29*, 3079–3104.

Gong, Z., Jin, L., Yu, X., Wang, B., Hu, S., Ruan, H., et al. (2023). Biodegradation of low density polyethylene by the fungus Cladosporium sp. recovered from a landfill site. *Journal of Fungi, 9*(6), 605.

Gowthami, A., Marjuk, M. S., Raju, P., Devi, K. N., Santhanam, P., Kumar, S. D., & Perumal, P. (2023). Biodegradation efficacy of selected marine microalgae against low-density polyethylene (LDPE): An environment friendly green approach. *Marine Pollution Bulletin, 190*, 114889.

Gu, J. D., Li, J. X., & Wang, Y. Y. (2005). Biochemical pathway and degradation of phthalate ester isomers by bacteria. *Water Science and Technology, 52*(8), 241–248.

Gui, Z., Liu, G., Liu, X., Cai, R., Liu, R., & Sun, C. (2023). A deep-sea bacterium is capable of degrading polyurethane. *Microbiology Spectrum, 11*, e00073–e00023.

Guo, H., Zheng, B., Jiang, D., & Qin, W. (2017). Overexpression of a laccase with dye decolorization activity from bacillus sp. induced in Escherichia coli. *Journal of Molecular Microbiology and Biotechnology, 27*(4), 217–227.

Han, Y. N., Wei, M., Han, F., Fang, C., Wang, D., Zhong, Y. J., et al. (2020). Greater biofilm formation and increased biodegradation of polyethylene film by a microbial consortium of *Arthrobacter* sp. and *Streptomyces* sp. *Microorganisms, 8*(12), 1979.

Hartmann, G. F., Ricachenevsky, F. K., Silveira, N. M., & Pita-Barbosa, A. (2022). Phytotoxic effects of plastic pollution in crops: What is the size of the problem? *Environmental Pollution, 292*, 118420.

Herrero Acero, E., Ribitsch, D., Dellacher, A., et al. (2013). Surface engineering of a cutinase from *Thermobifida cellulosilytica* for improved polyester hydrolysis. *Biotechnology and Bioengineering, 110*(10), 2581–2590.

Hou, L., Xi, J., Liu, J., Wang, P., Xu, T., Liu, T., et al. (2022). Biodegradability of polyethylene mulching film by two pseudomonas bacteria and their potential degradation mechanism. *Chemosphere, 286*, 131758.

Huerta Lwanga, E., Gertsen, H., Gooren, H., et al. (2016). Microplastics in the terrestrial ecosystem: Implications for *Lumbricus terrestris* (Oligochaeta, Lumbricidae). *Environmental Science & Technology, 50*(5), 2685–2691.

Ihssen, J., Reiss, R., Luchsinger, R., Thöny-Meyer, L., & Richter, M. (2015). Biochemical properties and yields of diverse bacterial laccase-like multicopper oxidases expressed in Escherichia coli. *Scientific Reports, 5*(1), 10465–10478.

Indumathi, A., & Gayathri, T. (2016). Plastic degrading ability of aspergillus oryzae isolated from the garbage dumping sites of Thanjavur, India. *International Journal of Current Microbiology and Applied Sciences., 8*(13), 2319–7706.

Jambeck, J. R., Geyer, R., Wilcox, C., et al. (2015). Plastic waste inputs from land into the ocean. *Science, 347*(6223), 768–771.

Janatunaim, R. Z., & Fibriani, A. (2020). Construction and cloning of plastic-degrading recombinant enzymes (MHETase). *Recent Patents on Biotechnology, 14*(3), 229–234.

Jeon, H. J., & Kim, M. N. (2015). Functional analysis of alkane hydroxylase system derived from *Pseudomonas aeruginosa* E7 for low molecular weight polyethylene biodegradation. *International Biodeterioration and Biodegradation, 103*, 141–146.

Jeyakumar, D., Chirsteen, J., & Doble, M. (2013). Synergistic effects of pretreatment and blending on fungi mediated biodegradation of polypropylenes. *Bioresource Technology, 148*, 78–85.

Johnnie, D. A., Issac, R., & Prabha, M. L. (2021). Bio efficacy assay of laccase isolated and characterized from Trichoderma viride in biodegradation of low density polyethylene (LDPE) and textile industrial effluent dyes. *Journal of Pure & Applied Microbiology, 15*(1), 410–420.

Joo, S., Cho, I. J., Seo, H., et al. (2018). Structural insight into molecular mechanism of poly (ethylene terephthalate) degradation. *Nature Communications, 9*(1), 382.

Jumper, J., Evans, R., Pritzel, A., et al. (2021). Highly accurate protein structure prediction with AlphaFold. *Nature, 596*(7873), 583–589.

Kale, S. K., Deshmukh, A. G., Dudhare, M. S., & Patil, V. B. (2015). Microbial degradation of plastic: A review. *Journal of Biochemical Technology, 6*(2), 952–961.

Karich, A., Ullrich, R., Scheibner, K., & Hofrichter, M. (2017). Fungal unspecific peroxygenases oxidize the majority of organic EPA priority pollutants. *Frontiers in Microbiology, 8*, 1463–1478.

Kawai, F., Oda, M., Tamashiro, T., et al. (2014). A novel ca 2+–activated, thermostabilized polyesterase capable of hydrolyzing polyethylene terephthalate from Saccharomonospora viridis AHK190. *Applied Microbiology and Biotechnology, 98*, 10053–10064.

Kay, M. J., Morton, L. H. G., & Prince, E. L. (1991). Bacterial degradation of polyester polyurethane. *International Biodeterioration, 27*(2), 205–222.

Khan, S., Ali, S. A., & Ali, A. S. (2023). Biodegradation of low density polyethylene (LDPE) by mesophilic fungus '*Penicillium citrinum*'isolated from soils of plastic waste dump yard, Bhopal, India. *Environmental Technology, 44*(15), 2300–2314.

Khatoon, N., Jamal, A., & Ali, M. I. (2019). Lignin peroxidase isoenzyme: A novel approach to biodegrade the toxic synthetic polymer waste. *Environmental Technology, 40*(11), 1366–1375.

Khoironi, A., Anggoro, S., & Sudarno. (2019). Evaluation of the interaction among microalgae Spirulina sp, plastics polyethylene terephthalate and polypropylene in freshwater environment. *Journal of Ecological Engineering, 20*(6), 161–173.

Khruengsai, S., Sripahco, T., & Pripdeevech, P. (2022). Biodegradation of polyester polyurethane by Embarria clematidis. *Frontiers in Microbiology, 13*, 874842.
Kim, J. H., Choi, S. H., Park, M. G., Park, D. H., Son, K. H., & Park, H. Y. (2022). Polyurethane biodegradation by Serratia sp. HY-72 isolated from the intestine of the Asian mantis *Hierodula patellifera. Frontiers in Microbiology, 13*, 1005415.
Kim, H. R., Lee, C., Shin, H., et al. (2023). Isolation of a polyethylene-degrading bacterium, *Acinetobacter guillouiae*, using a novel screening method based on a redox indicator. *Heliyon, 9*(5), e15731.
Kulkarni, S., Nene, S., & Joshi, K. (2017). Production of Hydrophobins from fungi. *Process Biochemistry, 61*, 1–11.
Kumar, R. V., Kanna, G. R., & Elumalai, S. (2017). Biodegradation of polyethylene by green photosynthetic microalgae. *Journal of Bioremediation & Biodegradation, 8*(381), 2.
Kumar, S., Das, M. P., Rebecca, L. J., & Sharmila, S. (2013). Isolation and identification of LDPE degrading fungi from municipal solid waste. *Journal of Chemical and Pharmaceutical Research, 5*(3), 78–81.
Kumari, A., & Chaudhary, D. R. (2020). Engineered microbes and evolving plastic bioremediation technology. In *Bioremediation of pollutants* (pp. 417–443). Elsevier.
Kurniawan, S. B., & Imron, M. F. (2019). The effect of tidal fluctuation on the accumulation of plastic debris in the Wonorejo River Estuary, Surabaya, Indonesia. *Environmental Technology & Innovation, 15*, 100420.
Kutmon, M., van Iersel, M. P., Bohler, A., et al. (2015). PathVisio 3: An extendable pathway analysis toolbox. *PLoS Computational Biology, 11*(2), e1004085.
Laroche, O., Pantos, O., Kingsbury, J. M., Zaiko, A., Wallbank, J., Lear, G., et al. (2023). A spatiotemporal analysis of marine diatom communities associated with pristine and aged plastics. *Biofouling, 39*, 1–17.
Liu, J., Zeng, Q., Lei, H., et al. (2023). Biodegradation of polyester polyurethane by Cladosporium sp. P7: Evaluating its degradation capacity and metabolic pathways. *Journal of Hazardous Materials, 448*, 130776.
Lin, Z., Jin, T., Xu, X., Yin, X., Zhang, D., Geng, M., et al. (2024). Screening and degradation characteristics of plastic-degrading microorganisms in film-mulched vegetable soil. *International Biodeterioration & Biodegradation, 186*, 105686.
Liu, K., Su, Z., Miao, S., et al. (2016). Enzymatic waterborne polyurethane towards a robust and environmentally friendly anti-biofouling coating. *RSC Advances, 6*(38), 31698–31704.
Lomwongsopon, P., & Varrone, C. (2022). Critical review on the progress of plastic bioupcycling technology as a potential solution for sustainable plastic waste management. *Polymers, 14*(22), 4996.
Loredo-Treviño, A., Gutiérrez-Sánchez, G., Rodríguez-Herrera, R., & Aguilar, C. N. (2012). Microbial enzymes involved in polyurethane biodegradation: A review. *Journal of Polymers and the Environment, 20*, 258–265.
Maeda, H., Yamagata, Y., Abe, K., et al. (2005). Purification and characterization of a biodegradable plasticdegrading enzyme from aspergillus oryzae. *Applied Microbiology and Biotechnology, 67*, 778–788.
Magnin, A., Pollet, E., Phalip, V., & Avérous, L. (2020). Evaluation of biological degradation of polyurethanes. *Biotechnology Advances, 39*, 107457.
Maroof, L., Iqbal, M., Farman, S., & Faisal, S. (2022). Biodegradation of low-density polyethylene (LDPE) bags by fungi isolated from waste disposal soil. *Applied and Environmental Soil Science, 2022*, 1–7.
Matthews, S., Belcher, J. D., Tee, K. L., et al. (2017). Catalytic determinants of alkene production by the cytochrome P450 peroxygenase OleTJE. *The Journal of Biological Chemistry, 292*, 5128–5143.
Mishra, A., Gupta, J., Kumari, T., Pal, R., & Thakur, I. S. (2021). Unravelling the attributes of novel cyanobacteria Jacksonvillea sp. ISTCYN1 by draft genome sequencing. *Bioresource Technology, 337*, 125473.

Mo, Y., Lao, H. I., Au, S. W., et al. (2022). Expression, secretion and functional characterization of three laccases in E. Coli. *Synthetic and Systems Biotechnology, 7*(1), 474–480.
Montazer, Z., Habibi-Najafi, M. B., Mohebbi, M., & Oromiehei, A. (2018). Microbial degradation of UV-pretreated low-density polyethylene films by novel polyethylene-degrading bacteria isolated from plastic-dump soil. *Journal of Polymers and the Environment, 26*, 3613–3625.
Montazer, Z., Habibi Najafi, M. B., & Levin, D. B. (2019). Microbial degradation of low-density polyethylene and synthesis of polyhydroxyalkanoate polymers. *Canadian Journal of Microbiology, 65*(3), 224–234.
Munir, E., Harefa, R. S. M., Priyani, N., & Suryanto, D. (2018, March). Plastic degrading fungi *Trichoderma viride* and *aspergillus nomius* isolated from local landfill soil in Medan. In *IOP conference series: Earth and environmental science* (Vol. 126, No. 1, p. 012145). IOP Publishing.
Ndahebwa Muhonja, C., Magoma, G., Imbuga, M., & Makonde, H. M. (2018). Molecular characterization of low-density polyethene (LDPE) degrading bacteria and fungi from Dandora dumpsite, Nairobi, Kenya. *International Journal of Microbiology, 2018.*
Nakajima-Kambe, T., Onuma, F., Kimpara, N., & Nakahara, T. (1995). Isolation and characterization of a bacterium which utilizes polyester polyurethane as a sole carbon and nitrogen source. *FEMS Microbiology Letters, 129*(1), 39–42.
Nakamiya, K., Ooi, T., & Kinoshita, S. (1997). Non-heme hydroquinone peroxidase from *Azotobacter beijerinckii* HM121. *Journal of Fermentation and Bioengineering, 84*(1), 14–21.
Nanda, S., Sahu, S., & Abraham, J. (2010). Studies on the biodegradation of natural and synthetic polyethylene by *Pseudomonas* spp. *Journal of Applied Sciences and Environmental Management, 14*(2).
Nandakumar, A., Chuah, J. A., & Sudesh, K. (2021). Bioplastics: A boon or bane? *Renewable and Sustainable Energy Reviews, 147*, 111237.
Oberbeckmann, S., Osborn, A. M., & Duhaime, M. B. (2016). Microbes on a bottle: Substrate, season and geography influence community composition of microbes colonizing marine plastic debris. *PLoS One, 11*(8), e0159289.
Ojha, N., Pradhan, N., Singh, S., et al. (2017). Evaluation of HDPE and LDPE degradation by fungus, implemented by statistical optimization. *Scientific Reports, 7*(1), 39515.
Olicón-Hernández, D. R., González-López, J., & Aranda, E. (2017). Overview on the biochemical potential of filamentous fungi to degrade pharmaceutical compounds. *Frontiers in Microbiology, 8*, 1792–1809.
Oluwoye, I., Machuca, L. L., Higgins, S., et al. (2023). Degradation and lifetime prediction of plastics in subsea and offshore infrastructures. *Science of the Total Environment*, 166719.
Orhan, Y., & Büyükgüngör, H. (2000). Enhancement of biodegradability of disposable polyethylene in controlled biological soil. *International Biodeterioration & Biodegradation, 45*(1–2), 49–55.
Osman, M., Satti, S. M., Luqman, A., Hasan, F., Shah, Z., & Shah, A. A. (2018). Degradation of polyester polyurethane by aspergillus sp. strain S45 isolated from soil. *Journal of Polymers and the Environment, 26*, 301–310.
Palm, G. J., Reisky, L., Böttcher, D., et al. (2019). Structure of the plastic-degrading *Ideonella sakaiensis* MHETase bound to a substrate. *Nature Communications, 10*(1), 1717.
Pathak, V. M. (2017). Review on the current status of polymer degradation: A microbial approach. *Bioresources and Bioprocessing, 4*(1), 1–31.
Patil, R., & Bagde, U. S. (2012). Isolation of polyvinyl chloride degrading bacterial strains from environmental samples using enrichment culture technique. *African Journal of Biotechnology, 11*(31), 7947–7956.
Penkhrue, W., Khanongnuch, C., Masaki, K., et al. (2015). Isolation and screening of biopolymer-degrading microorganisms from northern Thailand. *World Journal of Microbiology and Biotechnology, 31*, 1431–1442.

Pinnell, L. J., Conkle, J. L., & Turner, J. W. (2022). Microbial succession during the degradation of bioplastic in coastal marine sediment favors sulfate reducing microorganisms. *Frontiers in Marine Science, 9*, 945822.

Plastics Europe. Plastics – The facts. (2018). *Technical report, Plastic Europe, Brussels, Belgium*, https://www.plasticseurope.org/application/files/6315/4510/9658/Plastics_the_facts_2018_AF_web.pdf. Last accessed 28/07/2023.

Plastics Europe. (2021). Plastics – The facts 2021. Available at https://plasticseurope.org/

Pramila, R., & Ramesh, K. V. (2011). Biodegradation of low-density polyethylene (LDPE) by fungi isolated from marine water a SEM analysis. *African Journal of Microbiology Research, 5*(28), 5013–5018.

Priya, A., Dutta, K., & Daverey, A. (2022). A comprehensive biotechnological and molecular insight into plastic degradation by microbial community. *Journal of Chemical Technology & Biotechnology, 97*(2), 381–390.

Puranik, S., Shukla, L., Kundu, A., et al. (2023). Exploring potent fungal isolates from sanitary landfill soil for in vitro degradation of dibutyl phthalate. *Journal of Fungi, 9*(1), 125.

Rani, A., & Singh, P. (2017). Screening of polyethylene degrading fungi from polyethylene dump site. *International Journal of ChemTech Research, 10*(3), 699–704.

Renstad, R., Karlsson, S., & Albertsson, A. C. (1999). The influence of processing induced differences in molecular structure on the biological and non-biological degradation of poly (3-hydroxybutyrate-co-3-hydroxyvalerate), P (3-HB-co-3-HV). *Polymer Degradation and Stability, 63*(2), 201–211.

Ru, J., Huo, Y., & Yang, Y. (2020). Microbial degradation and valorization of plastic wastes. *Frontiers in Microbiology, 11*, 442.

Russell, J. R., Huang, J., Anand, P., et al. (2011). Biodegradation of polyester polyurethane by endophytic fungi. *Applied and Environmental Microbiology, 77*(17), 6076–6084.

Rüthi, J., Cerri, M., Brunner, I., et al. (2023). Discovery of plastic-degrading microbial strains isolated from the alpine and Arctic terrestrial plastisphere. *Frontiers in Microbiology, 14*, 1178474.

Sakti, A. D., Sembiring, E., Rohayani, P., et al. (2023). Identification of illegally dumped plastic waste in a highly polluted river in Indonesia using Sentinel-2 satellite imagery. *Scientific Reports, 13*(1), 5039.

Sakhalkar, S., & Mishra, R. L. (2013). Screening and identification of soil fungi with potential of plastic degrading ability. *Indian Journal of Applied Research, 3*(12), 1–3.

Sánchez, C. (2009). Lignocellulosic residues: Biodegradation and bioconversion by fungi. *Biotechnology Advances, 27*(2), 185–194.

Sangale, M. K., Shahnawaz, M., & Ade, A. B. (2019). Potential of fungi isolated from the dumping sites mangrove rhizosphere soil to degrade polythene. *Scientific Reports, 9*(1), 5390.

Sanniyasi, E., Gopal, R. K., Gunasekar, D. K., & Raj, P. P. (2021). Biodegradation of low-density polyethylene (LDPE) sheet by microalga, Uronema africanum Borge. *Scientific Reports, 11*(1), 17233.

Santo, M., Weitsman, R., & Sivan, A. (2013). The role of the copper-binding enzyme, laccase, in the biodegradation of polyethylene by the actinomycete *Rhodococcus ruber*. *International Biodeterioration and Biodegradation, 84*, 204–210.

Sarkhel, R., Sengupta, S., Das, P., & Bhowal, A. (2020). Comparative biodegradation study of polymer from plastic bottle waste using novel isolated bacteria and fungi from marine source. *Journal of Polymer Research, 27*, 1–8.

Schmit, J. P., & Mueller, G. M. (2007). An estimate of the lower limit of global fungal diversity. *Biodiversity and Conservation, 16*, 99–111.

Schmidt, J., Wei, R., Oeser, T., et al. (2017). Degradation of polyester polyurethane by bacterial polyester hydrolases. *Polymers, 9*(2), 65.

Schwartz, M., Perrot, T., Aubert, E., et al. (2018). Molecular recognition of wood polyphenols by phase II detoxification enzymes of the white rot Trametes versicolor. *Scientific Reports, 8*(1), 8472–8483.

Seo, H., Kim, S., Son, H. F., et al. (2019). Production of extracellular PETase from *Ideonella sakaiensis* using sec-dependent signal peptides in *E. coli*. *Biochemical and Biophysical Research Communications, 508*(1), 250–255.
Sevilla, M. E., Garcia, M. D., Perez-Castillo, Y., et al. (2023). Degradation of PET bottles by an engineered *Ideonella sakaiensis* PETase. *Polymers, 15*(7), 1779.
Shah, Z., Gulzar, M., Hasan, F., & Shah, A. A. (2016). Degradation of polyester polyurethane by an indigenously developed consortium of Pseudomonas and bacillus species isolated from soil. *Polymer Degradation and Stability, 134*, 349–356.
Sharma, J., Gurung, T., Upadhyay, A., Nandy, K., Agnihotri, P., & Mitra, A. K. (2014). Isolation and characterization of plastic degrading bacteria from soil collected from the dumping grounds of an industrial area. *International Journal of Advanced and Innovative Research, 3*(3), 225–232.
Sharma, N., Kumar, V., Maitra, S. S., Lakkaboyana, S. K., & Khantong, S. (2021). DBP biodegradation kinetics by *Acinetobacter* sp. 33F in pristine agricultural soil. *Environmental Technology & Innovation, 21*, 101240.
Sheik, S., Chandrashekar, K. R., Swaroop, K., & Somashekarappa, H. M. (2015). Biodegradation of gamma irradiated low density polyethylene and polypropylene by endophytic fungi. *International Biodeterioration & Biodegradation, 105*, 21–29.
Shi, L., Liu, H., Gao, S., Weng, Y., & Zhu, L. (2021). Enhanced extracellular production of is PETase in *Escherichia coli* via engineering of the *pel*B signal peptide. *Journal of Agricultural and Food Chemistry, 69*(7), 2245–2252.
Shimao, M., Saimoto, H., Kato, N., & Sakazawa, C. (1983). Properties and roles of bacterial symbionts of polyvinyl alcohol-utilizing mixed cultures. *Applied and Environmental Microbiology, 46*(3), 605–610.
Singh, G., Singh, A. K., & Bhatt, K. (2016). Biodegradation of polythenes by bacteria isolated from soil. *International Journal of Research and Development in Pharmacy & Life Sciences, 5*(2), 2056–2062.
Skariyachan, S., Megha, M., Kini, M. N., et al. (2015). Selection and screening of microbial consortia for efficient and ecofriendly degradation of plastic garbage collected from urban and rural areas of Bangalore, India. *Environmental Monitoring and Assessment, 187*, 4174.
Son, H. F., Cho, I. J., Joo, S., et al. (2019). Rational protein engineering of thermo-stable PETase from *Ideonella sakaiensis* for highly efficient PET degradation. *ACS Catalysis, 9*(4), 3519–3526.
Soud, S. A. (2019). Biodegradation of polyethylene LDPE plastic waste using locally isolated *Streptomyces* sp. *Journal of Pharmaceutical Sciences and Research, 11*(4), 1333–1339.
Sowmya, H. V., Ramalingappa, K., & M., & Thippeswamy, B. (2015). Degradation of polyethylene by Penicillium simplicissimum isolated from local dumpsite of Shivamogga district. *Environment, Development and Sustainability, 17*, 731–745.
Srikanth, M., Sandeep, T. S. R. S., Sucharitha, K., & Godi, S. (2022). Biodegradation of plastic polymers by fungi: A brief review. *Bioresources and Bioprocessing, 9*(1), 42.
Su, Y., Qi, H., Hou, Y., et al. (2022). Combined effects of microplastics and benzo [a] pyrene on the marine diatom *Chaetoceros muelleri*. *Frontiers in Marine Science, 8*, 779321.
Sulaiman, S., Yamato, S., Kanaya, E., et al. (2012). Isolation of a novel cutinase homolog with polyethylene terephthalate-degrading activity from leaf-branch compost by using a metagenomic approach. *Applied and Environmental Microbiology, 78*(5), 1556–1562.
Sun, Y., Wu, M., Zang, J., Du, L., Huang, M., Chen, C., & Wang, J. (2023). Plastisphere microbiome: Methodology, diversity, and functionality. *iMeta, 2*(2), e101.
Taghavi, N., Singhal, N., Zhuang, W. Q., & Baroutian, S. (2021). Degradation of plastic waste using stimulated and naturally occurring microbial strains. *Chemosphere, 263*, 127975.
Takamoto, T., Shirasaka, H., Uyama, H., & Kobayashi, S. (2001). Lipase-catalyzed hydrolytic degradation of polyurethane in organic solvent. *Chemistry Letters, 30*(6), 492–493.
Takei, D., Washio, K., & Morikawa, M. (2008). Identification of alkane hydroxylase genes in *Rhodococcus* sp. strain TMP2 that degrades a branched alkane. *Biotechnology Letters, 30*, 1447–1452.

Tanasupawat, S., Takehana, T., Yoshida, S., et al. (2016). *Ideonella sakaiensis* sp. nov., isolated from a microbial consortium that degrades poly (ethylene terephthalate). *International Journal of Systematic and Evolutionary Microbiology, 66*(8), 2813–2818.
Tokiwa, Y., & Suzuki, T. (1977). Hydrolysis of polyesters by lipases. *Nature, 270*(5632), 76–78.
Tseng, W. S., Lee, M. J., Wu, J. A., Kuo, S. L., Chang, S. L., Huang, S. J., & Liu, C. T. (2023). Poly (butylene adipate-co-terephthalate) biodegradation by *Purpureocillium lilacinum* strain BA1S. *Applied Microbiology and Biotechnology, 107*(19), 6057–6070.
UniProt Consortium. (2019). UniProt: A worldwide hub of protein knowledge. *Nucleic Acids Research, 47*(D1), D506–D515.
Usha, R., Sangeetha, T., & Palaniswamy, M. (2011). Screening of polyethylene degrading microorganisms from garbage soil. *Libyan Agriculture Research Center Journal International, 2*(4), 200–204.
Vaksmaa, A., Hernando-Morales, V., Zeghal, E., & Niemann, H. (2021). Microbial degradation of marine plastics: Current state and future prospects. *Biotechnology for sustainable environment, 1*, 111–154.
Vermeiren, P., Muñoz, C. C., & Ikejima, K. (2016). Sources and sinks of plastic debris in estuaries: A conceptual model integrating biological, physical and chemical distribution mechanisms. *Marine Pollution Bulletin, 113*(1–2), 7–16.
Vidiella, B., & Solé, R. (2022). Ecological firewalls for synthetic biology. *Iscience, 25*(7), 104658.
Viljakainen, V. R., & Hug, L. A. (2021). New approaches for the characterization of plastic-associated microbial communities and the discovery of plastic-degrading microorganisms and enzymes. *Computational and Structural Biotechnology Journal, 19*, 6191–6200.
Vingiani, G. M., Leone, S., De Luca, D., Borra, M., Dobson, A. D., Ianora, A., et al. (2022). First identification and characterization of detoxifying plastic-degrading DBP hydrolases in the marine diatom Cylindrotheca closterium. *Science of the Total Environment, 812*, 152535.
Wallace, N. E., Adams, M. C., Chafin, A. C., et al. (2020). The highly crystalline PET found in plastic water bottles does not support the growth of the PETase-producing bacterium *Ideonella sakaiensis*. *Environmental Microbiology Reports, 12*(5), 578–582.
Wang, W., Leung, A. O. W., Chu, L. H., et al. (2018). Phthalates contamination in China: Status, trends and human exposure-with an emphasis on oral intake. *Environmental Pollution, 238*, 771–782.
Wang, J., Youkharibache, P., Zhang, D., et al. (2020). iCn3D, a web-based 3D viewer for sharing 1D/2D/3D representations of biomolecular structures. *Bioinformatics, 36*(1), 131–135.
Wang, D., Zhang, P., Yan, M., et al. (2022). Degradation mechanism and properties of debris of photocatalytically degradable plastics LDPE-TiO2 vary with environments. *Polymer Degradation and Stability, 195*, 109806.
Watanabe, M., Kawai, F., Shibata, M., et al. (2003). Computational method for analysis of polyethylene biodegradation. *Journal of Computational and Applied Mathematics, 161*, 133–144.
Wei, R., & Wierckx, N. (2021). Microbial degradation of plastics. *Frontiers in Microbiology, 12*, 635621.
Wilkes, R. A., & Aristilde, L. (2017). Degradation and metabolism of synthetic plastics and associated products by pseudomonas sp.: Capabilities and challenges. *Journal of Applied Microbiology, 123*(3), 582–593.
Wong, J. K. H., Lee, K. K., Tang, K. H. D., & Yap, P. S. (2020). Microplastics in the freshwater and terrestrial environments: Prevalence, fates, impacts and sustainable solutions. *Science of the Total Environment, 719*, 137512.
Wright, R. J., Bosch, R., Gibson, M. I., & Christie-Oleza, J. A. (2020). Plasticizer degradation by marine bacterial isolates: A proteogenomic and metabolomic characterization. *Environmental Science & Technology, 54*(4), 2244–2256.
Wu, H., Liu, Q., Sun, W., Lu, Y., Qi, Y., & Zhang, H. (2023). Biodegradability of polyethylene mulch film by *Bacillus paramycoides*. *Chemosphere, 311*, 136978.

Wufuer, R., Li, W., Wang, S., & Duo, J. (2022). Isolation and degradation characteristics of PBAT film degrading bacteria. *International Journal of Environmental Research and Public Health, 19*(24), 17087.

Yadav, V., Dhanger, S., & Sharma, J. (2022). Microplastics accumulation in agricultural soil: Evidence for the presence, potential effects, extraction, and current bioremediation approaches. *Journal of Applied Biology and Biotechnology, 10*(2), 38–47.

Yoon, M. G., Jeon, J. H., & Kim, M. N. (2012). Biodegradation of polyethylene by a soil bacterium and *AlkB* cloned recombinant cell. *Journal of Bioremediation & Biodegradation, 3*, 145.

Yoshida, S., Hiraga, K., Takehana, T., et al. (2016). A bacterium that degrades and assimilates poly (ethylene terephthalate). *Science, 351*(6278), 1196–1199.

Yu, J., Sun, L., Ma, C., et al. (2016). Thermal degradation of PVC: A review. *Waste Management, 48*, 300–314.

Zahra, S., Abbas, S. S., Mahsa, M. T., & Mohsen, N. (2010). Biodegradation of low-density polyethylene (LDPE) by isolated fungi in solid waste medium. *Waste Management, 30*(3), 396–401.

Zeghal, E., Vaksmaa, A., Vielfaure, H., Boekhout, T., & Niemann, H. (2021). The potential role of marine fungi in plastic degradation–a review. *Frontiers in Marine Science, 8*, 738877–738894.

Zhang, F., Ding, Z., Gong, H., & Chi, J. (2019). Effects of microphytobenthos *Cylindrotheca closterium* on the fate of di-n-butyl phthalate in an aquatic microcosm. *Marine Pollution Bulletin, 140*, 101–106.

Zhang, Y., Lin, Y., Gou, H., et al. (2022a). Screening of polyethylene-degrading bacteria from *Rhyzopertha Dominica* and evaluation of its key enzymes degrading polyethylene. *Polymers, 14*(23), 5127.

Zhang, K., Hu, J., Yang, S., Xu, W., et al. (2022b). Biodegradation of polyester polyurethane by the marine fungus *Cladosporium halotolerans* 6UPA1. *Journal of Hazardous Materials, 437*, 129406.

Zhang, N., Ding, M., & Yuan, Y. (2022c). Current advances in biodegradation of polyolefins. *Microorganisms, 10*(8), 1537.

Zylstra, E. R. (2013). Accumulation of wind-dispersed trash in desert environments. *Journal of Arid Environments, 89*, 13–15.

Chapter 4
Microbial Degradation: Understanding the Mysteries of Polyethylene Terephthalate (PET) Degradation, "By Nature's Recyclers"

Ashish Kumar, Bharti Kukreti, Vindhya Bundela, Prerna Rani, and Ajay Veer Singh

Abstract Polyethylene terephthalate (PET), which is a widely used synthetic polymer, poses significant environmental challenges due to its resistance to degradation. The traditional methods of recycling PET waste associated with burning and landfilling prove challenging, due to the detrimental impacts on both terrestrial and aquatic organisms. However, there are potentially effective ways to combat plastic pollution since a variety of microbial species, such as bacteria and fungi, have demonstrated the capacity to enzymatically break down PET. The enzyme responsible for PET degradation includes cutinases and PETases, providing insight into the biochemical processes that help break down this resistant polymer. The understanding of recent advancements and obstacles in PET biodegradation is essential for formulating enduring remedies to alleviate plastic pollution. The purpose of this chapter is to give the state of knowledge in the subject, directing future studies and assisting in the creation of environmentally appropriate PET waste management techniques.

Keywords Polyethylene terephthalate · PETase · Biodegradability · Bioengineering · Microbial consortia

4.1 Introduction

The excessive use of plastic has significantly harmed ecosystems, resulting in substantial damage to the natural environment and its ecology. It is widely acknowledged that approximately 80% of the world's usable plastic consists of thermoplastic polymers, including polyethylene (PE), polyurethane (PUR), polyvinyl chloride

A. Kumar · B. Kukreti · V. Bundela · P. Rani · A. V. Singh (✉)
Biofortification Lab, Department of Microbiology, College of Basic Sciences and Humanities, G.B. Pant University of Agriculture and Technology, Pantnagar, Uttarakhand, India

R. Soni et al. (eds.), *Advanced Strategies for Biodegradation of Plastic Polymers*, https://doi.org/10.1007/978-3-031-55661-6_4

(PVC), polypropylene (PP), polystyrene (PS), and polyethylene terephthalate (PET). Within this group of plastics, polyethylene terephthalate (PET) predominates as the material of choice for the production of drinking bottles, foil, textile fibers, and food containers, owing to its exceptional mechanical and thermal properties (Danso et al., 2019). PET is a long linear polymer made up of repeating units of aromatic terephthalic acid (TPA) and ethylene glycol (EG). The nomenclature for the monomeric unit of PET is known as bis(2-hydroxyethyl) terephthalate (BHET) (Khairul Anuar et al., 2022). Globally, PET is extensively used in the preparation of a wide array of single-use plastic products. As highlighted by numerous major global organizations, the alarming state of plastic accumulation in ecosystems is expected to worsen unless measures are implemented to decrease plastic consumption and manufacturing (Hiraga et al., 2019). Traditional techniques for managing PET waste, such as mechanical recycling, landfilling, ignition, and compound hydrolysis, have several drawbacks, including a reduction in mechanical properties, the use of land resources, the production of toxic compounds, and significant energy consumption. Therefore, it is essential to devise innovative strategies for the upcycling of massive amounts of plastic garbage (Chen et al., 2022). There have already been many great applications created that draw inspiration from natural ecosystems. The biological techniques of PET waste degradation are effective and the best alternative way to overcome the chemical method to maintain environmental sustainability (Khairul Anuar et al., 2022). Even though PET is considered nonbiodegradable, multiple studies have demonstrated its degradation by microbes or their enzymes (Mohanan et al., 2020; Santacruz-Juarez et al., 2021; Ahmaditabatabaei et al., 2021). Among the extracellular enzymes secreted by microbes for plastic degradation, some have been identified so far, including cutinases, lipases, PETases, proteases, and esterases. They also produce hydrophilic intermediates to depolymerize PET and use it as a carbon source for their survival. Additionally, they utilize PET intermediates for metabolism and degradation processes (Ahmaditabatabaei et al., 2021; Bala et al., 2022). To get significant results, microbial breakdown can be enhanced and utilized together with physicochemical techniques. Overall, the degradation of PET by microorganisms holds potential as a viable solution to address the problem of plastic waste, offering a more sustainable approach for managing PET waste in the environment. However, continuous research and development is required if these techniques are to be enhanced and optimized for usage in real-world scenarios. In view of the above context, the current manuscript discusses about the methods of PET degradation, including the employment of various microbes and the mechanisms connected to them.

4.2 Strategy Used for PET Degradation

Since the year 1950, plastics have been produced in large quantities, leading to their pervasive use across a variety of industries and establishing them as an indispensable component of modern society (Thompson et al., 2009). Synthetic materials, notably polyethylene terephthalate (PET), provide economic advantages to packaging companies due to their widespread use in everyday life, ease of manufacturing, cost-effectiveness, durability, and resilience (Lebreton & Andrady, 2019). However, the consumption of synthetic plastic has risen as an immense ecological issue due to its barely breakdown nature, which remains in the environment for a century or more (Eriksen et al., 2014; Lechner et al., 2014). Pollution caused by the plastic exists in almost every niche of the planet, with profound impacts on ecological systems, particularly in aquatic ecosystems, affecting equally large and small fauna and flora (Seo & Park, 2020). Degradation is the term for any alteration of the physical or chemical properties and can result from chemical, physicochemical, or biological processes. Usually, the initial phase in any degradation pathway is the breaking of the polymeric substances by the physical action of a mechanical nature (Li et al. 2019a, b). Hydrolysis and oxidation are the two main chemical and biological processes that lead to polymer deterioration (Fotopoulou & Karapanagioti, 2019). When it comes to most types of degradation, Zheng et al. (2005) found that materials with pure carbon chain backbones are particularly resilient, but materials containing heteroatoms in the structure (such as polyesters and polyamines) exhibit higher vulnerability to degradation. PET serves as a prime illustration of a synthetic polymer. Despite the presence of ester links that are typically susceptible to degradation, it remains remarkably resilient due to its inherent aromatic nature. It is an aromatic polyester, and its breakdown challenges are attributed to the existence of covalent linkages (non-hydrolyzable) constituting a component called diethylene glycol terephthalate (Kawai et al., 2014a). The biodegradation of PET typically leads to the shortening of its chain length, resulting in a decrease in the polymer's molecular mass. Subsequently, this process generates oligomers, dimers, and water-soluble monomers that can then traverse microbial cell membranes, serving as a carbon source (Kawai et al., 2014b). Natural abiotic degradation of synthetic polymers occurs through mechanical breakdown induced by erosion, tidal forces, and waves, mirroring the gradual decomposition in the natural environment. A brittle material is produced by the effects of photooxidation and hydrolysis, and mechanical breakdown can readily follow, leading to the development of micro- and nanoplastic pieces. On the contrary, mechanical fragmentation does not alter the molecular mass of the polymer, whereas chemical fragmentation might (Lambert & Wagner, 2016).

In aerobic outdoor conditions, degradation due to sunlight is regarded as the most significant abiotic degradation pathway (Ali et al., 2021). The majority of synthetic polymers are vulnerable to deterioration brought on by visible and ultraviolet (UV) rays. Generally, the near-UV radiations (400–290 nm) govern the shelf life of polymeric materials in outdoor applications (Singh & Sharma, 2008). Upon exposure to near-ultraviolet light, PET undergoes photodegradation, leading to chain splitting via Norrish I and II reactions. The UV radiation provides activation energy to start the oxygen incorporation into the PET waste polymer chain. PET comprised alternating ethylene glycolate and terephthalate subunits, linked via ester linkages. According to Fotopoulou and Karapanagioti (2019), PET oxidation due to light involves photoproduct generation like hydroperoxide species, by the oxidation of the CH_2 groups next to the ester linkages via multiple pathways. According to Fairbrother et al. (2019), photodegradation of PET causes a direct cleavage of the ester link, resulting in the formation of terephthalic acid, esters, carbon monoxide/dioxide, carboxylic acids, and anhydrides. Additionally, cross-linking occurs, causing the polymer to lose its flexibility and develop an uneven surface (Fotopoulou & Karapanagioti, 2019). This cycle is repeated until the molecular weight of the polymer chains is sufficiently low so that microorganisms can quickly convert them into carbon dioxide. However, for complete decay, this process is estimated to necessitate a minimum of 50 years to reach full degradation (Thachnatharen et al., 2021).

Hydrolysis is another method of PET waste depolymerization, where it yields terephthalic acid (TPA) and ethylene glycol (EG) as a by-product. One of the most important phases in the abiotic degradation pathway is represented by this step. The presence of catalysts, such as the ions generated during this process, speeds up hydrolysis. The hydrolysis rate is determined by the vulnerability of polymeric chemical bonds to water attack and the degree of water migration across the polymeric substances (Padsalgikar, 2017). PET can hydrolyze in an alkaline, acidic, or natural environment (Sheel & Pant, 2019). During acid-base catalyzed reactions, the process begins by the attack of a nucleophile on a carbonyl group in ester or amide bonds (Hosseini et al., 2007). The residue of PET is reacted using NaOH or KOH to produce terephthalate salt, which is then further acidified to produce TPA monomer. In neutral hydrolysis, hot water or hot steam is employed to obtain EG and TPA as the primary products. Filtration is used to remove TPA from the post-reaction mixture, whereas distillation or extraction is used to recover an adequate quantity of diluted EG (Thachnatharen et al., 2021). There have been at least four different kinetic schemes proposed, and the mechanism underlying the hydrolysis of PET is not completely understood. It is hypothesized that hydrolysis takes place when water diffuses into the polymer's amorphous areas. According to Sammon et al. (2000), the rate will obviously be determined by the morphology and degree of crystallinity of the polymer as well as by temperature and relative humidity. Additionally, Sammon et al. (2000) investigated the impact of hydrolytic breakdown on the PET structure and observed a noteworthy transformation in the v(C-H) band. This change was attributed to the increase in hydrophilic end groups, resulting in a shift in the environment of aliphatic methyl groups and substantial hydrogen bonding.

Another strategy that is applied for synthetic polymer degradation is thermal degradation. It refers to the degradation of polymers induced by the input of high-temperature energy. At high temperatures, thermo-oxidative reactions can take place, leading to the fragmentation of polymer chains and the release of radicals (Pirsaheb et al., 2020). As per Venkatachalam et al. (2012), during the initial stage of PET degradation under elevated temperatures (250–350 °C), it is believed that random scission of the main chain ester bond occurs, resulting in the formation of vinyl esters and carboxyl end groups. This process involves β-scission, where the methylene group adjacent to the carbonyl groups is implicated, leading to the creation of double bonds at the end groups. The vinyl ester is then trans-esterified to produce the vinyl alcohol, which is subsequently tautomerized to produce acetaldehyde (East, 2005). Interestingly, the thermal degradation of PET allows obtaining additional degraded products, including ethylene, aromatic compounds, other aldehydes, cyclic oligomers, and carbon monoxide or dioxide (Das & Tiwari, 2019).

There is still a need for an environmentally friendly and highly efficient method to address synthetic polymer waste management. Numerous studies in this area (Zhang et al., 2020; Sarkhel et al., 2020; Mohanan et al., 2020; Palm et al., 2019) are focusing on the potential for microbial degradation of packaging materials. There is a great demand for eco-friendly recycling techniques to address the widespread synthetic polymer waste pollution of the environment. Microorganisms can be employed to biodegrade polymer trash in a cost-effective, environmentally acceptable manner through either their cultivation or the extraction of their enzymes (Restrepo-Florez et al., 2014). According to Francois-Heude et al. (2014), biodegradation is defined by a number of phases that can be distinguished by particular terminology. The biodeterioration process, which is the first stage of biodegradation, is carried out by a variety of microbial communities. Deterioration involves superficial degradation that alters the mechanical, physical, and chemical attributes of a given material (Lucas et al., 2008). The splitting of polymeric compounds into oligomers, dimers, or monomers via enzymes synthesized by microbes is the second step in the depolymerization process. The next stage is assimilation followed by the process known as mineralization. The term "assimilation" refers to the microbial metabolism's incorporation of molecules carried through the cytoplasm to create energy, new biomass, storage vesicles, and a variety of primary and secondary metabolites. Moreover, mineralization is the secretion of various salts, both simple and complex, as well as metabolites that enter the extracellular environment (Francois-Heude et al., 2014). Since biodegradation is more affordable and more effective and does not result in secondary pollutants like incineration and landfilling of PET waste, it is an ideal alternative to conventional waste management approaches (Michaud et al., 2007). Polymer-degrading microorganisms are a useful remedy for various types of environmental contaminants due to the metabolic diversity of these organisms (Iranzo et al., 2001). When PET waste is subjected to polymer-degrading microorganisms during biodegradation, this causes the organisms to attach and colonize the surface of the PET waste. Microbial hydrolysis involves two distinct phases. Initially, the enzyme adheres to the polymer substrate, and subsequently, it triggers the hydrolytic cleavage process. Once low-molecular-weight oligomers, dimers, and monomers are formed from the PET waste polymer, carbon dioxide and water are eventually produced (Zhang et al., 2004). The strategy of biodegrading synthetic polymer waste is widely adopted, with ongoing extensive research being conducted on a global scale.

4.3 Microbial-Based Mechanism Involved in PET Degradation

PET represents more than 10% of the global plastic market and is one of the most common types of polyester plastic used in a variety of industries. Therefore, it is essential to illustrate how to upcycle this material in an environmentally beneficial way. Numerous studies have shown that fungi and bacteria are efficient PET bio-degraders.

4.3.1 Enzyme-Based PET Degradation

Microbial populations can adapt to new environments and adhere to them, and they predominantly produce certain enzymes which enable them to survive on persistent polymer debris as their only source of carbon (Wang et al. 2020a, b). Numerous extracellular enzymes hydrolyze and oxidize polymers, causing the polymer chain to split into short-length polymers and small molecules (including oligomers, dimers and monomers), which are easily utilized by microbes (Chen et al., 2019). Additionally, microbes might assimilate these low-molecular-weight products as a part of their cellular metabolism (Wilkes et al., 2017). Over time, polymers undergo mineralization into H_2O and CO_2 through a combination of external and intracellular processes in aerobic environments. In anaerobic conditions, this transformation results in the production of CH_4, CO_2, organic acids, H_2O, and NH_4, thereby promoting the proliferation of microbial biomass. Hydrolyzable polymers, like PET, become more vulnerable to biodegradation due to the existence of multiple biodegradation pathways. These pathways include extracellular enzymes that play a role in breaking down cellulose and proteins (Chen et al., 2019). Recent studies have identified certain fungal and bacterial strains capable of biodegrading PET into smaller molecules like bis(2-hydroxyethyl) terephthalate (BHET) and mono(2-hydroxyethyl) terephthalate (MHET) (Wei et al., 2016). An enzyme known as PETase, which can hydrolyze PET, was, for the first time, identified in *Ideonella sakaiensis* which belongs to the family Comamonadaceae and genus *Ideonella* and has been found to be ubiquitous in nature (Yoshida et al., 2016; Danso et al., 2018). The bacterial strain *I. sakaiensis* harbors a dual enzyme system belonging to the tannase family, PETase (hydrolyzes polyethylene terephthalic acid) and MHETase (hydrolyzes mono-(2-hydroxyethyl) terephthalic acid), which works synergistically in the transformation of PET polyester into environmentally safe, simple, and reusable monomers. PETase breaks down PET polymer into MHET, which is further hydrolyzed into TPA and EG. The bacterium continued to use these end products as a source of nourishment (Austin et al., 2018). Microbes that have a metabolic system for these compounds can degrade these monomers. TPA gets converted to PCA (protocatechuate) using DCD (1,6-dihydroxycyclohexa-2,4-diene dicarboxylate) as an intermediate with the help of the tph enzyme (Salvador et al., 2019). In addition, bacteria like *Pseudomonas aeruginosa* and *P. putida* are able to transform EG to glyoxylate via periplasmic alcohol dehydrogenase enzyme, which is then directed straight into the Krebs cycle through several metabolic compounds like isocitrate or, in the case of *Escherichia coli*, propanediol oxidoreductase (Luo & Lee, 2017).

Moreover, it has been discovered that several enzymes, including nitro-benzyl-esterase, lipase, serine esterase, and cutinase, can hydrolyze PET. According to Müller et al. (2005), the enzymatic hydrolysis of PET using a cutinase like hydrolase produced by the soil actinobacterium *Thermobifida fusca* can break down up to 54% of the PET over 21 days at 55 °C. After that, Liebminger et al. (2007) demonstrated the activity of fungi that produced polyesterase and cutinase (*Penicillium citrinum*) and showed how well polyesterase breaks down PET. After that, Liebminger et al. (2007) demonstrated the activity of the fungus *Penicillium citrinum* that produced polyesterase and cutinase which hydrolyze 3PET [bis-(benzoyloxyethyl)terephthalate] polymer into BHET and MHET. Esterase is a member of the enzyme family capable of splitting short-chain acyl ester linkage present in PET monomers with observed surface modulation. According to research, PET degradation by these esterases was first reported in *Bacillus* and *Nocardia* (Sharon & Sharon, 2012). The interfacial activation phenomenon is a characteristic of lipases, which are well recognized for their hydrolysis of long-chain (greater than C10) water-insoluble triglycerides. According to reports, lipases can partially degrade PET materials by improving their wettability, dye ability, and absorbent qualities (Gupta et al., 2015). Numerous species, including *Thermomyces lanuginosus*, *Candida Antarctica*, and *Burkholderia* spp., have been reported for lipases production as PET whole-cell catalyst (Vertommen et al., 2005; Nechwatal et al., 2006; Eberl et al., 2009). Additionally, mycelia can secrete extracellular enzymes like depolymerase that break down polymers into simpler forms (Ameen et al., 2015; Ali et al., 2021). The large surface area of the fungal mycelium allows it to effectively access and destroy polymers (Sánchez, 2020). Notably, fungi tend to secrete a much higher concentration of enzymes compared to bacteria (Gangola et al., 2019). However, bacteria generally outperform fungi in terms of growth rate and adaptability to diverse environments due to their rapid reproduction and ability to thrive across a broad pH range. Fungi, on the other hand, typically prefer acidic pH conditions and exhibit a slower growth rate compared to bacteria. However, fungi have a higher metabolic rate than bacteria. To address challenges related to adaptability and growth rates, a common approach involves leveraging the enzymatic capabilities of both bacteria and fungi for biodegradation purposes. Another potential solution is to incorporate a fungal degradative gene into the bacterial genome.

4.3.2 Bioengineering Approach

Advancements in genetic engineering have opened exciting avenues for manipulating and enhancing the capabilities of microorganisms involved in PET degradation. Through the modification of genes responsible for PET breakdown, scientists can elevate enzyme production, enhance efficiency, and expand the range of substrates targeted (Dvořák et al., 2017). These synthetic biology techniques open the door to the development of microbial "bio-recyclers" that hold the potential to revolutionize the management of plastic waste entirely. Recent advancements in protein engineering have tackled a range of enzyme-related challenges through innovative methods. These methods encompass a range of strategies, including mutagenesis techniques such as random mutagenesis and site-directed mutagenesis, concurrent amino acid substitutions, genome editing, computational genomics, structural biology, and directed evolution (Kumar et al., 2018; Samak et al., 2020). This revolutionary approach may address the issue of the buildup of plastic waste while recovering the raw materials required for the manufacture of recovered plastic (Danso et al., 2018; Parisi et al., 2019). By identifying the genes for metabolizing the synthetic polymer and transforming them into suitable host organisms using appropriate vectors under the control of promoters, recombinant DNA technology was used to evolve the potential microbes (Hussain et al., 2018; Mandhare et al., 2018). There has been a lot of research toward developing novel enzymes that can break down PET or improve the degradability of existing ones. Since the *I. sakaiensis* enzyme *Is*PETase discovery, numerous research teams have investigated the structure and important catalytic residues (Han et al., 2017; Joo et al., 2018), which contributed to additional protein engineering efforts. Protein engineering through rational design modifies a protein of interest using prior knowledge of its structure and catalytic efficiency as well as computer simulation and modeling (Tiwari et al., 2012; Zhu et al., 2021). In the field of enzyme engineering, rational biocatalyst design is a promising future development. Recently, multiple algorithms, including homology modeling and molecular docking, were developed to consciously construct PET biocatalysts and modulate enzyme specificities For instance, Austin et al. (2018) conducted engineering experiments with *Is*PETase, resulting in increased activity and improved PET degradation. In order to redesign the *Is*PETase through a computational technique to maintain its stability in higher melting temperature, a 300-fold increase in the breakdown of semicrystalline PET is observed (Cui et al., 2021). Additionally, only by replacing the amino acid sequence with conserved sequences of leaf branch compost cutinases, in the active pocket of *Thermobifida fusca* enzyme, *Tf*Cut2 cutinase resulted in notable enhancement of PET degradation at elevated temperature (65 °C) (Wei et al., 2016). Similar to this, Tournier et al. (2020) improved thermostability and PET depolymerization by redesigning the PETase found in leaf-branch compost and afterward utilized the resultant monomers to resynthesize PET. Furthermore, Knott et al. (2020) in his research have designed a hybrid of the *I. sakaiensis* dual enzyme system (composed of PETase alongside the MHETase enzyme), which readily cleaves MHET into TA and EG, thereby enhancing PET biodegradation. Numerous initiatives also have been made to date to enhance PETase's catalytic performance in PET biodegradation. Austin et al. (2018) created the PETase$^{W159H/S238F}$ double mutant, which showed better PET degradation and increased substrate scope by constricting the binding cleft in the PETase structure.

Another effective method for minimizing substrate diffusion and mitigating the toxicity of reaction intermediates is the use of fusion enzymes (Sharshar et al., 2019). The combination of PET-hydrolyzing enzymes with signal peptides and short peptides that point recombinant proteins in the direction of the secretary pathway is an important method for increasing the quantity and catalytic activity of these enzymes. *Phaeodactylum tricornutum*, a photosynthetic diatom, was modified through protein engineering by Moog et al. (2019) to serve as a useful alternative chassis for PET biodegradation. For PETase expression in *P. tricornutum*, enzyme $PETase^{R280A}$ was coupled with an alkaline phosphatase (AP) signal peptide from *P. tricornutum* in a culture medium. The diatom's ability to secrete and produce $PETase^{R280A}$ was dramatically increased by this fusion resulting in an accelerated degradation rate and decreased crystallinity level of polyethylene terephthalate glycol (PETG) by 80-fold higher turnover compared with the bottle grade PET at 30 °C.

Moreover, in order to improve the thermodynamic properties of these polyester hydrolases, engineered posttranslational modifications (such as glycosylation, acylation, and methylation) might be applied (Shirke et al., 2018). In particular, bioprocesses involving engineered mesophilic organisms like *E. coli*, which can only grow to a maximum temperature of 48–50°C after evolutionary selection and at a fitness cost (Rudolph et al., 2010; Guyot et al., 2014), are incapable of hydrolysis at those temperatures. The methylotrophic yeast *Pichia pastoris* is an eukaryotic microorganism that may be easily genetically modified and grows at high cell densities. It is interesting to note that when recombinant LCC is expressed in *P. pastoris*, it yields properly folded glycosylated LCC, and this glycosylated LCC exhibits a higher thermostability by 10°C than the native LCC (Shirke et al., 2018). Similarly, in another study, it was found that the glycosylated cutinase of *Thermobifida cellulosilytica*, Thc_Cut1, mutant strains when engineered in *P. pastoris*, showed effective degradation of poly(butylene succinate) and poly(3-hydroxybutyrate-co-3-hydroxyvalerate) within 96 h (Gamerith et al., 2017). Thus, the above mentioned approaches made possible attempts to degrade polymers like PET (Fig. 4.1).

4.4 Microorganism Involved in Degradation of PET

4.4.1 Bacteria

In nature, microorganisms, primarily bacteria and certain fungi, are the primary degraders. These organisms are capable of adapting to new substrates because of their high metabolic potential and quick rates of replication. A potent method of environmental detoxification may involve the selection of highly efficient degradative microorganisms, followed by their adaptation to naturally persistent substance. Despite the favorable outcomes of many laboratory tests, only few of them are usable on a large scale (Miglani et al., 2022). *B. subtilis*, a gram-positive bacterium, is regarded as a potential bacterial suspension for biodegradation since it has a

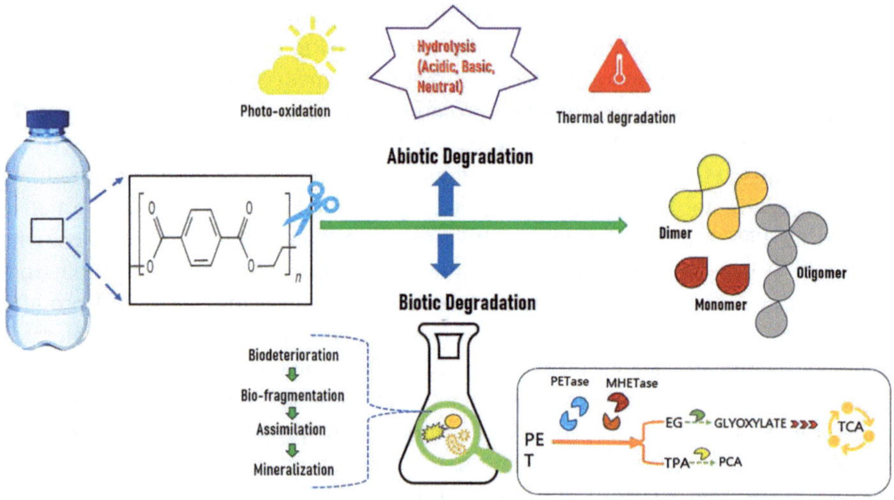

Fig. 4.1 Exploring the dual pathways of PET degradation, driven by abiotic and biotic forces

strong tolerance to hostile conditions and has been used to release proteins that can digest a variety of contaminants (Huang et al. 2018a, b). Due to its advantages of having a high production capacity, rapid expansion, and the devoid of an outer membrane, *B. subtilis* opposed to *E. coli*, which normally produce an inclusion body, is considered to be an appropriate microbial host for the creation of heterologous proteins (Arora, 2020). *I. sakaiensis*, a bacterium, has recently been found to solely utilize PET for carbon and energy in aerobic events. *I. sakaiensis* perform PET degradation by secreting two enzymes (PETase and MHETase), which can break down PET polymer into the environmentally safe monomers TPA and EG (Palm et al., 2019). Dhaka et al. (2022) reported PET plastic degradation at the rate of 65% by the three rhizospheric bacterial isolates *Priestia aryabhattai* VT 3.12, *Bacillus pseudomycoides* VT 3.15, and *Bacillus pumilus* VT 3.16.

4.4.2 Fungi

In situations where readily available carbon sources are limited, the significance of extracellular enzymes produced by fungi becomes more pronounced in influencing whether a fungal species can utilize plastic as an alternative carbon source. Fungi and their special enzyme systems can break down PET. Accumulation of a high proportion of aromatic terephthalate units restricts enzymatic breakdown and lowers chain mobility/flexibility in PET structures. Various fungi, however, have the ability to increase PET's susceptibility to destruction via the enzyme system (Gao et al., 2022; Khairul Anuar et al., 2022). A variety of fungi hydrolytic enzymes have so far been discovered and have many different applications. Among the array of enzymes, such as esterases, lipases, and cutinases, one notable application involves their capacity to hydrolyze the ester bonds linking the monomeric structural units of PET. This enzymatic action facilitates the degradation of PET into low-molecular-weight (MW) oligomers like BHET and MHET (Ahmaditabatabaei et al., 2021). By severing the tiny-chain acyl ester-ester connections present in PET monomers, esterases aid in the surface alteration of the target PET (Temporiti et al., 2022). Lipases are well known for partially catalyzing the hydrolysis of PET textiles by increasing their wettability, and this property is known as the interfacial activation phenomenon (Ekanayaka et al., 2022). Lipolytic esterolytic enzymes known as cutinases have an aggressive catalytic behavior toward PET breakdown (Arnling Baath et al., 2022). In addition to hydrolases, peroxidases that attack lignin can also break down plastics. Polyethylene (PE) and polyvinyl chloride (PVC) can be effectively broken down by fungi using laccases and peroxidases that they typically utilize to degrade lignin (Santacruz-Juarez et al., 2021).

4.4.3 Yeast

Numerous yeast strains have been identified as capable of degrading a wide range of plastics, including both biodegradable (Kitamoto et al., 2023) and nonbiodegradable plastics (Elsamahy et al., 2023). Additionally, researchers noted that yeast can ingest polyunsaturated fatty acids, which can be utilized to treat hydrophobic wastes and wastewater from the food industry that contains a lot of oils (Baghban et al., 2019). In light of the findings presented by Kosiorowska et al. (2022a), it is evident that employing *Y. lipolytica's* lipases or other esterases for the enzymatic depolymerization of PET holds substantial promise as a viable solution to the significant environmental problems stemming from improperly disposed plastic packaging. Yeast strains secrete various PET hydrolytic enzymes such as cutinase, lipases, esterases, and PETase (Carniel et al., 2017; Chen et al., 2020). These days, researchers have identified and modified these enzymes that degrade PET and enhance PET (polyethylene terephthalate) plastic degradation. Genetic engineering plays an important role in the bioremediation of PET, for example, bacterial PETase displays on the surface of yeast cells (*Pichia pastoris*) can degrade various commercially available highly crystalline PET and increase degradation of PET for about 36 times higher than purified PETase (Chen et al., 2020). Lipase isolated from *Candida antarctica* helps in the degradation of PET; it can also synergistically break down TPA in combination with cutinase isolated from *Humicola insolens* (Carniel et al., 2017).

4.4.4 Algae

Within sewage water, there is a documented tendency for algae to establish colonies on surfaces composed of synthetic materials like PET. Interestingly, these colonizing algae have been found to be nontoxic and less hazardous (Moog et al. 2019).

The initiation of PET biodegradation commences as algae adhere to the outer surface, with a critical role played by these organisms in secreting ligninolytic and exopolysaccharide enzymes essential for the biodegradation process (Kocer et al., 2021). The algal enzymes interact with polymers made of plastic causing chemical bonds to weaken, which starts the biodegradation process. Algae convert plastic waste into beneficial metabolites for the environment throughout the mineralization process, including CO_2 and H_2O, and are added to new biomass (Chia et al., 2020). The algal enzymes interface with the plastic surface's macromolecules to start the biodegradation process. Given the discovery that the algae species colonizing the PET surface exhibit higher protein and carbohydrate content along with enhanced specific growth rates, it becomes evident that PET serves as a nutrient source for these algae (Bher et al., 2022). Previous studies have revealed a range of biodegradation processes, including fouling, corrosion, hydrolysis, penetration, deterioration of leaching components, and the diffusion of pigment colors into polymers (Issac & Kandasubramanian, 2021). When considering algal deterioration, it is essential to account for key factors such as temperature, sunlight exposure, oxygen levels, and moisture presence (Nowicka-Krawczyk et al., 2022). In contrast to bacterial systems, which are known to produce endotoxins with pathogenic properties, considered as biological pollutants, microalgae emerge as a promising choice for PET biodegradation. This potential is underpinned by the fact that microalgae, unlike bacteria, do not generate endotoxins and do not rely on organic carbon sources for growth and development under photoautotrophic conditions (Chia et al., 2020).

4.4.5 Actinomycetes

Actinobacteria, as a collective group of organisms, hold substantial ecological significance and play pivotal roles in various biological processes. These include their contributions to the biogeochemical cycle, bioremediation, bioweathering, and the promotion of plant growth. Actinomycetes synthesize a diverse group of metabolites that are essential for the breakdown of plastics (Amobonye et al., 2021). Numerous studies have conclusively demonstrated the pivotal role of potent actinomycetes in bioremediation processes (Nazari et al., 2022; Oliveira et al., 2022). Additionally, adding compost to disturbed soil accelerates the bioremediation of contaminants like petroleum, pesticides, polycyclic aromatic hydrocarbons (PAH), heavy metals, and plastic (Huang et al. 2018a, b; Mawang et al., 2021). In the later phases of composting, actinobacteria grow due to the cooperative efforts of bacteria, fungi, and actinobacteria (Bhatti et al., 2017). In the wake of thermobiotic conditions fostered by preceding microorganisms, thermotolerant actinobacteria typically emerge as the predominant inhabitants. During the initial phase of the thermobiotic state, the compost is first colonized by thermotolerant actinobacteria, followed by the subsequent establishment of thermophilic actinobacteria (Amobonye et al., 2021). They are also thought to be capable of producing a bioemulsifier. When plastic polymers are broken down, microbes produce a chemical known as a bioemulsifier (Colin et al., 2013; Selim et al., 2021). The novel marine actinomycete strains *S. gougerotti*, *M. matsumotoense*, and *N. prasina* were found to be capable of biodegrading plastics (Oliveira et al., 2022). Another research on actinomycetes by Salam et al. (2014) isolated a novel anthracene-degrading strain *Microbacterium* sp. which is capable of degrading 90.12% of anthracene in 21 days.

4.4.6 PET Degradation by Genetically Modified Microorganism

An environmentally friendly strategy for plastic biodegradation, utilizing engineered enzymes, has emerged as a promising alternative for the future circular economy. Numerous genetic alterations have been performed to enhance microorganisms' inherent capacities for degrading PET, or by overexpressing the enzymes needed for this activity in the allogeneic host (Qi et al., 2021a). *E. coli* is a crucial model bacterium for the creation of recombinant proteins because of its distinct genetic history, easy-to-maintain growing conditions, and benefits from high density cultivation (Rosano & Ceccarelli, 2014). Yan et al. (2021) genetically modify a thermophilic anaerobic bacterium called *C. thermocellum* which enabled the bacterium to secrete a thermophilic cutinase (LCC), which can degrade PET at temperature up to 70 °C. *Yarrowia lipolytica* is a yeast that works well as a host for manufacturing heterologous proteins (Baghban et al., 2019). This established nonpathogenic unconventional yeast. *Y. lipolytica* is employed in the biodegradation of soil polluted with variety of hydrocarbons due to its capacity to assimilate a variety of hydrophobic sources of carbon, including n-alkanes (fuel), fatty acids (oils), or polyols (Madzak, 2021). Numerous studies on the prospective uses of *Y. lipolytica* in the environment have been conducted to date (Galvez-Lopez et al., 2019; Kosiorowska et al., 2022b; Sales et al., 2022). Even though microbial degradation is now too slow for industrial use, developments in protein/metabolic editing may ultimately provide a long-term solution to the breakdown of plastic waste.

4.4.7 Consortia-Based Degradation of PET

Consortia are a group of similar as well as different microorganisms which is used to perform similar activity by utilizing different metabolic functions. There are two types of consortia used for PET biodegradation: natural microbial consortia and synthetic microbial consortia (Liang et al., 2022). The majority of microbial consortia that can now break down PET are naturally microbial consortiums. Three *Pseudomonas* species and two *Bacillus* species were part of a consortium that was discovered by researchers to be able to lower the weight of granular PET. After being exposed to the consortium for 6 weeks, a 100 mg granule of PET weighed 3.15 mg less, indicating that the strains may work together to break down PET (Roberts et al., 2020). Synthetic biology now prioritizes the study of synthetic microbial consortia which mimic natural microbial consortia to carry out challenging biological activities (Liang et al., 2022). In particular, for the bioconversion of pollutants, it is critical to investigate possibilities and transform the functioning of microbial consortia members for specific techniques. Utilizing synthetic microbial consortia, previous investigations have demonstrated possibilities of biodegradation and bioconversion (Roberts et al., 2020). Qi et al. (2021b) evaluate PET degradation by using an artificial consortium which could degrade 13.6% of PET film in 7 days. When compared to pure culture, plastic degradation by synthetic microbial consortia has the following advantages: (i) diverse bacteria can perform PET biodegradation and bioconversion concurrently; (ii) different enzyme systems working together and the combination of distinct microorganisms' metabolic pathways can reduce the activity of products of degradation and increase the degradation rate; and (iii) the formation of synthetic microbial consortium is more capable and quicker than conventional metabolic engineering approaches (Table 4.1).

4.5 Factors Affecting PET Biodegradation

Biodegradability of the plastic can be affected by a range of factors, including the chemical composition, biological and physical characteristics of the polymers, and the presence microorganism. Polymer molecular mass and density are important factors holding a role in degradability (Tokiwa et al., 2009). Microbes found difficulty in utilizing high-molecular-weight polymers as substrates in comparison to the low-molecular-weight polymers. The arrangement and composition of monomers and the type of polymerization used can affect how susceptible the plastic is to enzymatic attack and microbial colonization (Shams et al., 2020). Plastic's surface qualities, such as roughness, porosity, and surface chemistry, can impact the rate of biodegradation and first-order structure, and high-order structure crystallinity is also crucial to the biodegradation process (Salam et al., 2021). Polyesters having side chains absorb less readily than those without. The degree of uniformity in the packing of polymer chains, known as crystallinity, can affect how easily enzymes and bacteria can access the polymer chains. Amorphous regions are generally more accessible for biodegradation than highly crystalline regions (Wang et al., 2021). Besides this, the functional groups, environmental factors, cross-linking and complexity in linear and branched structures, exposure to UV radiation, and mechanical forces such as abrasion, stretching, cooling, heating, and thawing can also accelerate the breakdown of plastics, making them more accessible to microbial action, and also the nature of polymers and physical characteristics such as pellets, films, or powder can greatly influence the rate and extent of biodegradation (Wang et al. 2020a, b; Koshti et al., 2018). The type of chemical bonds within the polymer chain

Table 4.1 List of different types of microorganisms and their enzymes for PET degradation

S. no.	Type of microorganism	Enzyme	References
1.	Bacteria		
	Ideonella sakaiensis 201-F6	PETase and MHETase	Yoshida et al. (2016)
	Escherichia coli BL21	PETase	Shi et al. (2021)
	E. coli	BIND-PETase	Zhu et al. (2022)
2.	Fungi		
	Fusarium solani	Cutinase	Tournier et al. (2020)
	Fusarium oxysporum	Cutinase	Kanelli et al. (2015)
	Penicillium restrictum and *P. simplicissimum*	Lipase	Moyses et al. (2021)
	Aspergillus tamarii and *Penicillium crustosum*	Lipase and cutinase	Anbalagan et al. (2021)
3.	Yeast		
	Pichia pastoris	PETase	Chen et al. (2020)
	Candida antarctica	Lipase	Carniel et al. (2017)
	Yarrowia lipolytica	Lipase	da Costa et al. (2020)
4.	Algae		
	Chlamydomonas reinhardtii CC-124	PETase	Kim et al. (2020)
	Phaeodactylum tricornutum	PETase	Moog et al. (2019)
5.	Actinomycetes		
	Thermobifida fusca KW3	Cutinase	Roth et al. (2014)
	Thermobifida cellulosilytica	Cutinase	Herrero Acero et al. (2011)
	Thermobifida halotolerans	Esterase	Ribitsch et al. (2012a)
	Saccharomonospora viridis AHK 190	Cutinase	Kawai et al. (2014a, b)
	Thermobifida alba	Cutinase	Ribitsch et al. (2012b)
6.	Consortium		
	Rhodococcus jostii, *Pseudomonas putida* and *Bacillus subtilis*	(PETase) and (MHETase)	Qi et al. (2021a, b)
	Bacillus thuringiensis strain C15, *Pseudomonas* sp. *strain B10*, *Pseudomonas* sp. *strain SWI36*, *Bacillus albus strain PFYN01*, and *Pseudomonas* sp. *strain SWI36*	Lipase, carboxylesterase, triacylglycerol esterase/lipase	Roberts et al. (2020)

plays a significant role in biodegradation. Certain bonds are more susceptible to microbial enzymatic activity than others. For instance, ester bonds (found in PET) are generally more labile (easily broken) compared to ether or amide bonds. The order of susceptibility (ester > ether > amide > urethane) reflects the relative ease of enzymatic cleavage for these bond types (Shams et al., 2020; Zeenat et al., 2021). Moreover, it is crucial to take into account other variables including pH levels, buffer solution concentration, plasticizer characteristics, and crystallinity degree. These variables can also influence the enzymatic hydrolysis of PET, by altering the enzyme activity or impeding access to the ester bond, which may therefore influence the effectiveness and pace of PET degradation (Maurya et al., 2020).

4.6 Current Research and Technology Advances

The biodegradation of PET by microorganisms has been the subject of ongoing and promising research and technological advancements. The discovery of novel and potentially PET-degrading microorganisms; the discovery of new enzymes; the use of microbial consortium and synergistic action, metagenomics, eliciting biochemical pathways involved in PET degradation, genetic engineering, waste management, and recycling; and the optimization of environmental factors for increasing the efficiency of PET degradation are the major advancements in the research field of plastic degradation. The biodegradation of PET involves assessing the capacity of microbes to decompose PET materials under controlled laboratory conditions. A variety of standard methods and guidelines has been established to evaluate the biodegradation of PET. Some of the commonly used methods include alterations in the polymer's surface topology, such as the existence of cracks, color, and crystallinity, and defragmentation by using electron microscopy and X-ray diffraction technique (Salam et al., 2021; Benavides et al., 2022). A change in the weight of the polymer is another way of assessing degradability. The magnitude of the biodegradation process can be better understood by combining a variety of advanced approaches. Scanning electron microscopy (SEM) enables high-resolution imaging and colonization of the surface morphology of PET samples; it is possible to determine the stages of cell division and biofilm formation before and after biodegradation as well as nutrient utilization by microorganisms on the PET surface for growth and development (Sarkhel et al., 2020). The chemical constitutions of PET sample changes are examined using Fourier transform infrared spectroscopy (FTIR). Changes in recognizable absorption bands can signal the development and demise of novel functional groups created during biodegradation (Khoironi et al., 2019). Atomic force microscopy (AFM) provides nanoscale imaging and topographical information about the PET surface. Differential scanning calorimetry (DSC) measures heat flow related with the changes in PET. Nuclear magnetic resonance spectroscopy and X- ray diffraction spectroscopy can be used to get information about the molecular structure and chemical composition of PET during degradation. Changes in crystallinity may be studied by X-ray diffraction (Benavides et al., 2022;

Salam et al., 2021). These techniques are used in many studies and show overall good approximation of the degradation rate of synthetic polymers. Torena et al. (2021) reported PET degradation by preparing microbial consortia. They assess degradation potential by standardized CO_2 evolution test further through SEM, DSC, FTIR, and viscometry analyses, and also HPLC analysis was done to determine by-products, e.g., BHET and TPA. In another study, they constructed a synthetic consortium of *Pseudomonas putida*, *Rhodococcus jostii*, and metabolically engineered *Bacillus subtilis* that secrete enzymes like PETase and MHETase and break down 13.6% of PET film in 7 days. Finally, they introduced *P. putida* to lower the inhibition due to ethylene glycol and 23.2% of mass loss was noted (Qi et al., 2021b). The disintegration of PET was measured by infrared spectroscopy and by scanning electron microscopy. These techniques reveal significant changes in the surface material of the plastic (Vague et al., 2019). Furthermore, in a research study, scientists identified bacteria and enzymes associated with PET degradation on a global scale. Their findings revealed that 25 terrestrial and 108 marine metagenomes collectively contained 349 PET hydrolases (Danso et al., 2018). Microbial consortium (*S. aurantiaca* + *B. subtilis* + *A. flavus* + *A. niger*) causes 28.78% weight loss of PET films and causes modification in surface carbonyl index and functional group detected by FTIR and field emission scanning electron microscopy (Maheswaran et al., 2023). Another study reported the isolation of a novel *Ideonella sakaiensis* 201-F6 bacterium capable of dissolving PET by enzymatic actions which break down PET into its monomers, terephthalic acid and ethylene glycol (Yoshida et al., 2016). Rhizospheric isolates *Priestia aryabhattai* VT 3.12 and *Bacillus pseudomycoides* VT 3.15 showed 65% of PET biodegradation assessed by HPLC and SEM. These microbes proved beneficial for remediating PET waste (Dhaka et al., 2022). Using a combination of standard assessment techniques and advanced research and methods can offer a comprehensive understanding of the biodegradation process and its effects on PET materials. This holistic approach enhances the reliability of the results and contributes to a more thorough evaluation of biodegradation.

4.7 Challenges and Limitations

Polyethylene terephthalate (PET) is a typical long polymer used in a variety of products, including packaging, textiles, and containers. Research and technology are advancing rapidly in PET hydrolysis by microorganisms, but there are still a number of significant obstacles to overcome in order to advance the field. Major challenges and limitations include the type of organisms, enzymes, intrinsic qualities of polymers, and selection on plastics prior to treatment. The inherent traits of PET polymers, including their hydrophobic and nonpolar nature, chemical inertness, poor surface wetting properties, and rigidity, pose inherent difficulties in terms of adhesion, coating, and biodegradation. Moreover, PET's physical characteristics, particularly its crystallinity and orientation, play a substantial role in determining an enzyme's effectiveness in breaking down its constituent elements (Webb et al., 2013). There is a necessity that enzymes should adhere to the plastic surface for better degradation. *Is*PETase is an enzyme specific to PET but shows poor catalytic activity and needs pretreatment for degradation. Another problem is uncertainty in the processing and pretreatments of different PETs. Furthermore, lack of knowledge and experience regarding technology, reaction engineering, design of the reactor, and enzyme production cost are other challenges related with degradation (Soong et al., 2022). Enzymes involved in PET degradation have certain limitations such as low catalytic turnover due to less active site, high temperature that alters enzyme activity, inhibition of catalytic reaction by MHET or intermediate metabolite, and interference of by-product during hydrolysis process (Ahmaditabatabaei et al., 2021). The current collection of studies indicates that only a few bacterial phyla possess the capacity to degrade PET, and also low gene frequencies indicate that this trait is limited and does not get an opportunity to spread in different populations (Zimmermann & Billig, 2011; Danso et al., 2018). Nevertheless, in the face of these challenges, several strategies hold promise for advancing PET biodegradation. These include harnessing high-temperature-tolerant enzymes, employing metagenomics to assess the uncultivable diversity of PET-degrading microbes, utilizing genetic modification techniques, customizing enzymes for enhanced activity, and incorporating additives and surfactants. These approaches collectively contribute to the acceleration of PET biodegradation.

4.8 Conclusion

It is abundantly clear that the persistent and rapid growth of the plastic industry has raised significant environmental concerns on a global scale, stemming from the staggering accumulation of PET waste in landfills, seas, and oceans year after year. The conventional methods of recycling PET trash (incineration and landfilling) continue to be challenging because of the disastrous consequences on plants, terrestrial and aquatic organisms, and humans. PET trash recycling calls for an effective and environmentally friendly method. A diversified group of microorganisms are accountable for degrading a variety of plastic polymers because these microorganisms must exist in various soils and water sources in order to survive. The environment in which plastic is present also affects how easily it can be broken down. The heterotrophic microbes use plastic as a carbon source or as a substrate. The main biodegradation process involves the attachment of microorganisms to the exterior surfaces, which is followed by their colonization at the exterior surface and the production of extracellular enzymes for the breakdown of large polymers of PET into monomers. To expedite the selection of pure microbial colonies for the purpose of breeding these PET-degrading microorganisms, it is imperative that we standardize and enhance the techniques for isolating such species. Equally critical is the assessment of the population and distribution of polymer-degrading microbes across diverse ecosystems. The cultivation of a diverse array of microorganisms, including bacteria, fungi, yeast, and actinomycetes, along with the strategic deployment of their genetically engineered robust enzyme systems, holds the key to establishing a sustainable PET recycling program. Notably, PET biodegradation achieved through synthetic microbial consortia has demonstrated superior efficiency compared to pure cultures. Furthermore, a comprehensive grasp of their molecular functions, facilitated by the elucidation of their crystal structures, promises to propel this research field into practical applications. In the collective pursuit of these goals, we can usher in a more sustainable and environmentally responsible era for PET waste management.

References

Ahmaditabatabaei, S., Kyazze, G., Iqbal, H. M. N., et al. (2021). Fungal enzymes as catalytic tools for polyethylene terephthalate (PET) degradation. *Journal of Fungi (Basel), 7*(11). https://doi.org/10.3390/jof7110931

Ali, S. S., Elsamahy, T., Koutra, E., et al. (2021). Degradation of conventional plastic wastes in the environment: A review on current status of knowledge and future perspectives of disposal. *Science of the Total Environment, 771*, 144719.

Ameen, F., Moslem, M., Hadi, S., et al. (2015). Biodegradation of low density polyethylene (LDPE) by mangrove fungi from the Red Sea Coast. *Progress in Rubber, Plastics and Recycling, 31*, 125–144. https://doi.org/10.1177/147776061503100204

Amobonye, A., Bhagwat, P., Singh, S., et al. (2021). Plastic biodegradation: Frontline microbes and their enzymes. *Science of the Total Environment, 759*, 143536. https://doi.org/10.1016/j.scitotenv.2020.143536

Anbalagan, S., Venkatakrishnan, H. R. R., Ravindran, J., et al. (2021). Hydrolytic degradation of polyethylene terephthalate by cutinase enzyme derived from fungal biomass–molecular characterization. *Biointerface Research in Applied Chemistry, 12*, 653–667. https://doi.org/10.33263/BRIAC121.653667

Arnling Baath, J., Novy, V., Carneir, L. V., et al. (2022). Structure-function analysis of two closely related cutinases from Thermobifida cellulosilytica. *Biotechnology and Bioengineering, 119*(2), 470–481. https://doi.org/10.1002/bit.27984

Arora, P. K. (2020). Bacilli-mediated degradation of xenobiotic compounds and heavy metals. *Frontiers in Bioengineering and Biotechnology, 8*, 570307.

Austin, H. P., Allen, M. D., Donohoe, B. S., et al. (2018). Characterization and engineering of a plastic-degrading aromatic polyesterase. *Proceedings of the National Academy of Sciences, 115*(19), E4350–E4357. https://doi.org/10.1073/pnas.1718804115

Baghban, R., Farajnia, S., Rajabibazl, M., et al. (2019). Yeast expression systems: Overview and recent advances. *Molecular Biotechnology, 61*(5), 365–384. https://doi.org/10.1007/s12033-019-00164-8

Bala, S., Garg, D., Thirumalesh, B. V., et al. (2022). Recent strategies for bioremediation of emerging pollutants: A review for a green and sustainable environment. *Toxics, 10*(8). https://doi.org/10.3390/toxics10080484

Benavides, C. D., Guzmán, M. P., Quijano, S. A., et al. (2022). Microbial degradation of polyethylene terephthalate: A systematic review. *SN Applied Sciences, 4*(10), 263.

Bhatti, A. A., Haq, S., & Bhat, R. A. (2017). Actinomycetes benefaction role in soil and plant health. *Microbial Pathogenesis, 111*, 458–467.

Bher, A., Mayekar, P. C., Auras, R. A., et al. (2022). Biodegradation of biodegradable polymers in mesophilic aerobic environments. *International Journal of Molecular Sciences, 23*(20).

Carniel, A., Valoni, É., Nicomedes, J., et al. (2017). Lipase from *Candida Antarctica* (CALB) and cutinase from *Humicolainsolens* act synergistically for PET hydrolysis to terephthalic acid. *Process Biochemistry, 59*, 84–90. https://doi.org/10.1016/j.procbio.2016.07.023
Chen, X., Xiong, X., Jiang, X., et al. (2019). Sinking of floating plastic debris caused by biofilm development in a freshwater lake. *Chemosphere, 222*, 856–864.
Chen, Z., Wang, Y., Cheng, Y., et al. (2020). Efficient biodegradation of highly crystallized polyethylene terephthalate through cell surface display of bacterial PETase. *Science of the Total Environment, 709*, 136138. https://doi.org/10.1016/j.scitotenv.2019.136138
Chen, Z., Duan, R., Xiao, Y., et al. (2022). Biodegradation of highly crystallized poly(ethylene terephthalate) through cell surface codisplay of bacterial PETase and hydrophobin. *Nature Communications, 13*(1), 7138. https://doi.org/10.1038/s41467-022-34908-z
Chia, W. Y., Ying Tang, D. Y., Khoo, K. S., et al. (2020). Nature's fight against plastic pollution: Algae for plastic biodegradation and bioplastics production. *Environmental Science and Ecotechnology, 4*, 100065. https://doi.org/10.1016/j.ese.2020.100065
Colin, V. L., Pereira, C. E., Villegas, L. B., et al. (2013). Production and partial characterization of bioemulsifier from a chromium-resistant actinobacteria. *Chemosphere, 90*(4), 1372–1378. https://doi.org/10.1016/j.chemosphere.2012.08.002
Cui, Y., Chen, Y., Liu, X., et al. (2021). Computational redesign of a PETase for plastic biodegradation under ambient condition by the GRAPE strategy. *ACS Catalysis, 11*(3), 1340–1350. https://doi.org/10.1021/acscatal.0c05126
da Costa, A. M., de Oliveira Lopes, V. R., et al. (2020). Poly (ethylene terephthalate) (PET) degradation by *Yarrowialipolytica*: Investigations on cell growth, enzyme production and monomers consumption. *Process Biochemistry, 95*, 81–90. https://doi.org/10.1016/j.procbio.2020.04.001
Danso, D., Schmeisser, C., Chow, J., et al. (2018). New insights into the function and global distribution of polyethylene terephthalate (PET)-degrading bacteria and enzymes in marine and terrestrial metagenomes. *Applied and Environmental Microbiology, 84*(8), 02773–02717.
Danso, D., Chow, J., & Streit, W. R. (2019). Plastics: Environmental and biotechnological perspectives on microbial degradation. *Applied and Environmental Microbiology, 85*(19).
Das, P., & Tiwari, P. (2019). Thermal degradation study of waste polyethylene terephthalate (PET) under inert and oxidative environments. *Thermochimica Acta, 679*, 178340. https://doi.org/10.1016/j.tca.2019.178340
Dhaka, V., Singh, S., Ramamurthy, P. C., et al. (2022). Biological degradation of polyethylene terephthalate by rhizobacteria. *Environmental Science and Pollution Research International, 30*, 116488. https://doi.org/10.1007/s11356-022-20324-9
Dvořák, P., Nikel, P. I., Damborský, J., et al. (2017). Bioremediation 3.0: Engineering pollutant-removing bacteria in the times of systemic biology. *Biotechnology Advances, 35*, 845–866. https://doi.org/10.1016/j.biotechadv.2017.08.001
East, A. J. (2005). Polyester fibres. In J. E. McIntyre (Ed.), *Synthethic fibres: Nylon, polyester, acrylic, polyolefin* (pp. 95–166). Woodhead Publishing Limited.
Eberl, A., Heumann, S., Brückner, T., et al. (2009). Enzymatic surface hydrolysis of poly (ethylene terephthalate) and bis (benzoyloxyethyl) terephthalate by lipase and cutinase in the presence of surface active molecules. *Journal of Biotechnology, 143*, 207–212. https://doi.org/10.1016/j.jbiotec.2009.07.008
Ekanayaka, A. H., Tibpromma, S., Dai, D., et al. (2022). A review of the fungi that degrade plastic. *Journal of Fungi (Basel), 8*(8). https://doi.org/10.3390/jof8080772
Elsamahy, T., Sun, J., Elsilk, S. E., et al. (2023). Biodegradation of low-density polyethylene plastic waste by a constructed tri-culture yeast consortium from wood-feeding termite: Degradation mechanism and pathway. *Journal of Hazardous Materials, 448*, 130944. https://doi.org/10.1016/j.jhazmat.2023.130944
Eriksen, M., Lebreton, L. C., Carson, H. S., et al. (2014). Plastic pollution in the world's oceans: More than 5 trillion plastic pieces weighing over 250,000 tons afloat at sea. *PLoS One, 9*, e111913. https://doi.org/10.1371/journal.pone.0111913

Fairbrother, A., Hsueh, H. C., Kim, J. H., et al. (2019). Temperature and light intensity effects on photodegradation of high-density polyethylene. *Polymer Degradation and Stability, 165*, 153–160. https://doi.org/10.1016/j.polymdegradstab.2019.05.002

Fotopoulou, K. N., & Karapanagioti, H. K. (2019). Degradation of various plastics in the environment. In H. Takada & H. Karapanagioti (Eds.), *Hazardous chemicals associated with plastics in the marine environment. The handbook of environmental chemistry* (Vol. 78, pp. 71–92). Springer.

Francois-Heude, A., Richard, E., Desnoux, E., et al. (2014). Influence of temperature, UV-light wavelength and intensity of polypropylene photo-thermal oxidation. *Polymer Degradation and Stability, 100*, 10–20. https://doi.org/10.1016/j.polymdegradstab.2013.12.038

Galvez-Lopez, D., Chavez-Melendez, B., Vazquez-Ovando, A., et al. (2019). The metabolism and genetic regulation of lipids in the oleaginous yeast *Yarrowia lipolytica*. *Brazilian Journal of Microbiology, 50*(1), 23–31. https://doi.org/10.1007/s42770-018-0004-7

Gamerith, C., Vastano, M., Ghorbanpour, S. M., et al. (2017). Enzymatic degradation of aromatic and aliphatic polyesters by *P. pastoris* expressed cutinase 1 from *Thermobifidacellulosilytica*. *Frontiers in Microbiology, 8*, 938. https://doi.org/10.3389/fmicb.2017.00938

Gangola, S., Joshi, S., Kumar, S., et al. (2019). Comparative analysis of fungal and bacterial enzymes in biodegradation of xenobiotic compounds. In P. Bhatt (Ed.), *Smart bioremediation technologies* (pp. 169–189). Academic. https://doi.org/10.1016/b978-0-12-818307-6.00010-x

Gao, R., Liu, R., & Sun, C. (2022). A marine fungus Alternaria alternata FB1 efficiently degrades polyethylene. *Journal of Hazardous Materials, 431*, 128617.

Gupta, D., Chaudhary, H., & Gupta, C. (2015). Topographical changes in polyester after chemical, physical and enzymatic hydrolysis. *Journal of the Textile Institute, 106*, 690–698.

Guyot, S., Pottier, L., Hartmann, A., et al. (2014). Extremely rapid acclimation of *Escherichia coli* to high temperature over a few generations of a fed-batch culture during slow warming. *MicrobiologyOpen, 3*, 52–63. https://doi.org/10.1002/mbo3.146

Han, X., Liu, W., Huang, J.-W., et al. (2017). Structural insight into catalytic mechanism of PET hydrolase. *Nature Communications, 8*(1), 2106. https://doi.org/10.1038/s41467-017-02255-z

Herrero Acero, E., Ribitsch, D., Steinkellner, G., et al. (2011). Enzymatic surface hydrolysis of PET: Effect of structural diversity on kinetic properties of cutinases from *Thermobifida*. *Macromolecules, 44*, 4632–4640. https://doi.org/10.1021/ma200949p

Hiraga, K., Taniguchi, I., Yoshida, et al. (2019). Biodegradation of waste PET: A sustainable solution for dealing with plastic pollution. *EMBO Reports, 20*(11), e49365. https://doi.org/10.15252/embr.201949365

Hosseini, S. S., Taheri, S., Zadhoush, A., et al. (2007). Hydrolytic degradation of poly(ethylene terephthalate). *Journal of Applied Polymer Science, 103*(4), 2304–2309. https://doi.org/10.1016/0014-3057(91)90237-I

Huang, X., Cao, L., Qin, Z., et al. (2018a). Tat-independent secretion of polyethylene terephthalate hydrolase PETase in *Bacillus subtilis* 168 mediated by its native signal peptide. *Journal of Agricultural and Food Chemistry, 66*(50), 13217–13227. https://doi.org/10.1021/acs.jafc.8b05038

Huang, Y., Xiao, L., Li, F., et al. (2018b). Microbial degradation of pesticide residues and an emphasis on the degradation of cypermethrin and 3-phenoxy benzoic acid: A review. *Molecules, 23*(9). https://doi.org/10.3390/molecules23092313

Hussain, I., Aleti, G., Naidu, R., et al. (2018). Microbe and plant assisted-remediation of organic xenobiotics and its enhancement by genetically modified organisms and recombinant technology: A review. *Science of the Total Environment*, 1582–1599. https://doi.org/10.1016/j.scitotenv.2018.02.037

Iranzo, M., Sainz-Pardo, I., Boluda, R., et al. (2001). The use of microorganisms in environmental remediation. *Annales de Microbiologie, 51*(2), 135–144.

Issac, M. N., & Kandasubramanian, B. (2021). Effect of microplastics in water and aquatic systems. *Environmental Science and Pollution Research International, 28*(16), 19544–19562.

Joo, S., Cho, I. J., Seo, H., et al. (2018). Structural insight into molecular mechanism of poly(ethylene terephthalate) degradation. *Nature Communications, 9*(1), 382. https://doi.org/10.1038/s41467-018-02881-1

Kanelli, M., Vasilakos, S., Nikolaivits, E., et al. (2015). Surface modification of poly (ethylene terephthalate) (PET) fibers by a cutinase from *Fusarium oxysporum. Process Biochemistry, 50*(11), 1885–1892. https://doi.org/10.1016/j.procbio.2015.08.013

Kawai, F., Oda, M., Tamashiro, T., et al. (2014a). A novel Ca 2+-activated, thermostabilized polyesterase capable of hydrolyzing polyethylene terephthalate from *Saccharomonosporaviridis* AHK190. *Applied Microbiology and Biotechnology, 98*, 10053–10064. https://doi.org/10.1007/s00253-014-5860-y

Kawai, F., Oda, M., Tamashiro, T., et al. (2014b). A novel Ca2+–activated, thermostabilized polyesterase capable of hydrolyzing polyethylene terephthalate from *Saccharomonospora viridis AHK190. Applied Microbiology and Biotechnology, 98*, 10053–10064. https://doi.org/10.1186/s13568-014-0044-9

Khairul Anuar, N. F. S., Huyop, F., Ur-Rehman, G., et al. (2022). An overview into polyethylene terephthalate (PET) hydrolases and efforts in tailoring enzymes for improved plastic degradation. *International Journal of Molecular Sciences, 23*(20). https://doi.org/10.3390/ijms232012644

Khoironi, A., Anggoro, S., & Sudarno. (2019). Evaluation of the interaction among microalgae spirulina sp, plastics polyethylene terephthalate and polypropylene in freshwater environment. *Journal of Ecological Engineering, 20*(6), 161–173.

Kim, J. W., Park, S. B., Tran, Q. G., et al. (2020). Functional expression of polyethylene terephthalate-degrading enzyme (PETase) in green microalgae. *Microb Cell Factories 19*(1), 1–9. https://doi.org/10.1186/s12934-020-01355-8

Kitamoto, H., Koitabashi, M., Sameshima-Yamashita, Y., et al. (2023). Accelerated degradation of plastic products via yeast enzyme treatment. *Scientific Reports, 13*(1), 2386. https://doi.org/10.1038/s41598-023-29414-1

Knott, B. C., Erickson, E., Allen, M. D., et al. (2020). Characterization and engineering of a two-enzyme system for plastics depolymerization. *Proceedings of the National Academy of Sciences, 117*(41), 25476–25485. https://doi.org/10.1073/pnas.2006753117

Kocer, A. T., Inan, B., Kaptan Usul, S., et al. (2021). Exopolysaccharides from microalgae: Production, characterization, optimization and techno-economic assessment. *Brazilian Journal of Microbiology, 52*(4), 1779–1790. https://doi.org/10.1007/s42770-021-00575-3

Koshti, R., Mehta, L., & Samarth, N. (2018). Biological recycling of polyethylene terephthalate: A mini-review. *Journal of Polymers and the Environment, 26*, 3520–3529.

Kosiorowska, K. E., Biniarz, P., Dobrowolski, A., et al. (2022a). Metabolic engineering of *Yarrowia lipolytica* for poly(ethylene terephthalate) degradation. *Science of the Total Environment, 831*, 154841. https://doi.org/10.1016/j.scitotenv.2022.154841

Kosiorowska, K. E., Moreno, A. D., Iglesias, R., et al. (2022b). Production of PETase by engineered *Yarrowia lipolytica* for efficient poly(ethylene terephthalate) biodegradation. *Science of the Total Environment, 846*, 157358. https://doi.org/10.1016/j.scitotenv.2022.157358

Kumar, N. M., Muthukumaran, C., Sharmila, G., et al. (2018). Genetically modified organisms and its impact on the enhancement of bioremediation. In S. Varjani, A. Agarwal, E. Gnansounou, & B. Gurunathan (Eds.), *Bioremediation: Applications for environmental protection and management* (pp. 53–76). Springer.

Lambert, S., & Wagner, M. (2016). Characterisation of nanoplastics during the degradation of polystyrene. *Chemosphere, 145*, 265–268. https://doi.org/10.1016/j.chemosphere.2015.11.078

Lebreton, L., & Andrady, A. (2019). Future scenarios of global plastic waste generation and disposal. *Palgrave Communications, 5*, 1–11.

Lechner, A., Keckeis, H., Lumesberger-Loisl, F., et al. (2014). The Danube so colourful: A potpourri of plastic litter outnumbers fish larvae in Europe's second largest river. *Environmental Pollution, 188*, 177–181. https://doi.org/10.1016/j.envpol.2014.02.006

Li, J., Wang, Y., Wang, X., et al. (2019a). Crystalline characteristics, mechanical properties, thermal degradation kinetics and hydration behavior of biodegradable fibers melt spun from polyoxymethylene/poly(L-lactic acid) blends. *Polymers, 11*(11), 1753. https://doi.org/10.3390/polym11111753

Li, K., Liu, Q., Fang, F., et al. (2019b). Microalgae-based wastewater treatment for nutrients recovery: A review. *Bioresource Technology, 291*, 121934. https://doi.org/10.1016/j.biortech.2019.121934

Liang, Y., Ma, A., & Zhuang, G. (2022). Construction of environmental synthetic microbial consortia: Based on engineering and ecological principles. *Frontiers in Microbiology, 13*, 829717.

Liebminger, S., Eberl, A., Sousa, F., et al. (2007). Hydrolysis of PET and bis-(benzoyloxyethyl) terephthalate with a new polyesterase from *Penicillium citrinum. Biocatalysis and Biotransformation, 25*(2–4), 171–177. https://doi.org/10.1080/10242420701379734

Lucas, N., Bienaime, C., Belloy, C., et al. (2008). Polymer biodegradation: Mechanisms and estimation techniques. *Chemosphere, 73*, 429–442. https://doi.org/10.1016/j.chemosphere.2008.06.064

Luo, Z. W., & Lee, S. Y. (2017). Biotransformation of p-xylene into terephthalic acid by engineered Escherichia coli. *Nature Communications, 8*, 1–8.

Madzak, C. (2021). Yarrowia lipolytica strains and their biotechnological applications: How natural biodiversity and metabolic engineering could contribute to cell factories improvement. *Journal of Fungi (Basel), 7*(7).

Maheswaran, B., Al-Ansari, M., Al-Humaid, L., et al. (2023). In vivo degradation of polyethylene terephthalate using microbial isolates from plastic polluted environment. *Chemosphere, 310*, 136757.

Mandhare, T., Nangare, P., Khuspe, P., et al. (2018). Recombinant DNA technology: A hopeful ray to improve life. *International Journal of Pharma Research and Health Sciences, 6*(2), 2346–2350. https://doi.org/10.21276/ijprhs.2018.02.02

Maurya, A., Bhattacharya, A., & Khare, S. K. (2020). Enzymatic remediation of polyethylene terephthalate (PET)–based polymers for effective management of plastic wastes: An overview. *Frontiers in Bioengineering and Biotechnology, 8*, 602325.

Mawang, C. I., Azman, A. S., Fuad, A. M., et al. (2021). Actinobacteria: An eco-friendly and promising technology for the bioaugmentation of contaminants. *Biotechnology Reports (Amsterdam, Netherlands), 32*, e00679. https://doi.org/10.1016/j.btre.2021.e00679

Michaud, L., Di Marco, G., Bruni, V., et al. (2007). Biodegradative potential and characterization of psychrotolerant polychlorinated biphenyl-degrading marine bacteria isolated from a coastal station in the Terra Nova Bay (Ross Sea, Antarctica). *Marine Pollution Bulletin, 54*(11), 1754–1761. https://doi.org/10.1016/j.marpolbul.2007.07.011

Miglani, R., Parveen, N., Kumar, A., et al. (2022). Degradation of xenobiotic pollutants: An environmentally sustainable approach. *Metabolites, 12*(9). https://doi.org/10.3390/metabo12090818

Mohanan, N., Montazer, Z., Sharma, P. K., et al. (2020). Microbial and enzymatic degradation of synthetic plastics. *Frontiers in Microbiology, 11*, 580709. https://doi.org/10.3389/fmicb.2020.580709

Moog, D., Schmitt, J., Senger, J., et al. (2019). Using a marine microalga as a chassis for polyethylene terephthalate (PET) degradation. *Microb Cell Fact 18*, 171. https://doi.org/10.1186/s12934-019-1220-z

Moyses, D. N., Teixeira, D. A., Waldow, V. A., et al. (2021). Fungal and enzymatic biodepolymerization of waste post-consumer poly (ethylene terephthalate) (PET) bottles using *Penicillium* species. *3 Biotech, 11*(10), 435. https://doi.org/10.1007/s13205-021-02988-1

Müller, R. J., Schrader, H., Profe, J., et al. (2005). Enzymatic degradation of poly (ethylene terephthalate): Rapid hydrolyse using a hydrolase from *T. fusca. Macromolecular Rapid Communications, 26*, 1400–1405. https://doi.org/10.1002/marc.200500410

Nazari, M. T., Simon, V., Machado, B. S., et al. (2022). Rhodococcus: A promising genus of actinomycetes for the bioremediation of organic and inorganic contaminants. *Journal of Environmental Management, 323*, 116220. https://doi.org/10.1016/j.jenvman.2022.116220

Nechwatal, A., Blokesch, A., Nicolai, M., et al. (2006). A contribution to the investigation of enzyme-catalysed hydrolysis of poly (ethylene terephthalate) oligomers. *Macromolecular Materials and Engineering, 291*, 1486–1494. https://doi.org/10.1002/mame.200600204

Nowicka-Krawczyk, P., Komar, M., & Gutarowska, B. (2022). Towards understanding the link between the deterioration of building materials and the nature of aerophytic green algae. *Science of the Total Environment, 802*, 149856.

Oliveira, J., Almeida, P. L., Sobral, R. G., et al. (2022). Marine-derived actinomycetes: Biodegradation of plastics and formation of PHA bioplastics-a circular bioeconomy approach. *Marine Drugs, 20*(12). https://doi.org/10.3390/md20120760

Padsalgikar, A. D. (2017). *Biological properties of plastics. Plastics in medical devices for cardiovascular applications*. Elsevier.

Palm, G. J., Reisky, L., Böttcher, D., et al. (2019). Structure of the plastic-degrading *Ideonella sakaiensis* MHETase bound to a substrate. *Nature Communications, 10*(1), 1717. https://doi.org/10.1038/s41467-019-09326-3

Parisi, D., Riley, C., Srivastava, A. S., et al. (2019). PET hydrolysing enzymes catalyse bioplastics precursor synthesis under aqueous conditions. *Green Chemistry, 21*, 3827–3833. https://doi.org/10.1039/C9GC01284B

Pirsaheb, M., Hossini, H., & Makhdoumi, P. (2020). Review of microplastic occurrence and toxicological effects in marine environment: Experimental evidence of inflammation. *Process Safety and Environmental Protection, 142*, 1–14.

Qi, X., Ma, Y., Chang, H., et al. (2021a). Evaluation of PET degradation using artificial microbial consortia. *Frontiers in Microbiology, 12*, 778828. https://doi.org/10.3389/fmicb.2021.778828

Qi, X., Yan, W., Cao, Z., et al. (2021b). Current advances in the biodegradation and bioconversion of polyethylene terephthalate. *Microorganisms, 10*(1). https://doi.org/10.3390/microorganisms10010039

Restrepo-Florez, J. M., Bassi, A., & Thompson, M. R. (2014). Microbial degradation and deterioration of polyethylene–A review. *International Biodeterioration and Biodegradation, 88*, 83–90.

Ribitsch, D., Acero, E. H., Greimel, K., et al. (2012a). A new esterase from *Thermobifida halotolerans* hydrolyses polyethylene terephthalate (PET) and polylactic acid (PLA). *Polymer, 4*(1), 617–629. https://doi.org/10.3390/polym4010617

Ribitsch, D., Acero, E. H., Greimel, K., et al. (2012b). Characterization of a new cutinase from *Thermobifida alba* for PET-surface hydrolysis. *Biocatalysis, 30*(1), 2–9. https://doi.org/10.3109/10242422.2012.644435

Roberts, C., Edwards, S., Vague, M., et al. (2020). Environmental consortium containing pseudomonas and bacillus species synergistically degrades polyethylene terephthalate plastic. *mSphere, 5*(6). https://doi.org/10.1128/mSphere.01151-20

Rosano, G. L., & Ceccarelli, E. A. (2014). Recombinant protein expression in *Escherichia coli*: Advances and challenges. *Frontiers in Microbiology, 5*, 172.

Roth, C., Wei, R., Oeser, T., et al. (2014). Structural and functional studies on a thermostable polyethylene terephthalate degrading hydrolase from *Thermobifidafusca*. *Applied Microbiology and Biotechnology, 98*, 7815–7823. https://doi.org/10.1007/s00253-014-5672-0

Rudolph, B., Gebendorfer, K. M., Buchner, J., et al. (2010). Evolution of Escherichia coli for growth at high temperatures. *The Journal of Biological Chemistry, 285*, 19029–19034. https://doi.org/10.1074/jbc.M110.103374

Salam, L. B., Obayori, O. S., & Olatoye, N. O. (2014). Biodegradation of anthracene by a novel actinomycete, *microbacterium* sp. isolated from tropical hydrocarbon-contaminated soil. *World Journal of Microbiology and Biotechnology, 30*(1), 335–341.

Salam, M. D., Varma, A., Prashar, R., et al. (2021). Review on efficacy of microbial degradation of polyethylene terephthalate and bio-upcycling as a part of plastic waste management. *Applied Ecology and Environmental Sciences, 9*, 695–703.

Sales, J. C. S., de Castro, A. M., et al. (2022). Post-consumer poly(ethylene terephthalate) (PET) depolymerization by *Yarrowia lipolytica*: A comparison between hydrolysis using cell-free enzymatic extracts and microbial submerged cultivation. *Molecules, 27*(21). https://doi.org/10.3390/molecules27217502

Salvador, M., Abdulmutalib, U., Gonzalez, J., et al. (2019). Genes for a circular and sustainable bio-PET economy. *Genes, 10*(5), 373. https://doi.org/10.3390/genes10050373

Samak, N. A., Jia, Y., Sharshar, M. M., et al. (2020). Recent advances in biocatalysts engineering for polyethylene terephthalate plastic waste green recycling. *Environment International, 145*, 106144. https://doi.org/10.1016/j.envint.2020.106144

Sammon, C., Yarwood, J., & Everall, N. (2000). An FT-IR study of the effect of hydrolytic degradation on the structure of thin PET films. *Polymer Degradation and Stability, 67*, 149–158.

Sánchez, C. (2020). Fungal potential for the degradation of petroleum-based polymers: An overview of macro- and microplastics biodegradation. *Biotechnology Advances, 40*, 107501.

Santacruz-Juarez, E., Buendia-Corona, R. E., Ramirez, R. E., et al. (2021). Fungal enzymes for the degradation of polyethylene: Molecular docking simulation and biodegradation pathway proposal. *Journal of Hazardous Materials, 411*, 125118. https://doi.org/10.1016/j.jhazmat.2021.125118

Sarkhel, R., Sengupta, S., Das, P., et al. (2020). Comparative biodegradation study of polymer from plastic bottle waste using novel isolated bacteria and fungi from marine source. *Journal of Polymer Research, 27*, 1–8. https://doi.org/10.1007/s10965-019-1973-4

Selim, M. S. M., Abdelhamid, S. A., & Mohamed, S. S. (2021). Secondary metabolites and biodiversity of actinomycetes. *Journal, Genetic Engineering & Biotechnology, 19*(1), 72.

Seo, S., & Park, Y. G. (2020). Destination of floating plastic debris released from ten major rivers around the Korean Peninsula. *Environment International, 138*, 105655.

Shams, M., Alam, I., & Chowdhury, I. (2020). Aggregation and stability of nanoscale plastics in aquatic environment. *Water Research, 171*, 115401.

Sharon, C., & Sharon, M. (2012). Studies on biodegradation of polyethylene terephthalate: A synthetic polymer. *Journal of Microbiology and Biotechnology Research, 2*, 248–257.

Sharshar, M. M., Samak, N. A., Hao, X., et al. (2019). Enhanced growth driven stepwise inducible expression system development in haloalkaliphilic desulfurizing *Thioalkali vibrioversutus*. *Bioresource Technology, 288*, 121486. https://doi.org/10.1016/j.biortech.2019.121486

Sheel, A., & Pant, D. (2019). Chemical depolymerization of PET bottles via glycolysis. In S. Thomas, A. Rane, K. Kanny, V. K. Abitha, & G. M. Thomas (Eds.), *Recycling of polyethylene terephthalate bottles* (pp. 61–84). William Andrew Publishing.

Shi, L., Liu, H., Gao, S., Weng, Y., & Zhu, L. (2021). Enhanced extracellular production of IsPETase in escherichia coli via engineering of the pelB signal peptide. *Journal of Agricultural and Food Chemistry, 69*(7), 2245–2252. https://doi.org/10.1021/acs.jafc.0c07469

Shirke, A. N., White, C., Englaender, J. A., et al. (2018). Stabilizing leaf and branch compost cutinase (LCC) with glycosylation: Mechanism and effect on PET hydrolysis. *Biochemistry, 57*, 1190–1200. https://doi.org/10.1021/acs.biochem.7b01189

Singh, B., & Sharma, N. (2008). Mechanistic implications of plastic degradation. *Polymer Degradation and Stability, 93*, 561–584.

Soong, Y. H. V., Sobkowicz, M. J., & Xie, D. (2022). Recent advances in biological recycling of polyethylene terephthalate (PET) plastic wastes. *Bioengineering, 9*(3), 98.

Temporiti, M. E. E., Nicola, L., Nielsen, E., et al. (2022). Fungal enzymes involved in plastics biodegradation. *Microorganisms, 10*(6). https://doi.org/10.3390/microorganisms10061180

Thachnatharen, N., Shahabuddin, S., & Sridewi, N. (2021). The waste management of polyethylene terephthalate (PET) plastic waste: A review. *IOP Conference Series: Materials Science and Engineering, 1127*(1), 012002.

Thompson, R. C., Moore, C. J., Vom Saal, F. S., et al. (2009). Plastics, the environment and human health: Current consensus and future trends. *Philosophical Transactions of the Royal Society. Series B, Biological Sciences, 364*, 2153–2166.

Tiwari, M. K., Singh, R., Singh, R. K., et al. (2012). Computational approaches for rational design of proteins with novel functionalities. *Computational and Structural Biotechnology Journal, 2*(3), e201204002. https://doi.org/10.5936/csbj.201209002

Tokiwa, Y., Calabia, B. P., Ugwu, C. U., et al. (2009). Biodegradability of plastics. *International Journal of Molecular Sciences, 10*(9), 3722–3742.

Torena, P., Alvarez-Cuenca, M., & Reza, M. (2021). Biodegradation of polyethylene terephthalate microplastics by bacterial communities from activated sludge. *Canadian Journal of Chemical Engineering, 99*, S69–S82.

Tournier, V., Topham, C. M., Gilles, A., et al. (2020). An engineered PET depolymerase to break down and recycle plastic bottles. *Nature, 580*(7802), 216–219. https://doi.org/10.1038/s41586-020-2149-4

Vague, M., Chan, G., Roberts, C., Swartz, N. A., et al. (2019). Pseudomonas isolates degrade and form biofilms on polyethylene terephthalate (PET) plastic. *BioRxiv, 647321*.

Venkatachalam, S., Nayak, G. S., Labde, V. J., et al. (2012). Degradation and recyclability of poly (ethylene terephthalate). In H. E. D. M. Saleh (Ed.), *Polyester* (pp. 75–99). InTech.

Vertommen, M., Nierstrasz, V., Van Der Veer, M., et al. (2005). Enzymatic surface modification of poly (ethylene terephthalate). *Journal of Biotechnology, 120*, 376–386. https://doi.org/10.1016/j.jbiotec.2005.06.015

Wang, J., Zhao, X., Wu, A., et al. (2020a). Aggregation and stability of sulfate-modified polystyrene nanoplastics in synthetic and natural waters. *Environmental Pollution, 114240*.

Wang, G. X., Huang, D., Ji, J. H., et al. (2020b). Seawater-degradable polymers-fighting the marine plastic pollution. *Advancement of Science, 8*, 2001121.

Wang, X., Bolan, N., Tsang, D. C. W., et al. (2021). A review of microplastics aggregation in aquatic environment: Influence factors, analytical methods, and environmental implications. *Journal of Hazardous Materials, 402*, 123496.

Webb, H. K., Arnott, J., Crawford, R. J., et al. (2013). Plastic degradation and its environmental implications with special reference to poly (ethylene terephthalate). *Polymers, 5*, 1–18. https://doi.org/10.3390/polym5010001

Wei, R., Oeser, T., Schmidt, J., et al. (2016). Engineered bacterial polyester hydrolases efficiently degrade polyethylene terephthalate due to relieved product inhibition. *Biotechnology and Bioengineering, 113*(8), 1658–1665. https://doi.org/10.1002/bit.25941

Wilkes, R. A., Aristilde, L., et al. (2017). Degradation and metabolism of synthetic plastics and associated products by *pseudomonas* sp.: capabilities and challenges. *Journal of Applied Microbiology, 123*(3), 582–593.

Yan, F., Wei, R., Cui, Q., et al. (2021). Thermophilic whole-cell degradation of polyethylene terephthalate using engineered *clostridium thermocellum*. *Microbial Biotechnology, 14*(2), 374–385. https://doi.org/10.1111/1751-7915.13580

Yoshida, S., Hiraga, K., Takehana, T., et al. (2016). A bacterium that degrades and assimilates poly(ethyleneterephthalate). *Science, 351*, 1196–1199. https://doi.org/10.1126/science.aad6359

Zeenat, A., Bukhari, D. A., Shamim, S., et al. (2021). Plastics degradation by microbes: A sustainable approach. *Journal of King Saud University – Science, 33*(6), 10.1016/j.jksus.2021.101538.

Zhang, J., Wang, X., Gong, J., et al. (2004). A study on the biodegradability of polyethylene terephthalate fiber and diethylene glycol terephthalate. *Journal of Applied Polymer Science, 93*(3), 1089–1096. https://doi.org/10.1002/app.20556

Zhang, J., Gao, D., Li, Q., et al. (2020). Biodegradation of polyethylene microplastic particles by the fungus aspergillus flavus from the guts of wax moth *galleria mellonella*. *Science of the Total Environment, 704*, 135931. https://doi.org/10.1016/j.scitotenv.2019.135931

Zheng, Y., Yanful, K., & Bassi, A. S. (2005). A review of plastic waste biodegradation. *Critical Reviews in Biotechnology, 25*, 243–250.

Zhu, B., Wang, D., & Wei, N. (2021). Enzyme discovery and engineering for sustainable plastic recycling. *Trends in Biotechnology*, 1–16.

Zhu, B., Ye, Q., Seo, Y., et al. (2022). Enzymatic degradation of polyethylene terephthalate plastics by bacterial Curli display PETase. *Environ Sci Technol Lett 9*(7), 650–657. https://doi.org/10.1021/acs.estlett.2c00332

Zimmermann, W., & Billig, S. (2011). Enzymes for the biofunctionalization of poly (ethylene terephthalate). Biofunctionalization of polymers and their applications. *Advances in Biochemical Engineering/Biotechnology, 125*, 97–120.

Chapter 5
Biodegradation of Polyurethane (PU) and Polyvinyl Chloride (PVC)

Manjul Gondwal, Akanksha Gwasikoti, Anjum Qureshi, Prem Prakash Solanki, Rakesh Kumar Verma, and Bhanu Pratap Singh Gautam

Abstract The accumulation of plastic trash has been recognized as a significant environmental issue that impacts all forms of life, natural ecosystems, and the global economy. In these circumstances, it is crucial to prioritize the search for environmentally friendly alternatives, such as biodegradation instead of conventional disposal methods. Currently, there is limited knowledge regarding the mechanisms and effectiveness of plastic biodegradation. The purpose of this study is to provide a concise overview of the biodegradation processes of polyurethane (PU) and polyvinyl chloride (PVC), highlighting their significance in terms of environmental sustainability and waste management. Biodegradation, which harnesses the power of microbes and enzymes, holds immense promise in tackling the environmental issues associated with synthetic polymers. Biodegradation refers to the natural process by which microbes and enzymes break down organic matter. To effectively develop strategies for managing plastic waste and promoting sustainability, it is essential to have a thorough understanding of the mechanisms and factors that affect the biodegradation of polyurethane (PU) and polyvinyl chloride (PVC).

Keywords Biodegradation · Polyurethane · Polyvinyl chloride · Microbial communities · Environmental implications

---

M. Gondwal · A. Gwasikoti · A. Qureshi · B. P. S. Gautam (✉)
Department of Chemistry, Laxman Singh Mahar Campus (affiliated to Soban Singh Jeena University, Almora), Pithoragarh, Uttarakhand, India

P. P. Solanki
Department of Chemistry, Institute of Science, Banaras Hindu University, Varanasi, Uttar Pradesh, India

R. K. Verma
Department of Zoology, Laxman Singh Mahar Campus (affiliated to Soban Singh Jeena University, Almora), Pithoragarh, Uttarakhand, India

R. Soni et al. (eds.), *Advanced Strategies for Biodegradation of Plastic Polymers*, https://doi.org/10.1007/978-3-031-55661-6_5

5.1 Introduction: The Biodegradation Pathways of Polyurethane and Polyvinyl Chloride

The study on the microbial degradation of polyurethane began in 1966 and 1968 (Cregut et al., 2013). According to Plastics Europe (2019), PVC was the third most widely used petroleum-based plastic in Europe in 2018, with a total volume of five million tons. Polyvinyl chloride (PVC) plastics are utilized in various industries such as food packaging, electronics, coatings, and medical devices. These applications utilize the advantageous qualities of PVC plastics, including their cost-effectiveness, durability, and mechanical attributes. The properties of a material are influenced by the specific type and quantity of plasticizers utilized (Bueno-Ferrer et al., 2010). The environmental impact of PVC plastics is a concern due to their high chlorine content and the necessary additives used in their manufacturing process. Additives and stabilizers, such as heavy metals or phthalates, play a vital role in enhancing the physical and chemical properties of end products, thereby ensuring their stability. It is crucial to acknowledge that these substances possess hazardous properties that can adversely affect human health (Glas et al., 2014) (Fig. 5.1).

The depolymerization and biodegradation of polyvinyl chloride (PVC) have been observed in *Tenebrio molitor* larvae. This study contributes to our understanding of the biodegradation of PVC, a significant polymer, alongside PS and PE (Peng et al., 2020).

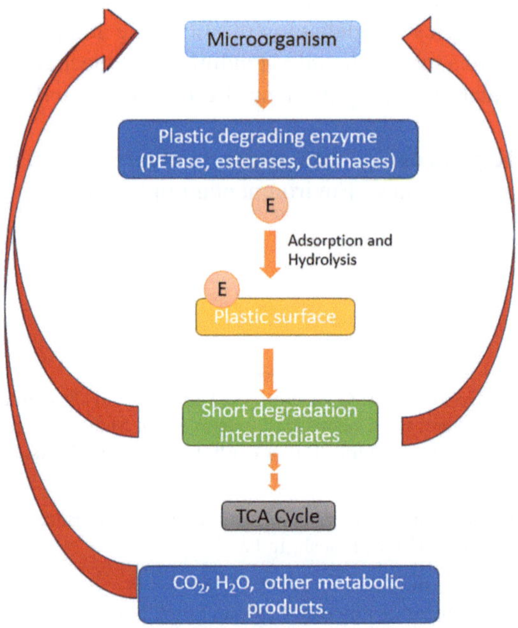

Fig. 5.1 General mechanism for biological degradation of plastics under aerobic conditions

The study aimed to examine the substantial biodegradation of untreated polyvinyl chloride (PVC) films that were plasticized. PBA-PU, a synthesized polymer, was employed to screen soil samples for microorganisms capable of degrading polyurethane (PU). *Cladosporium* sp. exhibited remarkable PU-degrading activity (Liu et al., 2023).

The bacteria capable of degrading polyester-based PUR polymers are primarily from the genera *Comamonas*, *Pseudomonas*, *Bacillus*, *Acinetobacter*, and *Corynebacterium*. Table 5.1 presents the information. *Pseudomonas putida* A12 is able to utilize Impranil DLN as its exclusive carbon source. Studies have demonstrated that this bacterium can degrade approximately 92% of the Impranil DLN substrate within a 4-day incubation period at 30 °C. The FTIR analysis provided additional evidence regarding the degradation process. The study demonstrated a decline in ester functional groups and the emergence of amide groups (Peng et al., 2014). The taxonomic term "*Comamonas* sp." is employed to denote an unspecified species within the *Comamonas* genus. The TB-35 strain effectively decomposed 50 mg of polyester-based PUR into diethylene glycol and adipic acid after being incubated at 30 °C for 7 days. The polymer served as the sole carbon source in this process. The degradation performance exhibited a 48% decrease in the absence of an additional nitrogen source (Nakajima-Kambe et al., 1995). *Bacillus subtilis* MZA-75 and *Pseudomonas aeruginosa* MZA-85 can both use polyester PUR film as their only carbon source for growth. Additionally, these microorganisms can completely convert the intermediates 1,4-butanediol (BDO) and adipic acid (AA) into carbon dioxide (CO_2) and water (H_2O). Furthermore, a study found that when cultivating both strains together in a mixed culture, the polyester PUR film experienced a 40% decrease in weight within a 30-day timeframe (Shah et al., 2016).

Microorganisms degrade PUR plastics primarily through the enzymatic activity of hydrolases, such as esterases, ureases, proteases, and amidases. Esterases, classified as EC 3.1 enzymes, play a crucial role in the degradation of polyester-based PUR plastics. These enzymes have a specific function in the hydrolysis of ester bonds present in the soft segments of plastics. Hydrolysis results in the liberation of carboxylic acid and alcohol end-groups. PudA is a membrane-bound enzyme classified as a phosphoric triester hydrolase (PUase). It is derived from the organism *C. acidovorans* TB-35 bacterium possessing a hydrophobic surface-binding domain (SBD) and a catalytic domain. The degradation of polyurethane (PUR) by this enzyme has been proposed to occur through a two-step process. First, it attaches to the hydrophobic PUR surface through its SBD (substrate-binding domain). Subsequently, it hydrolyzes the ester bonds in PUR. Insufficient research has been conducted on the degradation of polyurethane (PUR) by ureases (Liu et al., 2021).

Table 5.1 List of potential microorganisms for degrading PU and PVC

Polymer	Microorganism	Source of isolation	Conclusion	Reference
Polyurethane	**Fungi**			
	Pestalotiopsis microspora	Plastic-degrading fungi	Clearance of the medium was observed	Russell et al. (2011)
	Aspergillus tubingensis	Soil	Complete degradation	Khan et al. (2017)
	Cladosporium cladosporioides	Plastic debris floating at the shoreline	A visible halo on agar plates was seen (Brunner et al., 2018)	Brunner et al. (2018)
	Leptosphaeria sp.			
	Penicillium griseofulvum			
	Xepiculopsis graminea			
	Aspergillus fumigatus	Soil	On both LDPE and PU media, a halo zone and morphological deterioration were seen on the growing culture (Raghavendra et al., 2016)	Raghavendra et al. (2016)
	Bacteria			
	Corynebacterium sp.	Soil	Along with the CO_2 release, hydrolysis zones were seen	Ali et al. (2014)
	Bacillus sp.			
	Pseudomonas sp.			
	Micrococcus sp.			
	Arthrobacter sp.			

Polyvinyl chloride	**Fungi**			
	Phanerochaete chrysosporium PV1	PVC films buried in the soil	A morphological degeneration was detected	Sah et al. (2011)
	Lentinus tigrinus PV2			
	Aspergillus niger PV3			
	Aspergillus sydowii PV4			
	Bacteria			
	Acanthopleurobacter pedis	Soil	Morphological damage was found	Sah et al. (2011), Anwar et al. (2013)
	Microbacterium sp.			
	Pseudomonas putida			
	Pseudomonas putida			
	Pseudomonas otitidis			
	Acanthopleura bacterpedis	Based on their ability to break down PVC, the bacterial strains were chosen	The developed consortium has a considerably improved biodegradation capacity for PVC	Anwar et al. (2013), Darby and Kaplan (1968)

5.2 Environmental Impact of Biodegradation in Polyurethane and PVC

A significant environmental issue is the slow degradation or lack of biodegradability of organic materials in natural environments, such as plastics. Plastics, including various types such as nylon, polycarbonate, polyethylene terephthalate, polyethylene, polypropylene, polystyrene, polytetrafluoroethylene, polyurethane, and polyvinyl chloride, are widely used in our daily lives (Begum et al., 2015). Polyethylene is extensively used in various human activities due to its easy processing for producing items used in food packaging, textile packaging, laboratory instruments, and automotive components. Synthetic polymers are highly hydrophobic and have a large molecular weight. Annually, a significant quantity of plastic bags, estimated to be between 500 billion and 1 trillion, are commonly used worldwide. The global consumption of polyethylene is growing at an annual rate of 12%, leading to the production of around 140 million tons of synthetic polymers each year (Roy et al., 2008). The large amount of polyethylene present in the environment raises ecological concerns regarding its disposal. Polythene waste poses a significant threat to marine life. The consumption of specific substances can sometimes cause intestinal blockages in aquatic organisms, including fish, avian species, and marine mammals (Secchi & Zarzur, 1999; Denuncio et al., 2011). Plastic pollution is currently causing significant harm to the marine environment, affecting at least 267 species. The impacted species include a range of mammals, with sea turtles experiencing an 86% impact rate and seabirds experiencing a 44% impact rate. Polyethylene bag ingestion has been observed to cause mortality in terrestrial organisms, including cows (Coe et al., 1997). Polythene packaging, specifically plastic, constitutes a substantial portion, approximately 10%, of the overall municipal waste produced worldwide. While a fraction of this polythene waste undergoes recycling, the majority is disposed of in landfills, where it takes hundreds of years to degrade. The increasing utilization of polythene poses a significant challenge in terms of degradation (Moore, 2008). Polythene is widely recognized as the most common form of nondegradable solid waste and is currently considered a significant threat to the environment. The increasing usage of polythene presents a significant challenge due to its degradation. The ingestion or entanglement of plastics in the oceans results in the deaths of approximately one million birds and 10,000 marine animals each year. There has been a significant focus on the biodegradation of plastic waste and the use of microorganisms to break down polymers. The shift in focus is a result of the inefficiency of current chemical and physical disposal methods used for these pollutants, as well as the associated environmental concerns (Begum et al., 2015).

5.3 Biodegradation of Polyurethane and PVC: Sustainable Waste Management

Plastic biodegradation entails a sequence of biochemical reactions that degrade polymeric materials into smaller molecules, such as oligomers or monomers. Subsequently, these smaller molecules can participate in the biogeochemical cycle. Enzymes from various classes are essential catalysts in the biodegradation process. Biodegradation can be categorized into four distinct stages. The breakdown of organic materials involves four primary processes (Skariyachan et al., 2022):

(a) Biodeterioration
(b) Biofragmentation
(c) Assimilation
(d) Mineralization

Biodeterioration involves the formation of carbonyl groups. This phenomenon is caused by the action of oxidative enzymes produced by microorganisms or by external factors such as sunlight, particularly ultraviolet radiation. Polymer degradation encompasses three primary processes (Fig. 5.2).

The degradation of polymers involves three primary processes:

1. Oxidation
2. Biofragmentation
3. Bioassimilation

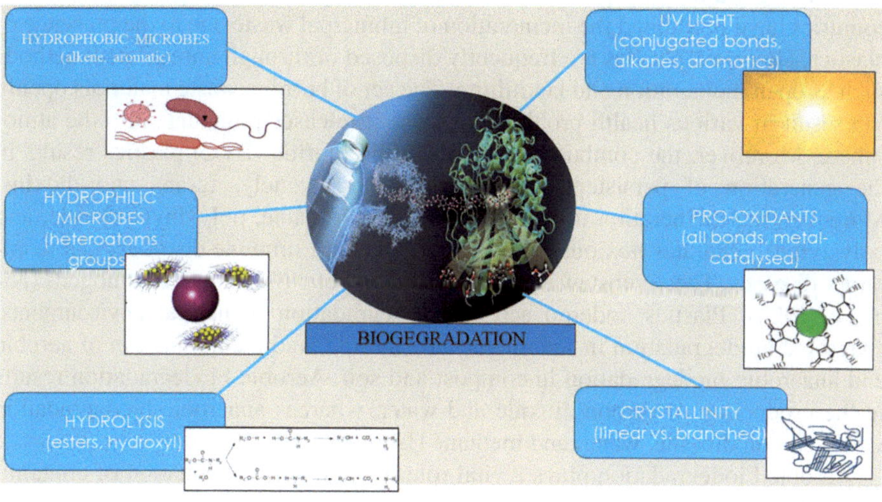

Fig. 5.2 Key properties and processes that induce biodegradation of different chemical groups

Oxidation leads to a decrease in the number of carbonyl groups and the formation of carboxylic acids. Biofragmentation refers to the enzymatic hydrolysis and/or fragmentation of polymer carbon chains, resulting in the release of intermediate products. Bioassimilation refers to the process in which bacteria or fungi metabolize small hydrocarbon fragments that are released during biofragmentation. Mineralization, the fourth step in the process, entails the intracellular transport of hydrolysis products within the cell wall. Subsequently, the hydrolysis products undergo conversion into microbial biomass within the cell, resulting in the emission of carbon dioxide and water (Mohanan et al., 2020). The initial stage of microbial degradation involves the establishment of microbial species on the plastic surface. Colonization leads to consortia development and subsequent biofilm formation, resulting in significant polymer surface degradation. Microbial adhesion to polymer surfaces entails the production of proteins and polysaccharides. These substances infiltrate the pores of the material and cause alterations in its dimensions. Biodeterioration refers to the adherence and activation of microorganisms on plastic surfaces. In this phase, the plastic polymer experiences surface degradation, leading to changes in its physicochemical properties (Ali et al., 2021a). Fragmentation is a lytic process that is essential for the degradation of polymers. This process results in the creation of monomers, dimers, and/or oligomers. Microorganisms employ diverse mechanisms to degrade polymers. These mechanisms entail the release of particular enzymes, including oxidoreductases and hydrolases, as well as the production of free radicals. The chemical bonds involved in plastic formation exhibit similarities to the bonds found in natural polymers, such as lignocellulosic compounds. Hence, it is plausible to propose that many enzymes capable of breaking down lignocellulose also possess the capacity to degrade plastic polymers. Several countries have prohibited the incineration of municipal waste due to the presence of plastics. However, plastics are frequently disposed of through unregulated methods such as open incineration and landfilling. The act of burning waste materials openly can result in various health problems because it releases pollutants into the atmosphere. Moreover, the combustion of polyvinyl chloride (PVC) plastics results in the generation of persistent organic pollutants, namely, furans and dioxins. Moreover, the incineration of polyethylene, polyurethane, polyvinyl chloride, and polystyrene generates noxious irritants that can cause immune disorders and respiratory diseases. These substances are classified as potential human carcinogens (Nir et al., 1993). Plastics undergo aerobic biodegradation in natural environments, anaerobic biodegradation in sediments and landfills, and a combination of aerobic and anaerobic biodegradation in compost and soil. Aerobic biodegradation results in the production of carbon dioxide and water, whereas anaerobic biodegradation yields carbon dioxide, water, and methane (Ishigaki et al., 2004).

Aerobic biodegradation plays a vital role in the natural attenuation of contaminants at hazardous waste sites. Aerobic microorganisms utilize oxygen as an electron acceptor for the purpose of metabolizing organic compounds into smaller molecules. The by-products of this process are CO_2 and H_2O. Extracellular and intracellular depolymerase enzymes are essential for the biological degradation of polymers. Microbial exoenzymes facilitate the degradation of complex polymers, resulting in the formation of shorter chains or smaller molecules such as oligomers, dimers, and monomers. The molecules are sufficiently small to dissolve in water, enabling them to traverse the semipermeable outer membranes of bacteria. Hence, bacteria can utilize these molecules as carbon and energy sources. The initial phase of polymer degradation is referred to as depolymerization. The term "mineralization" refers to the process in which inorganic entities such as CO_2, H_2O, or CH_4 are produced. Polyvinyl chloride (PVC) is a durable thermoplastic material renowned for its resistance to abrasion and chemical exposure. Furthermore, it exhibits low moisture absorption. While there is extensive research on the thermal and photodegradation of PVC, there is limited literature on its biodegradation (Alshehrei, 2017). The low molecular weight of PVC renders it susceptible to biodegradation by white-rot fungi. Polyvinyl chloride (PVC) can depolymerize and biodegrade within a specific temperature range (Peng et al., 2020).

5.4 Microbial Degradation of Polyurethane and PVC in Natural Environments

Researchers in the scientific community are actively seeking microorganisms that can degrade polyurethanes due to their resistance to degradation by microbial catalysts. In vitro assays were used to evaluate the biodegradation of polyurethane. These assays have yielded valuable insights into the mechanisms utilized by microorganisms for biodegradation. Moreover, the literature has reported further evidence of biodegradation, such as the decrease in polyurethane mass caused by microbial consumption. Microbial degradation has been shown to negatively impact the mechanical properties of polyurethane foams, particularly in terms of elongation and tensile strength. Infrared spectroscopy was used to detect changes in the spectral fingerprints of polyurethane foams, which were found to be indicative of biodegradation (Dutta et al., 2010; Sarkar et al., 2006). Various microorganisms have been found to accelerate the biodegradation of biodegradable plastics in landfills. These microorganisms include bacteria like *Pseudomonas* sp., nylon-eating bacteria, and flavobacteria. They aid in the degradation of nylon through the action of the nylonase enzyme (Negoro, 2000).

Enzymatic degradation of plastic through hydrolysis involves enzymes binding to the polymer, enabling the hydrolytic breakdown of polymers into smaller units like oligomers, dimers, and monomers. Subsequently, these smaller entities undergo mineralization, resulting in the formation of carbon dioxide (CO_2) and water (H_2O). Previous research has highlighted the ability of fungi, bacteria, and algae to break down polymers. These organisms have been observed to thrive on easily degradable polymers like cellulose and lignin (Ali & Sun, 2015) (Fig. 5.3).

The mycelium of fungi has a large surface area, allowing it to efficiently penetrate and break down conventional plastic polymers. In addition, mycelia possess the capacity to release extracellular enzymes, including depolymerase, that aid in the decomposition of polymers into smaller entities like oligomers, dimers, and

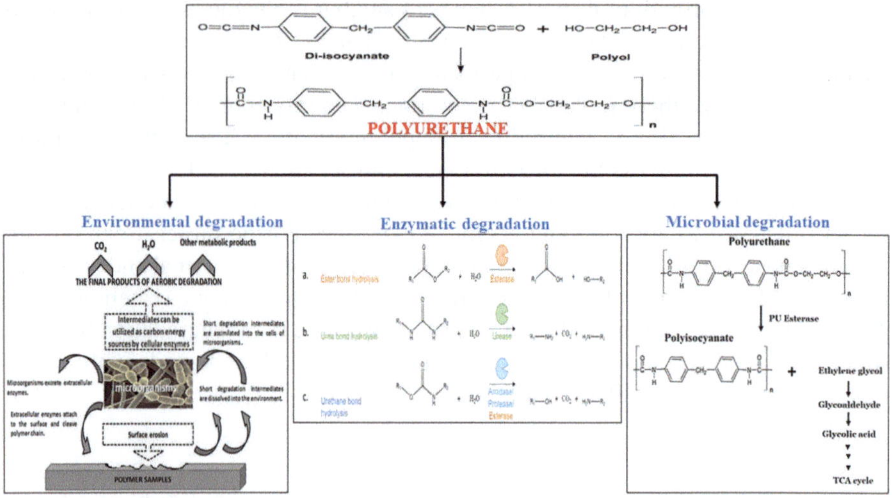

Fig. 5.3 Different types of biodegradation processes used for polyurethane

monomers. Fungi have a higher concentration of secreted enzymes compared to bacteria (Gangola et al., 2019). Subsequently, intracellular enzymes present in fungi can uptake and degrade the generated monomers into minerals. White-rot and brown-rot fungi are capable of efficiently facilitating the biodegradation of plastic and other polymeric materials. This is due to their ability to produce a range of extracellular enzymes, such as lignin peroxidase, manganese peroxidase, versatile peroxidase, and multi-copper oxidase laccase. Enzymes are essential in the breakdown of lignin, converting it into carbon dioxide (CO_2) and water (H_2O) (Ali et al., 2021b). Table 5.1 provides a comprehensive compilation of fungal strains that have shown effectiveness in breaking down plastic waste materials. Our investigation revealed the presence of *Aspergillus* spp. and multiple *Penicillium* species. Several strains, including *Aspergillus flavus*, have shown promise in the biodegradation of plastics. *Aspergillus flavus* is particularly notable for its frequent use and high efficiency in plastic degradation. Fungal strains capable of plastic degradation have been found in both contaminated soil and seawater. Similarly, different bacterial strains can inhabit and metabolize plastic compounds in soil or water contaminated with plastic. Several studies have indicated that using specific bacteria to degrade plastic shows potential as an effective bioremediation method (Yoshida et al., 2016). Our literature review found that *Bacillus* spp. and *Pseudomonas* spp. were tested, along with different species of *Streptomyces*, exhibiting significant effectiveness in dealing with different plastic polymers. Nevertheless, *Bacillus cereus* emerged as the dominant species, with plastic-contaminated soil serving as the primary medium for its isolation. However, the rate of fungal degradation surpasses that of bacteria. It is important to note that the degradation of plastic by fungi necessitates more stable conditions compared to bacterial degradation (Artigas Alejo, 2008) (Fig. 5.4).

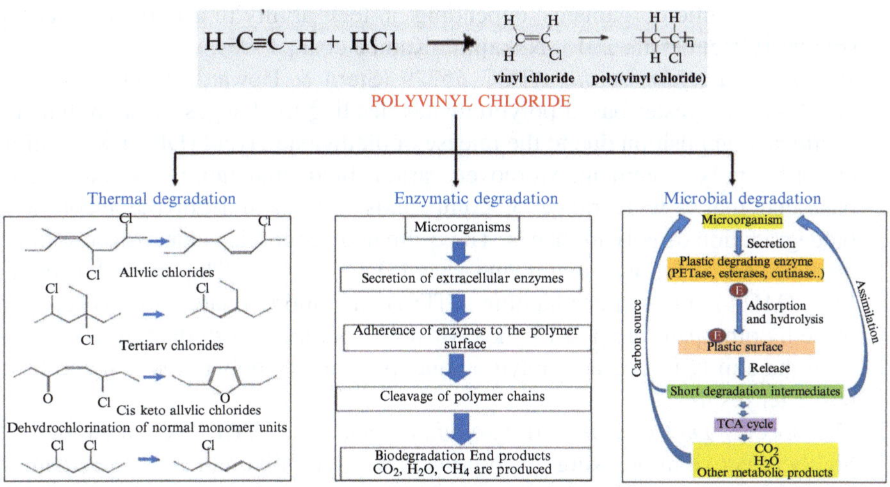

Fig. 5.4 Different types of biodegradation processes used for polyvinyl chloride

5.5 Challenges and Prospects in Biodegrading Polyurethane and PVC for Eco-friendly Technologies

Most PUR (polyurethane) polymer materials are thermosetting polymers. There are two types of polyurethane: ester-type polyurethane (ES-PU) and ether-type polyurethane (ET-PU). Studies have shown that polyether-based polyurethane (PUR) materials have a reduced risk of fungal infestation compared to polyester-based PUR materials (Darby & Kaplan, 1968). The study showed that *Comamonas acidovorans* strain TB-35 can effectively degrade ES-PU derived from polydiethylene adipate with molecular weights of 2500 and 2690, as well as TDI. The degradation of ES-PU is facilitated by the hydrolytic enzyme derived from *C. Acidovorans* TB-35 strain demonstrates the ability to synthesize diethylene glycol and adipic acid (Nakajima-Kambe et al., 1995). The degradation of ES-PU, which is synthesized using TDI, PCL-diol (Mn 1250), and ethylenediamine, is observed to be caused by cholesterol esterase. This leads to the release of hard segment constituents (Fukushima et al., 2010).

1. The use of solid-rich medium in biodegradation assays allows for direct interaction between the degrader and the substrate. The *Acidovorans* strain showed significantly higher rates of polyurethane degradation compared to previously observed rates by environmental inoculums. The *acidovorans bacterium* can degrade 2–3 g of polyurethane within 4–6 days. The observed degradation activity in this study is approximately 400 times higher than previously reported rates of polyurethane degradation by naturally occurring microbial communities. During polyurethane biodegradation, microorganisms utilize diethylene glycol (DEG) and trimethylolpropane as carbon sources. These molecules are then bio-

degraded by microorganisms, depending on their ability to use them as carbon sources (Cregut et al., 2013; Nakajima-Kambe et al., 1995).

2. *Pseudomonas chlororaphis* ATCC 55729 (Stern & Howard, 2000) was incubated with polyester-based polyurethanes, leading to changes in the polymer's chemical composition due to the release of diethylene glycol (DEG), a constituent of the polyol segment. Moreover, various biodegradation assays have demonstrated that a wide range of compounds can result from the incomplete biodegradation of polyurethanes. The estimated economic values of these intermediates are as follows: diethylene glycol (DEG) at 0.98 €/kg, methylenedianiline (MDA) or toluenediamine (TDA) at approximately 2 €/kg, and trimethylolpropane at up to 15 €/kg. Most microorganisms studied in this review have the ability to degrade polyurethane, particularly polyurethane made from polyester resins.
3. The species *Dietzamaris* and *Exophiala jeanselmei* have been identified as highly effective in breaking down polyether-based polyurethanes. *Dietzamaris*, formerly known as *Rhodococcus equi* TB-60, belongs to the *Actinobacteria* phylum, while *Exophiala jeanselmei* is a fungus classified under the *Ascomycota* phylum (Akutsu-Shigeno et al., 2006; Owen et al., 1996). Different methods have been used to evaluate the degradation of polyurethane, resulting in different results depending on the type of polyurethane (polyether or polyester-based), its chemical composition (including polyol and isocyanate composition), and its ability to be broken down by living organisms (Huang & Roby, 1986). Polyester-based polyurethane has been found to be susceptible to microbial degradation under aerobic conditions. The physical and chemical changes observed during the biodegradation assays provide evidence of the compounds' biodegradation, leading to a significant decrease in weight (Dutta et al., 2010; Filip, 1978; Iannace et al., 1999). Assays used to evaluate the biodegradability of polyurethane compounds, specifically those that measure carbon dioxide (CO_2) production following the OECD Sturm procedure (standard 301b), indicate that polyurethanes have limited biodegradability. Polyurethanes should be classified as non-readily biodegradable substances or inherently persistent entities in the environment. Only 51.1% of the tested polyurethanes showed susceptibility to biodegradation after an 84-day incubation period, with an estimated degradation rate of around 10 mg/l (Iannace et al., 1999). The assay requires a minimum threshold of 60% release of the molecule as carbon dioxide within 28 days for successful completion. Polyurethanes derived from polyethers demonstrate significant resistance to microbial attacks, showing no instances of biodegradation or only minimal degradation under aerobic and anaerobic conditions. One potential reason for this observation is that polyether resins contain chemical ether bonds that are resistant to microbial enzyme activity, while polyester resins have ester bonds that are more susceptible to cleavage by hydrolases.
5. The *Ascomycota* phylum includes fungal genera such as *Aspergillus*, *Pestalotiopsis*, and *Gliocladium* (Russell et al., 2011; Cangemi et al., 2008; Kay et al., 1991). Filamentous fungi are well-known for their ability to break down diverse organic compounds. The mentioned abilities depend on effectively colo-

nizing the substrate and secreting numerous enzymes. Several fungal species isolated from soil samples have shown potential in biodegrading polyurethane foams. Fungal biodegradation differs from bacterial biodegradation in terms of its distinct characteristics, as it does not involve modifying the mechanical resistance of polyurethane foams. Bacterial degradation significantly compromises the structural integrity of polyurethanes, while fungal attacks on polyurethane foams cause more subtle alterations to their properties (Wilcox et al., 2015).

5.6 Microbial Communities for Polyurethane and PVC Biodegradation

Artificially selected microbial communities have emerged as a promising approach for the removal of pollutants through bioremediation and biodegradation. This method offers several advantages, including its potential for rapid and cost-effective pollutant removal, its environmentally friendly nature, and its social acceptability. This holds true even in the absence of prior knowledge regarding the particular species and pathways implicated in the process of degradation. Moreover, the incorporation of natural polymers, such as starch, into plastic blends has been demonstrated to enhance their biodegradability (Borchert et al., 2021). Enhancing biodegradation is associated with a process that involves the rapid enzymatic hydrolysis of starch. This process induces porosity in the polymer, rendering it susceptible to degradation by both biotic and abiotic factors (Wool et al., 2000). In their study, Karimi and Biria (2019) observed the biodegradation process in LDPE-starch blend samples. The researchers utilized an aqueous solution of alpha-amylase to accomplish this. Amylase, an enzyme, was found to display co-metabolic activity by hydrolyzing starch (also called Provisional 15) and PE molecules. Furthermore, the incorporation of hydrophilic starch into hydrophobic PE 640 significantly improved the biodegradability of both polymers.

Urease is the enzyme that catalyzes the hydrolysis of urea linkages in specific poly(urea-urethane) polymers. This process results in the liberation of two amine molecules and carbon dioxide gas. There is a lack of literature on the degradation of polyurethane (PUR) by ureases. The limited availability of reports is due to the inherent challenge of hydrolyzing urea bonds (Phua et al., 1987). Urea bonds exhibit greater resistance to degradation in comparison to ester bonds. Ureases (EC 3.5.1.5) exhibit enzymatic activity toward polyether urea (PUR). The primary cause of PUR degradation is the hydrolysis of urea bonds. Two bottlenecks impede the enzymatic degradation of polyurethane (PUR), as recently identified by researchers. The majority of identified PUR hydrolases primarily focus on hydrolyzing ester bonds present in the soft segments of polyester-based PUR materials. No authentic urethanase enzyme capable of directly degrading the urethane bonds present in PUR polymers has been reported (Wei et al., 2020). Future research should prioritize the investigation of enzymes with "polyurethanase" activity. The integration of this method with existing polyester hydrolases will effectively tackle the difficulties related to the enzymatic breakdown of polyurethane (PUR). Moreover, the hydrolyzable chemical bonds present in PUR polymers exhibit limited enzyme accessibility due to their insoluble hydrophobic nature. The process can be enhanced by incorporating polymer-binding domains with well-defined specificity (Van Tilbeurgh et al., 1986). Functional polymer-binding peptides, known as anchor peptides, are utilized for the specific binding of cells to polymer surfaces. This process enhances the accessibility of polymers that are insoluble (Dedisch et al., 2020). A screening system has been created to aid in the directed evolution of the polypropylene (PP)-binding peptide LCI. This system employs cell surface display of an esterase derived from *Escherichia coli*. This approach has led to the identification of a mutant variant of LCI that exhibits a 12-fold higher binding affinity to PP compared to the original, wild-type LCI. Additionally, reports have discussed the utilization of ultrahigh-throughput screening approaches (Apitius et al., 2019) and rational engineering techniques to advance the development of anchor peptides. The enzyme Tcur1278, obtained from *Thermomonospora curvata* DSM 43183 and recognized for its hydrolytic activity on PET (Wei et al., 2014), has been genetically combined with the anchor peptide tachystatin A2. The enzyme fusion has significantly enhanced the hydrolytic activity on polyester PUR nanoparticles (Impranil DLN SD) compared to the unmodified enzyme. This improvement is likely a result of enhanced substrate adsorption (Islam et al., 2019).

The enzymes involved in the biodegradation of polyurethane are found in two primary compartments and can be classified as either secreted or membrane-bound during the initial stage of degradation. The model's validity has been partially confirmed through the use of recombinant expression techniques for studying extracellular enzymes and biochemical assays for examining membrane-bound enzymes. Furthermore, a comprehensive explanation of the biodegradation process has been provided. The initial stage of polyurethane biodegradation occurs when the enzyme, anchored to the membrane, attaches to the polyurethane surface. Subsequently, the urethane bond undergoes hydrolysis facilitated by the membrane-bound enzyme. This process results in the liberation of monomers and the constituent elements of the polyurethane material (Howard et al., 2012). The aforementioned procedure enables microorganisms to generate a significant quantity of compounds resulting from the degradation of polyurethane in their surrounding environment. The limited effectiveness of freely released enzymes in binding to the bulk substrate is attributed to a reduced interaction between the enzyme and substrate. This is because polyurethanes, being insoluble substrates, pose a challenge for enzyme-substrate interaction. Nevertheless, these enzymes can still contribute to the biodegradation process by physically eroding the polyurethane surface. The objective of this procedure is to enhance the interaction between microorganisms and polyurethane, thereby promoting the biodegradation process through increased microbial adhesion. In this model, unbound enzymes collaborate as cooperative entities, with the membrane playing a vital role.

Binding-capable enzymes are crucial for the degradation of polyurethanes. The membrane-bound enzyme plays a vital role in facilitating enzyme concentration near the substrate in microorganisms. This concentration improves enzyme effectiveness. Prior research has observed the presence of membrane-bound enzymes attaching to a substrate surface before beginning enzymatic degradation. This phenomenon has been observed in enzymes that degrade insoluble substrates. In 1999, Nakajima-Kambe et al. proposed a mechanism to explain the operational role of membrane-bound polyurethanase in acidovorans. The membrane-bound enzyme is crucial for the microorganism's utilization of polyurethane as a substrate, enabling its thriving. The significance of freely released polyurethanases in the environment may be underestimated, but their contribution remains crucial. The membrane-bound enzyme efficiently degrades polyurethane, leading to significant accumulation of polyurethane monomers bound to the enzymes. Moreover, it is noteworthy that these enzymes possess a comparatively limited surface area, which facilitates effective retrieval. Extracellular polyurethanases play a crucial role in the efficient degradation of polymers. They have a vital function in finalizing the degradation process initiated by enzymes bound to the membrane (Cregut et al., 2013).

The diversity of plastic-degrading microorganisms is influenced by various factors, including environmental conditions, temperature, pH, and other relevant variables. Soil moisture content has a notable influence on microorganism growth in environmental settings. Soil moisture is a crucial factor that initiates microbial activity. An increase in moisture content is likely to enhance the hydrolytic cleavage of microbes. Temperature and pH are influential factors in the degradation process. The enzymatic degradation capacity decreases with increasing temperature. Polymers with high melting points exhibit reduced susceptibility to degradation. pH variations affect the hydrolytic reaction rate. pH fluctuations can have a substantial impact on microbial growth and degradation rates (Wilcox et al., 2015).

5.7 Enzymes and Microbial Agents in Polyurethane and PVC Biodegradation

Enzymes responsible for polyurethane biodegradation are located in two main compartments and can be either secreted or membrane-bound during the initial phase of the process. Polyurethane biodegradation begins with the enzyme, attached to the membrane, binding to the surface of the material. The urethane bond is hydrolyzed by a membrane-bound enzyme, leading to the release of monomers and components of the polyurethane material (Howard et al., 2012). This process facilitates the production of numerous compounds by microorganisms through the degradation of polyurethane in their immediate environment. The insolubility of polyurethanes poses a challenge for efficient enzyme-substrate binding when free enzymes are directly added to the medium. This is mainly due to the restricted interaction between the enzymes and the substrate in large quantities. Despite this limitation, enzymes can still actively engage in the biodegradation process by mechanically breaking down the surface of polyurethane. This procedure increases the contact area between microorganisms and polyurethane, promoting faster biodegradation by improving microbial adhesion. In this model, the free enzymes work together, while the membrane-bound enzymes play a crucial role in polyurethane degradation. The microorganism benefits from having a membrane-bound enzyme because it can concentrate some of its enzymes near the substrate, leading to improved efficiency of the released free enzymes (Cregut et al., 2013).

Exiguobacterium sp. is a bacterium capable of polystyrene degradation. The YT2 protein is obtained from *Tenebrio molitor* and *Acinetobacter* sp. Prior research has documented the existence of *Tribolium castaneum* (Wang et al., 2020). The *Bacillus* sp. bacterial strain has the capability to degrade polyethylene. YP1 and *Enterobacter asburiae* YT1 were isolated from the insects *Plodia interpunctella* and *Acinetobacter* sp. The NyZ450 strain and *Bacillus* species are of interest. The strain NyZ451, derived from the bacterium *T. molitor* has been previously identified in prior research. The fungal strain *Aspergillus flavus* G10, isolated from *Gryllus bimaculatus*, has been found to have the capability to degrade PS-based polyurethane (PU) in the field of PU-related research. This study examines *Xanthomonas* sp. The HY-71 strain, capable of degrading three types of polyurethane (polyacrylic-, polystyrene-, and polyethylene-based), was isolated from the intestinal tract of mature Japanese carpenter bees. This report presents the preliminary findings on the degradation and utilization of polyurethane (PU) by *Xanthomonas* species isolated from insects' gastrointestinal tract. The strain HY-71 was evaluated for its capacity to produce extracellular enzymes responsible for the degradation of polyurethane (PU) components. Strain HY-71 exhibited the ability to produce extracellular enzymes, as indicated by the presence of a transparent halo zone resulting from protease activity and the formation of a calcium complex due to esterase activity around the bacterial growth (Kim et al., 2022) (Fig. 5.5).

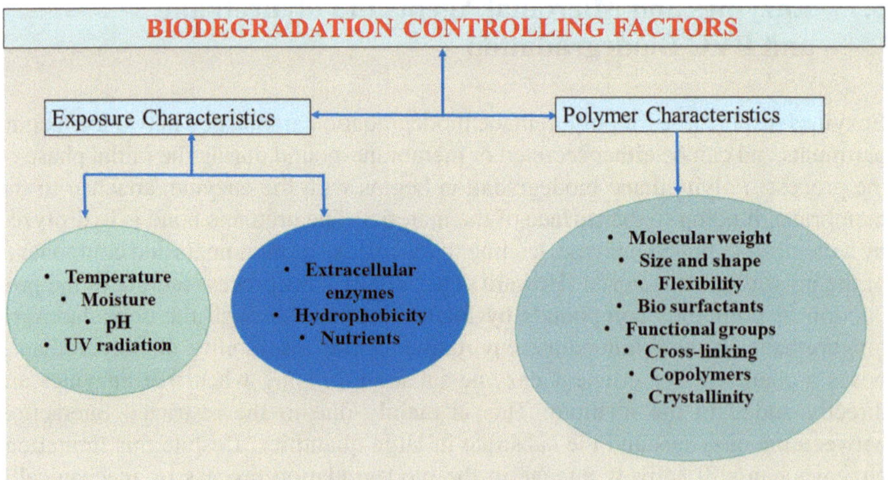

Fig. 5.5 Factors affecting the rate of polymer biodegradation

5.8 Bioremediation and Biotechnological Applications for Polyurethane and PVC Biodegradation

Scientists have been actively investigating microorganisms that can degrade polyurethanases, despite the presence of microbial catalysts, because of their ability to survive in various environmental conditions. Polyurethane biodegradation was assessed using in vitro assays. These assays have provided valuable insights into the mechanisms employed by microorganisms in the biodegradation process. Furthermore, the literature extensively documents additional observations supporting the occurrence of biodegradation. These observations demonstrate the decrease in polyurethane mass due to microbial consumption. Research has demonstrated that microbial degradation can adversely affect the mechanical properties of polyurethane foams, specifically in relation to elongation and tensile strength. In addition, infrared spectroscopy was employed to identify alterations in the spectral characteristics of polyurethane foams, offering reliable proof of biodegradation (Kim et al., 2022).

Fungal biodegradation differs from bacterial biodegradation by specifically targeting the mechanical resistance of polyurethane foams. Bacterial degradation has a significant impact on the structural integrity of polyurethanes, whereas fungal attacks on polyurethane foams result in more subtle changes to their properties (Oprea, 2010). Based on the aforementioned observation, it seems that the degradation of the substrate is more likely attributed to overall deterioration rather than a deliberate and targeted assault. While this method may facilitate complete biodegradation of polyurethane, its economic feasibility for disposal is questionable. The study of biodegradation mechanisms has primarily focused on proteobacterial strains, particularly those within the γ-*Proteobacteria* class. Examples of these strains include *Delftia acidovorans* (formerly known as *Comamonas acidovorans* TB-35), *Pseudomonas fluorescens*, and *Pseudomonas chlororaphis* ATCC 55729 (Nakajima-Kambe et al., 1995). Microorganisms utilize polyurethanes as a carbon source for their growth and metabolic requirements. These phenomena are mainly observed in fungi, but they can also occur in *Drosophila*. *Acidovorans* can function as a nitrogen source in the presence of polyurethane, in addition to its primary role. In many instances, bacteria use polyurethane as a carbon and nitrogen source (Nakajima-Kambe et al., 1995; Ibrahim et al., 2011), leading to the reduction or elimination of the degradation process. Yeast extracts were incorporated into the growth medium as a nitrogen source to enhance the biodegradation of polyurethane by bacteria (Ruiz et al., 1999). Microorganisms utilize diethylene glycol (DEG) and trimethylolpropane (CH_3CH_2C) as carbon sources during the polyurethane biodegradation process. Subsequently, microorganisms facilitate the breakdown of these molecules via microbial biodegradation. This process depends on microorganisms' capacity to utilize carbon sources as a potential energy source (Nakajima-Kambe et al., 1995). Stern and Howard (2000) have shown that *Pseudomonas chlororaphis* ATCC 55729 can degrade polyester-based polyurethanes (Nakajima-Kambe et al., 1995). Observation of the incubation of the microorganism with polyester-based polyurethanes revealed alterations in the polymer's chemical composition. This can be attributed to the release of DEG, a constituent of the polyol region. Additionally, various biodegradation assays have shown that polyurethanes can undergo incomplete biodegradation, resulting in the production of a diverse array of compounds.

Two recently discovered species are *Dietzamaris* and *Exophiala jeanselmei*. *Dietzamaris*, formerly referred to as *Rhodococcus equi* TB-60, belongs to the *Actinobacteria* phylum. *Exophiala jeanselmei* is an *Ascomycota* fungus. These species have demonstrated remarkable proficiency in degrading polyether-based polyurethanes (Akutsu-Shigeno et al., 2006). Researchers have recently investigated the genetic manipulations related to polyurethanases produced by *Pseudomonas chlororaphis*, *Pseudomonas aeruginosa*, and *Delftia*. Researchers have found that acidovorans bacteria can facilitate the development of a *Saccharomyces cerevisiae* strain capable of degrading polyurethanes. The statement provided is insufficient and lacks necessary details for rephrasing in an academic style. The user did not provide any text to rewrite. The application of yeast-arming technologies in engineering has led to the development of *cerevisiae*, suggesting potential for the advancement of degradation strategies. These strategies employ genetically modified strains that possess specific genetic tools for targeting polyurethanes (Shibasaki et al., 2009). New and targeted strategies are being developed within specific organisms to address the previously ineffective traditional biochemical enzymatic degradation of polyurethane. These approaches provide alternative methods for producing commercially valuable by-products and promoting the degradation of pollutants. Shibasaki et al. (2009) successfully engineered a specific strain of *Saccharomyces cerevisiae*, a yeast species. This strain has been genetically engineered to express lipases throughout its entire cellular membrane. The aim of this modification was to enhance the degradation of polyester-polyurethane. This study demonstrated the feasibility of this approach for the first time (Cregut et al., 2013; Shibasaki et al., 2009).

5.9 Addressing Environmental Concerns: Biodegradation of Polyurethane and PVC

According to the 2020 estimates, approximately 195 countries generated a total of 400 million metric tons of plastic waste. Around 8.8 million metric tons were deposited into the Earth's oceans (Koller & Braunegg, 2018). The United Nations Centre for Regional Development (UNCRD) has identified several Southeast Asian countries, namely, China, Indonesia, the Philippines, Vietnam, Sri Lanka, Thailand, Malaysia, and Bangladesh, as having exhibited notable mismanagement in their management of plastic waste. The report highlights that approximately 88% of plastic waste generated in these nations ultimately enters water bodies. The global plastic production from 1950 to 2019 amounted to 8100 million metric tons (Mt). The global production of plastics has experienced substantial growth, rising from two million metric tons (Mt) in 1950 to 450 million metric tons (Mt) in 2019. The current global production capacity of bio-based or biodegradable plastics is four million metric tons (Mt). Monomers derived from fossil hydrocarbons, such as ethylene and propylene, are commonly used in plastic production. The nonbiodegradable nature of commonly used plastics leads to their accumulation in landfills and disposal in bodies of water. The ongoing presence of this issue in the environment poses a substantial threat to both the environment and its diverse ecosystems. The global prevalence and widespread occurrence of plastic pollution in the ocean is a significant concern. Research studies have estimated that the ocean contains approximately 580,000 pieces of plastic debris per square kilometer. The United Nations convention on biological diversity's recent evaluation report examines the impact of marine plastic waste on various species. The group encompasses over 600 organisms, spanning from microorganisms to whales. The primary means of exerting influence is by consuming these waste materials. Various procedures and methods are being employed to address this issue. Currently, numerous local industries engage in the production of various degradable plastic bags. These bags can be classified as either fully degradable or partially degradable, achieved through the inclusion of additives during the manufacturing process. Carbohydrates have been extensively utilized in diverse industries for the production of biodegradable plastics that can degrade under ultraviolet (UV) radiation or through microbial activity. The extensive accumulation of plastics in landfills, in oceans, or through incineration poses a significant threat to the environment and ecosystems. The combustion of plastic materials emits noxious gases, such as furans and dioxins. These substances possess greenhouse properties and contribute to ozone layer depletion. Dioxin is widely acknowledged as a significant contributor to human health issues and soil contamination (Wilcox et al., 2015). Living organisms are experiencing significant water pollution caused by the improper disposal of plastic waste in various bodies of water such as lakes, rivers, ponds, and similar habitats. The lakes in Dhaka, an urban metropolis, exemplify the detrimental effects of plastic pollution. Visitors frequently discard plastic bottles, cans, bags, and other plastic items into the lakes. The introduction of plastic waste into aquatic environments has adverse effects on hydrodynamics, fish reproduction, and essential organisms. Additionally, the presence of polymers in marine environments has the potential to exacerbate global warming. The presence of a shaded canopy created by these organisms inhibits plankton growth. Temperature and photooxidation are the main drivers of plastic fragmentation into microplastics (MP) on land surfaces. Microplastic particles, when fragmented, have the potential to migrate into deeper soil layers through the activities of plants and soil organisms. This process may lead to groundwater contamination and soil degradation (Scheurer & Bigalke, 2018).

5.10 Sustainable Degradation of Polyurethane Waste Through Microbial and Enzymatic Processes

The degradation capacity of the identified microorganisms and enzymes is currently limited. To establish an effective biodegradation and recycling method for polyurethane (PUR) waste, various challenges must be confronted. It is essential to develop high-throughput and standardized screening methods. Accurate detection of polyether-PUR-degrading microbes is essential. Further research is necessary to fully understand the biocatalytic degradation mechanism of polyurethane (PUR), particularly in relation to polyether-based PUR plastics. Gaining this understanding enhances the optimization of protein and strain engineering techniques. This will enhance the efficiency of polyurethane (PUR) degradation, thereby satisfying the demands of industrial applications (Liu et al., 2021). The utilization of a specific enzyme mixture containing polyesterases and urethane-degrading enzymes has the potential to significantly enhance the degradation of various polyurethane (PUR) waste within a bioreactor system. In the present context, it is recommended to employ diverse protein engineering strategies, including the newly developed machine-learning-based techniques. Additionally, a promising approach is to develop engineering techniques that can incorporate PUR-specific binding domains into the microbial degradation process of solid PUR. Improving the accessibility of the material to biocatalysts would aid in its degradation. It is essential to prioritize the development of an efficient and cost-effective method for separating degradation products from waste polyurethane (PUR). Enhancing the biocatalytic efficiency is crucial for small molecules, such as particular alcohols and acids. Prompt removal of toxic substances, such as diamines, is crucial for successful enzymatic degradation. In summary, it is essential to systematically establish and engineer microbial metabolic pathways for the purpose of converting monomers or oligomers into valuable products. The strain capable of thriving on PUR oligomers and monomers exhibits significant potential as a chassis cell for upcycling (Espinosa et al., 2020). Despite the existing technical challenges in the downstream processing of polyurethane (PUR) degradation products, the wide range of small molecules obtained presents considerable opportunities for their conversion into valuable chemicals. Efficient production of value-added chemicals necessitates the optimization of metabolic pathways and synthesis modules in genetically modified microorganisms. The Plastics Microbial Biodegradation Database (PMBD) compiles 949 documented relationships between microorganisms and plastics.

Furthermore, it provides details on 79 genes implicated in the biodegradation of plastics. Advancements in identifying and modifying microbes and enzymes that possess the ability to selectively degrade particular polymer types will enhance the recycling of mixed plastic waste fractions. Currently, these fractions are disposed of either through incineration or landfilling. Kremser et al. have developed a novel process that integrates bioleaching and enzyme-based techniques for the recycling of multilayer packaging. This exemplifies their exceptional work. The study effectively demonstrated a novel technique for extracting uncontaminated polyethylene (PE) and aluminum hydroxide from beverage cartons. The study's findings have significant implications for the potential application of biological recycling methods in multilayer materials. The findings suggest potential for developing new biocatalysts and bioprocesses. These advancements have the potential to enable cost-effective and efficient conversion and decomposition of plastics, facilitating their widespread reuse and environmentally sustainable disposal (Kumar et al., 2022).

5.11 Conclusion

The study of biodegradation of polyurethane (PU) and polyvinyl chloride (PVC) is of great importance due to its significant implications for environmental sustainability and waste management. Both polyurethane (PU) and polyvinyl chloride (PVC) are extensively utilized in multiple industries, making them major contributors to the accumulation of plastic waste and environmental pollution. Biodegradation has the potential to provide a promising solution for addressing these environmental concerns. Microbial and enzymatic processes are essential for breaking down PU and PVC, facilitating their transformation into simpler substances that are biologically compatible. These processes offer an environmentally sustainable solution by converting nonbiodegradable plastics into materials that can be naturally assimilated by ecosystems.

Nevertheless, there are still challenges to overcome in the field, including slow degradation rates and the requirement for ideal environmental conditions. To tackle these challenges, it is necessary to conduct additional research in order to improve the rates of biodegradation. This research should focus on identifying microbial strains and enzymes that are effective in breaking down the targeted substances. Additionally, it is important to optimize the conditions under which degradation occurs to ensure maximum efficiency. A comprehensive understanding of the mechanisms behind PU and PVC biodegradation, as well as the factors that influence these processes, is essential for the development of effective waste management strategies and the advancement of a circular economy. Future research should prioritize optimizing biodegradation processes and exploring innovative approaches to fully harness the potential of biodegradation for sustainable management of plastic waste. Furthermore, the field of biodegradation will be significantly advanced, and a more sustainable future will be realized through the use of interdisciplinary collaboration and innovative technologies.

References

Akutsu-Shigeno, Y., Adachi, Y., Yamada, C., Toyoshima, K., Nomura, N., Uchiyama, H., & Nakajima-Kambe, T. (2006). Isolation of a bacterium that degrades urethane compounds and characterization of its urethane hydrolase. *Applied Microbiology and Biotechnology, 70*, 422–429.

Ali, S. S., & Sun, J. (2015). Physico-chemical pretreatment and fungal biotreatment for park wastes and cattle dung for biogas production. *SpringerPlus, 4*, 1–14.

Ali, M. I., Ahmed, S., Robson, G., Javed, I., Ali, N., Atiq, N., & Hameed, A. (2014). Isolation and molecular characterization of polyvinyl chloride (PVC) plastic degrading fungal isolates. *Journal of Basic Microbiology, 54*(1), 18–27.

Ali, S. S., Elsamahy, T., Koutra, E., Kornaros, M., El-Sheekh, M., Abdelkarim, E. A., et al. (2021a). Degradation of conventional plastic wastes in the environment: A review on current status of knowledge and future perspectives of disposal. *Science of the Total Environment, 771*, 144719.

Ali, S. S., Al-Tohamy, R., Koutra, E., El-Naggar, A. H., Kornaros, M., & Sun, J. (2021b). Valorizing lignin-like dyes and textile dyeing wastewater by a newly constructed lipid-producing and lignin modifying oleaginous yeast consortium valued for biodiesel and bioremediation. *Journal of Hazardous Materials, 403*, 123575.

Alshehrei, F. (2017). Biodegradation of synthetic and natural plastic by microorganisms. *Journal of Applied & Environmental Microbiology, 5*(1), 8–19.

Anwar, M., Negi, H., Zaidi, M., Haider, G., Gupta, S., & Goel, R. (2013). Biodeterioration studies of thermoplastics in nature using indigenous bacterial consortium. *Brazilian Archives of Biology and Technology, 56*, 475–484.

Apitius, L., Rübsam, K., Jakesch, C., Jakob, F., & Schwaneberg, U. (2019). Ultrahigh-throughput screening system for directed polymer binding peptide evolution. *Biotechnology and Bioengineering, 116*(8), 1856–1867.

Artigas Alejo, J. (2008). *The role of fungi and bacteria on the organic matter decomposition process in streams: Interaction and relevance in biofilms*. Universitat de Girona.

Begum, M. A., Varalakshmi, B., & Umamagheswari, K. (2015). Biodegradation of polythene bag using bacteria isolated from soil. *International Journal of Current Microbiology and Applied Sciences, 4*(11), 674–680.

Borchert, E., Hammerschmidt, K., Hentschel, U., & Deines, P. (2021). Enhancing microbial pollutant degradation by integrating eco-evolutionary principles with environmental biotechnology. *Trends in Microbiology, 29*(10), 908–918.

Brunner, I., Fischer, M., Rüthi, J., Stierli, B., & Frey, B. (2018). Ability of fungi isolated from plastic debris floating in the shoreline of a lake to degrade plastics. *PLoS One, 13*(8), e0202047.

Bueno-Ferrer, C., Garrigós, M. C., & Jiménez, A. (2010). Characterization and thermal stability of poly(vinyl chloride) plasticized with epoxidized soybean oil for food packaging. *Polymer Degradation and Stability, 95*(11), 2207–2212.

Cangemi, J. M., Santos, A. M. D., Neto, S. C., & Chierice, G. O. (2008). Biodegradation of polyurethane derived from castor oil. *Polímeros, 18*, 201–206.

Coe, J. M., Andersson, S., & Rogers, D. B. (1997). Marine debris in the Caribbean region. In *Marine debris: Sources, impacts, and solutions* (pp. 25–33). Springer New York.

Cregut, M., Bedas, M., Durand, M., & Thouand, G. (2013). New insights into polyurethane biodegradation and realistic prospects for the development of a sustainable waste recycling process. *Biotechnology Advances, 31*(8), 1634–1647.

Darby, R. T., & Kaplan, A. M. (1968). Fungal susceptibility of polyurethanes. *Applied Microbiology, 16*(6), 900–905.

Dedisch, S., Wiens, A., Davari, M. D., Söder, D., Rodriguez-Emmenegger, C., Jakob, F., & Schwaneberg, U. (2020). Matter-tag: A universal immobilization platform for enzymes on polymers, metals, and silicon-based materials. *Biotechnology and Bioengineering, 117*(1), 49–61.

Denuncio, P., Bastida, R., Dassis, M., Giardino, G., Gerpe, M., & Rodríguez, D. (2011). Plastic ingestion in Franciscana dolphins, Pontoporia blainvillei (Gervais and d'Orbigny, 1844), from Argentina. *Marine Pollution Bulletin, 62*(8), 1836–1841.

Dutta, S., Karak, N., Saikia, J. P., & Konwar, B. K. (2010). Biodegradation of epoxy and MF modified polyurethane films derived from a sustainable resource. *Journal of Polymers and the Environment, 18*, 167–176.

Espinosa, M. J. C., Blanco, A. C., Schmidgall, T., Atanasoff-Kardjalieff, A. K., Kappelmeyer, U., Tischler, D., & Eberlein, C. (2020). Toward biorecycling: Isolation of a soil bacterium that grows on a polyurethane oligomer and monomer. *Frontiers in Microbiology, 11*, 404.

Filip, Z. (1978). Decomposition of polyurethane in a garbage landfill leakage water and by soil microorganisms. *European Journal of Applied Microbiology and Biotechnology, 5*, 225–231.

Fukushima, K., Abbate, C., Tabuani, D., Gennari, M., Rizzarelli, P., & Camino, G. (2010). Biodegradation trend of poly (ε-caprolactone) and nanocomposites. *Materials Science and Engineering: C, 30*(4), 566–574.

Gangola, S., Joshi, S., Kumar, S., & Pandey, S. C. (2019). Comparative analysis of fungal and bacterial enzymes in biodegradation of xenobiotic compounds. In *Smart bioremediation technologies* (pp. 169–189). Academic Press.

Glas, D., Hulsbosch, J., Dubois, P., Binnemans, K., & De Vos, D. (2014). End-of-life treatment of poly(vinyl chloride) and chlorinated polyethylene by dehydrochlorination in ionic liquids. *ChemSusChem, 7*(2), 610–617.

Howard, G. T., Norton, W. N., & Burks, T. (2012). Growth of Acinetobacter gerneri P7 on polyurethane and the purification and characterization of a polyurethanase enzyme. *Biodegradation, 23*, 561–573.

Huang, S. J., & Roby, M. S. (1986). Biodegradable polymers poly(amide-urethanes) [1]. *Journal of Bioactive and Compatible Polymers, 1*(1), 61–71.

Iannace, S., Nocilla, G., & Nicolais, L. (1999). Biocomposites based on sea algae fibers and biodegradable thermoplastic matrices. *Journal of Applied Polymer Science, 73*(4), 583–592.

Ibrahim, I. N., Maraqa, A., Hameed, K. M., Saadoun, I. M., & Maswadeh, H. M. (2011). Assessment of potential plastic-degrading fungi in Jordanian habitats. *Turkish Journal of Biology, 35*(5), 551–557.

Ishigaki, T., Sugano, W., Nakanishi, A., Tateda, M., Ike, M., & Fujita, M. (2004). The degradability of biodegradable plastics in aerobic and anaerobic waste landfill model reactors. *Chemosphere, 54*(3), 225–233.

Islam, S., Apitius, L., Jakob, F., & Schwaneberg, U. (2019). Targeting microplastic particles in the void of diluted suspensions. *Environment International, 123*, 428–435.

Karimi, M., & Biria, D. (2019). The promiscuous activity of alpha-amylase in biodegradation of low-density polyethylene in a polymer-starch blend. *Scientific Reports, 9*(1), 2612.

Kay, M. J., Morton, L. H. G., & Prince, E. L. (1991). Bacterial degradation of polyester polyurethane. *International Biodeterioration, 27*(2), 205–222.

Khan, S., Nadir, S., Shah, Z. U., Shah, A. A., Karunarathna, S. C., Xu, J., et al. (2017). Biodegradation of polyester polyurethane by Aspergillus tubingensis. *Environmental Pollution, 225*, 469–480.

Kim, J. H., Choi, S. H., Park, M. G., Park, D. H., Son, K. H., & Park, H. Y. (2022). Biodegradation of polyurethane by Japanese carpenter bee gut-associated symbionts Xanthomonas sp. HY-71, and its potential application on bioconversion. *Environmental Technology & Innovation, 28*, 102822.

Koller, M., & Braunegg, G. (2018). Advanced approaches to produce polyhydroxyalkanoate (PHA) biopolyesters in a sustainable and economic fashion. *The EuroBiotech Journal, 2*(2), 89–103.

Kumar, V., Mitra, D., Rani, A., Suyal, D. C., Gautam, B. P. S., Jain, L., & Soni, R. (2022). *Bio-inoculants for biodegradation and bioconversion of agrowaste: Status and prospects* (pp. 351–367). Emerging Trends and Strategies.

Liu, J., He, J., Xue, R., Xu, B., Qian, X., Xin, F., Blank, L. M., Zhou, J., Wei, R., Dong, W., & Jiang, M. (2021). Biodegradation and up-cycling of polyurethanes: Progress, challenges, and prospects. *Biotechnology Advances, 48*, 107730.

Liu, J., Zeng, Q., Lei, H., Xin, K., Xu, A., Wei, R., Li, D., Zhou, J., Dong, W., & Jiang, M. (2023). Biodegradation of polyester polyurethane by Cladosporium sp. P7: Evaluating its degradation capacity and metabolic pathways. *Journal of Hazardous Materials, 448*, 130776.

Mohanan, N., Montazer, Z., Sharma, P. K., & Levin, D. B. (2020). Microbial and enzymatic degradation of synthetic plastics. *Frontiers in Microbiology, 11*, 580709.

Moore, C. J. (2008). Synthetic polymers in the marine environment: A rapidly increasing, long-term threat. *Environmental Research, 108*(2), 131–139.

Nakajima-Kambe, T., Onuma, F., Kimpara, N., & Nakahara, T. (1995). Isolation and characterization of a bacterium which utilizes polyester polyurethane as a sole carbon and nitrogen source. *FEMS Microbiology Letters, 129*(1), 39–42.

Negoro, S. (2000). Biodegradation of nylon oligomers. *Applied Microbiology & Biotechnology, 54*(4), 461–466.

Nir, M. M., Miltz, J., & Ram, A. (1993). Update on plastics and the environment: Progress and trends. *Plastics Engineering (USA), 49*(3), 75–93.

Oprea, S. (2010). Dependence of fungal biodegradation of PEG/castor oil-based polyurethane elastomers on the hard-segment structure. *Polymer Degradation and Stability, 95*(12), 2396–2404.

Owen, S., Otani, T., Masaoka, S., & Ohe, T. (1996). The biodegradation of low-molecular-weight urethane compounds by a strain of Exophiala jeanselmei. *Bioscience, Biotechnology, and Biochemistry, 60*(2), 244–248.

Peng, Y., Shih, Y., Lai, Y., Liu, Y., Liu, Y., & Lin, N. (2014). Degradation of polyurethane by bacterium isolated from soil and assessment of polyurethanolytic activity of a Pseudomonas putida strain. *Environmental Science and Pollution Research, 21*(16), 9529–9537.

Peng, B., Chen, Z., Chen, J., Yu, H., Zhou, X., Criddle, C. S., Wu, W., & Zhang, Y. (2020). Biodegradation of Polyvinyl Chloride (PVC) in Tenebrio molitor (Coleoptera: Tenebrionidae) larvae. *Environment International, 145*, 106106.

Phua, S. K., Castillo, E., Anderson, J. M., & Hiltner, A. (1987). Biodegradation of a polyurethane in vitro. *Journal of Biomedical Materials Research, 21*(2), 231–246.

Raghavendra, V. B., Uzma, M., & Govindappa, M. (2016). Screening and identification of polyurethane (PU) and low density polyethylene (LDPE) degrading soil fungi isolated from municipal solid waste. *International Journal of Current Research, 8*(07), 34753–34761.

Roy, P. K., Titus, S., Surekha, P., Tulsi, E., Deshmukh, C., & Rajagopal, C. (2008). Degradation of abiotically aged LDPE films containing pro-oxidant by bacterial consortium. *Polymer Degradation and Stability, 93*(10), 1917–1922.

Ruiz, C., Main, T., Hilliard, N. P., & Howard, G. T. (1999). Purification and characterization of twopolyurethanase enzymes from Pseudomonas chlororaphis. *International Biodeterioration & Biodegradation, 43*(1–2), 43–47.

Russell, J. R., Huang, J., Anand, P., Kucera, K., Sandoval, A. G., Dantzler, K. W., et al. (2011). Biodegradation of polyester polyurethane by endophytic fungi. *Applied and Environmental Microbiology, 77*(17), 6076–6084.

Sah, A., Negi, H., Kapri, A., Anwar, S., & Goel, R. (2011). Comparative shelf life and efficacy of LDPE and PVC degrading bacterial consortia under bioformulation. *Ekologija, 57*(2). https://doi.org/10.6001/ekologija.v57i2.1885

Sarkar, S., Singha, P. K., Dey, S., Mohanty, M., & Adhikari, B. (2006). Synthesis, characterization, and cytotoxicity analysis of a biodegradable polyurethane. *Materials and Manufacturing Processes, 21*(3), 291–296.

Scheurer, M., & Bigalke, M. (2018). Microplastics in Swiss floodplain soils. *Environmental Science & Technology, 52*(6), 3591–3598.

Secchi, E. R., & Zarzur, S. (1999). Plastic debris ingested by a Blainville's beaked whale, Mesoplodon densirostris, washed ashore in Brazil. *Aquatic Mammals, 25*(1), 21–24.

Shah, Z. A., Gulzar, M., Hasan, F., & Shah, A. A. (2016). Degradation of polyester polyurethane by an indigenously developed consortium of Pseudomonas and Bacillus species isolated from soil. *Polymer Degradation and Stability, 134*, 349–356.

Shibasaki, S., Kawabata, A., Tanino, T., Kondo, A., Ueda, M., & Tanaka, M. (2009). Evaluation of the biodegradability of polyurethane and its derivatives by using lipase-displaying arming yeast. *Biocontrol Science, 14*(4), 171–175.

Skariyachan, S., Taskeen, N., Kishore, A. P., & Krishna, B. V. (2022). Recent advances in plastic degradation – From microbial consortia-based methods to data sciences and computational biology driven approaches. *Journal of Hazardous Materials, 426*, 128086.

Stern, R. V., & Howard, G. T. (2000). The polyester polyurethanase gene (pueA) from Pseudomonas chlororaphis encodes a lipase. *FEMS Microbiology Letters, 185*(2), 163–168.

Van Tilbeurgh, H., Tomme, P., Claeyssens, M., Bhikhabhai, R., & Pettersson, G. (1986). Limited proteolysis of the cellobiohydrolase I from Trichoderma reesei: Separation of functional domains. *FEBS Letters, 204*(2), 223–227.

Wang, Z., Xin, X., Shi, X., & Zhang, Y. (2020). A polystyrene-degrading Acinetobacter bacterium isolated from the larvae of Tribolium castaneum. *Science of the Total Environment, 726*, 138564.

Wei, R., Oeser, T., Then, J., Kühn, N., Barth, M., Schmidt, J., & Zimmermann, W. (2014). Functional characterization and structural modeling of synthetic polyester-degrading hydrolases from Thermomonospora curvata. *AMB Express, 4*, 1–10.

Wei, R., Tiso, T., Bertling, J., O'Connor, K., Blank, L. M., & Bornscheuer, U. T. (2020). Possibilities and limitations of biotechnological plastic degradation and recycling. *Nature Catalysis, 3*(11), 867–871.

Wilcox, C., Van Sebille, E., & Hardesty, B. D. (2015). Threat of plastic pollution to seabirds is global, pervasive, and increasing. *Proceedings of the National Academy of Sciences, 112*(38), 11899–11904.

Wool, R. P., Raghavan, D., Wagner, G. C., & Billieux, S. (2000). Biodegradation dynamics of polymer–starch composites. *Journal of Applied Polymer Science, 77*(8), 1643–1657.

Yoshida, S., Hiraga, K., Takehana, T., Taniguchi, I., Yamaji, H., Maeda, Y., et al. (2016). A bacterium that degrades and assimilates poly (ethylene terephthalate). *Science, 351*(6278), 1196–1199.

Chapter 6
Approaches to Degrading Polystyrene (PS) Using Diverse Microorganisms

Hemant Dasila, Bharti Kukreti, Seema Bisht, Damini Maithani, Viabhav Kumar Upadhayay, and Shubham Pandey

Abstract Use of plastics in the modern world has significantly increased and its persistent nature in an environment is still a major area of concern. Polystyrene (PS) is a synthetic derivative of plastic which is composed of styrene monomer units, and the presence of phenol group in the styrene monomer contributes to its stability. PS is used in large quantities as it is used for various packaging purposes in industry, but the major area of concern is its persistent nature in an environment. In addition to various physical and chemical techniques, the biodegradation method using microorganisms has also been employed. Numerous bacterial and fungal groups have been reported to possess potential to biodegrade PS. Microorganism-mediated PS degradation involves destruction of the core styrene structure which results in lowering molecular mass. Microorganism-mediated enzymatic degradation of PS is also one of the popular methods of biodegradation. Different techniques like scanning electron microscopy, transmission electron microscopy, biogas production, and determination of molar mass have been used to detect the biodegradation of PS.

Keywords Polystyrene · Scanning electron microscopy · Biodegradation · Biogas

H. Dasila (✉)
Department of Microbiology, Akal College of Basic Sciences, Eternal University, Sirmaur, Himachal Pradesh, India

B. Kukreti · S. Bisht · S. Pandey
Department of Microbiology, College of Basic Sciences, G.B.P.U.A.&T., Pantnagar, Uttarakhand, India

D. Maithani
Department of Microbiology, School of Bioengineering and Bioscience, Lovely Professional University, Phagwara, Punjab, India

V. K. Upadhayay
Department of Microbiology, College of Basic Sciences & Humanities, Dr. Rajendra Prasad Central University, Pusa, Samastipur, Bihar, India

R. Soni et al. (eds.), *Advanced Strategies for Biodegradation of Plastic Polymers*, https://doi.org/10.1007/978-3-031-55661-6_6

6.1 Introduction

The demand for plastic materials for day-to-day usage keeps on increasing, and it would be very fair to accept the fact that the modern world without plastics would not exist. One major disadvantage of using plastic is its persistence in an environment. Plastic and its derivatives are resistant to natural biodegradation processes, and the physical methods that are employed to degrade them are not only expensive but also not safe for the environment. Production of plastics is easy and cheap; thus, there is sharp increase in the global production of plastics in earth (Geyer et al., 2017; Plakunov et al., 2020; Lau et al., 2020). A major significant category which caused a spike in plastic production is the disposable items made of plastics. In 2015, the annual production of plastics worldwide was 350 million tons which is very significant from the year 1950, in which the plastic production was only 2 million tons (Geyer et al., 2017). In 2019, the production rate was increased up to 368 million tons. Overall, from the year 1950 to 2015, there was 8300 million tons of plastics produced.

Polystyrene (PS) is a synthetic derivative of plastic composed of a styrene monomer which is an aromatic hydrocarbon in nature [$(C_8H_8)n$]. PS has a high degree of polymerization and could be in the range from 600 to 2500 monomer units (Fig. 6.1). The presence of phenol makes styrene to have a stable crystalline structure. The properties of PS depend upon the positioning of the phenyl group. A standard PS polymer that is recommended for commercialization purposes has a phenyl group which is arranged in both sides of the chain, and this arrangement prevents PS crystallization.

PS is a rigid polymer with high optical transmission and weak mechanical strength. PS has a low density with good dielectric properties along with good cold resistance of up to −40 °C. PS is a homo-polymer with a molecular weight of 100,000–400,000 g/mol. PS is used for manufacturing different products like packaging stuffs, disposable tableware, disposable cups, cutlery, food containers, and compact disks (Yang et al., 2015b). Due to such wider use of plastics and its derivative like PS, plastics have accumulated in large amount in the environment causing different diseases that are related to human health, polluting the environment, and

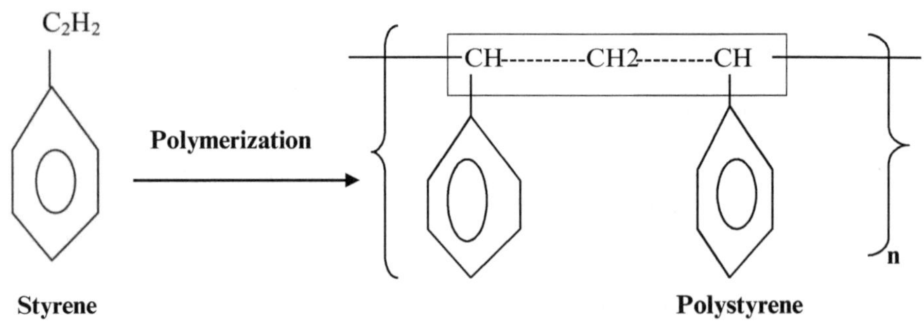

Fig. 6.1 Polystyrene polymerization from a styrene unit

making significant changes in the soil ecosystem. The PS material can be recycled but most of the PS foam ends in landfill. The major problem of recycling PS is its transportation cost, and because of this, the recycling rate of PS is now increasing in some developed countries like the United States.

6.2 Polystyrene's Persistent Nature

PS is generally a very durable and a prominent nonbiodegradable thermoplastic. PS biodegradation actually occurs in the environment but at a very slow rate, which results in the long persistence of PS in the form of solid waste (Kotova et al., 2021). Ho et al. (2018a, b) reported the same in which PS biodegradation in soil is less than 1%, and even after 90 days, there was no such significant increase in the degradation rate. A similar report has been observed in which a PS sheet was buried in soil for about 32 years and yet there was no significant degradation in PS (Otake et al., 1995). The hydrophobic character in PS is the most significant reason for its resistance in an environment. The molecular weight of a compound affects the hydrophobicity of PS. The hydrophobic pattern in PS also affects the microbial attachment proficiency. PS generally has a high molecular weight and a poor water-soluble agent which account for its tough resistance against microorganisms to utilize it as a source of carbon and for energy purposes. Biological processes via release of extracellular enzymes can occur outside the cell, but these enzymes are too large to penetrate inside the polymer structure (Palmisano & Pettigrew, 1992). During manufacturing of PS, the use of different chemical like flame retardants, processing lubricant, stabilizers, and antioxidant (bisphenol A) provides resistance to PS against chemical oxidation and biodegradation, and this contributes significantly to persistent nature of PS. Other additives that are used as stabilizing materials in thermoplastic preparation include antimicrobial agents, dyes, and pigments (Saron & Felisberti, 2006). Some reports suggest that the use silver nanoparticles has been done for its antibacterial properties, and similar types of nanoparticles are being used in plastic manufacturing which offer resistance against natural plastic biodegradation processes mediated by bacteria (Ho et al., 2018a, b). The list of some important additives which provides stability to PS is given in Table 6.1. Microbial degradation is also reported in electronic waste which offers the same kind of resistance against degradation (Dasila et al., 2023c; Debbarma et al., 2023).

Table 6.1 List of chemical additives used to provide stability to thermoplastics

S. no.	Additive	Function
1.	Tinuvin P	UV stabilizer
2.	Acrawax	Processing lubricant
3.	Decabromodiphenyl oxide	Heat resistant
4.	Saytex (BCL-462)	Heat resistant
5.	Irganox 1076	Antioxidant
6.	Zinc stearate	Processing lubricant

6.3 PS Degradation Studies

Biodegradation of PS has been previously studied, and some serious efforts have been done to investigate the effect of microorganisms on PS and thermoplastic degradation. However, biodegradation of PS is complex and not fully understood. Shimpi et al. (2012) reported the biodegradation of PS using *Pseudomonas aeruginosa*. Isolated *Rhodococcus ruber* strain was reported to biodegrade three different forms of PS (PS flakes, PS powder, and microtiter plates developed from PS). Most of the bacterial isolates first adhere to the PS surface and then form a thick biofilm around the surface which results in the reduction of the weight of PS, suggesting bacterial-mediated biodegradation. Motta et al. (2009) used *Curvularia* species to determine the rate of biodegradation of atactic PS. Reports suggest that hyphae adhere to the PS surface followed by the penetration of the PS surface, and dusting of spores has also been observed in the entire treated sample after 9 weeks. The positive outcome of using this approach is that tracking of metabolic pathway is easy which can be different in different environmental conditions. However, there is one disadvantage too, which ignores the maximum possibility of PS biodegradation via cooperative process, i.e., consortium-mediated PS biodegradation. All these studies have contributed to the fact that microorganisms can be very useful tools to biodegrade PS and other thermoplastics.

PS biodegradation studies have also been observed underwater. In water, PS grafted polymer was synthesized and subsequently added in water, furthering the biodegradation rate, which has been observed by measuring the mass. Reduction in the mass of PS from 45.8% to 93.1% after 27 days indicates the PS biodegradation in water (Nikolic et al., 2014). Different techniques have been used to determine the PS biodegradation like Fourier transform infrared (FTIR) spectroscopy, scanning electron microscopy (SEM), thermogravimetric analysis (TGA), nuclear magnetic resonance (NMR) spectroscopy, gas chromatography/mass spectrometry, and X-ray photoelectron spectroscopy. Degradation in the average molecular weight is a clear indication of PS degradation. In addition to that is the change in molecular pattern which is a measurement of bulk deterioration of polymeric surface. A similar study on biodegradation of high-grade PS was done with starch as a carrier material and was put under the 3 L vessel for 12 weeks in aerobic condition with proper aeration. The results showed that inclusion of starch as a carrier material accelerated the structural molecular changes. Therefore, the approach of PS biodegradation using real environmental conditions such as river water, soil, and activated sludge is a unique method. In general, the rate of indigenous biodegradation is slow, but using different additives like starch can accelerate the biodegradation process. To promote the biodegradation rate of PS in an environment, modifying PS by blending or grafting using copolymers and addition of prooxidant such as metal salt like manganese, iron, and cobalt increases the rate of PS oxidation in an indigenous environment. Prooxidant also promotes PS biodegradation rate by adding an oxygen-containing group that can be further used by microorganisms for PS biodegradation. The rate

Table 6.2 List of PS types and its biodegradation causative agent

S. no.	PS type	Microorganisms used	Techniques for monitoring	References
1.	High-impact PS	*Enterobacter* and *Citrobacter*	Fourier transform infrared (FTIR) spectroscopy, scanning electron microscopy (SEM), and thermogravimetric analysis (TGA).	Park and Kim (2019)
2.	High-impact PS	*Bacillus* spp. and *Pseudomonas* spp.	FTIR, TGA, and nuclear magnetic resonance (NMR) spectroscopy	Sarmah and Rout (2019)
3.	PS (disposable plates)	*Pseudomonas aeruginosa*, *Staphylococcus aureus*, and *Aspergillus niger*	Weight loss percentage	Mohan et al. (2016)
4.	PS films of 0.02 mm thickness	*Exiguobacterium* sp. strain YT2	Gas chromatography/mass spectrometry (GC/MS), SEM, and X-ray photoelectron spectroscopy	Skariyachan et al. (2018)
5.	Modified PS	*Pseudomonas aeruginosa*	FTIR	Delacuvellerie et al. (2019)

of PS biodegradation also depends upon the type of microorganisms used, and the combination of Gram-positive and Gram-negative bacteria is also used to biodegrade PS. It has been reported that Gram-positive cocci-shaped (*Bacillus*) and Gram-negative rod-shaped (*Pseudomonas*) bacteria reduce PS mass by 30% in garbage soil (Montazer et al., 2020; Park & Kim, 2019; Delacuvellerie et al., 2019). The list of PS biodegrading associated microorganisms with their respective monitoring techniques is shown in Table 6.2.

6.4 Analytical Method for Monitoring PS Degradation

There are different analytical methods that are used to monitor the degradation pattern in PS structure. Some of the prominent methods are summarized below

6.4.1 Changes in Morphology Properties

Morphology in the PS structure can be monitored via observing the surface color and roughening of the surface using techniques like scanning electron microscopy (SEM) and atomic force microscopy (AFM) (Joshi et al., 2022). Changes in the polyethylene structure by SEM and AFM have been observed using the two strains

Bacillus sp. YP1 and *Enterobacter asburiae* YT1 isolated from the worm's gut (Yang et al., 2014). To understand the basic mechanisms, several techniques are used like X-ray powder diffraction (XRD), transmission electron microscopy (TEM), and scanning force microscopy (SFM) (Lucas et al., 2008).

6.4.2 Changes in Molar Mass and Weight Loss

Plastic degradation measurements have been regularly monitored via measuring weight loss. Broadening of molecular weight distribution occurred during plastic degradation. To analyze the molecular weight distribution and the average molecular weight (Mw) during plastic degradation, gel permeation chromatography (GPC) is performed. GPC provides great insight of the break in chain length (Raddadi & Fava, 2019). Mass spectrometry (MS) is the technique used to detect the metabolic intermediate that occurs during plastic biodegradation (Kundu et al., 2014).

6.4.3 Analyzing Functional Group Dynamics

To measure the chemical modification in the functional group during plastic biodegradation, Fourier transform infrared spectroscopy (FTIR) is performed (Rajandas et al., 2012). A 1710–1715 cm^{-1} increase in absorbance has been observed during polyethylene (PE) biodegradation (Ghatge et al., 2020). However, one significant disadvantage of using FTIR is that new functional groups could be generated due to the microbial attachment on the surface or due to the release of chemical additives. In addition to that, X-ray photoelectron spectroscopy (XPS) has been frequently used to identify new functional groups that have been added during plastic degradation (Tennakoon et al., 2020). For instance, at 288.2 eV on XPS spectra, carbonyl functional group has been observed during polyethylene biodegradation (Morgen et al., 2020).

6.4.4 Determination of Biogas Production

Respirometry or gas chromatography, which is performed for the production of biogas during PS biodegradation, is considered as an important technique to measure the degradation rate of plastics (Kundu et al., 2014). The release of carbon dioxide during PS biodegradation can be entrapped using a sodium hydroxide solution to form sodium carbonate, which is a measure of total organic carbon (Meyer-Cifuentes et al., 2020).

6.5 Free-Living Microorganism's Diversity Mediating Plastic Biodegradation

Biodegradation of plastic is not an easy process and it takes time. However, potential microorganisms have been isolated from time to time from different sites like soil, compost, activated sludge, and seawater (Montazer et al., 2020). For instance, strains of *Acinetobacter pittii*, *Pseudomonas putida*, and *Micrococcus luteus* have been isolated from soil and reported to show bio-fragmentation via utilizing plastics as a whole carbon and energy source (Montazer et al., 2019). Members of more than 20 different bacteria genera have been reported to degrade plastics and its synthetic derivatives like polyethylene (Ghatge et al., 2020). *Pseudomonas* strains are metabolically very versatile in nature, which make them potential candidates for biodegradation and prominent biofertilizers (Dasila et al., 2023a, b). Biodegradation of plastic derivates, i.e., polyvinyl chloride (PVC), has been reported, in which the potential degraders were fungi. Fungi (*Aspergillus niger*, *A. sydowii*, *Phanerochaete chrysosporium*, and *Lentinus tigrinus*) have been reported to degrade PVC, which is later confirmed by SEM data and by measuring the fungal biomass (Ali et al., 2014). Diversity in the fungal groups with potential degrading abilities depends upon the enrichment of the plastic substrate in the source environment. For example, soil contaminated with PVC plastic derivatives has the potential fungi degraders like *Mucor* sp. and *Penicillium* sp., which is later confirmed by infrared spectrometry (Pardo-Rodríguez & Zorro-Mateus, 2021). Bio-augmentation is a similar type of method of bioremediation in which soil is applied with substrate of interest where a potential bio-degrader using the same substrate for carbon and energy source results in its biodegradation.

PS biodegradation has also been studied on larvae of two mealworm species, i.e., *T. obscures* and *T. molitor* (Yang et al., 2015a; Peng et al., 2019). Using radioisotopic analysis showed that the PS foam was metabolized to carbon dioxide using a radioisotopic-labeled PS carbon (13C). The PS reduction observed was found to be 47.7% and 49.2% of the remaining PS was excreted, and some minor parts were incorporated into the lipid portion. GC-MS analysis results showed that certain intermediate metabolites were produced which are known to be associated with PS degradation. During succession of larval stage, it was observed that initially the predominant larvae were enterobacteria and lactobacteria which were replaced by *enterococci* and spiroplasma, respectively, during further succession. Thirty bacterial isolates associated with gut microbiota were subsequently identified using molecular tools and were incubated in the presence of PS for 28 days. During the incubation period, it was observed that there is significant increase in the strains of *Enterobacter hormaechei*, *Bacillus cereus*, *Chryseobacterium* sp., *Exiguobacterium*, and *E. faecalis* (Kotova et al., 2021). Bioinoculants are not only restricted to PS biodegradation, but they have been known to biodegrade more complex substances like E-waste (Dasila et al., 2023c).

6.6 Studies on PS Degradation Mediated by Microorganisms

The carbon-carbon structure in PS renders it resistant to natural biodegradation. Furthermore, given that plastics have only recently entered the environment within the last few decades, specialized enzymatic mechanisms within microbial hosts for breaking them down may not have been developed. While certain microorganisms capable of biodegrading polystyrene (PS) have been identified, the precise mechanism governing PS degradation, as well as the enzyme and biodegradation pathways that are responsible, remain largely unexplored. The integration of both biotic and abiotic elements is seen as a viable approach for decomposing PS polymers as well (Zhang et al., 2022a, b). Some researchers have put forth potential enzymes for PS degradation, like esterases, oxidative enzymes, and lipases. Notably, the strength of PS's C–C bonds diminishes in either acidic or alkaline conditions, suggesting that alkane hydroxylases might cleave the primary C–C bonds of PS in such circumstances (Xu et al., 2019). However, it is worth noting that the presence of bulky styrene groups within the active site of these enzymes could impede the enzymatic cleavage of C–C bonds. In a study by Przemieniecki et al. (2020), various enzymes, namely, naphthol-AS-BI-phosphohydrolase, β-galactosidase, β-glucuronidase, acid phosphatase, alkaline phosphatase, and leucine arylamidase, exhibited elevated functions within the bacterial community of the gut of *Tenebrio molitor* (mealworms). This observation suggests that nature may harbor depolymerase enzymes with known enzymatic classes capable of acting on extended alkane chains or polymers that include aromatic components, hinting at the potential existence of enzymatic capability for breaking down PS polymers in natural environments.

The depolymerization of polystyrene (PS) yields its constituent monomer, known as styrene, which can be readily subjected to biodegradation by various microorganisms, including *Pseudomonas*, *Rhodococcus*, *Xanthobacter*, *Nocardia*, and brown-rot fungus (*Gloeophyllum trabeum*) through specific styrene catabolism pathways (Parthasarathy et al., 2022). Styrene undergoes aerobic degradation via two primary routes, one targeting the vinyl moiety and the other entailing a direct cleavage of the ring. When in anaerobic conditions, styrene undergoes a conversion process into phenylacetic acid (PAA) (O'Leary et al., 2002), which is subsequently converted into benzoic acid (Grbicgalic et al., 1990), providing a potential avenue for complete PS biodegradation, contingent upon the identification and optimization of the depolymerization enzyme responsible. In the context of *Pseudomonas*, the degradation of styrene is primarily achieved through two well-defined pathways: the upper pathway and the lower pathway (Johnston et al., 2018). In the upper sequence, styrene gets oxidized through the activity of styrene monooxygenase, resulting in the creation of styrene oxide, which necessitates the involvement of reducing electrons. Following this, styrene oxide undergoes isomerization to become phenylacetaldehyde via the actions of styrene oxide isomerases. Phenylacetaldehyde is subsequently subjected to oxidation by phenylacetaldehyde dehydrogenase, ultimately producing phenylacetates (Fig. 6.2). The lower route encompasses the binding of phenylacetates with coenzyme A, resulting in the formation of coenzyme phenylethyl, which is then processed by

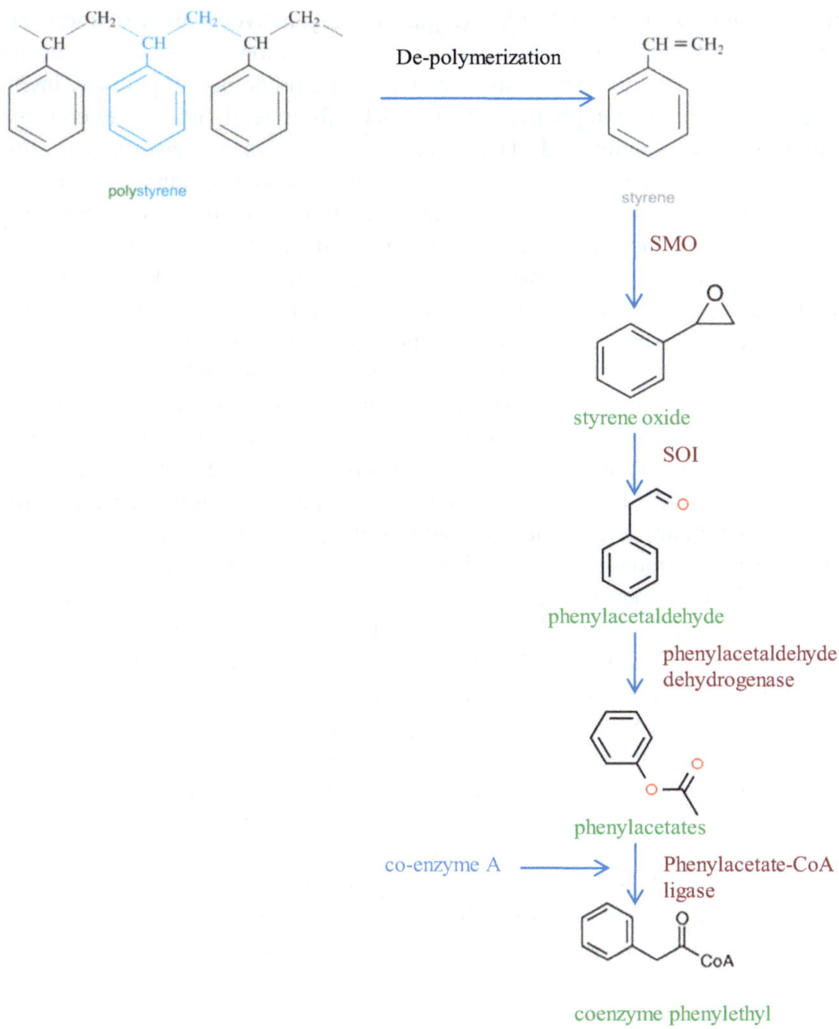

Fig. 6.2 Enzymatic degradation mechanisms of PS mediated by *Pseudomonas* strain

specialized enzymes to generate acetyl-CoA inside the tricarboxylic acid cycle (Yanto, 2019).

The biodegradation of polystyrene (PS) is a multifaceted process that initiates once microorganisms establish colonies on PS particles and secrete enzymes that facilitate the fragmentation of the polymer into smaller molecular units, including monomers, oligomers, and dimmers (Miloloža et al., 2022; Ho et al., 2018a, b). Interestingly, certain microorganisms can utilize styrene itself as a source of carbon for their growth. For instance, *Rhodococcus ruber* has been demonstrated to form biofilms on PS surfaces (Mor & Sivan, 2008) and exhibit partial degradation of the polymer. The rate of biodegradation is contingent upon factors such as the thickness of the PS material and its molecular weight. Notably, a diverse array of microorganisms possesses the capacity to facilitate the biodegradation of styrene. Various pathways exist for the catabolism of styrene, although a predominant pathway involves the oxidation of styrene to phenylacetate, which subsequently undergoes conversion through the tricarboxylic acid (TCA) cycle. Microorganisms have the capacity to incorporate polystyrene (PS) into their metabolic processes because they can utilize styrene as a source of carbon. Additionally, the molecular structure of polystyrene closely bears a likeness to lignin, which, in conjunction with other complex aromatic hydrocarbons such as chrysene, pyrene, phenanthrene, and benzo[a]pyrene, can undergo degradation through the action of ligninolytic enzymes (Yanto, 2019). Ligninolytic enzymes are oxidative enzymes that lack specificity and are triggered by mediators that are not proteins, and they have a pivotal role in breaking down lignin. Various enzymatic pathways have been proposed for the biodegradation of PS, involving enzymes such as laccases, oxidoreductases, hydroquinone, alkane hydroxylases, lipases, and P450 monooxygenases (Liyuan & Erica, 2021; Wei & Zimmermann, 2017). These pathways encompass the conversion of PS into carboxylic acids, which are further metabolized through β-oxidation and the Krebs cycle. One suggested degradation pathway involves the conversion of styrene into styrene epoxide through the catalytic action of SMO (styrene monooxygenase). Subsequently, this styrene epoxide undergoes further transformation into 4-maleylacetoacetate, facilitated by a series of enzymes such as SOI (styrene oxide isomerase), PAALDH (phenylacetaldehyde dehydrogenase), PAAH (phenylacetate hydroxylase), HPAAH (2-hydroxyphenylacetate hydroxylase), and HGADO (homogentisate 1,2-dioxygenase). In the β-oxidation pathway, 4-maleylacetoacetate is transformed into acetyl-CoA, which then becomes integrated into the TCA cycle, contributing to important biosynthesis pathways at the core of cellular processes (Ho et al., 2018a, b; Ru et al., 2020; Ali et al., 2021; Miloloža et al., 2022; Zhang et al., 2022a, b). Alternatively, in a second proposed biodegradation pathway for polystyrene (PS), styrene undergoes hydroxylation through the action of SDO (styrene dioxygenase) on its aromatic ring, yielding the formation of styrene cis-glycol. This styrene cis-glycol compound can undergo further metabolic processes to convert into acetyl-CoA through a series of enzymatic reactions involving CGDH (cis-glycol dehydrogenase), CDO (catechol 2,3-dioxygenase), HMASALDH (2-hydroxymuconic acid semialdehyde hydrolase), HPDEH (2-hydroxypenta-2,4-dienoate hydratase), HOA (4-hydroxy-2-oxovalerate aldolase), and the PDHC (pyruvate dehydrogenase complex). In a similar manner, acetyl-CoA subsequently enters the TCA cycle, contributing to processes like biomass synthesis or the accumulation of other metabolites, such as PHAs (polyhydroxyalkanoates) (Ho et al., 2018a, b; Ru et al., 2020; Ali et al., 2021; Miloloža et al., 2022; Zhang et al., 2022a, b).

An alternative proposed biodegradation route for polystyrene (PS) begins with the initial breaking of the substituent in PS polymers. Enzymes like aromatic ring hydroxylases and monooxygenases are potential contenders for dismantling the aromatic ring component of PS. Nevertheless, the precise details of this degradation pathway and the specific enzymes involved remain incompletely understood.

Tischler et al. (2009) have illuminated an additional oxygenation of the substituent moiety process within *Rhodococcus opacus* 1CP. This process converts styrene into styrene oxide through the action of the enzyme SMO (styrene monooxygenase). Subsequently, an isomerization step facilitated by SOI (styrene oxide isomerase) takes place, resulting in the production of phenylacetaldehyde. Additional chemical changes to styrene encompass its conversion via reduction to ethylbenzene, the introduction of hydroxyl groups to the side chain, resulting in the formation of either 1-phenylethanol or 2-phenylethanol, and its transformation into phenylacetic or benzoic acids via anaerobic degradation routes. In the case of *Rhodococcus rhodochrous* NCIMB 13259, it employs ring dioxygenation to break down styrene, generating vinylcatechol, which is subsequently processed into acetaldehyde and pyruvate (Warhurst et al., 1994). These intricate pathways and mechanisms collectively highlight the diverse strategies employed by various microorganisms to catabolize styrene, a crucial step in the potential biodegradation of PS. Understanding these processes at the molecular level and identifying key enzymes involved in depolymerizing PS can aid in the advancement of more effective and environmentally friendly approaches for managing plastic waste.

6.7 Mechanism of Polystyrene Biodegradation

Plastic products have revolutionized our daily lives due to their light weight, exceptional chemical stability, high durability, and cost-effectiveness (Gu et al., 2020). These products can decompose into microparticles (<5 mm) through the collaborative effects of abiotic (including chemical and physical) and biotic mechanism (involves living organisms) in the natural environment termed as microplastics (MPs) (Thompson et al., 2004). Worldwide, efforts have been made by all countries to mitigate and reduce MP pollution at its source. Polystyrene (PS), whether in the form of microparticles or nanoparticles, has been identified as toxic to animals, humans, and plants. Aquatic animals, particularly, have been extensively studied for their sensitivity to PS toxicity (Wang et al., 2019; Zhu et al., 2020; Qiang & Cheng, 2021). Therefore, PS biodegradation is crucial due to the environmental and health concerns linked with the widespread utilization and disposal of polystyrene products. Two predominant approaches for PS decomposition are advanced oxidation processes (AOPs) and biological degradation. These mechanisms break down the polymer's chemical bonds into smaller molecules that can eventually be transformed into useful products or even completely mineralized into CO_2 and H_2O (Silva et al., 2018). The adaptation of microorganisms to MP environments has been seen to enhance the biodegradation process. A wide array of organisms, covering diverse groups and species, engage in degradation studies. These studies have highlighted the significance of gut microflora (bacteria and fungi) in facilitating PS degradation within the digestive tracts of these organisms (Peng et al., 2020; Cucini et al., 2020; Billen et al., 2020). For example, gut flora of insect larvae including strains of *Bacillus*, *Klebsiella*, and *Enterobacter* have shown the ability to degrade

PS (Yang et al., 2014), raising hopes for the degradation of other plastic wastes. An enzyme-mediated mechanism involving isolation of *Pseudomonas* strain DSM 50,071 from beetle larval gut known as *Zophobas atratus* has also been reported, with serine hydrolase (SH) being the key player in PS degradation (Kim et al., 2020). The simplest approach of PS polymer degradation is believed to be through the interaction of abiotic and biotic elements (Qin et al., 2021). PS, with its aromatic durable thermoplastic nature and C–C backbone, naturally degrades at a slow pace (Ho et al., 2018a, b). Its hydrophobic properties hinder dissolution in water, thereby limiting interaction with enzymes. PS biodegradation is initiated when microorganisms colonize the PS surface, rendering it more hydrophilic and secreting enzymes which results in the production of relatively small molecular fragments known as oligomers and monomers. Mor and Sivan (2008) reported that when the Gram-positive bacterium *Rhodococcus ruber* forms a slimy layer on the PS surface, its partial degradation starts, while a filamentous fungus (*Curvularia*) had to attached its hyphae and penetrate the PS structure for its degradation (Motta et al., 2009). The enzymes produced by microorganisms precisely target the molecular structure of MPs, breaking them down into monomers that can be utilized for the generation of energy and nutrients for microbes (Islam et al., 2019; Ganesh Kumar et al., 2020; Kumar et al., 2022). The destructive metabolism of styrene, a component of PS, involves various enzymatic pathways, such as its oxidation, leading to the generation of phenylacetate, which further enters the tricarboxylic acid cycle (TCA) or citric acid cycle (Danso et al., 2019).

In 1997, the initial discovery of enzyme involved in PS degradation was documented. This enzyme, hydroquinone peroxidase sourced from the bacterium *Azotobacter beijerinckii* HM121, is asserted to possess the capability to break down PS. The transformation of water-insoluble PS into a smaller, water-soluble molecule involves the use of dichloromethane. Subsequently, hydroquinone peroxidase is responsible for degrading the water-soluble PS (Nakamiya et al., 1997). Over 20 years later, a study presented findings on the degradation of PS, which encompassed a series of enzymatic reactions (Ho et al., 2018a, b; Danso et al., 2019). Degradation is initiated from splitting of either the primary or the adjacent chain, leading to distinct degradation pathways (Hou & Majumder, 2021). Cleaving the primary chain might produce diverse ring compounds, including toluene and styrene. Subsequently, they follow two destructive metabolic routes which use them as a source of carbon. Thus, PS mineralization process has been observed as a substrate in the TCA cycle (Wu et al., 2023). The process commences with the hydrolytic breakdown of the PS backbone into styrene monomers. PS is transformed into styrene epoxide in a reaction catalyzed by styrene monooxygenase (SMO) (Oelschlägel et al., 2018). Subsequently, it is subjected to isomerization, becoming 3-phenylacetaldehyde due to the involvement of styrene oxide isomerase (SOI). Phenylacetaldehyde dehydrogenase converts 3-phenyl acetaldehyde into 4-phenylacetic acid. Afterward, the conversion of 4-phenylacetic acid to 5-phenylacetyl coenzyme A takes place with the assistance of an enzyme belonging to the ligase family known as phenylacetyl CoA ligase enzyme. Through β-oxidation, 5-phenylacetyl coenzyme A yields acetyl-CoA, an intermediate of citric acid cycle

(Ho et al., 2018a, b; Danso et al., 2019). Another study has reported about an alternate pathway wherein styrene epoxide is transformed into 4-maleylacetoacetate. This conversion is facilitated by various enzymes including styrene oxide isomerase (SOI) and phenylacetate hydroxylase (PAAH), 2-hydroxyphenylacetate hydroxylase (HPAAH), phenylacetaldehyde dehydrogenase (PAALDH), and homogentisate 1,2-dioxygenase (HGADO) (Ru et al., 2020). Following this, the beta oxidation of 4-maleylacetoacetate leads to the generation of acetyl-CoA, and subsequently, the TCA cycle serves as the hub for cellular metabolism (Zhang et al., 2022a, b).

In certain investigations, an alternate pathway was suggested, wherein the ring of styrene undergoes hydroxylation by an oxidoreductase enzyme, i.e., styrene dioxygenase (SDO), resulting in the formation of styrene cis-glycol (Ru et al., 2020). This compound can then be further converted into acetyl-CoA through a sequence of enzymatic reactions involving cis-glycol dehydrogenase, catechol 2,3-dioxygenase, 2-hydroxymuconic acid semialdehyde hydrolase, 2-hydroxypenta-2,4-dienoate hydratase, 4-hydroxy-2-oxovalerate aldolase, and the pyruvate dehydrogenase complex (Miloloža et al., 2022). Furthermore, acetyl-CoA will subsequently go into citric acid cycle, contributing to either biomass generation or the accumulation of various metabolites including polyhydroxyalkanoates (Oelschlägel et al., 2018; Ru et al., 2020; Ali et al., 2021; Zhang et al., 2022a, b). Additionally, laccase, a copper-binding enzyme, and oxidoreductases from *Rhodococcus ruber* (C208) have been demonstrated to participate in the process of PS biodegradation (Santo et al., 2013). Aside from laccase, alkane hydroxylase and hydroquinone peroxidases from the *Pseudomonas aeruginosa* strain E7 have also been documented in this context (Kotova et al., 2021). Xu et al. (2019) utilized calculations of quantum mechanics for illustrating the enzymatic PS degradation process using oxidoreductase. They specifically highlight the involvement of cytochrome P450 in a chemical reaction that breaks saturated hydrocarbon bonds, as indicated by Matthews et al. (2017). These enzymes stand out as promising players for PS ring degradation, although the comprehensive degradation route and the involvement of specific enzymes have yet to be fully disclosed.

In recent times, several researches have highlighted the capacity of microorganisms obtained from various sources, such as sea, waste disposal sites, plastic-infested soils, and also the digestive systems of insects, to degrade polystyrene (PS). These studies have explored a range of PS materials, including pure PS as well as modified forms and PS blended with other polymers. Ward et al. (2006), in their study, observed that the strain of *Pseudomonas putida* CA-3 utilized oil derived from thermal degradation of styrene as a source of nutrient and energy to produce storage molecules, including PHAs (Ward et al., 2006). Interestingly, styrene itself has been employed as a source of carbon for certain microorganism's cellular growth (Mor & Sivan, 2008; Motta et al., 2009). Strains of *Bacillus*, *Sphingobacterium*, and *Xanthomonas* demonstrated effective degradation of PS ranging from 40% to 56% efficiency (Eisaku et al., 2003). In another study, Krueger et al. (2015) documented the depolymerization of poly(sodium 4-styrenesulfonate) (PSS) which decreases to up to 50% within 20 days due to the actions of the white-rot fungus *Gloeophyllum Trabeum*. Furthermore, in another research, authors revealed four bacterial strains, namely, *Alcaligenes*, *Citrobacter sedlakii*, *Brevundimonas diminuta*, and *Enterobacter*, that are responsible for PS depolymerization, with *Enterobacter* achieving a maximum depolymerization rate of 12.4% within over a span of 1 month (Sekhar et al., 2016). Researchers identified a *Pseudomonas aeruginosa* strain that has the capacity to break down altered PS (Shimpi et al., 2012), while investigations were conducted on *Curvularia* species to assess their potential in degrading atactic PS without any preceding treatment (Motta et al., 2009). A strain of *P. aeruginosa* was identified as having the capability to degrade altered PS (Shimpi et al., 2012), while investigations were conducted on *Curvularia* sp. to assess their potential in completely invading the surface of oxidized PS for its degradation (Motta et al., 2009). In various research, the role of fungus has also been reported to degrade PS. Certain fungi, such as *Pleurotus ostreatus*, *Phanerochaete chrysosporium*, *Trametes versicolor*, and *Gloeophyllum trabeum*, displayed their ability to depolymerize polystyrene in conjunction with polyphenolic materials (Krueger et al., 2015). Additionally, species like *Fusarium*, *Aspergillus*, and *Penicillium* were recognized for their ability to break down PS foam (Umamaheswari & Murali, 2013). *Geomyces* and *Mortierella* species have also been implicated in PS degradation (Oviedo-Anchundia et al., 2021).

6.8 Conclusion

Polystyrene (PS) is a synthetic form of plastic which offers some serious resistance to degradation. Different physical and chemical methods have been employed, but these methods are not only harmful to the environment but also expensive. Thus, to degrade PS, an alternative strategy is needed and degradation via microorganisms is one of them. PS biodegradation rate differs from microorganisms to microorganisms. However, there are different factors like growth rate, preferred carbon/energy source, and nutritional requirements which affect microorganisms' biodegradation rate. There is huge diversity of microorganisms in terms of their abundance and function which are capable of degrading PS. Biodegradation methods like bio-augmentation and bio-stimulation provide a natural way to speed up the degradation rate in a natural environment. To assess the biodegradation rate, different chemical techniques like SEM-XRD, TEM, FTIR, IR spectroscopy, GC-MS, and AFM have been utilized. These tools provide insight mechanisms of microorganism-mediated PS biodegradation.

References

Ali, M. I., Ahmed, S., Robson, G., et al. (2014). Isolation and molecular characterization of polyvinyl chloride (PVC) plastic degrading fungal isolates. *Journal of Basic Microbiology, 54*(1), 18–27.

Ali, S. S., Elsamahy, T., Al-Tohamy, R., et al. (2021). Plastic wastes biodegradation: Mechanisms, challenges and future prospects. *Science of the Total Environment, 780*, 146590. https://doi.org/10.1016/j.scitotenv.2021.146590

Billen, P., Khalifa, L., Van Gerven, F., et al. (2020). Technological application potential of polyethylene and polystyrene biodegradation by macro-organisms such as mealworms and wax moth larvae. *Science of the Total Environment, 735*, 139521. https://doi.org/10.1016/j.scitotenv.2020.139521

Cucini, C., Leo, C., Vitale, M., et al. (2020). Bacterial and fungal diversity in the gut of polystyrene-fed *Alphitobius diaperinus* (Insecta: Coleoptera). *Animal Gene, 17*, 200109. https://doi.org/10.1016/j.angen.2020.200109

Danso, D., Chow, J., & Streita, W. R. (2019). Plastics: Environmental and biotechnological perspectives on microbial degradation. *Applied and Environmental Microbiology, 85*(19), e01095–e01119.

Dasila, H., Sah, V. K., Jaggi, V., et al. (2023a). Untapped indigenous PSB potential from forest ecosystem for enhancing soil enzyme activity and agronomic traits of different wheat genotypes. *Vegetos*, 1–14.

Dasila, H., Sah, V. K., Jaggi, V., et al. (2023b). Cold-tolerant phosphate-solubilizing Pseudomonas strains promote wheat growth and yield by improving soil phosphorous (P) nutrition status. *Frontiers in Microbiology, 14*, 1135693.

Dasila, H., Maithani, D., Srivastava, P., et al. (2023c). Bioremediation: A sustainable way for E-waste management. In *Microbial technology for sustainable E-waste management* (pp. 113–126). Springer International Publishing.

Debbarma, P., Suyal, D. C., Kumar, S., et al. (2023). Bioremediation of E-waste through microbial exopolysaccharides: A perspective. In *Microbial technology for sustainable E-waste management* (pp. 245–257). Springer International Publishing.

Delacuvellerie, A., Cyriaque, V., Gobert, S., et al. (2019). The plastisphere in marine ecosystem hosts potential specific microbial degraders including Alcanivorax borkumensis as a key player for the low-density polyethylene degradation. *Journal of Hazardous Materials, 380*, 120899. https://doi.org/10.1016/j.jhazmat

Eisaku, O., Linn, K., Takeshi, E., et al. (2003). Isolation and characterization of polystyrene degrading microorganisms for zero emission treatment of expanded polystyrene. *Environmental Engineering Research, 40*, 373–379. https://doi.org/10.11532/proes1992.40.373

Ganesh Kumar, A., Anjana, K., Hinduja, M., et al. (2020). Review on plastic wastes in marine environment: Biodegradation and biotechnological solutions. *Marine Pollution Bulletin, 150*, 110733. https://doi.org/10.1016/j.marpolbul.2019.110733

Geyer, R., Jambeck, J., & Law, K. L. (2017). Production, use, and fate of all plastics ever made. *Science Advances, 3*, 1700782.

Ghatge, S., Yang, Y., Ahn, J. A., et al. (2020). Biodegradation of polyethylene: A brief review. *Applied Biological Chemistry, 63*(1), 1–14. https://doi.org/10.1186/s13765-020-00511-3

Grbicgalle, D., Churchmaneisel, N., Mrakovic, I., et al. (1990). Microbial transformation of styrene by anaerobic consortia. *The Journal of Applied Bacteriology, 69*, 247–260.

Gu, Y., Zhao, J., & Johnson, J. A. (2020). Polymer networks: From plastics and gels to porous frameworks. *Angewandte Chemie, International Edition, 59*(13), 5022–5049.

Ho, B. T., Roberts, T. K., & Lucas, S. (2018a). An overview on biodegradation of polystyrene and modified polystyrene: The microbial approach. *Critical Reviews in Biotechnology, 38*(2), 308–320.

Ho, B. T., Roberts, T. K., & Lucas, S. (2018b). An overview on biodegradation of polystyrene and modified polystyrene: The microbial approach. *Critical Reviews in Biotechnology, 38*, 1–13.

Hou, L., & Majumder, E. L. (2021). Potential for and distribution of enzymatic biodegradation of polystyrene by environmental microorganisms. *Materials (Basel), 14*(3), 503.

Islam, S., Apitius, L., Jakob, F., et al. (2019). Targeting microplastic particles in the void of diluted suspensions. *Environment International, 123*, 428–435. https://doi.org/10.1016/j.envint.2018.12.029

Johnston, B., Radecka, I., Hill, D., et al. (2018). The microbial production of polyhydroxyalkanoates from waste polystyrene fragments attained using oxidative degradation. *Polymers, 10*(9), 957. https://doi.org/10.3390/polym10090957

Joshi, G., Goswami, P., Verma, P., et al. (2022). Unraveling the plastic degradation potentials of the plastisphere-associated marine bacterial consortium as a key player for the low-density polyethylene degradation. *Journal of Hazardous Materials, 425*, 128005.

Kim, H. R., Lee, H. M., Yu, H. C., et al. (2020). Biodegradation of polystyrene by *Pseudomonas* sp. isolated from the gut of superworms (larvae of *Zophobas atratus*). *Environmental Science & Technology, 54*, 6987–6996. https://doi.org/10.1021/acs.est.0c01495

Kotova, I. B., Taktarova, Y. V., Tsavkelova, E. A., et al. (2021). Microbial degradation of plastics and approaches to make it more efficient. *Microbiology, 90*, 671–701. https://doi.org/10.1134/S0026261721060084

Krueger, M. C., Harms, H., & Schlosser, D. (2015). Prospects for microbiological solutions to environmental pollution with plastics. *Applied Microbiology and Biotechnology, 99*, 8857–8874. https://doi.org/10.1007/s00253-015-6879-4

Kumar, V., Sharma, N., Duhan, L., et al. (2022). Microbial engineering strategies for synthetic microplastics clean up: A review on recent approaches. *Environmental Toxicology and Pharmacology, 98*, 104045. https://doi.org/10.1016/j.etap.2022.104045

Kundu, D., Hazra, C., Chatterjee, A., et al. (2014). Biopolymer and biosurfactant-graft-calcium sulfate/polystyrene nanocomposites: Thermophysical, mechanical and biodegradation studies. *Polymer Degradation and Stability, 10*(7), 37–52.

Lau, W. W. Y., Shiran, Y., Bailey, R. M., et al. (2020). Evaluating scenarios toward zero plastic pollution. *Science, 369*(6510), 1455–1461.

Liyuan, H., & Erica, L. W. M. (2021). Potential for and distribution of enzymatic biodegradation of polystyrene by environmental microorganisms. *Materials, 14*, 503. https://doi.org/10.3390/ma14030503

Lucas, N., Bienaime, C., Belloy, C., Queneudec, M., Silvestre, F., & Nava-Saucedo, J. E. (2008). Polymer biodegradation: Mechanisms and estimation techniques – A review. *Chemosphere, 73*(4), 429–442.

Matthews, S., Belcher, J. D., Tee, K. L., et al. (2017). Catalytic determinants of alkene production by the cytochrome P_{450} peroxygenase OleT_{JE}. *The Journal of Biological Chemistry, 292*, 5128–5143.

Meyer-Cifuentes, I. E., Werner, J., Jehmlich, N., et al. (2020). Synergistic biodegradation of aromatic-aliphatic copolyester plastic by a marine microbial consortium. *Nature Communications, 11*(1), 5790.

Miloloža, M., Ukić, Š., Cvetnić, M., et al. (2022). Optimization of polystyrene biodegradation by *Bacillus cereus* and *Pseudomonas alcaligenes* using full factorial design. *Polymers, 14*(20), 4299. https://doi.org/10.3390/polym14204299

Mohan, A. J., Sekhar, V. C., Bhaskar, T., et al. (2016). Microbial assisted high impact polystyrene (HIPS) degradation. *Bioresource technology 213*, 204–207.

Montazer, Z., Habibi Najafi, M. B., & Levin, D. B. (2019). Microbial degradation of low-density polyethylene and synthesis of polyhydroxyalkanoate polymers. *Canadian Journal of Microbiology, 65*(3), 224–234.

Montazer, Z., Habibi Najafi, M. B., & Levin, D. B. (2020). Challenges with verifying microbial degradation of polyethylene. *Polymers, 12*(1), 123. https://doi.org/10.3390/polym12010123

Mor, R., & Sivan, A. (2008). Biofilm formation and partial biodegradation of polystyrene by the actinomycete *Rhodococcus ruber*: Biodegradation of polystyrene. *Biodegradation, 19*, 851–858. https://doi.org/10.1007/s10532-008-9188-0

Morgen, T. O., Baur, M., Göttker-Schnetmann, I., et al. (2020). Photodegradable branched polyethylenes from carbon monoxide copolymerization under benign conditions. *Nature Communications, 11*(1), 3693.

Motta, O., Proto, A., De Carlo, F., et al. (2009). Utilization of chemically oxidized polystyrene as co-substrate by filamentous fungi. *International Journal of Hygiene and Environmental Health, 212*, 61–66.

Nakamiya, K., Sakasita, G., Ooi, T., et al. (1997). Enzymatic degradation of polystyrene by hydroquinone peroxidase of *Azotobacter beijerinckii* HM121. *Journal of Bioscience and Bioengineering, 84*, 480–482. https://doi.org/10.1016/S0922-338X(97)82013-2

Nikolic, V., Velickovic, S., & Popovic, A. (2014). Biodegradation of polystyrene-graft-starch copolymers in three different types of soil. *Environmental Science and Pollution Research, 21*, 9877–9886.

O'Leary, N. D., O'Connor, K. E., & Dobson, A. D. W. (2002). Biochemistry, genetics and physiology of microbial styrene degradation. *FEMS Microbiology Reviews, 26*, 403–417. https://doi.org/10.1111/j.1574-6976.2002.tb00622.x

Oelschlägel, M., Zimmerling, J., & Tischler, D. (2018). A review: The styrene metabolizing cascade of side-chain oxygenation as biotechnological basis to gain various valuable compounds. *Frontiers in Microbiology, 9*, 490.

Otake, Y., Kobayashi, T., Asabe, H., et al. (1995). Biodegradation of low-density polyethylene, polystyrene, polyvinyl chloride, and urea formaldehyde resin buried under soil for over 32 years. *Journal of Applied Polymer Science, 56*, 1789–1796.

Oviedo-Anchundia, R., del Castillo, D. S., Naranjo-Moran, J., et al. (2021). Analysis of the degradation of polyethylene, polystyrene and polyurethane mediated by three flamentous fungi isolated from the Antarctica. *African Journal of Biotechnology, 20*(2), 66–76. https://doi.org/10.5897/AJB2020.17200

Palmisano, A. C., & Pettigrew, C. A. (1992). Biodegradability of plastics. *Bioscience, 42*(9), 680–685.

Pardo-Rodríguez, M. L., & Zorro-Mateus, P. J. P. (2021). Biodegradation of polyvinyl chloride by *Mucor sp* and *Penicillium sp* isolated from soil. *Revista de Investigación, Desarrollo e Innovación, 11*(2), 387–399.

Park, S. Y., & Kim, C. G. (2019). Biodegradation of micro-polyethylene particles by bacterial colonization of a mixed microbial consortium isolated from a landfill site. *Chemosphere, 222*, 527–533.

Parthasarathy, A., Miranda, R. R., Eddingsaas, N. C., Chu, J., Freezman, I. M., Tyler, A. C., & Hudson, A. O. (2022). Polystyrene degradation by Exiguobacterium sp. RIT 594: Preliminary evidence for a pathway containing an atypical oxygenase. *Microorganisms, 10*(8), 1619. https://doi.org/10.3390/microorganisms10081619

Peng, B. Y., Su, Y., Chen, Z., et al. (2019). Biodegradation of polystyrene by dark (*Tenebrio obscurus*) and yellow (*Tenebrio molitor*) mealworms (Coleoptera: Tenebrionidae). *Environmental Science & Technology, 53*(9), 5256–5265. https://doi.org/10.1021/acs.est.8b06963

Peng, B. Y., Li, Y., Fan, R., et al. (2020). Biodegradation of low-density polyethylene and polystyrene in superworms, larvae of *Zophobas atratus* (Coleoptera: Tenebrionidae): Broad and limited extent depolymerization. *Environmental Pollution, 266*, 115206. https://doi.org/10.1016/j.envpol.2020.115206

Plakunov, V. K., Gannesen, A. V., Mart'yanov, S. V., et al. (2020). Biocorrosion of synthetic plastics: Degradation mechanisms and methods of protection. *Microbiology (Moscow), 89*, 647–659.

Przemieniecki, S. W., Kosewska, A., Ciesielski, S., et al. (2020). Changes in the gut microbiome and enzymatic profile of Tenebrio molitor larvae biodegrading cellulose, polyethylene and polystyrene waste. *Environmental Pollution, 256*, 113265. https://doi.org/10.1016/j.envpol.2019.113265

Qiang, L., & Cheng, J. (2021). Exposure to polystyrene microplastics impairs gonads of zebrafsh (*Danio rerio*). *Chemosphere, 263*, 128161.

Qin, Z. H., Mou, J. H., Chao, C. Y. H., et al. (2021). Biotechnology of plastic waste degradation, recycling, and valorization: Current advances and future perspectives. *ChemSusChem, 14*(19), 4103–4114. https://doi.org/10.1002/cssc.202100752

Raddadi, N., & Fava, F. (2019). Biodegradation of oil-based plastics in the environment: Existing knowledge and needs of research and innovation. *Science of the Total Environment, 67*(9), 148–158.

Rajandas, H., Parimannan, S., Sathasivam, K., et al. (2012). A novel FTIR-ATR spectroscopy based technique for the estimation of low-density polyethylene biodegradation. *Polymer Testing, 31*(8), 1094–1099.

Ru, J., Huo, Y., Yang, Y., et al. (2020). Microbial degradation and valorization of plastic wastes. *Frontiers in Microbiology, 11*, 442. https://doi.org/10.3389/fmicb.2020.00442

Santo, M., Weitsman, R., & Sivan, A. (2013). The role of the copper-binding enzyme, laccase, in the biodegradation of polyethylene by the actinomycete *Rhodococcus ruber*. *International Biodeterioration and Biodegradation, 84*, 204–210. https://doi.org/10.1016/j.ibiod.2012.03

Sarmah, P., & Rout, J., (2019). Cyanobacterial degradation of low-density polyethylene (LDPE) by Nostoc carneum isolated from submerged polyethylene surface in domestic sewage water. *Energy, Ecology and Environment 4*(5), 240–252.

Saron, C., & Felisberti, M. I. (2006). Influence of colorants on the degradation and stabilization of polymers. *Química Nova, 29*, 124–128.

Sekhar, V. C., Nampoothiri, K. M., Mohan, A. J., et al. (2016). Microbial degradation of high impact polystyrene (HIPS), an e-plastic with decabromodiphenyl oxide and antimony trioxide. *Journal of Hazardous Materials, 318*, 347–354. https://doi.org/10.1016/j.jhazmat.2016.07.008

Shimpi, N., Borane, M., Mishra, S., et al. (2012). Biodegradation of polystyrene (PS)-poly (lactic acid)(PLA) nanocomposites using *Pseudomonas aeruginosa*. *Macromolecular Research, 20*, 181–187. https://doi.org/10.1007/s13233-012-0026-1

Silva, A. B., Costa, M. F., & Duarte, A. C. (2018). Biotechnology advances for dealing with environmental pollution by micro(nano)plastics: Lessons on theory and practices. *Current Opinion in Environmental Science & Health, 1*, 30–35.

Skariyachan, S., Patil, A. A., Shankar, A., et al. (2018). Enhanced polymer degradation of polyethylene and polypropylene by novel thermophilic consortia of Brevibacillus sps. and Aneurinibacillus sp. screened from waste management landfills and sewage treatment plants. *Polymer Degradation and Stability 149*, 52–68.

Tennakoon, A., Wu, X., Paterson, A. L., et al. (2020). Catalytic upcycling of high-density polyethylene via a processive mechanism. *Nature Catalysis 3*(11), 893–901.

Thompson, R. C., Olsen, Y., Mitchell, R. P., et al. (2004). Lost at sea: Where is all the plastic? *Science, 304*, 838. https://doi.org/10.1126/science.1094559

Tischler, D., Eulberg, D., Lakner, S., et al. (2009). Identification of a novel self-sufficient styrene monooxygenase from *Rhodococcus Opacus 1CP*. *Journal of Bacteriology, 191*(15), 4996–5009.

Umamaheswari, S., & Murali, M. (2013). FTIR spectroscopic study of fungal degradation of poly(ethylene terephthalate) and polystyrene foam. *Elixir Chemical Engineering, 64*, 19159–19164.

Wang, J., Li, Y., Lu, L., et al. (2019). Polystyrene microplastics cause tissue damages, sex-specific reproductive disruption and transgenerational effects in marine medaka (*Oryzias melastigma*). *Environmental Pollution, 254*, 113024. https://doi.org/10.1016/j.envpol.2019.113024

Ward, P. G., Goff, M., Donner, M., et al. (2006). A two-step chemo-biotechnological conversion of polystyrene to a biodegradable thermoplastic. *Environmental Science & Technology, 40*, 2433–2437. https://doi.org/10.1021/es0517668

Warhurst, A. M., Clarke, K. F., Hill, R. A., et al. (1994). Metabolism of styrene by *Rhodococcus rhodochrous* NCIMB 13259. *Applied and Environmental Microbiology, 60*(4), 1137–1145.

Wei, R., & Zimmermann, W. (2017). Microbial enzymes for the recycling of recalcitrant petroleum-based plastics: How far are we? *Microbial Biotechnology, 10*(6), 1308–1322. https://doi.org/10.1111/1751-7915.12710

Wu, F., Guo, Z., Cui, K., et al. (2023). Insights into characteristics of white rot fungus during environmental plastics adhesion and degradation mechanism of plastics. *Journal of Hazardous Materials, 448*, 130878. https://doi.org/10.1016/j.jhazmat.2023.130878

Xu, J., Cui, Z., Nie, K., et al. (2019). A quantum mechanism study of the C-C bond cleavage to predict the bio-catalytic polyethylene degradation. *Frontiers in Microbiology, 10*, 489. https://doi.org/10.3389/fmicb.2019.00489

Yang, J., Yang, Y., Wu, W. M., et al. (2014). Evidence of polyethylene biodegradation by bacterial strains from the guts of plastic-eating waxworms. *Environmental Science & Technology, 48*(23), 13776–13784. https://doi.org/10.1021/es504038a

Yang, Y., Yang, J., Wu, W. M., et al. (2015a). Biodegradation and mineralization of polystyrene by plastic-eating mealworms: Part 1. Chemical and physical characterization and isotopic tests. *Environmental Science & Technology, 49*, 12080–12086.

Yang, Y., Yang, J., Wu, W. M., et al. (2015b). Biodegradation and mineralization of polystyrene by plastic-eating mealworms: Role of gut microorganisms. *Environmental Science & Technology, 49*, 12087–12093.

Yanto, D. H. Y. (2019). *IOP Conference Series: Earth and Environmental Science, 308*, 012001.

Zhang, H., Kong, D., Wang, L., et al. (2022a). Degradation of UV-pretreated polyolefins by latex clearing protein from Streptomyces sp. Strain K30. *Science of the Total Environment, 806*, 150779.

Zhang, Y., Pedersen, J. N., Eser, B. E., et al. (2022b). Biodegradation of polyethylene and polystyrene: From microbial deterioration to enzyme discovery. *Biotechnology Advances, 60*, 107991. https://doi.org/10.1016/j.biotechadv.2022.107991

Zhu, M., Chernick, M., Rittschof, D., et al. (2020). Chronic dietary exposure to polystyrene microplastics in maturing Japanese medaka (*Oryzias latipes*). *Aquatic Toxicology, 220*, 105396. https://doi.org/10.1016/j.aquatox.2019.105396

Chapter 7
Diversified Analytical Methods Used to Analyze Plastic Biodegradation

Javier E. Vilasó-Cadre, Lázaro Adrián González-Fernández, Nahum Andrés Medellín-Castillo, and Iván A. Reyes-Domínguez

Abstract Analytical methods for microplastics are essential for understanding and monitoring the degradation processes of these micromaterials, with some techniques even being able to identify them in living organisms. This chapter presents the fundamentals and applications of the main methods for the analysis of microplastics with a focus on degradation processes in environmental samples. Optical, electron, and atomic force microscopy methods are described. Spectrometric, thermal analysis, and chromatographic methods are also described. This wide variety of methods must be used together to obtain an integral qualitative and quantitative characterization. Hybrid techniques are of great interest as they show the best performance. Research on analytical methods for microplastics is still an area of opportunity, as there are still many practical limitations.

Keywords Plastic biodegradation · Analytical methods · Microscopy · Spectrometry · Thermal analysis · Chromatographic methods

J. E. Vilasó-Cadre
Institute of Metallurgy, Autonomous University of San Luis Potosí, San Luis Potosí, Mexico

L. A. González-Fernández (✉)
Multidisciplinary Graduate Program in Environmental Sciences,
Autonomous University of San Luis Potosí, San Luis Potosí, Mexico

N. A. Medellín-Castillo
Multidisciplinary Graduate Program in Environmental Sciences,
Autonomous University of San Luis Potosí, San Luis Potosí, Mexico

Center for Research and Graduate Studies, Faculty of Engineering, San Luis Potosí, Mexico

I. A. Reyes-Domínguez
Institute of Metallurgy, Autonomous University of San Luis Potosí, San Luis Potosí, Mexico

National Council of Humanities, Science and Technology of Mexico, Mexico City, Mexico

R. Soni et al. (eds.), *Advanced Strategies for Biodegradation of Plastic Polymers*, https://doi.org/10.1007/978-3-031-55661-6_7

7.1 Introduction

Both research and technologies implemented for the degradation of microplastics always require an analytical monitoring method. Monitoring starts with the characterization of the microplastics in the sample and the subsequent determination of the degradation effects and often the products of this process.

There are a variety of methods for identifying and quantifying microplastics. Among the most commonly used for identification are microscopic methods, e.g., optical, electronic, and atomic force. Clearly, the more advanced methods allow for a more comprehensive characterization of these micro- and nanomaterials, but they are not specifically suited for quantification. Other methods are more focused on characterization only, such as thermal analysis. Spectroscopic methods also have primarily qualitative applications in the analysis of microplastics. Chromatographic methods are ideal for the quantification of both microplastics and their degradation products. There are hybrid techniques that give the best results because they are obviously more advanced, but they are also more expensive.

The reality is that the existing methods are very flexible, and it is risky to categorize them definitively as quantitative or qualitative. This is partly because the skills and experience of the analyst play a fundamental role in the analysis of microplastics.

In this chapter, we present the methods used for the qualitative and quantitative analysis of microplastics and their degradation products during the processes aimed at eliminating from the environment this type of material, which, due to its persistence and toxicity, is one of the most important pollutants today.

7.2 Microscopy Techniques

Microscopy is a method based on the magnification of a sample to study its microscopic characteristics, reaching the nanometer or even atomic level in some techniques. There are several types of microscopies, which in this text, we will classify as optical, electron, and atomic force microscopy. Some microscopy techniques can be hybridized with others or with spectroscopic methods. The most common microscopy techniques such as classical optical, polarized light, fluorescence, scanning, and transmission electron microscopy have been used for the analysis of microplastics. Hybrid optical and electron techniques have also been used. Among them, however, scanning electron microscopy is the most widely used to analyze these micromaterials.

7.2.1 Optical Microscopy

In general, optical microscopy is the observation of specimens by magnification using optical lenses and a source of visible and, in some techniques, ultraviolet radiation. The image observed in a microscope can be two-dimensional or three-dimensional, and in the latter case, it is referred to as a stereoscopic microscope. Conventional optical microscopes allow higher magnification than a stereoscope. In stereoscopic microscopes, no sample preparation is required; in the case of liquids, a microplastic separation step is performed, which is generally common for the analysis of these materials by any method.

In the case of microplastic analysis, morphological description with the stereomicroscope typically has a detection limit of 50 μm, but characterization of particles smaller than 100 μm may be limited. The use of digital cameras attached to the stereoscope is one of the main advantages of this method, as it facilitates morphological description or size measurement. Normally, microscopic methods are not suitable for quantification of microplastics, but only for morphological identification and characterization. The identification of very small microplastic particles is not accurate because it is subject to possible confusion with other materials, and under- or overestimation may occur, and here the experience of the analyst in identification plays a very important role. Some criteria that can be used to classify particles under stereoscopic observation as microplastics are absence of cellular structure, stable thickness along its entire length, and homogeneous color. Suspected particles can be examined with a hot needle to observe morphological changes common in thermoplastic materials (Kalaronis et al., 2022).

Another important optical microscopy technique is polarized light. Polarized light is that in which the electromagnetic waves all oscillate in only one plane. The polarized light microscope contains two polarizing prisms, one of which is placed between the light source and the specimen to allow illumination with polarized light. The second polarizer, called the analyzer, is placed between the specimen and the ocular lens, and its function is to be able to extinguish the light transmitted by the first polarizer. Cross-polarizers allow the observation of interference phenomena of polarized light due to the birefringence of anisotropic samples, and this is used to identify materials, including some microplastics. Polyethylene microparticles, for example, can be better identified in a polarized light microscope than in a stereoscope because they can exhibit color interference patterns. Cellulose microfibers appear bright when viewed with crossed prisms. Polyamide microfibers also show strong birefringence when viewed with crossed polarizers while rotating the specimen. Pleochroism can sometimes be observed in some microplastics (Gwinnett et al., 2021; Labbe et al., 2020).

Fluorescence microscopy is also an optical technique in which the sample is illuminated with radiation of a specific wavelength at which the electrons in the molecules are excited; the excited electrons emit photons when they return to their basal state. This phenomenon is called molecular fluorescence. In this fluorescence technique, images are observed based on the molecular emission of radiation, either autogenously or because they have been previously stained with a substance that, when bound, allows the observation of molecular fluorescence. For the observation of microplastics, the coupling of the monochromatic source, which can be an LED or a laser, to a stereoscope is simple. Labbe et al. (2020) provide details on how to do this. In cases where such coupling is done, it is necessary to use a light filter in front of the ocular lenses that blocks the scattered excitation light.

Nile Red is the dye of choice for observing fluorescence in microplastics, with green, yellow, or red emission depending on the hydrophobicity of the polymer. Safranine T and fluorescein isophosphate (FITC) are also used to identify these micromaterials, with excitation at 520 nm and 488 nm, respectively (Lv et al., 2019). Rose Bengal can also be used. It is common for dyes to give false positives in the identification of microplastics due to organic and biological matter, for example, Nile Red has the ability to bind to cellular lipid materials. Therefore, for more accurate results, identification by fluorescence microscopy should be preceded by a digestion step to remove organic matter or separation by flotation. Digestion can be performed in an acidic medium, in a basic medium, with 30% hydrogen peroxide, or with Fenton's reagent. However, extreme conditions of acidity or basicity degrade the polymers and can damage the morphology of the microparticles. Among the microplastics that can be best analyzed by fluorescence microscopy are polyethylene terephthalate (PET), polypropylene (PP), polystyrene (PS), polyethylene (PE), nylon (PA), polyester (PES), and elastane (polyurethane) (Kalaronis et al., 2022).

Optical microscopy is a cheap, simple, and above all versatile tool for the analysis of microplastics during studies and the implementation of degradation technologies, but it has less detection capacity and accuracy than other more powerful tools such as scanning electron microscopy; for this reason, for a more reliable analysis by optical microscopy, several of the described techniques should be used.

7.2.2 Electron Microscopy

There are two types of electron microscopy: scanning and transmission. The difference with optical microscopy is that the latter uses light and optical lenses, while electron microscopy uses an accelerated electron beam focused on the sample by electromagnetic lenses in a vacuum column. Electron microscopy allows to obtain images with higher resolution and therefore higher magnification than optical microscopy, mainly due to the fact that in the latter method, the resolution is limited to the wavelength of light, which is much larger than that of an electron beam accelerated by a potential difference.

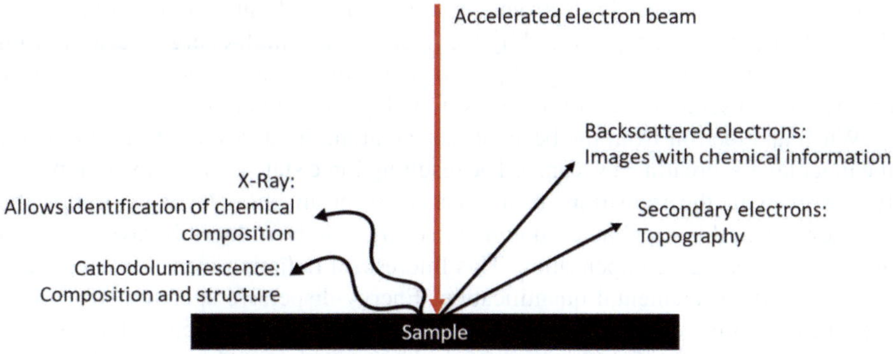

Fig. 7.1 Phenomena used by the SEM to examine the sample when an accelerated electron beam is incident on its constituent materials. The straight arrows represent electron phenomena, while the wavy arrows represent electromagnetic waves

Scanning electron microscopy (SEM) consists of obtaining images and chemical information of a sample by the effects produced on conductive materials by an accelerated electron beam. The electron source can be a cathode consisting of a tungsten or LaB_6 filament or a tungsten single crystal. The electrons are accelerated by a potential difference between a cathode and an anode, commonly between 5 and 30 kV. The accelerated electrons travel through a column containing the electromagnets, which act as electromagnetic lenses, and finally reach the sample. Figure 7.1 shows the interaction phenomena that occur when an accelerated electron beam strikes a material, which are detected in the SEM to obtain topographical, compositional, and even structural information. Other phenomena such as Auger electron emission also occur, but most SEM applications use secondary electrons, backscattered electrons, and X-rays. The SEM can only examine conductive samples, so for nonconductive materials, a preparation is made in which a layer of carbon or gold is deposited. The resolution of an electron microscope depends on many factors, such as the acceleration voltage of the electrons or the type of electron emission source, but for SEM, it is proposed that the resolution is close to 1 nm, although this is not generally achieved.

Secondary electrons are the most commonly used for SEM micrographs. They are produced when an electron in the beam passes close to the nucleus of an atom in the sample and provides enough energy for an inner electron to leave the nuclear shell. These electrons are low energy, less than 5 keV, so they must come from atoms near the surface of the sample, and it is precisely this that allows topographic micrographs to be observed when a detector for these electrons is used in the microscope.

Backscattered electrons are those in the electron beam that are repelled by the nucleus of the atoms in the sample. The energy of these electrons depends on the atomic number, resulting in micrographs that show differences in brightness depending on the composition.

Cathodoluminescence is electromagnetic radiation or light, ranging from the visible to the near infrared, produced by the interaction of high-energy electrons with a luminescent material. This signal provides information about crystalline defects, energy gaps in semiconductors, composition, dopants, and more.

When an electron from the beam strikes an atom, it causes an electron to leave the inner layers, creating a vacancy. The resulting ionic state is unstable, minimizing its energy with the transition of an electron from an outer layer to occupy the vacancy. When this happens, a photon with energy on the order of X-rays is emitted as a result of energy compensation. This fluorescent radiation from the sample can then be used for elemental quantification. Energy-dispersive spectrometry (EDS) classifies the emitted X-rays according to their energy, while wavelength-dispersive spectrometry (WDS) classifies them according to their wavelength. EDS is more commonly used because the analysis is faster, but the results are semiquantitative under common conditions. WDS is a more time-consuming method because it is based on X-ray diffraction, but it is more accurate because of its higher resolution.

SEM is the most widely used method for the analysis of microplastics (Kalaronis et al., 2022). In addition to the ability to obtain images at higher magnifications, SEM has an important advantage in that no treatment of the sample is required other than the carbon or gold coating necessary to make the surface conductive, but no digestion to remove organic matter, for example, as needed for fluorescence microscopy analysis. EDS analysis of morphology and composition allows identification of polymer microparticles. In addition, EDS allows the identification of metals bound to microplastics. However, EDS does not provide an accurate indication of whether the analyzed particle is a polymeric material.

An advanced technique that provides higher-resolution micrographs than a conventional SEM is that using a field emission electron source, abbreviated as FE-SEM. Field emission produces a highly focused electron beam that improves spatial resolution. This electron generation technique uses low voltages, typically between 0.02 and 5 kV. This minimizes the effect of charging and damage to nonconducting materials that often occurs with high voltages in the SEM. These features make this technique ideal for use in the analysis of microplastics due to the organic nature of these materials.

The surface history of microplastics is important in SEM analysis because the surface undergoes significant changes depending on the environmental conditions. For example, the same surface morphology is not observed for microparticles from river or seawater as for those from wastewater. According to Kalaronis et al. (2022), microplastics extracted from river, sea, and lake waters are observed with cracked, rough, and edged surfaces, whereas microplastics extracted from wastewater are observed with smooth and aged surfaces. Obviously, this is not generalizable, but they are elements that allow to have criteria for identification by SEM when the elemental composition by EDS is also considered. EDS results should show a high content of carbon and oxygen. But as described, the presence of metals may also be possible in microplastics, although in smaller quantities.

On the other hand, in the transmission electron microscopy (TEM), the image is obtained by detecting the electrons passing through the sample by means of a fluorescent screen. The electron sources are the same as those described for the SEM, and a vacuum column with electromagnetic lenses that focus the electron beam is also used, although there are some differences in the arrangement of the lenses and the internal structure of the column. The accelerating voltage used in the TEM, typically between 60 and 300 kV, is higher than that used in the SEM. The TEM allows higher magnification and resolution (<0.1 nm) than the SEM, in addition to providing structural information through electron diffraction or the high-resolution images themselves. EDS analysis is also possible with the TEM.

TEM images are two-dimensional due to the fact that a micrograph is obtained as a result of the transmission of electrons, as distinct from SEM, where images are obtained by electrons scattered or emitted from the surface, so that micrographs show the topography of the sample.

TEM is the method par excellence for characterizing nanomaterials because it allows higher magnifications than SEM. This would lead to the assumption that TEM is the appropriate method to study nanoplastics, but this is not the case. The TEM is an instrument with practical limitations for studying micro- and nanoplastics due to the amorphous nature of these materials and their organic elemental composition. What is usually done is staining with heavy elements that enhance the interaction with the electron beam. For this reason, published work using TEM for the analysis of microplastics is scarce compared to that of using SEM. Nevertheless, it is a valid technique for this application as long as the sample is properly treated, and the heavy element staining does not alter the chemical structure of the polymer. For crystalline or predominantly crystalline nanoplastics, these treatments may not be necessary.

One of the major advantages of the TEM is its ability to perform electron energy loss spectroscopy (EELS) for polymer analysis. This spectroscopic technique analyzes the electrons inelastically scattered by the sample, providing high-resolution information on atomic composition, chemical bonding, and many other properties. Complementing EDS with EELS is a good strategy to study materials by TEM, including micro- and nanoplastics, especially in degradation studies because it allows the study of chemical changes.

The hybrid scanning transmission electron microscopy (STEM) technique has a great potential for the analysis of polymeric particles, especially nanoplastics, due to the higher resolution achieved compared to a conventional SEM. STEM combines the principles of SEM and TEM to produce a transmission image by scanning the electron beam across the sample. It achieves better resolution than a SEM without the high energy used in a TEM. Although a SEM can be modified to become a STEM, the greatest benefits are achieved when the TEM is adapted for use as a STEM. In fact, many modern TEMs can be used in STEM mode.

Electron microscopy techniques are very versatile for the analysis of microplastics in the degradation processes of these materials. The combination of optical and electron microscopy techniques can lead to a much more accurate analysis than using each technique separately. However, in microscopy, the experience of the analyst is critical to obtaining reliable results, as it relies heavily on the ability to identify based on experience. Electron microscopy in general is a very powerful method for the analysis of micro- and nanomaterials, but its main disadvantage is the cost of the microscopes as well as the materials and equipment required for sample preparation.

7.2.3 Atomic Force Microscopy

Atomic force microscopy (AFM) is a powerful method that allows topographic imaging, property study, and surface manipulation by the interaction of forces between a very thin tip of a mechanical probe and the sample surface. The mechanical probe is called a cantilever. Figure 7.2 shows a representation of the principle of AFM. The sample is deposited on a substrate which can be a mica, glass, or silicon and is moved by a piezoelectric. As the tip approaches a critical distance with respect to the sample, a deflection of the cantilever occurs due to attractive forces, and when close enough, repulsive forces will also begin to dominate. The interaction force between the tip of the cantilever and the sample surface is measured by means of a laser incident on the probe and whose reflection beam is detected. The cantilever can be made of various materials, the most common being silicon nitride, silicon dioxide, and silicon.

The thinner the tip, the higher the image resolution and the more accurate the force measurement. Typically, this tip has nanometer dimensions ranging from 20 nm to a little less than 1 nm. The AFM can be operated in three main modes: contact mode, contact dynamic mode, and noncontact dynamic mode. In the first, the tip of the cantilever is in constant contact with the sample surface; in the second, known as tapping, the cantilever oscillates vertically at high frequency, moving away from and toward the surface; in the third, the tip moves toward and away from the surface, but there is no contact. In the AFM, the entire piezoelectric motion system of the scanner and cantilever, as well as the detection system, is controlled by a computer. The resolution of the AFM is less than 1 nm, allowing visualization

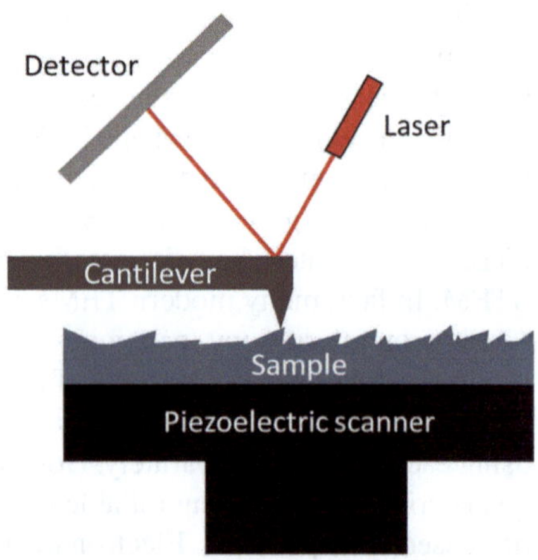

Fig. 7.2 Representation of the principle of operation of an atomic force microscope

and manipulation at the atomic scale. In contrast to SEM, AFM not only provides high-resolution images of the surface but also allows the characterization of surface properties such as stiffness, adhesiveness, hydrophobicity, conductivity, magnetization, work function, and others (Fu et al., 2020; Mariano et al., 2021).

For AFM analysis of microplastics, samples can be immobilized on adhesive pads or tapes and mounted on steel mounting disks (Fu et al., 2020). Since AFM allows not only to study surface morphology but also to determine surface physical properties, it is an ideal method for micro- and nanoplastic degradation studies, since a change in properties is not necessarily reflected in surface morphology. This is why AFM is considered a key method for understanding the degradation of these polymers.

AFM has been used to study the properties and degradation of microplastics in marine ecosystems such as polyvinyl chloride (PVC), PE, or PP (Dussud et al., 2018; Kumari et al., 2019; Melo-Agustín et al., 2022). One of the main limitations of AFM is that slow scanning speeds are generally required to obtain high-resolution images, although some advanced AFM techniques allow faster scanning (Fu et al., 2020).

AFM can be hybridized with Raman and infrared spectroscopy, which allows obtaining chemical information about the sample and not only physical as in the traditional technique, something that has been greatly exploited to study microplastics because in the analysis of these materials, chemical information is key to control and understand degradation processes or binding with other contaminants such as metals.

7.2.4 Raman and FTIR Microspectroscopy

Both Raman spectroscopy and FTIR are of the vibrational type. Raman spectroscopy is based on the inelastic scattering of monochromatic light, usually a laser. It allows to obtain a spectrum of peaks and bands where some vibrational modes are identified. FTIR is also a vibrational spectroscopy method that uses infrared radiation to obtain an interferogram and applies Fourier transform to bring it into the frequency domain. FTIR is considered a more powerful method than Raman spectroscopy, but the latter often shows peaks that are weak in FTIR, so the two are complementary. There are hybrid techniques between optical microscopy and Raman and FTIR spectroscopy, known as micro-Raman and micro-FTIR, respectively. They combine an optical microscope with a spectrometer to obtain Raman or FTIR spectra on microparticles smaller than 10 μm. This facilitates the identification of microplastics and the chemical changes that can occur as a result of their degradation.

According to Fu et al. (2020), micro-Raman has been increasingly applied in the identification of microplastics in samples of seawater, freshwater, wastewater, aquatic organisms, food, beverages, and cosmetics. However, micro-FTIR has also been widely applied in the analysis of microplastics in surface water, sediment, wastewater, leachate, atmosphere, food, and even biological organisms.

7.3 Respirometric Methods

Respirometric methods are a set of techniques used to measure the rate of respiration in living organisms or biological systems. These methods involve monitoring the consumption of oxygen (O_2) or production of carbon dioxide (CO_2) as a means to quantify the metabolic activity of cells or organisms (King & Dutka, 2019).

There are several respirometric methods commonly employed, including the following:

1. Closed respirometry: In this method, the organism or sample is placed in a closed system, such as a respirometer, where changes in gas composition can be measured over time. The decrease in oxygen concentration or increase in carbon dioxide concentration indicates the rate of respiration (Archiza et al., 2017).
2. Open respirometry: This method involves measuring the exchange of gases between the organism or sample and its surrounding environment. This is typically done using gas analyzers or gas exchange chambers. Oxygen consumption or carbon dioxide production is determined by comparing the gas concentrations before and after exposure to the organism or sample (Nowack et al., 2020).
3. Microrespirometry: This technique is used to measure respiration rates at a small scale, such as individual cells or tiny organisms. It involves using microsensors or microrespirometers to directly measure changes in oxygen or carbon dioxide levels at a microscopic level (DeLiberto et al., 2020).
4. Mass-specific respirometry: This method calculates respiration rates per unit of body mass, allowing for comparisons across different sizes or species. It helps to account for variations in metabolic activity related to body size or weight (Patel et al., 2018).

Respirometric methods are widely used in diverse fields, including physiology, ecology, biochemistry, and environmental science, to study cellular metabolism, energy budgets, and the impact of environmental factors on living organisms.

Respirometric methods play a crucial role in assessing the biodegradation potential of plastics. These methods involve measuring the oxygen consumption or carbon dioxide production of microbial organisms when exposed to plastic samples. Here is a concise overview of respirometric methods used in plastics biodegradation research:

1. Closed respirometry: Closed respirometers are commonly used for measuring the oxygen uptake rate (OUR) during plastic biodegradation. Plastic samples are incubated with microbial cultures in a closed system, and the decrease in oxygen concentration over time is measured. The OUR is then calculated based on the rate of oxygen depletion (Kim, 2023).
2. Manometric respirometry: In this method, plastic samples are placed in an airtight container connected to a pressure transducer. As microorganisms degrade the plastic, they produce carbon dioxide, causing an increase in pressure. By monitoring the pressure changes, researchers can estimate the rate of plastic biodegradation (Strebl et al., 2023).

3. Off-gas analysis: Off-gas analysis involves monitoring the composition of gases released during plastic biodegradation. This method can be used to measure carbon dioxide production or methane emission, which can indicate the extent of polymer degradation (Macfarlane, 2017).
4. Stoichiometric analysis: Stoichiometric analysis focuses on determining the microbial metabolic pathway during plastic degradation. It involves measuring the ratio of oxygen consumed to carbon dioxide produced, allowing researchers to understand the efficiency of biodegradation processes (Mainardis et al., 2021).

It is worth noting that respirometric methods provide valuable insights into the biodegradation potential of plastics under controlled laboratory conditions. However, they may not fully represent real-world degradation scenarios. Therefore, a combination of respirometric studies, field trials, and other analytical techniques is often employed to obtain a comprehensive understanding of plastic biodegradation.

7.4 Spectroscopic Methods

7.4.1 Infrared Spectroscopy

FTIR (Fourier transform infrared) spectroscopy is a powerful analytical technique used to identify and characterize chemical compounds based on their unique infrared absorption patterns. FTIR spectroscopy utilizes infrared radiation, which lies in the electromagnetic spectrum between visible light and microwaves. This radiation is characterized by its wavelengths ranging from approximately 0.7 to 300 micrometers (Smith, 2011).

The sample to be analyzed is typically prepared as a thin film or thoroughly ground into a fine powder. It is important to ensure that the sample thickness is consistent to obtain accurate and reproducible results. As the infrared radiation interacts with the sample, the molecules in the sample absorb specific wavelengths of the infrared light. This absorption occurs due to the vibrational and rotational motions of the atoms within the sample's chemical bonds.

The transmitted or absorbed infrared radiation is detected by an FTIR spectrophotometer. The resulting data is represented as a plot of intensity (transmittance or absorbance) on the y-axis versus wavenumber (the reciprocal of the wavelength) on the x-axis. This plot is called an infrared spectrum. FTIR spectra provide information about the types of chemical functional groups present in the sample. Each functional group exhibits characteristic absorption bands in the infrared spectrum, allowing for identification and characterization of the sample's chemical composition (Simonescu, 2012).

To identify an unknown sample, its FTIR spectrum can be compared to reference spectra in databases to find matches. This can assist in the identification of the sample's components or chemical structure. FTIR spectroscopy can also be used for quantitative analysis. By measuring the intensity of specific absorption bands, the concentration of a particular component in a sample can be determined (Lin et al., 2021).

FTIR spectroscopy is widely used in various scientific fields, including chemistry, materials science, pharmaceuticals, and forensic analysis. Its ability to quickly provide valuable information about molecular composition makes it a versatile and powerful analytical tool.

FTIR spectroscopy can be utilized as a valuable tool for studying the biodegradation of plastics. FTIR allows for the analysis of plastic samples before and after biodegradation to detect any chemical changes that occur. The technique is sensitive to various functional groups present in plastics, such as carbonyl (C=O), hydroxyl (O-H), and amide (C=O-N) groups. By comparing the infrared spectra of the plastic sample before and after biodegradation, one can determine if there have been any alterations in chemical structure due to microbial activity (J.-L. Xu et al., 2019).

FTIR can also be used to quantitatively assess the degradation of specific functional groups in plastics. By monitoring the intensity of specific absorption bands related to targeted chemical bonds, researchers can track the degradation progress and determine the extent of biodegradation (Jung et al., 2021).

The analysis of FTIR spectra can help identify the by-products or metabolites formed during plastic biodegradation. The appearance of new peaks or significant changes in existing peaks in the spectra may indicate the formation of degradation products, offering valuable insights into the biodegradation pathways (Khan et al., 2023).

By performing periodic FTIR measurements during the biodegradation process, researchers can track the changes in functional groups over time. This allows for the evaluation of the kinetics of biodegradation and provides a deeper understanding of the degradation mechanism (Singh et al., 2020).

FTIR spectroscopy can also facilitate comparative studies between different types of plastics to determine their relative biodegradability. By analyzing the FTIR spectra of various plastics before and after biodegradation, researchers can assess which materials are more prone to degradation and identify potential factors influencing their biodegradability (Sarkhel et al., 2020).

In conclusion, FTIR spectroscopy serves as a valuable analytical technique for studying plastic biodegradation, providing important information about chemical changes, degradation products, and degradation kinetics. It can aid in the development of more sustainable and biodegradable plastics by allowing researchers to assess and optimize their biodegradability.

7.4.2 Nuclear Magnetic Resonance

Nuclear magnetic resonance (NMR) spectroscopy is a powerful analytical technique used to study the structure, dynamics, and interactions of molecules. It is based on the principle that certain atomic nuclei possess a magnetic property called spin, and when placed in a strong magnetic field, they can absorb and emit electromagnetic radiation at specific frequencies. The fundamental aspects of NMR spectroscopy are the followings (James, 1998):

- Nuclear spins: Nuclei with an odd number of protons or neutrons have a nonzero nuclear spin and can be studied using NMR. Common nuclei used in NMR spectroscopy include hydrogen-1 (^{1}H), carbon-13 (^{13}C), nitrogen-15 (^{15}N), and phosphorus-31 (^{31}P).
- Chemical shift: In an NMR spectrum, the position of a signal is given in terms of chemical shift, measured in parts per million (ppm). Chemical shift represents the magnetic environment experienced by the nucleus, influenced by its electron density and neighboring atoms. It provides information about the chemical structure of the molecule.
- Signal intensity: The intensity of an NMR signal is proportional to the number of nuclei giving rise to the signal. It reflects the abundance of the particular nucleus in the sample and can be used for quantitative analysis.
- Spin-spin coupling: Spin-spin coupling occurs when the spins of neighboring nuclei interact with each other. It leads to the splitting of NMR signals into multiple peaks, known as multiplets. Spin-spin coupling provides valuable information about the connectivity and proximity of atoms in a molecule.
- Relaxation processes: NMR signals arise from the excitation and relaxation of nuclear spins. Two relaxation processes, called T1 (spin-lattice relaxation) and T2 (spin-spin relaxation), determine the lifetimes of excited states. T1 reflects the rate at which spins return to thermal equilibrium, while T2 describes the dephasing of spins due to interactions with their surroundings.
- Two-dimensional NMR (2D-NMR): 2D-NMR spectroscopy combines information from two different NMR experiments to provide more detailed structural information. It is commonly used to resolve overlapping signals, determine spatial connectivity, and study molecular dynamics.

NMR spectroscopy finds applications in various scientific fields, including chemistry, biochemistry, materials science, and medicine. It enables the determination of molecular structures, identification of unknown compounds, analysis of reaction mechanisms, and investigation of protein dynamics, among other applications.

NMR spectroscopy is a valuable tool for studying plastic biodegradation. NMR can provide insights into the changes in the molecular structure of plastics during the degradation process. By analyzing the molecular structure of plastics before and after biodegradation, NMR can reveal changes in the structure of the plastic. For example, NMR spectroscopy can detect chemical modifications in plastics during biodegradation, such as the formation of carbonyl groups in the oxidation pathway during polystyrene biodegradation (Sintim et al., 2020).

In addition, NMR can confirm the formation of specific chemical groups during plastic biodegradation. By using NMR spectroscopy, researchers can gain valuable insights into the molecular changes that occur in plastics during biodegradation. This information can help understand the mechanisms of biodegradation (Moon et al., 2023), assess the efficiency of degradation processes (Nikafshar & Nejad, 2022), and potentially guide the development of more sustainable plastic materials (Avella et al., 2005).

Besides to the applications mentioned earlier, NMR spectroscopy can also be used to study the biodegradation of plastics in marine environments. For example, a study used NMR spectroscopy to analyze the chemical changes in PET films exposed to seawater for up to 12 months. The study found that the PET films underwent significant chemical changes during biodegradation, including the formation of carbonyl groups and the cleavage of ester bonds (Niaounakis et al., 2019).

To conclude, NMR spectroscopy is a powerful tool for studying plastic biodegradation. By providing insights into the molecular changes that occur during biodegradation, NMR can help researchers understand the mechanisms of biodegradation, assess the efficiency of degradation processes, and potentially guide the development of more sustainable plastic materials.

7.4.3 X-Ray Diffraction

X-ray diffraction (XRD) is a technique used to study the crystal structure of materials. The principle behind XRD analysis is based on the interaction of X-rays with the atoms in a sample. When X-rays are directed at a sample, they interact with the atoms in the sample, and the X-rays that are scattered by the atoms in the sample are detected by a detector. The detector records the intensity of the scattered X-rays at different angles, and the intensity of the scattered X-rays is related to the arrangement of atoms in the sample (Chatterjee, 2000).

The data obtained from the detector is analyzed to determine the crystal structure of the sample. XRD analysis is one of the keystones of materials science and materials research, and it has a wide range of applications in different fields such as geology, chemistry, and materials science. XRD can be used to identify the crystal structure of a material, determine the orientation of crystals in a sample, and study the effects of temperature and pressure on crystal structures. By providing information on the crystal structure of materials, XRD can help researchers understand the properties and behavior of materials and potentially guide the development of new materials with desired properties (Ali et al., 2022).

In the realm of plastic biodegradation, XRD serves as a fundamental technique for investigating the crystalline structure of materials. Its application allows for the elucidation of any modifications that may transpire within the material's structure during the biodegradation process. During plastic biodegradation, the polymers that make up the plastic can break down into smaller fragments due to the action of microbes or environmental factors (Sarkhel et al., 2020).

These fragments may exhibit different crystal structures compared to the original plastic. By subjecting the degraded plastic to XRD analysis, researchers can identify any changes in the material's crystal structure. This can help determine if there are new crystalline phases formed or if the overall crystallinity of the material has changed. XRD analysis can also be used to compare the crystal structures of different types of plastics before and after degradation, allowing for a better understanding of the degradation behavior of different materials (Abubaker & Taha, 2021).

XRD can be used to analyze changes in the crystalline structure of plastic materials as they degrade over time. By measuring diffraction patterns, researchers can monitor the crystalline phases present in the plastic and assess how they are affected by biodegradation processes. XRD can also help determine the rate at which plastics degrade. By studying the changes in X-ray diffraction patterns over time, researchers can gather valuable data on the degradation kinetics of plastic materials, which can be used to better understand the biodegradation process (Achilias et al., 2011).

XRD can also aid in identifying the by-products that result from plastic biodegradation. By analyzing the diffraction patterns of degraded plastic samples, researchers can identify new crystalline phases or changes in existing ones, providing insights into the breakdown of the plastic structure. Additionally, XRD can be used to compare the biodegradation efficacy of different plastic materials or additives. By studying the changes in diffraction patterns for various samples under controlled degradation conditions, researchers can assess the impact of different factors on the rate and extent of biodegradation.

In addition, XRD can contribute to determining the environmental impact of plastic biodegradation processes. By examining the diffraction patterns of degraded plastics in different environments, researchers can gain insights into how the biodegradation process is influenced by external factors, such as temperature, humidity, or microbial activity (Gao et al., 2022).

To summarize, XRD serves as a valuable tool in the study of plastic biodegradation, as it unveils molecular and atomic alterations within the material. This knowledge aids in evaluating the extent of degradation and potentially facilitates the development of more sustainable plastic materials.

7.4.4 X-Ray Fluorescence

The fundamental principle of X-ray fluorescence (XRF) spectroscopy is based on the interaction of X-rays with matter. When a sample is exposed to high-energy X-rays, the atoms in the sample absorb the X-ray energy and become excited. This causes the inner-shell electrons to transition to higher energy levels. Once the inner-shell electrons are excited, they quickly return to their ground state, releasing the excess energy in the form of X-rays. These emitted X-rays have characteristic energies that are specific to the elements present in the sample (Russ, 2013).

By measuring the energy and intensity of the X-rays emitted, XRF spectroscopy can identify the elements present in the sample and quantify their concentrations. This is done by comparing the measured X-ray wavelengths or energies to known standards or databases.

XRF can be performed using two main types of excitation sources: X-ray tube-based (wavelength-dispersive XRF, or WDXRF) and radioactive source-based (energy-dispersive XRF, or EDXRF). Both methods follow the same principles of X-ray excitation and detection (Georgakopoulou et al., 2017).

XRF is a powerful technique used for qualitative and quantitative analysis of various materials, such as metals, minerals, ceramics, and environmental samples. It is widely employed in fields ranging from geology and archaeology to materials science and quality control (Shackley, 2011).

X-ray fluorescence (XRF) spectroscopy can be applied in various ways to study plastic biodegradation. XRF can identify and quantify the elemental composition of degraded plastic samples. By analyzing the X-ray emissions from the sample, researchers can determine the presence and concentration of elements, such as carbon, oxygen, nitrogen, and other additives or contaminants that may be associated with biodegradation products (Finzi-Quintão et al., 2016).

XRF can also track alterations in the elemental composition of plastics as they biodegrade. By comparing the XRF spectra of original and degraded plastic samples, researchers can observe any elemental changes that occur during the biodegradation process. This information can provide insights into the breakdown and transformation of plastic materials (Zhang et al., 2021).

XRF can additionally help identify specific elements or compounds that result from the biodegradation of plastics. By analyzing the X-ray emissions from degraded plastic samples, researchers can detect the presence of new elements or compounds that may be formed during the biodegradation process. This knowledge contributes to a deeper understanding of the biodegradation mechanisms (Baidurah, 2022).

This technique can aid in assessing the environmental impact of plastic biodegradation. By studying the elemental composition of degraded plastic samples collected from different environments, researchers can gain insights into any potential leaching or migration of elements from the degraded plastics. This information is valuable for understanding the potential ecological implications of plastic biodegradation (Turner, 2016).

XRF can also be used to assess the degradation of additives commonly found in plastics, such as stabilizers, plasticizers, or flame retardants. By analyzing the changes in elemental composition or concentration of these additives using XRF, researchers can gain insights into their stability and fate during biodegradation (Zhang et al., 2021).

This technique can help in understanding the interactions between microorganisms and plastic during biodegradation. By mapping the elemental distribution on the surface of a plastic sample using XRF microscopy, researchers can study microbial colonization patterns and identify any elemental changes associated with microbial activity (Baidurah, 2022).

XRF analysis offers insights into the molecular breakdown of polymers during biodegradation. Through the examination of elemental composition or concentration shifts in polymer samples at different degradation stages, researchers can uncover specific degradation mechanisms.

Moreover, XRF proves valuable in assessing plastic biodegradation across varying environmental conditions, including temperature, humidity, and microbial activity. By scrutinizing elemental variations in plastic samples subjected to diverse conditions via XRF, researchers can compare degradation effectiveness and rates across different scenarios.

Furthermore, XRF finds application in traceability studies for identifying the source or origin of plastic materials. By scrutinizing the elemental composition of plastic samples using XRF, researchers can establish a distinctive signature for distinct batches or sources. This facilitates effective date coding and traceback analysis, serving quality control objectives (Zhang et al., 2021).

As can be seen, XRF spectroscopy is a valuable tool for investigating plastic biodegradation, providing information on elemental composition, changes in plastic structure, identification of degradation products, and environmental impact. It offers a nondestructive and relatively quick analytical method for studying the chemical transformations taking place during the biodegradation process.

7.5 Thermal Methods

7.5.1 Differential Scanning Calorimetry

The fundamental principle of differential scanning calorimetry (DSC) is to measure the heat flow associated with physical and chemical processes occurring in a sample as a function of temperature. DSC works by comparing the heat flow into or out of a sample with a reference material as both are subjected to a controlled temperature program. The sample and reference are placed in separate compartments of a calorimeter, and any temperature difference between them is continuously measured (Freire, 1995).

As the temperature is increased or decreased, various thermal transitions and reactions can occur in the sample, such as phase changes, melting, crystallization, and chemical reactions. These processes involve energy exchanges, which result in heat flow differences between the sample and reference. DSC measures the heat flow difference between the sample and reference and plots it against temperature, generating a thermogram or DSC curve. The DSC curve provides information about the energy changes associated with the processes occurring in the sample (Farah et al., 2018).

Key parameters obtained from a DSC curve include (Menczel et al., 2023) the following:

- Melting point: The temperature at which a solid material changes to a liquid. This is indicated by an endothermic peak in the DSC curve.
- Crystallization point: The temperature at which a supercooled liquid or amorphous material transforms to a crystalline state. This is indicated by an exothermic peak in the DSC curve.

- Glass transition temperature: The temperature at which an amorphous material undergoes a transition from a rigid, glassy state to a rubbery, more mobile state. This is indicated by an endothermic step or gradual change in the DSC curve.
- Enthalpy changes: The area under the peaks in the DSC curve provides information about the energy absorbed or released during a thermal event, such as melting or crystallization.

DSC is widely used in various fields, including materials science, polymers, pharmaceuticals, and food science, to investigate the thermal properties, phase transitions, chemical reactions, and stability of materials. It can also be applied in the study of plastic biodegradation in several ways.

DSC can identify and characterize changes in the thermal behavior of plastic materials during biodegradation. By examining DSC curves of original and degraded plastic samples, researchers can observe shifts in melting temperatures, crystallization behavior, and glass transition temperatures. These changes in thermal behavior can provide insights into the structural alterations and degradation mechanisms taking place in the plastic (Cafiero et al., 2015).

DSC can also be used to determine the kinetics of plastic biodegradation. By subjecting plastic samples to controlled temperature programs and monitoring the heat flows during degradation, researchers can quantify the rate of degradation. This information is valuable for understanding the degradation kinetics of plastic materials and predicting their durability under different environmental conditions (Vogel et al., 2021).

DSC can additionally help evaluate the impact of additives on the biodegradation behavior of plastics. By comparing the DSC curves of plastic samples with different additives or additive concentrations, researchers can assess how the presence of additives affects the thermal properties and degradation characteristics of the plastics. This knowledge is useful for optimizing the formulation of biodegradable plastics and understanding the role of additives in their degradation (Šudomová et al., 2023).

This technique can be used to determine the extent of compostability or biodegradability of plastic materials. By analyzing the DSC curves of plastic samples before and after composting, researchers can observe changes in thermal behavior that indicate the degree of degradation. This enables the assessment of a plastic material's suitability for composting applications and provides insights into its environmental impact (Delacuvellerie et al., 2021).

DSC can additionally help study the interactions between microorganisms and plastics during biodegradation. By monitoring the heat flow changes during the degradation of plastic samples in the presence of specific microorganisms, researchers can gain insights into the microbe-induced degradation processes. This information contributes to understanding the role of microorganisms in plastic biodegradation and can aid in the development of bio-based strategies for plastic waste management (Osman et al., 2018; Vogel et al., 2021).

All these examples highlight the versatility of DSC in investigating various aspects of plastic biodegradation, including thermal behavior changes, degradation kinetics, additive effects, compostability assessment, and microbial interactions. DSC provides valuable insights into the physical and chemical transformations occurring during the biodegradation of plastic materials.

7.5.2 *Differential Thermal Analysis*

Differential thermal analysis (DTA) is a technique used in materials science to analyze the thermal behavior of a sample as it is subjected to controlled temperature changes. It measures the temperature difference between the sample and a reference material as the two are simultaneously heated or cooled. During a DTA experiment, both the sample and reference material are placed in separate crucibles and heated or cooled at a constant rate. As the temperature changes, the difference in heat flow between the sample and reference material is measured, recorded, and plotted on a graph (Xu et al., 2005).

The fundamentals of differential thermal analysis (DTA) involve the measurement and analysis of temperature differences between a sample and a reference material as they undergo controlled heating or cooling. The key components of DTA are (Wagner, 2017) the following:

- Sample and reference: DTA requires both a sample of interest and a reference material. The sample can be a solid, liquid, or gas, while the reference material is typically an inert substance with a known thermal behavior.
- Temperature control: DTA experiments involve subjecting the sample and reference material to controlled temperature changes. This can be achieved by using a heating or cooling device that maintains a constant and programmable temperature ramp.
- Heat flow measurement: DTA measures the difference in heat flow between the sample and reference material as they are heated or cooled. This is done through thermocouples or other temperature sensors that monitor the temperature changes in both the sample and reference material.
- DTA curve: The DTA data is plotted as a graph known as a DTA curve. This curve shows the temperature difference between the sample and reference material as a function of time or temperature. It provides insights into thermal events such as phase transitions, chemical reactions, and thermal stability.
- Interpretation: Analyzing the DTA curve requires understanding the behavior of materials under different thermal conditions. Thermal events such as endothermic (heat absorption) or exothermic (heat release) peaks indicate specific processes occurring in the sample, such as melting, crystallization, or chemical reactions.

DTA can be used to investigate various properties of materials, such as phase transitions, thermal stability, thermal decomposition, and chemical reactions. By analyzing the DTA curves, researchers can gather information about the sample's melting point, crystallization behavior, heat capacity, and other thermal characteristics. DTA is a useful tool in fields like metallurgy, ceramics, polymers, and pharmaceuticals, allowing scientists to better understand the thermal properties and behavior of materials. It can be a valuable tool for studying the biodegradation behavior of plastic materials (Matjašič et al., 2021).

For example, DTA can help determine the temperature range at which plastic biodegradation occurs. By subjecting the plastic sample to temperature ramps, any thermal events associated with degradation, such as weight loss or changes in heat flow, can be observed and analyzed (Din et al., 2020).

DTA can also provide insights into the kinetics of plastic biodegradation. By varying the temperature ramp rates and analyzing the resulting DTA curves, researchers can determine the rate at which the material degrades and identify any temperature-dependent factors influencing the process (Das & Tiwari, 2017).

DTA, as the other techniques abovementioned, can be used to compare the biodegradation behavior of different plastic materials. By subjecting various plastics to the same thermal conditions and analyzing their DTA curves, researchers can evaluate their relative biodegradability and compare their thermal stability (Chamas et al., 2020).

Additionally, DTA can aid in understanding the mechanisms of plastic biodegradation. By analyzing the DTA curves and correlating them with other characterization techniques, researchers can gain insights into the thermal degradation pathways, such as depolymerization, chain scission, or other chemical reactions taking place during biodegradation (S. S. Ali et al., 2021).

DTA can also be used to study the effects of additives on plastic biodegradation. By conducting DTA experiments on plastic samples with different additives or additives at varying concentrations, researchers can identify any changes in thermal behavior and assess the impact of additives on the degradation process.

In conclusion, DTA can provide valuable information about the thermal behavior and degradation kinetics of plastic materials during biodegradation processes. This knowledge can contribute to the development of more sustainable plastics and improve our understanding of their environmental impact.

7.5.3 Thermogravimetric Analysis

Thermogravimetric analysis (TGA) is a technique used to analyze the thermal behavior of materials by measuring their weight changes as they undergo controlled temperature changes. It provides valuable information about the composition, stability, and decomposition characteristics of a material when subjected to heating or cooling. In a typical TGA experiment, a small sample of the material of interest is placed in a controlled atmospheric environment (usually a nonreactive gas like nitrogen). The sample is then subjected to a programmed temperature increase or decrease, known as a temperature ramp, using a furnace or other heating/cooling apparatus (Loganathan et al., 2017).

As the temperature changes, the TGA apparatus continuously measures the weight of the sample. This weight measurement allows for the detection of any weight loss or gain that occurs as the material undergoes thermal events such as decomposition, evaporation, oxidation, or other chemical reactions. The weight changes are typically recorded as a function of temperature or time. The recorded data is plotted on a graph, known as a TGA curve, which displays the percentage weight loss or gain of the material as a function of temperature or time. Peaks and plateaus on the TGA curve correspond to specific thermal events or reactions, providing insight into the material's stability, composition, and degradation behavior.

TGA can be coupled with other analysis techniques such as mass spectrometry (TGA-MS) or infrared spectroscopy (TGA-FTIR) to identify the evolved gases or analyze the chemical composition of the sample during the thermal process (Kök et al., 2017).

The essentials of TGA are as follows (Saadatkhah et al., 2020):

- Sample measurement: TGA involves weighing a small sample of the material of interest and placing it in a controlled atmospheric environment. The sample can be a solid, liquid, or gas.
- Temperature control: TGA experiments require precise temperature control. The sample is heated or cooled at a controlled rate using a furnace or other heating/cooling devices. The temperature ramp is typically programmed to increase or decrease linearly.
- Weight measurement: During heating or cooling, the TGA apparatus continuously measures the weight of the sample as it undergoes thermal changes. This is usually done using a highly sensitive microbalance or load cell.
- TGA curve: The weight change of the sample is plotted on a graph known as a TGA curve, which shows the percentage weight loss or gain as a function of temperature or time. The TGA curve provides information about the thermal stability, decomposition, or other reactions occurring in the material.
- Interpretation: The TGA curve can be analyzed to understand the thermal behavior of the material. The TGA curve may exhibit characteristic features such as weight loss or gain peaks, plateaus, or continuous weight changes, which correspond to various thermal events like degradation, evaporation, sublimation, oxidation, or decomposition.
- Additional analysis: TGA results can be further analyzed by coupling TGA with other techniques such as mass spectrometry (TGA-MS) or infrared spectroscopy (TGA-FTIR). This allows for the identification of evolved gases and the characterization of reaction products during the thermal process.

TGA is widely used in fields such as materials science, chemistry, polymers, and pharmaceuticals to study thermal stability, degradation behavior, and composition changes in various materials. It provides valuable insights into the thermal characteristics of substances, aiding in the development of new materials, evaluation of stability, and understanding of decomposition mechanisms (Escalante et al., 2022).

TGA can be applied in the study of plastic biodegradation to assess the thermal behavior and decomposition characteristics of plastic materials under controlled temperature conditions. TGA can help determine the temperature range at which plastic biodegradation occurs. By subjecting the plastic sample to a temperature ramp, any weight loss associated with degradation can be observed and quantified. The temperature at which significant weight loss occurs can provide insights into the thermal stability and degradation temperature of the plastic material (Bhandari et al., 2021).

TGA can also be used to study the kinetics of plastic biodegradation. By varying the temperature ramp rates and analyzing the resulting TGA curves, researchers can determine the rate at which the material degrades and observe any temperature-dependent factors influencing the degradation process. This information is valuable for understanding the degradation kinetics and designing more sustainable plastics (Escalante et al., 2022).

Additionally, TGA can provide information about the decomposition products generated during plastic biodegradation. By analyzing the evolved gases using techniques such as TGA-FTIR or TGA-MS, researchers can identify the types of gases released and gain insights into the degradation pathways and mechanisms (Eimontas et al., 2021).

TGA can also be used to compare the biodegradation behavior of different plastic materials. By conducting TGA experiments on different plastics under the same thermal conditions, researchers can assess their relative biodegradability and compare their thermal stability. This knowledge can guide the selection of appropriate plastics for specific applications with a focus on biodegradability (Patnaik et al., 2020).

In addition, TGA can be used to study the effects of additives on plastic biodegradation. By conducting TGA experiments on plastic samples with different types or concentrations of additives, researchers can assess how additives influence the thermal stability and biodegradation behavior of the plastic material (Arráez et al., 2019).

As can be noticed, TGA provides valuable insights into the thermal behavior and decomposition characteristics of plastic materials during biodegradation processes. This information aids in understanding their environmental fate, developing more sustainable plastics, and assessing the potential impact of additives or manufacturing processes on biodegradability.

7.6 Chromatographic Methods

Optical, electron, and atomic force microscopy methods are very useful for identifying and characterizing the properties of microplastics with a focus on degradation studies and technologies but are limited in their quantitative application. Chromatographic methods are primarily useful for the quantification of these micromaterials.

Both gas chromatography (GC) and high-performance liquid chromatography (HPLC) are useful for the quantification of microplastics. Regarding degradation and biodegradation, these chromatographic methods are of particular interest because they allow the determination of decomposition products, thus providing more chemical information about the process. Chromatographic methods are separation methods in themselves, but they have a quantitative application because there is always a detection system for the analytes that allows obtaining chromatograms that are used for quantification. A chromatogram is a plot of the detector response as a function of time.

In gas chromatography, the sample is volatilized, and separation occurs in a gas mobile phase that transports the analytes through the chromatographic column. The stationary phase can be solid or liquid, but the latter is more commonly used, where the liquid is immobilized in an inert solid. Several types of detectors are used in GC, the most common being flame ionization, thermionic, thermal conductivity, and atomic emission. However, the best analytical results are obtained when the chromatograph is coupled to a mass spectrometry, a hybrid technique often abbreviated as GC/MS. Mass spectrometry fragments molecules and detects them based on the mass-to-charge ratio of the fragments.

GC/MS has been used for the determination of microplastics. It is particularly important in the analysis of degradation products of polymers such as PP, PE, and PS. GC/MS analysis can be performed for microplastics in river, lake, and seawater, soils, and sediments. GC/MS can be combined with a pyrolysis step (PGC/MS). Pyrolysis uses high temperatures to break down the polymeric molecules of microplastics into smaller molecules that can be analyzed by GC/MS. The advantage of PGC/MS is that it allows the analysis of more complex molecules and provides a more complete structural characterization. PGC/MS can be coupled with a thermal desorption device, which allows better analysis of organic additives in plastics. All of these gas chromatography techniques are particularly important when trying to study the chemical effects that can be produced by the degradation processes of micropolymers, because they allow a comprehensive analysis not only of the microplastics themselves but also of the molecules associated with their degradation.

On the other hand, in HPLC, the mobile phase is a liquid that moves the analytes at high pressure through a stationary phase where they are retained and separated. Likewise, the method uses a detector to record a response signal from the analytes and quantify them through chromatograms. In this method, solubility and polarity play a fundamental role in the separation process. HPLC detectors are generally UV-visible, refractometers, fluorescence detectors, and electrochemical detectors. However, as in GC, to improve detection by eliminating the effect of interferences, it is possible to couple liquid chromatography to mass spectrometry (LC-MS).

Hybrid LC/MS is also particularly useful for the analysis of microplastic degradation products. When polymer degradation is used as a sample treatment, it is possible to quantify microplastics by analyzing the depolymerization products. For example, Wang et al. (2017) applied thermal basic hydrolysis to depolymerize two polymers: polycarbonate (PC) and polyethylene terephthalate (PET). They used LC-MS/MS to quantify bisphenol A and para-phthalic acid, which allowed them to determine the amounts of PC and PET in environmental samples. The use of an additional MS analyzer further improves detection.

In general, in both gas and liquid chromatography, analyte identification is based on retention times obtained from chromatograms, while quantification is based on integration of chromatographic peaks. Chromatographic methods have very low detection limits and are very useful for the quantitative study of microplastics and their degradation products, which is not possible with microscopic methods.

7.7 Conclusions

There is a wide variety of instrumental methods for the analysis of microplastics during degradation processes. Most of them are aimed at the identification and qualitative characterization of these materials. Chromatographic methods are the ones that can provide the best quantitative results. Hybrid techniques between microscopic methods, between microscopic and spectroscopic methods, and between chromatographic methods and mass spectrometry can offer the best analytical performance. However, simple methods can also be very powerful if the analyst is experienced in their use and application. A combination of analytical techniques is the best strategy for a comprehensive characterization of microplastics. The analysis must also pay close attention to the methods used to separate these materials and often to the digestion of interfering materials such as organic matter. Many of the methods for the analysis of microplastics can be improved and some are still an area of analytical research to achieve results with better accuracy and thus avoid under- or overestimation of these micromaterials in samples of mainly environmental interest. In addition, the study of their degradation products, although possible, is still limited and therefore represents another challenge for analytical research.

References

Abubaker, S. A., & Taha, A. H. (2021). Identification and characterization of different types of plastics wastes using X-ray diffraction and X-ray fluorescence techniques. *ARO-The Scientific Journal of Koya University, 9*(2), 22–25.

Achilias, D. S., Panayotidou, E., & Zuburtikudis, I. (2011). Thermal degradation kinetics and isoconversional analysis of biodegradable poly (3-hydroxybutyrate)/organomodified montmorillonite nanocomposites. *Thermochimica Acta, 514*(1–2), 58–66.

Ali, S. S., Elsamahy, T., Al-Tohamy, R., Zhu, D., Mahmoud, Y. A.-G., Koutra, E., Metwally, M. A., Kornaros, M., & Sun, J. (2021). Plastic wastes biodegradation: Mechanisms, challenges and future prospects. *Science of the Total Environment, 780*, 146590.

Ali, A., Chiang, Y. W., & Santos, R. M. (2022). X-ray diffraction techniques for mineral characterization: A review for engineers of the fundamentals, applications, and research directions. *Minerals, 12*(2), 205.

Archiza, B., Welch, J. F., & Sheel, A. W. (2017). Classical experiments in whole-body metabolism: Closed-circuit respirometry. *European Journal of Applied Physiology, 117*, 1929–1937.

Arráez, F. J., Arnal, M. L., & Müller, A. J. (2019). Thermal degradation of high-impact polystyrene with pro-oxidant additives. *Polymer Bulletin, 76*, 1489–1515.

Avella, M., De Vlieger, J. J., Errico, M. E., Fischer, S., Vacca, P., & Volpe, M. G. (2005). Biodegradable starch/clay nanocomposite films for food packaging applications. *Food Chemistry, 93*(3), 467–474.

Baidurah, S. (2022). Methods of analyses for biodegradable polymers: A review. *Polymers, 14*(22), 4928.

Bhandari, N. L., Bhandari, G., Bista, S., Pokhrel, B., Bist, K., & Dhakal, K. N. (2021). Degradation of fundamental polymers/plastics used in daily life: A review. *Bibechana, 18*(1), 240–253.

Cafiero, L., Fabbri, D., Trinca, E., Tuffi, R., & Vecchio Ciprioti, S. (2015). Thermal and spectroscopic (TG/DSC–FTIR) characterization of mixed plastics for materials and energy recovery under pyrolytic conditions. *Journal of Thermal Analysis and Calorimetry, 121*, 1111–1119.

Chamas, A., Moon, H., Zheng, J., Qiu, Y., Tabassum, T., Jang, J. H., Abu-Omar, M., Scott, S. L., & Suh, S. (2020). Degradation rates of plastics in the environment. *ACS Sustainable Chemistry & Engineering, 8*(9), 3494–3511.

Chatterjee, A. K. (2000). X-ray diffraction. In *Handbook of analytical techniques in concrete science and technology* (pp. 275–332).

Das, P., & Tiwari, P. (2017). Thermal degradation kinetics of plastics and model selection. *Thermochimica Acta, 654*, 191–202.

Delacuvellerie, A., Benali, S., Cyriaque, V., Moins, S., Raquez, J.-M., Gobert, S., & Wattiez, R. (2021). Microbial biofilm composition and polymer degradation of compostable and non-compostable plastics immersed in the marine environment. *Journal of Hazardous Materials, 419*, 126526.

DeLiberto, A. N., Drown, M. K., Oleksiak, M. F., & Crawford, D. L. (2020). Measuring complex phenotypes: A flexible high-throughput design for micro-respirometry. *BioRxiv*, 2003–2020.

Din, M. I., Ghaffar, T., Najeeb, J., Hussain, Z., Khalid, R., & Zahid, H. (2020). Potential perspectives of biodegradable plastics for food packaging application-review of properties and recent developments. *Food Additives & Contaminants: Part A, 37*(4), 665–680.

Dussud, C., Meistertzheim, A. L., Conan, P., Pujo-Pay, M., George, M., Fabre, P., Coudane, J., Higgs, P., Elineau, A., Pedrotti, M. L., Gorsky, G., & Ghiglione, J. F. (2018). Evidence of niche partitioning among bacteria living on plastics, organic particles and surrounding seawaters. *Environmental Pollution, 236*, 807–816. https://doi.org/10.1016/J.ENVPOL.2017.12.027

Eimontas, J., Striūgas, N., Abdelnaby, M. A., & Yousef, S. (2021). Catalytic pyrolysis kinetic behavior and TG-FTIR-GC–MS analysis of metallized food packaging plastics with different concentrations of ZSM-5 zeolite catalyst. *Polymers, 13*(5), 702.

Escalante, J., Chen, W.-H., Tabatabaei, M., Hoang, A. T., Kwon, E. E., Lin, K.-Y. A., & Saravanakumar, A. (2022). Pyrolysis of lignocellulosic, algal, plastic, and other biomass wastes for biofuel production and circular bioeconomy: A review of thermogravimetric analysis (TGA) approach. *Renewable and Sustainable Energy Reviews, 169*, 112914.

Farah, J. S., Silva, M. C., Cruz, A. G., & Calado, V. (2018). Differential calorimetry scanning: Current background and application in authenticity of dairy products. *Current Opinion in Food Science, 22*, 88–94.

Finzi-Quintão, C. M., Novack, K. M., & Bernardes-Silva, A. C. (2016). Identification of biodegradable and oxo-biodegradable plastic bags samples composition. *Macromolecular Symposia, 367*(1), 9–17.

Freire, E. (1995). Differential scanning calorimetry. In *Protein stability and folding: Theory and practice* (pp. 191–218). Springer.

Fu, W., Min, J., Jiang, W., Li, Y., & Zhang, W. (2020). Separation, characterization and identification of microplastics and nanoplastics in the environment. *Science of the Total Environment, 721*, 137561. https://doi.org/10.1016/J.SCITOTENV.2020.137561

Gao, R., Liu, R., & Sun, C. (2022). A marine fungus Alternaria alternata FB1 efficiently degrades polyethylene. *Journal of Hazardous Materials, 431*, 128617.

Georgakopoulou, M., Hein, A., Müller, N. S., & Kiriatzi, E. (2017). Development and calibration of a WDXRF routine applied to provenance studies on archaeological ceramics. *X-Ray Spectrometry, 46*(3), 186–199.

Gwinnett, C. M. B., Osborne, A. O., & Jackson, A. R. W. (2021). The application of tape lifting for microplastic pollution monitoring. *Environmental Advances, 5*, 100066. https://doi.org/10.1016/J.ENVADV.2021.100066

James, T. L. (1998). *Fundamentals of NMR* (Online Textbook) (pp. 1–31). Department of Pharmaceutical Chemistry, University of California.

Jung, S., Cho, S.-H., Kim, K.-H., & Kwon, E. E. (2021). Progress in quantitative analysis of microplastics in the environment: A review. *Chemical Engineering Journal, 422*, 130154.

Kalaronis, D., Ainali, N. M., Evgenidou, E., Kyzas, G. Z., Yang, X., Bikiaris, D. N., & Lambropoulou, D. A. (2022). Microscopic techniques as means for the determination of microplastics and nanoplastics in the aquatic environment: A concise review. *Green Analytical Chemistry, 3*, 100036. https://doi.org/10.1016/J.GREEAC.2022.100036

Khan, S., Ali, S. A., & Ali, A. S. (2023). Biodegradation of low density polyethylene (LDPE) by mesophilic fungus 'Penicillium citrinum' isolated from soils of plastic waste dump yard, Bhopal, India. *Environmental Technology, 44*(15), 2300–2314.

Kim, D. I. (2023). Metabolic rates of Japanese anchovy (Engraulis japonicus) during early development using a novel modified respirometry method. *Animals, 13*(6), 1035.

King, E. F., & Dutka, B. J. (2019). Respirometric techniques. In *Toxicity testing using microorganisms* (pp. 75–113). CRC Press.

Kök, M. V., Varfolomeev, M. A., & Nurgaliev, D. K. (2017). Crude oil characterization using tga-dta, tga-ftir and tga-ms techniques. *Journal of Petroleum Science and Engineering, 154*, 537–542.

Kumari, A., Chaudhary, D. R., & Jha, B. (2019). Destabilization of polyethylene and polyvinylchloride structure by marine bacterial strain. *Environmental Science and Pollution Research, 26*, 1507–1516. https://doi.org/10.1007/S11356-018-3465-1

Labbe, A. B., Bagshaw, C. R., & Uttal, L. (2020). Inexpensive adaptations of basic microscopes for the identification of microplastic contamination using polarization and Nile red fluorescence detection. *Journal of Chemical Education, 97*(11), 4026–4032. https://doi.org/10.1021/ACS.JCHEMED.0C00518

Lin, H., Bean, S. R., Tilley, M., Peiris, K. H. S., & Brabec, D. (2021). Qualitative and quantitative analysis of sorghum grain composition including protein and tannins using ATR-FTIR spectroscopy. *Food Analytical Methods, 14*, 268–279.

Loganathan, S., Valapa, R. B., Mishra, R. K., Pugazhenthi, G., & Thomas, S. (2017). Thermogravimetric analysis for characterization of nanomaterials. In *Thermal and rheological measurement techniques for nanomaterials characterization* (pp. 67–108). Elsevier.

Lv, L., Qu, J., Yu, Z., Chen, D., Zhou, C., Hong, P., Sun, S., & Li, C. (2019). A simple method for detecting and quantifying microplastics utilizing fluorescent dyes - Safranine T, fluorescein isophosphate, Nile red based on thermal expansion and contraction property. *Environmental Pollution, 255*, 113283. https://doi.org/10.1016/J.ENVPOL.2019.113283

Macfarlane, D. J. (2017). Open-circuit respirometry: A historical review of portable gas analysis systems. *European Journal of Applied Physiology, 117*(12), 2369–2386.

Mainardis, M., Buttazzoni, M., Cottes, M., Moretti, A., & Goi, D. (2021). Respirometry tests in wastewater treatment: Why and how? A critical review. *Science of the Total Environment, 793*, 148607.

Mariano, S., Tacconi, S., Fidaleo, M., Rossi, M., & Dini, L. (2021). Micro and nanoplastics identification: Classic methods and innovative detection techniques. *Frontiers in Toxicology, 3*, 636640. https://doi.org/10.3389/FTOX.2021.636640

Matjašič, T., Simčič, T., Medveššek, N., Bajt, O., Dreo, T., & Mori, N. (2021). Critical evaluation of biodegradation studies on synthetic plastics through a systematic literature review. *Science of the Total Environment, 752*, 141959.

Melo-Agustín, P., Kozak, E. R., de Jesús Perea-Flores, M., & Mendoza-Pérez, J. A. (2022). Identification of microplastics and associated contaminants using ultra high resolution microscopic and spectroscopic techniques. *Science of the Total Environment, 828*, 154434. https://doi.org/10.1016/J.SCITOTENV.2022.154434

Menczel, J. D., Andre, R., Kohl, W. S., Krongauz, V. V., Lőrinczy, D., Reading, M., & Grebowicz, J. (2023). Fundamentals of DSC. In *The handbook of differential scanning calorimetry* (pp. 1–189). Elsevier.

Moon, Y., Hwang, R. Y., Park, S., & Han, O. H. (2023). 1H NMR spectroscopy of degraded perfluorosulfonic acid membranes: A simple methodology for detecting onset of degradation. *Journal of Electroanalytical Chemistry, 932*, 117268.

Niaounakis, M., Kontou, E., Pispas, S., Kafetzi, M., & Giaouzi, D. (2019). Aging of packaging films in the marine environment. *Polymer Engineering & Science, 59*(s2), E432–E441.

Nikafshar, S., & Nejad, M. (2022). Evaluating efficacy of different UV-stabilizers/absorbers in reducing UV-degradation of lignin. *Holzforschung, 76*(3), 235–244.

Nowack, J., Dill, V., & Dausmann, K. H. (2020). Open-flow respirometry under field conditions: How does the airflow through the nest influence our results? *Journal of Thermal Biology, 92*, 102667.

Osman, M., Satti, S. M., Luqman, A., Hasan, F., Shah, Z., & Shah, A. A. (2018). Degradation of polyester polyurethane by aspergillus sp. strain S45 isolated from soil. *Journal of Polymers and the Environment, 26*, 301–310.

Patel, H., Kerndt, C. C., & Bhardwaj, A. (2018). *Physiology, respiratory quotient*. StatPearls Publishing.

Patnaik, S., Kumar, S., & Panda, A. K. (2020). Thermal degradation of eco-friendly alternative plastics: Kinetics and thermodynamics analysis. *Environmental Science and Pollution Research, 27*(13), 14991–15000.

Russ, J. C. (2013). *Fundamentals of energy dispersive X-ray analysis: Butterworths monographs in materials*. Butterworth-Heinemann.

Saadatkhah, N., Carillo Garcia, A., Ackermann, S., Leclerc, P., Latifi, M., Samih, S., Patience, G. S., & Chaouki, J. (2020). Experimental methods in chemical engineering: Thermogravimetric analysis—TGA. *The Canadian Journal of Chemical Engineering, 98*(1), 34–43.

Sarkhel, R., Sengupta, S., Das, P., & Bhowal, A. (2020). Comparative biodegradation study of polymer from plastic bottle waste using novel isolated bacteria and fungi from marine source. *Journal of Polymer Research, 27*, 1–8.

Shackley, M. S. (2011). An introduction to X-ray fluorescence (XRF) analysis in archaeology. In *X-ray fluorescence spectrometry (XRF) in geoarchaeology* (pp. 7–44). Springer.

Simonescu, C. M. (2012). Application of FTIR spectroscopy in environmental studies. *Advanced Aspects of Spectroscopy, 29*(1), 77–86.

Singh, S., Kumar, V., Singla, S., Sharma, M., Singh, D. P., Prasad, R., Thakur, V. K., & Singh, J. (2020). Kinetic study of the biodegradation of acephate by indigenous soil bacterial isolates in the presence of humic acid and metal ions. *Biomolecules, 10*(3), 433.

Sintim, H. Y., Bary, A. I., Hayes, D. G., Wadsworth, L. C., Anunciado, M. B., English, M. E., Bandopadhyay, S., Schaeffer, S. M., DeBruyn, J. M., & Miles, C. A. (2020). In situ degradation of biodegradable plastic mulch films in compost and agricultural soils. *Science of the Total Environment, 727*, 138668.

Smith, B. C. (2011). *Fundamentals of Fourier transform infrared spectroscopy*. CRC Press.

Strebl, M. G., Bruns, M. P., & Virtanen, S. (2023). Respirometric in situ methods for real-time monitoring of corrosion rates: Part III. Deconvolution of electrochemical polarization curves. *Journal of the Electrochemical Society, 170*(6), 61503.

Šudomová, L., Doležalová Weissmannová, H., Steinmetz, Z., Řezáčová, V., & Kučerík, J. (2023). A differential scanning calorimetry (DSC) approach for assessing the quality of polyethylene terephthalate (PET) waste for physical recycling: A proof-of-concept study. *Journal of Thermal Analysis and Calorimetry, 148*, 1–13.

Turner, A. (2016). Heavy metals, metalloids and other hazardous elements in marine plastic litter. *Marine Pollution Bulletin, 111*(1–2), 136–142.

Vogel, K., Wei, R., Pfaff, L., Breite, D., Al-Fathi, H., Ortmann, C., Estrela-Lopis, I., Venus, T., Schulze, A., & Harms, H. (2021). Enzymatic degradation of polyethylene terephthalate nanoplastics analyzed in real time by isothermal titration calorimetry. *Science of the Total Environment, 773*, 145111.

Wagner, M. (2017). *Thermal analysis in practice: Fundamental aspects*. Carl Hanser Verlag GmbH Co KG.

Wang, L., Zhang, J., Hou, S., & Sun, H. (2017). A simple method for quantifying polycarbonate and polyethylene terephthalate microplastics in environmental samples by liquid chromatography-tandem mass spectrometry. *Environmental Science and Technology Letters, 4*(12), 530–534. https://doi.org/10.1021/ACS.ESTLETT.7B00454

Xu, W., Li, S., Whitely, N., & Pan, W.-P. (2005). Fundamentals of TGA and SDT.

Xu, J.-L., Thomas, K. V., Luo, Z., & Gowen, A. A. (2019). FTIR and Raman imaging for microplastics analysis: State of the art, challenges and prospects. *TrAC Trends in Analytical Chemistry, 119*, 115629.

Zhang, K., Hamidian, A. H., Tubić, A., Zhang, Y., Fang, J. K. H., Wu, C., & Lam, P. K. S. (2021). Understanding plastic degradation and microplastic formation in the environment: A review. *Environmental Pollution, 274*, 116554.

Chapter 8
Genetically Engineered Plastic Munching Microbes: Recent Advancements and Perspectives

Neha Saini, Diksha Bharti, Shivani Singh, Ruksana Banoo, Vindhya Bundela, Pooja Nain, and Manvika Sahgal

Abstract The discovery of the plastic-degrading microbes has unveiled new avenues for addressing the global plastic pollution crisis and possess the potential to revolutionize waste management practices. Microorganisms possess substantial potential to facilitate a more sustainable plastic economy through processes like biodegradation and enzymatic recycling of polymers. With the advent of gene editing tools, the field of molecular biology introduced innovative methods to precisely integrate foreign genes into specific genomic locations in various microorganisms, addressing the numerous challenges linked with existing methods. Recently, gene targeting methods involving zinc finger nucleases (ZFNs), transcription activator-like effector nucleases (TALENs), and clustered regularly interspaced short palindrome repeats (CRISPR) have gained prominence. This chapter concentrates on diverse strategies for production of degrading enzymes, synthetic biology, metabolic engineering, and gene targeting, which offer promising solutions to plastic waste management.

Keywords Synthetic polymer · Biodegradation · Enzymatic catalysis · Genetically engineered · Microorganisms · Recombinant DNA technology · Polyethylene terephthalate

N. Saini · D. Bharti · S. Singh · R. Banoo · V. Bundela · M. Sahgal (✉)
Department of Microbiology, G.B. Pant University of Agriculture & Technology, Pantnagar, Uttarakhand, India

P. Nain
Department of Soil Science, College of Agriculture, G.B. Pant University of Agriculture & Technology, Pantnagar, Uttarakhand, India

R. Soni et al. (eds.), *Advanced Strategies for Biodegradation of Plastic Polymers*, https://doi.org/10.1007/978-3-031-55661-6_8

8.1 Introduction

Global plastic output is expected to reach 390.7 million metric tons in 2021, representing a 4% annual rise (Statista Research Department, 2023). Xia (2022) reported that plastic generation will increase by around 1800 million tons by 2050. According to the Central Pollution Control Board (CPCB) report (2019–2020), India produces 3.5 million tons of plastic waste in a year. Only 30% of it is recycled. Almost two-thirds of plastic trash originates from plastics having a lifetime of less than 5 years, and roughly 12% comes from consumer items. If current consumption trends continue, the polymer industry may consume up to 20% of the total fossil fuel by 2050 (Godfrey, 2019; CIEL, 2020).

Plastics are synthetic polymer-containing materials that are extensively utilized in packaging (40%), electronics and automobiles (6%), building and construction (19%), and agricultural industries (3%) (Plastics Europe, 2022; Huda & Widiastuti, 2021). It is a significant global concern due to the properties of plastic such as long life span, durability, flexibility, inertness, cheapness, and high use of single-use plastic products such as bottles, plastic bags, and packaging (Sanniyasi et al., 2021). After the coronavirus pandemic 2019, the use of single-use plastic wares such as gloves, masks, tissues, and surgical instruments has increased (Morgana et al., 2021; Yang et al., 2022). Poor management, low biodegradability, and its persistent nature leads to the presence of plastic in the environment for prolonged periods (Anand et al., 2023).

Plastic pollution is classified into three types: macroplastics (such as plastic chairs, plastic shopping bags, and vehicle parts), mesoplastics (such as fishing nets), and microplastics (such as plastic beads, pellets, and shattered plastic particles). Polyethylene (PE), high-density polyethylene (HDPE) and low-density polyethylene (LDPE), polystyrene (PS), polypropylene (PP), polyurethane (PUR), polyvinyl chloride (PVC), and polyethylene terephthalate (PET) are examples of synthetic polymers (Tokiwa et al., 2009) which account for 80% of the total worldwide plastic usage (Plastics Europe, 2019).

Conventional methods that are widely used for disposing of plastic trash are landfilling, incineration, recycling, and reusing (Ru et al., 2020). Landfilling and incineration are not considered safe and eco-friendly due to the accumulation of hazardous waste in the soil which leads to low soil fertility and the emission of toxic gases, i.e., greenhouse gases, into the environment causing global warming (Kale et al., 2015). Cycling often entails complex processes that result in low-value products with undesirable properties. These methods are ineffective, unsustainable, and non-eco-friendly and produces hazardous secondary products during plastic waste management. So, there is need for potential and eco-friendly method of plastic degradation.

Recently, biological methods have emerged as promising and environmentally friendly approaches that involve biodegradation and bioremediation. They are economically cheaper and complex processes that involve the following steps:

biodeterioration, depolymerization, and mineralization. Plastics, characterized by a carbon-carbon backbone structure, serve as carbon sources utilized by microbes such as bacteria, actinomycetes, fungi, and algae. Microbes involve the cleavage of chemical bonds and generate the energy by assimilation of electrons.

In this a microorganism biologically transforms complex form of polymers by the production of extracellular enzymes and utilizes it for the production of its biomass (Haider et al., 2019). The biodegradation rate of these synthetic plastics is also affected by physicochemical circumstances such as light, temperature, oxygen, and specific microbes (Rujnic-Sokele & Pilipovic, 2017). Currently, scientists have identified 27 enzymes capable of degrading synthetic polymers (Danso et al., 2019), most of them belonging to the serine hydrolase category and include cutinases, lipases, and carboxylesterases. *I. sakaiensis* produces the enzymes PETase and MHETase (Yoshida et al., 2016; Furukawa et al., 2019), which can efficiently break down polyethylene terephthalate and polyurethane.

A prominent milestone in the field of plastic degradation is the revelation of *Ideonella sakaiensis* 201-F6 bacterium that was initially discovered in 2016 in Japan. *Ideonella sakaiensis* was the first known plastic-eating bacteria that could destroy polyethylene terephthalate (PET) polymers by using them as a carbon and a source of energy for their metabolism processes. Prominent bacteria involved in plastic biodegradation include *Arthrobacter*, *Pseudomonas*, *Bacillus*, *Micrococcus*, *Corynebacterium*, *Nocardia*, *Escherichia coli*, and *Micrococcus* (Arman Alim et al., 2022). Moreover, various types of fungi, such as *Fusarium* spp., *Aspergillus* spp., and *Penicillium* spp., have been identified as contributors to plastic degradation (Ma et al., 2022). A number of algal species also are known to degrade polyethylene including the genera *Oscillatoria*, *Spirulina*, *Lyngbya*, *Nostoc*, *Chlorella*, *Hydrocoleum*, *Pithophora*, *Plasmodium*, *Phormidium*, *Anomoeoneis*, *Stigeoclonium*, and *Nitzschia* have been discovered to colonize low-density polyethene bags in water (Bhuyar et al., 2019).

Some issues in the biodegradation of plastic arise due to the nature of plastic such as hydrophobicity, size, and chemical nature (Sivan et al., 2006). Now there is an urgent need to promote and increase the efficiency and effectiveness of plastic-degrading microorganisms by the use of modern biotechnological methods. The modern biotechnological method involves recombinant DNA technology, gene modification, and gene cloning in order to enhance the biodegradation efficiency of microbes. These techniques alter the specificity and affinity of enzymes and construct a novel pathway within microbes to degrade various plastics (Kumar et al., 2020). In this chapter, we have provided a comprehensive overview of recent advancements in biotechnological strategies aimed at addressing plastic pollution. These include enzyme modification, genetic and metabolic engineering, gene editing tools, and the integration of smart technologies like computational biology, nanotechnology, and omics technologies into the biodegradation process (Fig. 8.1).

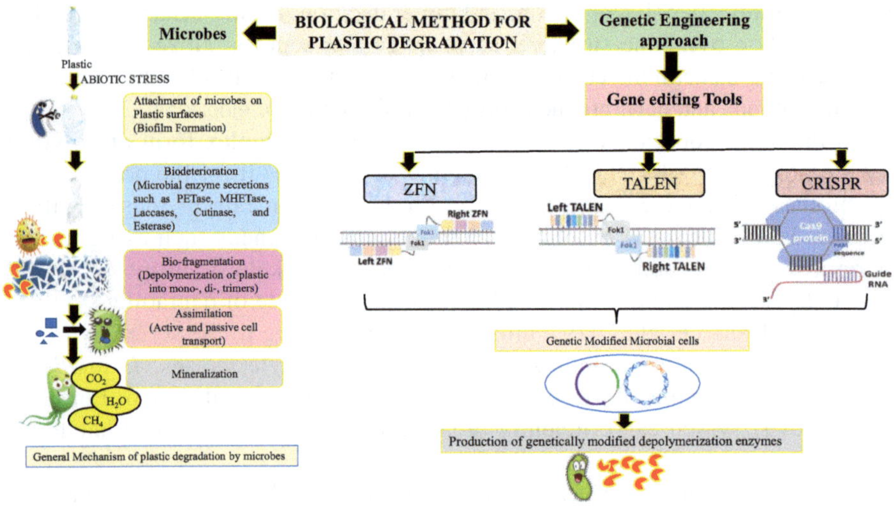

Fig. 8.1 Integration of advanced gene editing techniques in the general biological mechanism of plastic degradation

8.2 Genetic and Metabolic Engineering

8.2.1 Genetic Engineering

Jack Williamson coined the term "genetic engineering" in 1951, which is also known as "genetic modification" or biotechnology, a cutting-edge field at the intersection of biology, genetics, and technology. It involves the deliberate alteration of an organism's genetic material, typically its DNA, to introduce specific and desired traits or characteristics. It is a fundamental genetic process that involves the transfer of genetic material between two DNA molecules with similar or identical sequences (Hachisuka et al., 2021). Plasmids are frequently utilized as a genetic engineering vehicle for microbial system manipulation, despite being unstable for large-scale applications that require the deletion, insertion, overexpression, or mutation of the desired genes. Enzyme-based processes in molecular biology offer precise control and modification at various levels, including promoter and ribosome binding strengths, messenger RNA (mRNA), and protein synthesis. The Cre/lox recombinase system facilitates targeted gene insertion and modulation of gene expression levels. These techniques, along with PCR, antisense RNA, and site-directed mutagenesis, have been employed in degrading plastic polymers. Furthermore, various genetic recombination techniques have been employed to the wild-type bacterial strain, resulting in strains exhibiting significantly accelerated activity. By construction of the pRT2210 plasmid (pSEVA441-EGFP-PrpsL-PelB-IsPETase-FLAG tag plasmid), *Pseudomonas putida* transformed into a polyethylene terephthalate (PET)-eating bacteria that are capable of producing PETase enzymes that may have functions against PET plastic in saltwater (Matar, 2020). Recombinant DNA technology and gene editing have played a pivotal role in enhancing microbial capabilities to break down hazardous plastic waste. The primary goal is to isolate, modify, and incorporate genes related to degradation pathways into genetically modified organisms including *E. coli* and *S. cerevisiae* BY 4741 which have been engineered to produce enzymes such as manganese-dependent peroxidase, which has key role in polyethylene degradation (Sharma, 2018).

Genetically engineered microorganisms (GEMs) are also known as genetically modified organisms (GMOs) whose genes are manipulated by using genetic engineering. These encompass various microorganisms such as bacteria, fungi, and yeast. The in vitro techniques have been used to alter the genetic makeup of these microorganisms. This transformation process involves the introduction of specific genes into a single microorganism, resulting in the creation of multiple microorganisms with modified genetic characteristics (Liu et al., 2019; Sanghvi et al., 2020) and optimizing their biodegradation potential and allowing them to target specific contaminants. The utilization of GMOs in plastic degradation has the potential to offer more sustainable and eco-friendly solution to address the problem of plastic degradation. Researchers have utilized genetic engineering techniques to modify bacteria like *E. coli* as a host organism in order to produce enzymes capable of degrading plastics owing to its ease of manipulation rapid growth and high level of protein expression. For example, genetically engineered strains of *E. coli* BL21 and *P. chrysosporium* have been used for the production of laccase enzyme that belongs to a class of oxidoreductase and plays a role in breaking down various compounds, including lignin and some types of plastics (Jaiswal et al., 2020). Additionally, *E. coli* cells were altered and genetically modified to produce BIND-PETase. This enzyme is engineered from curli, which is a protein involved in the formation of bacterial biofilms, and it possesses the capability to depolymerize PET plastics (Zhu et al., 2022).

One notable breakthrough occurred at the University of Portsmouth in the UK, where researchers harnessed genetic recombination to express the functional enzymes IsPETase and MHETase. IsPETase enzymes were introduced into *Chlamydomonas reinhardtii* using a recombinant vector termed pBR9_PETase_Cre and MHETase into *Escherichia coli* (*E. coli*) BL21 utilizing the vector pUCIDT plasmid, which includes an indigenous signal peptide derived from *I. sakaiensis*. These enzymes recombined to produce a "super enzyme" which, compared to the previously existing strains, could break down PET plastic six times more quickly (Mathew et al., 2021; Kaur et al., 2023). This genetic modification improves the efficiency, stability, and specificity of the substrate and the longevity of plastic degradation processes as compared to naturally occurring plastic-degrading enzymes. These advancements present promising solutions for addressing environmental pollution and waste management challenges.

8.2.2 *Gene Editing Techniques*

8.2.2.1 Homologous Recombination-Based Gene Editing

Homologous recombination (HR)-based systems are early gene editing methods that have played a significant role in eukaryotic gene editing (Capecchi, 1989). These systems aim to introduce foreign genes into recipient cells and replace target genes through site-specific recombinases, achieving precise gene modifications. Notable examples are the Lambda-red system (*Escherichia coli*), the Cre-loxP (*Saccharomyces cerevisiae*), and FLP-FRT systems that are applicable to fungi.

Lambda-Red System

The Lambda-red system was introduced in 1998 by Murphy for gene modification only in *E. coli*. It functions independently of the host's recombinant machinery and employs three bacteriophage-derived proteins (Exo, Gam, and Beta) delivered through vector plasmids to facilitate gene modification. Exonuclease degrades double-stranded DNA (dsDNA) to create single-stranded DNA (ssDNA) fragments with a 3′ sticky end. Gram protein shields linear DNA from endogenous *E. coli* nucleases, and Beta annealing protein protects the ssDNA generated by Exo (Dillingham & Kowalczykowski, 2008). This system is independent of restriction sites and is flexible for gene editing, but it is specific to *E. coli*. Additionally, it has adapted to achieve gene editing in *Aspergillus* and can be combined with recent technologies like CRISPR.

Cre-loxP

The Cre-loxP system is based on DNA sequence-specific recombination enzyme Cre, discovered in P1 bacteriophages in 1981. The Cre recombinase acts on DNA substrates containing loxP sites, leading to various genomic modifications like deletions, inversions, translocations, and cassette exchanges. To employ the Cre-loxP system, loxP sites and a marker gene must first be integrated into the host genome, followed by the expression of Cre to accomplish gene modification. This was successfully applied in various organisms, including *Aspergillus*, *Saccharomyces*, *Trichoderma*, *Lecanicillium*, *Neurospora*, and others, making it a powerful tool for multigene manipulation and gene fragment integration (Xu et al., 2019; Moon et al., 2022).

FLP-FRT System

The FLP-FRT system is a similar approach like Cre-loxP (Sadowski, 1995), which originates from *S. cerevisiae* and uses flippase (FLP) to recognize pairs of FLP recombinase target (FRT) sequences flanking a genomic region. Like Cre-loxP, it also relies on HR with short homologous arms for nonspecific gene knock-in or knockout. While initially applied in *Penicillium chrysogenum* and *Sordaria macrospora* (Kopke et al., 2010), the system has developed into recombination-based genetic engineering and is widely used across various organisms, including yeast, filamentous fungi, plants, and mammals (Zeng et al., 2022) However, it is important to observe that these systems can only act on specific sequences recognized by recombinases and may have a higher probability of misintegration.

8.2.2.2 Nuclease-Based Technologies

Compared to traditional homology-directed repair (HR) gene editing techniques, nuclease-based gene editing methods rely on the utilization of designed nucleases for more precise targeting of specific sequences. These nucleases are capable of accurately acting on the desired DNA sequences reducing the likelihood of random insertions and, importantly, enabling the identification and modification of arbitrary gene sequences. Zinc finger nucleases (ZFNs), transcription activator-like effector nucleases (TALENs), and RNA-guided engineered nucleases (RGENs) are three major types of engineered nucleases that exhibit site specificity. They function by introducing double-strand breaks (DSBs) in the DNA, which can subsequently trigger either homology-directed repair (HDR) or nonhomologous end joining (NHEJ) mechanisms. Among them, CRISPR-Cas system has gained particular prominence as an effective tool for genomic exploration (Pant et al., 2022).

Zinc Finger Nucleases (ZFNs)

Zinc finger nucleases (ZFNs) are a type of engineered protein fused with a nonspecific FokI endonuclease catalytic domain and 3–6 Cys2-His2 zinc finger (ZF) DNA-binding domains that comprise approximately 30 amino acids. This DNA-binding domain is responsible for recognizing and engaging with a specific sequence of three nucleotides in target DNA. Since unique DNA sequences could be the focus of the ZF domain engineering, a double-strand break (DSB) can be induced at the target site when two ZFNs are designed to target contiguous sequences on a DNA double strand. Insertions or deletions at the target site may result from the repair of the DSB by the cell's internal repair mechanisms, such as nonhomologous end joining (NHEJ), which can cause gene disruption or knockout. This modular binding property allows ZFNs to identify and bind to DNA sequences that are 9–18 base pairs in length. By designing ZFNs with the appropriate DNA-binding modules, researchers can effectively target and induce double-strand breaks at desired locations in the chromosome. The capability of ZFNs to generate site-specific DNA breaks has revolutionized genetic engineering and genome editing. While ZFN technology has been utilized relatively successfully in organisms like *A. nidulans* (Fayyaz et al., 2020), its extensive use has been limited. This limitation is primarily due to the complex and time-consuming process of constructing ZF domains and verifying their specificity and cytotoxicity, further hindering their adoption in research.

Transcription Activator-Like Effector Nucleases (TALENs)

Transcription activator-like effector nucleases are composed of a nuclease domain, frequently FokI, and a DNA-binding domain derived from transcription activator-like effectors (TALEs). TALEs are naturally generated proteins from the *Xanthomonas* species of plant pathogenic bacteria (Grau et al., 2013). TALENs are intended to recognize particular DNA sequences, just like ZFNs. A DSB (double-strand break) can be produced by directing TALENs to particular sequences on two strands of the DNA double helix. If donor DNA template is readily available, repair of the DSB can result in either precise alteration through homology-directed repair (HDR) or gene disruption by nonhomologous end joining. TALENs comprise DNA-binding domains made up of multiple 33–35 amino acid repeat domains, all of them recognizing one base pair. As a result, recognition of C, T, A, and G only requires the occurrence of a minimum of four different types of DNA-binding module. Furthermore, compared to triplet-restricted ZFNs, the single-base identification of TALE-DNA-binding domain repeats offered more design flexibility (Joung & Sander, 2013).

Currently, TALEN technology has found success in generating stable genetic modifications in human cells and model organisms, including *Drosophila*, *Caenorhabditis elegans*, *zebrafish*, *mice*, and *Arabidopsis thaliana* (Fayyaz et al., 2020). In *A. oryzae*, it has been established that TALENs can delete large fragments of target DNA sequences (Mizutani et al., 2017). However, it is important to note that each base within the target sequence requires a specific TALE recognition module, making the construction process of TALENs labor-intensive.

RNA-Guided Engineered Nucleases (RGENs) or CRISPR-Cas Systems

RGENs, also known as CRISPR-Cas systems ("clusters of regularly interspaced short palindromic repeats'), utilize a short guide RNA (sgRNA) to target a specific DNA sequence. This discovery unfolded in *Escherichia coli*, where an unusual repetitive DNA sequence was identified in 1987 by Y. Ishino on a gene responsible for isoenzyme conversion of alkaline phosphatase. These repetitive sequences eventually came to be known as CRISPR. CRISPR regions had "spacer" DNA sequences interspersed among the repeats. These spacers turned out to be genetic sequences serving as an acquired immunity system. CRISPR/Cas are the most reliable and highly specific tools for precision of genome editing and engineering (Ishino et al., 2018).

CRISPR-Cas systems are classified into two classes depending on the complexity of Cas proteins and the organization of effector domains. Class 1 includes Type I, Type III, and Type IV, and they require the Cas complex to recognize and degrade foreign DNA. Class 2, on the other hand, comprises three types (II, V, and VI) and is further divided into ten subtypes. Class 2 systems are distinct in that they only need a single Cas protein guided by RNA to perform interference functions and have been extensively used in gene editing (Koonin et al., 2017). Class 2 Type II CRISPR-Cas9 systems are well-studied; Cas9 is used with RuvC and HNH nuclease domains guided by sgRNA to create blunt end DNA double-strand breaks. Class 2 Type V Cas12a (Cpf1) does not require tracrRNA, uses crRNA alone, and generates sticky end double-strand breaks. Cas12a has potential for multiple targeting and RNA editing. Class 2 Type VI Cas13 edits RNA and shows research promise. The functionality of CRISPR/Cas9 proteins is notably specific in their interaction with foreign DNA (Fig. 8.2).

The process begins with the assembly of a pre-CRISPR RNA (pre-CrRNA) molecule, which occurs through the addition of tracrRNA, and the Cas9 protein forms a Cas complex or CrRNA-tracrRNA-Cas9 complex (Jiang & Doudna, 2017). This Cas complex is recognized as a particular genetic trait known as the protospacer adjacent motif (PAM). Normally, the PAM is situated close to the target of interest or the foreign DNA. Once the Cas complex identifies this PAM sequence, it initiates a meticulous and precise cleavage of the foreign DNA. This recognition event is key, as it directs the system to the location where the nucleic acid sequence needs to be excised (Tarasava et al., 2018). Repair of the DSB by nonhomologous end joining can lead to gene disruption, while repair via homology-directed repair can be harnessed to insert, correct, or replace specific DNA sequences. CRISPR-Cas9 has become immensely popular because of its ease of design and versatility in genome editing applications (Table 8.1).

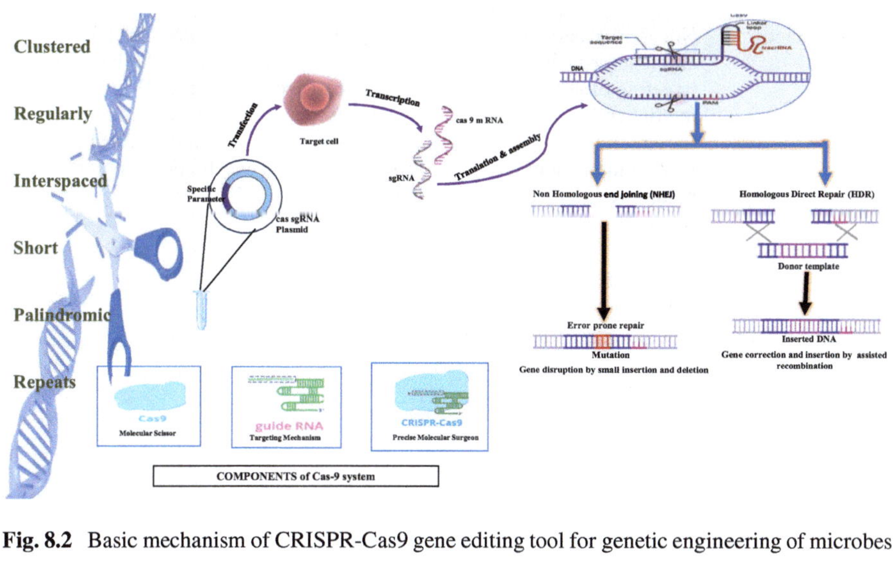

Fig. 8.2 Basic mechanism of CRISPR-Cas9 gene editing tool for genetic engineering of microbes

Table 8.1 Differences between ZFN, TALEN, and CRISPR/Cas9

Characteristics	ZFN	TALEN	CRISPR/Cas9
Origin and occurrence	1996, all eukaryotes	2009, *Xanthomonas*	2012, *Streptococcus pyrogens*
Abbreviation	Zinc finger nucleases	Transcription activator-like nucleases	Clustered regulatory interspaced short palindromic repeats
Gene editing	Man-made	Man-made	Natural defense mechanism in bacteria
Composition	Zinc finger domain and Fok1 endonuclease	TALE protein (bacterial) and Fok1 endonuclease	Two RNAs (trans-activating crRNA and a single-guide RNA) and Cas proteins
Precision	Moderate	Higher	Higher
Effectiveness	Ordinary	Ordinary	Higher
Immune reaction	Low	Low	High
Length of target DNA sequence	9–36 bp	30–40 bp	22 bp and PAM sequence
Mechanism of action (breakdown of DNA)	Targeted protein-DNA and double-strand break	Targeted protein-DNA and double-strand break	Targeted RNA-DNA and single- or double-strand break
Construction	Difficult	Easy	Easy
Off-target effect due to mutagenesis	Difficult to determine and not known	Difficult to determine and not known	Easy to determine and known
Multiplexing capability	Not	Not	Yes
Advantages	Less than 1 kb protein size permit packing in a single AAV	Precise recognition by 1 bp Rather simple target recognition region	Modify DNA with higher precision Permit multiplexing
Disadvantages	Not appropriate for more efficient throughput targeting of gene, required multiple of three sequences for targeting Inconvenient cloning requires extra linker sequences to attach modules Designing is hard Not stable	Packaging into virus is challenging due to large size of protein Specificity may be affected by repetitive sequences Target sites need a 5′ thymine base	Cas9 nuclease is relatively large to pack into AAV Off-targeting is more Limited PAM motif in human genome

Abbreviation: *AAV* adeno-associated virus vector, *PAM* protospacer adjacent motif

8.3 Metabolic Engineering of Enzymes

8.3.1 Enzymes Involved in Degradation of Synthetic Polymers

Enzymes serve as biological catalysts for the microbial degradation of plastic because they enable microorganisms for direct uptake and greatly speeding up the high-molecular-weight conversion of plastic polymers into valuable end products. The mechanism responsible for plastic degradation relies on enzymes found within microbial cells, either inside or outside the cell. These enzymes effectively dismantle the polymer chains, incorporating them into the microbial cells and ultimately releasing metabolic byproducts such as carbon dioxide (CO_2), water (H_2O), methane (CH_4), and nitrogen (N_2) (Amobonye et al., 2021). They can act either at the ends of the polymer chains (exolytic or exo-attack) or within the polymer chains (endolytic or endo-attack). When enzymes act from the outside (exo-attack), they produce smaller oligomers or monomers, which are eaten by microorganisms. In contrast, when enzymes act within the polymer (endo-attack), they significantly reduce the molecular weight of polymers, making the resulting products less likely to be assimilated without further degradation. For example, brominated high-impact plastic can be successfully broken down by extracellular depolymerization of *Pseudomonas* sp., and internal polyethylene glycol (PEG) dehydrogenase from *P. stutzeri* is used to change PEG into glyoxylic acid. Currently, scientists have identified 27 enzymes capable of degrading synthetic polymers (Danso et al., 2019), most of them belonging to the serine hydrolase category, including cutinases, lipases, and carboxylesterases.

8.3.2 Hydroxylases (Alkane Hydroxylases)

Alkane hydroxylases play a crucial role in breaking down various plastics like polyethylene, polystyrene, polypropylene, and polyvinyl chloride due to their carbon-containing structures (Pinto et al., 2022). Specifically, AlkB enzymes play a vital role in breaking down linear alkanes found in plastics. They achieve this through terminal or subterminal (ω-1) hydroxylation, resulting in the production of 1-alkanol or 2-alkanol, respectively. Alkanols are further oxidized by various enzymes leading to the formation of alkanoic acid (Inderthal et al., 2021). In efforts to enhance plastic degradation, the co-expression of AlkB enzymes with specific redox partners from *Pseudomonas aeruginosa* in *E. coli* has demonstrated significant potential. This cooperative action leads to the conversion of plastics, particularly polyethylene, into carbon dioxide (Park & Park, 2018). Notably, *P. aeruginosa* possesses two alkane hydroxylases, AlkB1 and AlkB2, with AlkB2 exhibiting greater effectiveness in breaking down low-molecular-weight polyethylene (LMWPE) compared to AlkB1 (Ali et al., 2023).

Moreover, bacteria that lack AlkB enzymes but contain the cytochrome P450 enzyme have also shown promise in plastic degradation (Li et al., 2020). By engineering the active site of enzyme in genetically modified bacteria like *E. coli*, *Pseudomonas putida*, or *Ralstonia*, along with the incorporation of essential enzymes such as P450, Adh, BVMO, and the esterase CYP153 (cytochrome P450 alkane hydroxylases) emerging as promising aerobic alkane degradation candidates for plastic degradation (Liang et al., 2016; Mohanan et al., 2020).

8.3.3 Hydrolases

Hydrolases like esterases, lipases, and cutinases are involved in breaking down plastic polymers. Esterases catalyze the cleavage of ester bonds in plastics, yielding alcohols, phenols, and acids. *Pseudomonas* sp. AKS2, for instance, produces an esterase capable of metabolizing polyethylene succinate (PES) into succinic acid, an essential component of the tricarboxylic acid (TCA) cycle (Abdelghafour et al., 2022). Lipases break down the triglyceride ester linkages into free fatty acids and glycerol. Some bacteria, like *Bacillus subtilis*, *Pseudomonas aeruginosa*, and *Burkholderia cepacia*, produce extracellular lipases that can degrade specific polyester polymers, including poly(ε-caprolactone) (PCL) and poly-4-hydroxyalkanoate (P(4HB) polymers. Lipases from *Aspergillus oryzae* and *Candida antarctica* have been synergistically utilized to hydrolyze PET into terephthalic acid (TPA), enhancing PET degradation (Maurya et al., 2020). But the naturally occurring hydrolases have some limitations such as substrate specificity, limited efficiency, environmental impact, stability, and longevity, so overcoming these limitations is necessary in engineering these enzymes.

8.3.4 PETase and MHETase Systems

In the bacterium *Ideonella sakaiensis*, two enzymes, PETase and MHETase, have been discovered (Yoshida et al., 2016). PETase (EC 3.1.1.101) belongs to the superfamily of α/β hydrolase and catalyzes the depolymerization of PET plastics by hydrolyzing the ester linkages within the polymer. Polyethylene terephthalate is degraded by both aerobes and anaerobes. The aerobic microbes, such as *Thermobifida*, *Saccharomonospora*, *Streptomyces*, *Stenotrophomonas pavanii*, *Thermobifida fusca*, and *Ideonella sakaiensis* (Farzi et al., 2019; Huang et al., 2022), and anaerobes such as *Pelotomaculum* and *Syntrophorhabdus*, are involved in the degradation of PET. This process generates dimeric bis(2-hydroxyethyl) (BHET) as the minor product and mono(2-hydroxyethyl) (MHET) as the major product. BHET molecules can further undergo degradation to convert into monomeric MHET. It has conserved active site catalytic triad of Ser-His-Asp and an optimal substrate binding site capable of accommodating four mono(2-hydroxyethyl) terephthalate (MHET) moieties of PET (Joo et al., 2018). PETase demonstrates significantly higher activity on PET compared to other enzymes but lower activity on p-nitrophenol-linked aliphatic esters.

MHETase (EC 3.1.1.102) belongs to the α/β hydrolase family that functions in cooperation with IsPETase, which is a two-enzyme system, to efficiently break down PET, releasing ethylene glycol (EG) and terephthalic acid (TPA). The intracellular mechanism of plastic degradation by microbes, especially with regard to the abovementioned enzymes, is still an area of ongoing research and requires further investigation (Knott et al., 2020). Utilizing genetic engineering tools, scientists have successfully generated mutant variants of the IsPETase and MHETase enzymes with notable enhancements in stability, catalytic efficiency, and substrate specificity (Kan et al., 2021).

8.3.5 Ligninolytic Enzymes

A group of white-rot fungi, i.e., *Phanerochaete*, *Pleurotus*, *Ganoderma*, *Bjerkandera*, *Dichomitus*, and *Trametes* (Koller et al., 2023; Voběrková et al., 2018), also plays a crucial role in the degradation of synthetic plastic polymers. Lignin peroxidases (LiP), laccase (Lac), manganese-dependent peroxidases (MnP), and versatile peroxidase (VP) are the four primary enzymes that make up the ligninolytic system in white-rot fungi. These enzymes exhibit remarkable oxidizing capabilities and versatility, e.g., *Pleurotus ostreatus* plays a significant role in the degradation of LDPE (Gómez-Méndez et al., 2018).

Laccases, also known as Lac (EC 1.10.3.2), are extracellular glycoproteins capable of oxidizing a wide range of compounds, including phenols, aromatic amines, and even non-phenolic substances when aided by mediators, for example, ABTS (2,2′-azinobis-(3-ethylbenzothiazoline-6-sulfonic acid), TEMPO (2,2,6,6-tetrameth yl-1-piperidinyloxy), and HBT (1-hydroxy-benzotriazole) (Singh & Gupta, 2020). Some microorganisms such as *Bacillus*, *Rhodococcus ruber*, *Psychrobacter*, *Pleurotus ostreatus*, and *Aspergillus flavus* produce laccase that breaks down polyethylene (Zhang et al., 2020; Shilpa & Meena, 2022).

Lignin peroxidase, Lip (EC 1.11.1.14), is an extracellular heme glycoprotein renowned for its role in lignin degradation. It is extensively studied and characterized in the fungus *Phanerochaete chrysosporium*. It oxidizes non-phenolic aromatic compounds, often utilizing hydrogen peroxide and mediators like veratryl alcohol to broaden its substrate. Veratryl alcohol is a fungal secondary metabolite known to enhance the enzyme's action (Zacchi et al., 2000).

Manganese peroxidase (MnP; EC 1.11.1.13) is a heme glycoprotein extracellular isoenzyme secreted by white-rot fungi with a notable ability to oxidize Mn^{2+} to Mn^{3+} during substrate oxidation, facilitating the degradation of both phenolic and non-phenolic compounds (Chowdhary et al., 2019; Kumar & Chandra, 2020). It has been extensively studied in the model fungus *Phanerochaete chrysosporium*, which serves as a key organism for lignin and xenobiotic biodegradation research (Palma et al., 2000). Versatile peroxidase (VP; EC 1.11.1.16), found primarily in *Pleurotus* and *Bjerkandera* species, combines features of LiP and MnP, allowing it to efficiently oxidize compounds with varying redox potentials, including plastics. VP holds significant promise for degrading persistent pollutants, showcasing the potential of white-rot fungi in combating plastic pollution (Kumar & Chandra, 2020).

8.3.6 Cutinases

Cutinases (EC 3.1.1.74), similar to PETase, belong to the category of α/β hydrolases and share a classical catalytic triad consisting of serine (Ser), histidine (His), and aspartic acid (Asp). These enzymes are naturally synthesized by plant pathogens to break down polyesters found in cutin and suberin layers but exhibit the capability to catalyze reactions involving various polyesters, triacylglycerols, waxes, and plastics (Nyyssölä & Ahlgren, 2019). Fungal cutinases have been isolated from diverse species such as *Penicillium citrinum*, *Aspergillus fumigatus*, *Aspergillus nidulans*, *Aspergillus oryzae*, *Fusarium solani* f.sp. *pisi*, *Fusarium oxysporum*, *Humicola insolens*, *Saccharomonospora viridis*, and *Moniliophthora roreri* (Ping et al., 2017; Tan et al., 2021) demonstrating the ability to hydrolyze low-crystallinity PET, polyethylene succinate (PES), polystyrene (PS), polycaprolactone (PCL), and polyethylene (PE). Bacterial cutinases, like those from *Thermobifida cellulosilytica* and *Thermobifida fusca*, can break down PET, reducing its crystallinity. Thc_Cut1 and Thc_Cut2 from T. *cellulosilytica*, as well as Thf42_Cut1 from *T. fusca*, have demonstrated hydrolytic activity on PET, releasing important breakdown products like MHET and TPA. To enhance efficiency, site-directed mutagenesis has been proposed to alter the Thc_Cut1 and Thc_Cut2 and improve its hydrolytic activity (Carr et al., 2020). When the active site of Thc_Cut1 was a double-mutated Q132A/T101A, it exhibited a 1.6-fold increase in hydrolytic activity of PET. In another study, the active site of TfCut2 was modified resulting in mutants like G62A and G62A/I213S. G62A had a fourfold higher hydrolytic rate and better substrate binding than the wild-type enzyme (Wei et al., 2016). Cut190 (S226P/R228S) double mutant from *Saccharomonospora viridis* AHK190, for PET degradation, showed improved activity and thermostability in contrast to the wild-type enzyme. The presence of calcium ions further enhanced its performance (Kawai et al., 2014).

Moreover, initiatives are being undertaken to engineer microorganisms like *Clostridium thermocellum* with PET-degrading cutinases, expanding their potential applications in plastic waste management. These findings highlight the versatility of cutinases in plastic degradation, underscoring their significance in addressing plastic pollution (Kaushal et al., 2021) (Table 8.2).

Table 8.2 Microorganisms and their mode of action in degrading different types of synthetic plastic

S. no.	Microorganism	Mode of action	Plastic degradation	References
1.	*Pseudomonas aeruginosa E7*	Alkane hydroxylase enzyme	Low-molecular-weight polyethylene	Jeon and Kim (2015)
2.	*Pseudomonas* sp.	Oxidative or hydrolytic enzymes	Polyethylene	Wilkes and Aristilde (2017)
3.	*Ideonella sakaiensis* 201-F6	PETase enzyme	Poly(ethylene terephthalate) (PET)	Austin et al. (2018)
4.	*Cladosporium cladosporioides*, *Xepiculopsis graminea*, *Penicillium griseofulvum*, and *Leptosphaeria* sp.	Extracellular enzyme diffused from hyphae into media and by oxidative reaction	Polyurethane	Brunner et al. (2018)
5.	*Ideonella sakaiensis*	PETase and MHETase enzyme	Polyethylene terephthalate (PET)	Palm et al. (2019)
6.	*Thermomonospora curvata*	Cutinase enzyme	Polyesters (PET), polypropylene, polyurethane.	Islam et al. (2019)
7.	*Aspergillus tamarii*	Lipases	Polyethylene terephthalate	Anbalagan et al. (2021)
8.	Amycolatopsis mediterranei	Cutinase	Poly(ε-caprolactone)	Tan et al. (2021)
9.	*Moraxella* sp. TA144	Psychrophilic esterase enzyme	Polyethylene terephthalate, polyurethane, Polyhydroxy butyrate, polybutylene succinates, polylactic acid	Nikolaivits et al. (2022)
10.	*Citrobacter freundii* and *Bacillus* sp.		Polyethylene	Ali et al. (2023)
11.	*Cladosporium* sp. CPEF-6	Laccase enzyme	Low-density polyethylene (LDPE)	Gong et al. (2023)
12.	*Chlamydomonas reinhardtii*	PETase enzyme	Polyethylene terephthalate (PET)	Di Rocco et al. (2023)
13.	*Candidatus* Bathyarchaeota archaeon	PET46 (putative archeal PET- degrading esterase)	Polyethylene terephthalate (PET)	Perez-Garcia et al. (2023)

8.4 Engineering of Metabolic Enzymes

8.4.1 PETase Engineering

Efforts to improve the thermal stability and activity of native PETase from *Ideonella sakaiensis* have involved strategic mutagenesis. One significant mutation, IsPETaseR280A, replaced arginine (R) at position 280 with alanine (A), resulting in a 22.4% increase in TPA and MHET release compared to wild-type PETase. Son et al. (2019) further enhanced PETase by introducing mutations S121D/D186H and S121E/D186H, creating variants with improved thermal stability through additional hydrogen bonds. Combining the R280A mutation with these stabilizing mutations in the triple mutant (S121E/D186H/R280A) significantly improved PET degradation by 13.9-fold compared to the native enzyme. Dai et al. (2021) introduced hydrophobic substrate-binding domains, including cellulose-binding domain (CBM), to PETase's C-terminus. Among the mutants, IsPETaseEHA_CBM showed a 2.28-fold increase in PET degradation product concentration compared to the original mutant, enhancing PETase's binding to PET plastic. Ma et al. (2018) targeted six key residues near the substrate-binding groove, resulting in mutants with increased PET affinity, notably I179F with the highest catalytic activity. Liu et al. (2018) improved PETase catalytic efficiency through structure-guided mutagenesis, leading to increased activity in mutants W130H and S209F. Austin et al. (2018) narrowed the binding cleft through mutagenesis, yielding the double-mutant S238F/W159H, which improved PET degradation and demonstrated the ability to degrade PEF. Additionally, photosynthetic microalgae *Phaeodactylum tricornutum* were engineered to produce hybrid PETase, enabling PET and PETG degradation at moderate temperatures (21 °C) and yielding the valuable breakdown products MHET and TPA (Moog et al., 2019). These mutagenetic approaches collectively contribute to the advancement of PETase variants with enhanced plastic-degrading capabilities and stability.

8.4.2 MHETase Engineering

The metabolic engineering of MHETase is not as advanced as PETase. However, there have been some noteworthy studies involving modifications to the amino acid sequence of MHETase. Palm et al. (2019) made an amino acid change within the enzyme's active site. Specifically, they focused on the amino acid Phe495, which plays an important role in substrate binding. By replacing Phe495 with alanine (A), they created the F495A protein variant. Their findings revealed that the mutant had a turnover rate for MHET that was more than two times lower compared to the wild-type enzyme, with a rate of about 5 s^{-1} (Palm et al., 2019). Another study, conducted by Sagong et al. (2020), aimed to engineer MHETase to enhance its ability to degrade BHET. The research indicated that MHETase could bind to BHET as a substrate, but its hydrolytic activity was quite low. Through targeted mutagenesis, they identified the importance of hydrophobic residues Leu254, Trp397, Phe415, and Phe495 in substrate binding and enzymatic catalysis. They performed several mutations, primarily centered around phenylalanine position 424 (F424). Three mutants (F424N, F424V, and F424I) exhibited a significant increase in activity against BHET, threefold higher than the native MHETase. Additionally, a point mutation at position 411, changing arginine to lysine (R411K), resulted in a 1.7-fold increase in activity against BHET compared to the wild-type enzyme. Building on these single mutants, further mutagenesis was carried out, leading to the creation of double protein variants (R411K/F424N, R422K/F424V, and R411K/F424I). These double mutants displayed relative activities against BHET of 8.7%, 10.5%, and 11.1%, respectively, compared to the native MHETase, which had only 1% relative activity against this substrate. Subsequent studies on MHETase mutants were based on earlier work by Palm et al. (2019), which identified the importance of the S416A mutation. Researchers then developed a triple mutant, R411K/S416A/F424I, which is shown to be 15.3-fold more active against BHET than the wild-type MHETase, significantly improving its ability to degrade PET films (Sagong et al., 2020).

8.4.3 Modification of Cutinases

FsCut, a cutinase enzyme derived from *Fusarium solani* f.sp. *pisi*, was genetically modified and subsequently heterologous expression in *Escherichia coli*, to enhance its activity to degrade solid polybutylene adipate terephthalate (PBAT) discs and utilize polybutylene succinate (PBS) as the sole carbon source in the growth medium (Santos-Beneit et al., 2023). High-level expression of leaf-branch compost cutinase (LCC) was achieved by inserting a signal peptide sequence and a constitutive promoter in *Clostridium thermocellum.* This engineered LCC exhibited improved PET film degradation, with a maximum weight loss of approximately 62% after 14 days of incubation at 60 °C (Yan et al., 2021).

8.4.4 Engineered Lipases

In the current investigation, the intracellular encoding genes of lipases, LIP1 (EY04_08410) and LIP2 (EY04_09635), both originated from *Pseudomonas chlororaphis* PA23. These genes were cloned and subsequently expressed within *Escherichia coli* BL21 (DE3). Both enzymes exhibited broad substrate specificity, including the ability to hydrolyze polyhydroxyalkanoates (PHAs), polylactic acid (PLA), and para-nitrophenyl (pNP) alkanoates. Gel permeation chromatography (GPC) analysis demonstrated a reduction in polymer molecular weight when incubated with LIP1 and LIP2. Additionally, these enzymes displayed some polymer-degrading activity on petroleum-based polymers like poly(ε-caprolactone) (PCL) and polyethylene succinate (PES), suggesting their potential use in biodegrading synthetic polyester plastics (Mohanan et al., 2022). The potential of these enzymes holds promise for future plastic biodegradation efforts due to their ability to break down various plastic polymers.

8.4.5 Engineering of Chimeric Enzymes

Chimeric enzymes, also referred to as fusion proteins, formed by combining genes from unrelated sources, initially encode distinct proteins. These hybrid enzymes inherit the characteristics of all the proteins they originate from, resulting in a single, versatile molecule. They hold great promising potential for the biodegradation of plastics, where multiple enzyme systems can work in synergy to break down polymers effectively. This synergy arises from the complementary properties of these enzymes in terms of their catalytic patterns and substrate specificity (Liu et al., 2019).

The study by Barth et al. (2016) demonstrated the successful use of immobilized TfCa in combination with LC-cutinase (LCC) or TfCut2 from *Thermobifida fusca* KW3 to enhance the biodegradation of PET. This approach effectively addressed the issue of MHET accumulation and significantly improved the hydrolytic efficiency, making it a promising strategy for tackling PET waste. In a study by Gamerith et al. (2017), they fused a hydrophobic-binding module of polyhydroxybutyrate depolymerase (PA_PBM) from *Alcaligenes faecalis* to polyamidase (PA) from *Nocardia farcinica* IMA 10152A through a C-terminal fusion approach. This fusion enzyme, referred to as PA_PBM, exhibited increased activity when applied to commercial polyurethane copolymers. This enhanced activity was apparent from the release of 4,4′-diaminodiphenylmethane (MDA) and various oligomers (Gamerith et al., 2017). Liu et al. (2018) created an artificial chimeric enzyme by combining lipase (Lip) from *Thermomyces lanuginosus* and cutinase (Cut) from *Thielavia terrestris* NRRL 8126. This chimeric enzyme, known as Lip-Cut, was constructed through end-to-end fusion and overexpression in *Pichia pastoris*. Lip-Cut exhibited enhanced degradation capabilities, particularly for poly(ε-caprolactone) (PCL). In fact, the weight loss of PCL films when treated with Lip-Cut was 13.3 times higher at 6 hours compared to Lip, 11.8 times higher compared to Cut, and 5.7 times higher compared to a mixture of Lip and Cut. GC-MS analysis identified 6-hydroxyhexanoic acid and 3-caprolactone as the primary products of PCL hydrolysis by Lip-Cut. Knott et al. (2020) developed a two-enzyme system by creating chimeric proteins that combined MHETase and PETase for enhanced PET degradation. Different linker lengths (8, 12, or 20 amino acids) were used to join MHETase's C-terminus to PETase's N-terminus. These chimeras significantly improved PET and MHET turnover compared to free enzymes. The chimeras increased TPA release threefold. Notably, constructs with PETase at the C-terminus and MHETase at the N-terminus could not express the protein.

Chen et al. (2021) enhanced PETase by adding glutamic acid (E) and lysine (K) to its C-terminus, creating PETase-EK fusion proteins of varying sizes (5, 10, and 30 kDa). These mutants displayed improved thermal stability, retaining 80% activity after 6 hours at 40 °C compared to the native enzyme's 65%. All mutants showed significantly increased PET degradation on both amorphous and bottle films of PET, with PETase-EK30 delivering the best results after 6 days at 40 °C. In essence, chimeric enzymes offer a promising solution for tackling plastic pollution through enhanced enzymatic degradation processes.

8.4.6 PET46 Enzyme

Recently, researchers have discovered an innovative PET-degrading enzyme called PET46 from a deep-sea microorganism using metagenomics methods. They identified the gene, a synthesized enzyme in *E. coli*. PET46 has unique properties, including the ability to degrade both long- and short-chain PET molecules continuously. It employs a distinct mechanism of substrate binding compared to known PET-degrading enzymes, featuring a crucial "lid" structure. PET46 shares similarities with ferulic acid esterase, suggesting potential applications in breaking down lignin in plant cell walls and degrading plastics. PET46 is exceptionally efficient at 70 °C, making it promising for biotechnology applications in addressing marine and terrestrial plastic pollution (Chow et al., 2023) (Table 8.3).

8.5 Other Advanced Techniques Involved in Plastic Mulching

8.5.1 Bioinformatic Approach

8.5.1.1 Omics Tools

Due to the rapid development of high-throughput sequencing, several omics tools such as genomics, transcriptomics, proteomics, and metabolomics have been increasingly applied in the study of the microbial ecology of plastic mulching and its biodegradation mechanisms (Rüthi et al., 2023). In plastic degradation pathways, novel genes or enzymes are uncovered through the genomic study of the microbial community of plastic mulching.

Table 8.3 Genetically modified microorganisms in munching of synthetic plastic polymers

S. no.	Microorganism	Modification	Plastic degradation	References
1.	*Fusarium solani* f.sp. *pisi*	Site-directed mutagenesis to create more space in the active site of the cutinase enzyme	Polyethylene terephthalate (PET)	Araujo et al. (2007)
2.	*Bacillus subtilis*	Integration of *Thermobifida fusca* cutinase TfCut2 gene	Polyethylene terephthalate (PET), polyethylene (PE), and polypropylene	Arutchelvi et al. (2008)
3.	*Escherichia coli Rosetta-gami B* (de3)	Integration of cutinase gene (cut190) from *Saccharomonospora viridis* AHK190	PET degradation	Kawai et al. (2014)
4.	*Thermobifida fusca*	Addition of Ca^{2+} or Mg^{2+} cations to stabilize esterases TfH, BTA2, Tfu0882, TfCut1, and TfCut2 gene for enhancing enzyme functionality and stability	Polyethylene terephthalate (PET)	Then et al. (2015)
5.	*Ideonella sakaiensis*	Arginine (R) in position 280 was replaced by alanine (A) in the PETase enzyme, enhancing its activity	Polyethylene terephthalate (PET)	Joo et al. (2018)
6.	*Pichia pastoris*	Integrations of lipase (LiP) gene from *Thermomyces lanuginosus* and cutinase (Cut) gene from *Thielavia terrestris* NRRL 8126 which were cloned end to end to form bifunctional lipase-cutinase enzyme	Poly(ε- caprolactone) PCL	Liu et al. (2018)
7.	*Phaeodactylum tricornutum*	Polyethylene terephthalate-hydrolyzing enzyme gene incorporated	Polyethylene terephthalate (PET)	Moog et al. (2019)
8.	*Clostridium thermocellum*	The thermophilic cutinase (LCC) was incorporated which was isolated from plant compost metagenome	Polyethylene terephthalate (PET)	Yan et al. (2021)
9.	*Thermobifida cellulosilytica* dsm44535	Site-directed mutagenesis on the gene-encoding cutinases	Polyethylene terephthalate (PET)	Herrero Acero et al. (2013)
10.	*Escherichia coli*	Intracellular lipases LIP1 and LIP2 from *Pseudomonas chlororaphis* PA23 were expressed into *E. coli*	PHAs, polylactic acid (PLA), and para-nitrophenyl (pNP) alkanoates	Mohanan et al. (2022)

8.5.1.2 Genomics

For the genotyping fingerprinting of the microorganisms, various PCR-based techniques, such as amplified fragment length polymorphisms, randomly amplified polymorphic DNA analysis, and terminal restriction fragment length polymorphisms, are used. More recently, the known enzymes encoded by several gene sequences to degrade polyurethane plastics were retrieved from landfill-derived metagenomes (Gaytán et al., 2019). For example, the environmental metagenomes were used to exhibit the hydrolytic activity for different polyesters such as poly(lactic acid), poly(ε-caprolactone), and poly(butylene succinate-co-adipate) and the screening of enzymes belonging to the esterase families by using function-based agar plate assays (Hajighasemi et al., 2018). Similarly, for the functional analysis, the disulfide bonds containing plastic-degrading enzymes in the host cells of *Pichia pastoris* were used, and these enzymes are inappropriately expressed in *E. coli* (Urbanek et al., 2020).

8.5.1.3 Transcriptomics

Transcriptomics is another technique in which the enzymes associated with the degradation mechanisms and the expression of the upregulated genes were elucidated. For example, the size and gene annotation quality of current databases of known plastic mulch-degrading enzymes are limiting the transcriptomics process. Gravouil et al. (2017) analyzed the expression of several functional genes in species of the *Rhodococcus* genus through the combining study of transcriptomics and lipidomics and identified the pathways involved in PE consumption. Gao and Sun (2021) used differential gene expression analysis for the degradation of poly(ethylene terephthalate) (PET) in three bacterial species (*Exiguobacterium* sp., *Halomonas* sp., *Ochrobactrum* sp.) and identified the genes for energy production and cell growth.

8.5.1.4 Proteomics

It is a posttranslational modification that involves interaction, functionality, and sequence analysis of a protein. The microarray or transcriptomic profiling techniques are used for proteomics analysis. Subsequently, these techniques can be applicable for the identification of novel functional genes, metabolic pathways, and novel proteomes in plastic mulch-degrading organisms. For example, the fungus *Knufia chersonesos* secreted enzymes to hydrolyze polybutylene adipate terephthalate (PBAT) which were screened by comparatively analyzing the exoproteome. In another study, a variety of natural and synthetic polyesters were hydrolyzed by a polyhydroxybutyrate depolymerase enzyme produced by *Alcanivorax* sp. through a comparative exoproteomic approach (Zadjelovic et al., 2020).

8.5.1.5 Metabolomics

This technique focuses on whole metabolic pathways of an organism (Dangi et al., 2019). Plastic mulch is supposed to be hydrolyzed by the exogenous response of microbe's enzymes. Further, the formation of many oxidizing side groups increases the availability for microbial assimilation and reduces the surface hydrophobicity through depolymerization of polymer chains. Ru et al. (2020) described that the isolated cutinase enzyme from the compost metagenomic library of *Pseudomonas mendocina*, *Fusarium solani*, and *Thermomyces insolens* has the ability to hydrolyze both the aliphatic and aromatic ester bonds found in polyesters by increasing the surface hydrophilicity and effectively involving in the degradation of PET.

8.5.2 Nanotechnology

Nanotechnology offers an eco-friendly, cost-effective solution to plastic degradation by enhancing barrier properties through nanoclay particles. These nanocomposites can be composted after use, returning them to the soil. Chandran et al. (2023) reported the degradation of polyethylene using silver nanoparticles from *Aspergillus oryzae* in 5 weeks. Additionally, nano-barium titanate nanoparticles facilitate bacterial consortium growth for efficient plastic biodegradation, with the preferred particle size being 38 nm (Pandey et al., 2015).

8.5.3 Molecular Modeling and Computational Technology or Chemistry

Computational tools and machine learning techniques play a pivotal role in plastic degradation research and development. Databases like UM-BBD, MetaCyc, and BioCyc provide essential information on metabolic pathways and genes involved in plastic biodegradation. These tools enable the prediction of synthetic polymer degradation pathways (Tourova et al., 2020). A greedy accumulated strategy for protein engineering (GRAPE) was utilized to enhance PETase from *Ideonella sakaiensis* (Cui et al., 2021).

Meng et al. (2021) research employed computational tools to identify a potentially beneficial mutation combination in PETase, resulting in a double mutant (changes from tryptophan (W) to histidine (H) at position 159 and from phenylalanine (F) to tyrosine (Y) at position 229), with enhanced thermal stability and significantly increased catalytic activity in degrading PET plastics.

8.5.4 *Synthetic Biology*

In synthetic biology, engineers employ biological techniques to design complex systems, such as those for plastic mulching. These systems sustain a vast population of microorganisms. By redesigning *E. coli* metabolic pathways with novel enzymes like methanol dehydrogenase and C-C bonding enzymes, efficient methanol assimilation is achieved. Synthetic pathways are developed to create key enzymes like P450, Adh, BVMO, and esterase, which are crucial for plastic-degrading bacteria. Recent research, such as by Yeom et al. (2021) and Tiso et al. (2020), demonstrates the potential of synthetic biology in engineering bacteria for efficient plastic degradation, with advanced enzymatic engineering techniques and genetic biosensors enhancing their functionality and facilitating microorganism screening.

8.6 Prospective Applications of GMO and Engineered Proteins

Genetic engineering emerged as a promising approach to tackle the complex task of plastic biodegradation. Since a single microorganism often lacks the capacity to perform all the necessary steps in plastic degradation, researchers have harnessed genetic engineering tools to transmit multiple genes accountable for different metabolic pathways into various microbial species. This approach enhances efficiency, eco-friendliness, and the availability of genetically modified organisms (GMOs) for plastic biodegradation. GMOs have gained considerable attention because of their ability to efficiently degrade various pollutants, including plastics, under diverse environmental conditions. For instance, researchers have engineered novel biosynthetic pathways in *E. coli* to convert plastic-derived monomers like terephthalic acid into valuable compounds like vanillin (Sadler & Wallace, 2021). Genetic modification has also facilitated the insertion of bacterial genes responsible for polyhydroxybutyrate (PHB) production into microalgae and macroalgae, reducing the cost of PHB production for plastic manufacturing.

Similarly, the modification of *Chlamydomonas reinhardtii* strain with two genes (phbB and phbC) from *Ralstonia eutropha* was used for the production of PHB plastic (Chaogang et al., 2010). Ghosh et al. (2019) reported that the genetically engineered microorganisms *Staphylococcus* sp. (52%) and *Pseudomonas* sp. (11%) have the efficiency to degrade PE. Additionally, the *E. coli* BL-21 synthesizes LC-cutinase during the PET degradation process which hydrolyzes the polymer to yield terephthalate and ethylene glycol. In the pursuit of sustainable plastic mulch production, the biorefinery concept, coupled with GMOs, holds significant promise. These GMOs can contribute not only to plastic waste disposal but also to the synthesis of biodegradable plastics, addressing the pressing issue of plastic pollution.

The use of engineered enzymes and microorganisms to break down plastics holds great potential for reducing the environmental impact of plastic waste. However, it is important to note that practical applications and scaling up such technologies for real-world plastic recycling may require further research, optimization, and regulatory considerations.

8.7 Limitation

While genetic engineering tools offer exciting possibilities for enhancing the performance of microbes in plastic mulching and biodegradation, there are several limitations considered as the utilization of GMOs in agriculture, including plastic mulching, raises ethical and social concerns regarding consumer acceptance, intellectual property rights, and potential unintended consequences. Additionally, genetically modified organisms (GMOs) involved in plastic biodegradation might introduce allergenic components, affecting both workers and consumers during production and disposal. The genetically engineered bacteria will have a long-term impact on the atmosphere and human beings. Escaping engineered microbes could interact with nontarget organisms, raising ecological and biosafety concerns. Costly development and maintenance of engineered strains, along with incomplete understanding of plastic degradation pathways, hinder widespread adoption, especially in small-scale agriculture. Despite progress, our understanding of the genetic and metabolic pathways involved in plastic degradation is incomplete, making it challenging to design effective genetic modifications. Furthermore, the lack of specificity in genetic tools may inadvertently disrupt nonplastic environmental components, compounding ecological disturbance.

8.8 Conclusion and Future Prospects

The widespread use of plastics coupled with limited recycling options has resulted in the accumulation of millions of metric tons of plastic waste in terrestrial and marine environments. Plastic biodegradation faces significant challenges owing to the high hydrophobicity, crystallinity, strong chemical bonds, and high molecular weight of these materials (Urbanek et al., 2018). This includes exploring the most promising biodegradation processes for synthetic plastics, synthesizing and applying highly efficient plastic-degrading enzymes. Fortunately, recent discoveries of microorganisms and their enzymes to degrade plastics possess a promising solution to address this ever-growing environmental challenge. The enzymes produced by these organisms have become the focus of extensive scientific investigations, seeking a deeper understanding of the mechanisms underlying plastic degradation. Genetic engineering approaches are currently being employed to enhance these enzymes, focusing on improving their catalytic efficiency and stability. Published findings indicate that naturally isolated microorganisms have limited plastic degradation capabilities. Therefore, improving enzymatic production and enhancing enzymatic activity, particularly when targeting specific materials with higher selectivity, are critical for advancing the rate of plastic biodegradation. The application of metabolic engineering holds significant promise in achieving these objectives. The use of molecular engineering tools in contemporary biotechnology allows for the creation of networks connecting genes and pathways. ZFNs, TALENs, and CRISPR-Cas9 (RGENs) are all programmable nucleases capable of generating site-specific DSBs in the genome. The choice of which technology to use depends on factors like target specificity, ease of design, and delivery methods, and CRISPR-Cas9 has become the most widely embraced genome editing tool due to its accessibility and efficiency. The integration of genomic, proteomic, and metabolomic data provides comprehensive insights into living systems and can strategically enhance the efficiency of microbial systems in plastic degradation. Synthetic biology, an emerging field combining biological and engineering techniques, offers computational tools to tackle metabolic pathways and regulatory networks comprehensively. This approach signifies a substantial step toward more effective and sustainable strategies for addressing plastic pollution.

In essence, the optimization of PET-degrading enzymes through genetic engineering signifies a promising era in combatting plastic pollution. Our understanding of the genetic complexity of these "plastic-munching microbes" enables us to find effective solutions to address the most pressing environmental challenges that our planet faces related to plastic waste.

References

Abdelghafour, M. M., Orbán, Á., Deák, Á., et al. (2022). Biocompatible poly (ethylene succinate) polyester with molecular weight dependent drug release properties. *International Journal of Pharmaceutics, 618*, 121653.

Ali, S., Bukhari, D. A., & Rehman, A. (2023). Call for biotechnological approach to degrade plastic in the era of COVID-19 pandemic. *Saudi Journal of Biological Sciences, 30*, 103583.

Amobonye, A., Bhagwat, P., et al. (2021). Plastic biodegradation: Frontline microbes and their enzymes. *Science of the Total Environment, 759*, 143536.

Anand, U., Dey, S., et al. (2023). Biotechnological methods to remove microplastics: A review. *Environmental Chemistry Letters, 21*, 1787.

Anbalagan, S., Venkatakrishnan, H. R. R., Ravindran, J., et al. (2021). Hydrolytic degradation of polyethylene terephthalate by cutinase enzyme derived from fungal biomass–molecular characterization. *Biointerface Research in Applied Chemistry, 12*, 653–667. https://doi.org/10.33263/BRIAC121.653667

Araújo, R., Silva, C. O., Neill, A., et al. (2007). Tailoring cutinase activity towards polyethylene terephthalate and polyamide 6, 6 fibers. *Journal of Biotechnology, 128*(4), 849–857. https://doi.org/10.1016/j.jbiotec.2006.12.028

Arman Alim, A. A., Mohammad Shirajuddin, S. S., & Anuar, F. H. (2022). A review of nonbiodegradable and biodegradable composites for food packaging application. *Journal of Chemistry*. https://doi.org/10.1155/2022/7670819

Arutchelvi, J., Sudhakar, M., Arkatkar, A., et al. (2008). Biodegradation of polyethylene and polypropylene. *Indian Journal of Biotechnology, 7*, 9–22. http://nopr.niscpr.res.in/handle/123456789/7326

Austin, H. P., Allen, M. D., Donohoe, B. S., et al. (2018). Characterization and engineering of a plastic-degrading aromatic polyesterase. *Proceedings of the National Academy of Sciences of the United States of America, 115*(19), E4350–E4357.

Barth, M., Honak, A., Oeser, T., et al. (2016). A dual enzyme system composed of a polyester hydrolase and a carboxylesterase enhances the biocatalytic degradation of polyethylene terephthalate films. *Biotechnology Journal, 11*(8), 1082–1087. https://doi.org/10.1002/biot.201600008

Bhuyar, P., Muniyasamy, S., & Govindan, N. (2019). Biodegradation of plastic waste by using microalgae and their toxins. *Journal of Biotechnology & Biomaterials, 8*(55), 10–3389.

Brunner, I., Fischer, M., Rüthi, J., et al. (2018). Ability of fungi isolated from plastic debris floating in the shoreline of a lake to degrade plastics. *PLoS One, 13*(8), e0202047. https://doi.org/10.1371/journal.pone.0202047

Capecchi, M. R. (1989). Altering the genome by homologous recombination. *Science, 244*(4910), 1288–1292.

Carr, C. M., Clarke, D. J., & Dobson, A. D. (2020). Microbial polyethylene terephthalate hydrolases: Current and future perspectives. *Frontiers in Microbiology, 11*, 571265.

Chandran, A., Iadecola, T., Khemani, V., et al. (2023). Quantum many-body scars: A quasiparticle perspective. *Annual Review of Condensed Matter Physics, 14*, 443–469.

Chaogang, W., Zhangli, H., Anping, L., et al. (2010). Biosynthesis of poly-3-hydroxybutyrate (phb) in the transgenic green alga *Chlamydomonas reinhardtii* 1. *Journal of Phycology, 46*(2), 396–402.

Chen, K., Hu, Y., Dong, X., et al. (2021). Molecular insights into the enhanced performance of EKylated PETase toward PET degradation. *ACS Catalysis, 11*(12), 7358–7370. https://doi.org/10.1021/acscatal.1c01062

Chow, J., Perez-Garcia, P., et al. (2023). Microbial enzymes will offer limited solutions to the global plastic pollution crisis. *Microbial Biotechnology, 16*(2), 195–217.

Chowdhary, P., Shukla, G., Raj, G., et al. (2019). Microbial manganese peroxidase: A ligninolytic enzyme and its ample opportunities in research. *SN Applied Sciences, 1*, 1–12. https://doi.org/10.1007/s42452-018-0046-3

CIEL. (2020). *Plastic Global law & Policy*. Center for International Environmental Law Control Pollution Board report. Guidelines for Assessment of Environment Compensation to be levied for Violation of Plastic Waste Management. https://cpcb.nic.in/uploads/plasticwaste/EC_Regime_PWM.pdf

Cui, Y., Chen, Y., Liu, X., et al. (2021). Computational redesign of a PETase for plastic biodegradation under ambient condition by the GRAPE strategy. *ACS Catalysis, 11*(3), 1340–1350.

Dai, L., Qu, Y., Huang, J. W., et al. (2021). Enhancing PET hydrolytic enzyme activity by fusion of the cellulose–binding domain of cellobiohydrolase I from *Trichoderma reesei*. *Journal of Biotechnology, 334*, 47–50.

Dangi, A. K., Sharma, B., Hill, R. T., et al. (2019). Bioremediation through microbes: Systems biology and metabolic engineering approach. *Critical Reviews in Biotechnology, 39*(1), 79–98.

Danso, D., Chow, J., & Streit, W. R. (2019). Plastics: Environmental and biotechnological perspectives on microbial degradation. *Applied and Environmental Microbiology, 85*(19), e01095–e01019.

Di Rocco, G., Taunt, H. N., Berto, M., et al. (2023). A PETase enzyme synthesised in the chloroplast of the microalga *Chlamydomonas reinhardtii* is active against post-consumer plastics. *Scientific Reports, 13*(1), 10028. https://doi.org/10.1038/s41598-023-37227-5

Dillingham, M. S., & Kowalczykowski, S. C. (2008). RecBCD enzyme and the repair of double-stranded DNA breaks. *Microbiology and Molecular Biology Reviews, 72*(4), 642–671.

Farzi, A., Dehnad, A., & Fotouhi, A. F. (2019). Biodegradation of polyethylene terephthalate waste using Streptomyces species and kinetic modeling of the process. *Biocatalysis and Agricultural Biotechnology, 17*, 25–31.

Fayyaz, M., Chew, K. W., Show, P. L., et al. (2020). Genetic engineering of microalgae for enhanced biorefinery capabilities. *Biotechnology Advances, 43*, 107554.
Furukawa, M., Kawakami, N., Tomizawa, A., et al. (2019). Efficient degradation of poly (ethylene terephthalate) with Thermobifida fusca cutinase exhibiting improved catalytic activity generated using mutagenesis and additive-based approaches. *Scientific Reports, 9*(1), 16038. https://doi.org/10.1038/s41598-019-52379-z
Gamerith, C., Vastano, M., Ghorbanpour, S. M., et al. (2017). Enzymatic degradation of aromatic and aliphatic polyesters by P. pastoris expressed cutinase 1 from Thermobifida cellulosilytica. *Frontiers in Microbiology, 8*, 938. https://doi.org/10.3389/fmicb.2017.00938
Gao, R., & Sun, C. (2021). A marine bacterial community capable of degrading poly (ethylene terephthalate) and polyethylene. *Journal of Hazardous Materials, 416*, 125928.
Gaytán, I., Sánchez-Reyes, A., Burelo, M., et al. (2019). Degradation of recalcitrant polyurethane and xenobiotic additives by a selected landfill microbial community and its biodegradative potential revealed by proximity ligation-based metagenomic analysis. *Frontiers in Microbiology, 10*, 2986.
Ghosh, S., Qureshi, A., & Purohit, H. J. (2019). Microbial degradation of plastics: Biofilms and degradation pathways. *Contaminants in Agriculture and Environment: Health Risks and Remediation, 1*, 184–199.
Godfrey, L. (2019). Waste plastic, the challenge facing developing countries – Ban it, change it, collect it. *Recycling, 4*(1), 3.
Gómez-Méndez, L. D., Moreno-Bayona, D. A., Poutou-Pinales, R. A., et al. (2018). Biodeterioration of plasma pretreated LDPE sheets by *Pleurotus ostreatus*. *PLoS One, 13*(9), e0203786.
Gong, Z., Jin, L., Yu, X., et al. (2023). Biodegradation of low-density polyethylene by the fungus *Cladosporium* sp. recovered from a landfill site. *Journal of Fungi, 9*(6), 605. https://doi.org/10.3390/jof9060605
Grau, J., Boch, J., & Posch, S. (2013). TALENoffer: Genome-wide TALEN off-target prediction. *Bioinformatics, 29*(22), 2931–2932.
Gravouil, K., Ferru-Clément, R., Colas, S., et al. (2017). Transcriptomics and lipidomics of the environmental strain *Rhodococcus ruber* point out consumption pathways and potential metabolic bottlenecks for polyethylene degradation. *Environmental Science & Technology, 51*(9), 5172–5181.
Hachisuka, S. I., Nishii, T., & Yoshida, S. (2021). Development of a targeted gene disruption system in the poly (ethylene terephthalate)-degrading bacterium *Ideonella sakaiensis* and its applications to PETase and MHETase genes. *Applied and Environmental Microbiology, 87*(18), e00020–e00021.
Haider, T. P., Völker, C., Kramm, J., et al. (2019). Plastics of the future? The impact of biodegradable polymers on the environment and on society. *Angewandte Chemie International, 58*(1), 50–62.
Hajighasemi, M., Tchigvintsev, A., Nocek, B., et al. (2018). Screening and characterization of novel polyesterases from environmental metagenomes with high hydrolytic activity against synthetic polyesters. *Environmental Science & Technology, 52*(21), 12388–12401.
Herrero Acero, E., Ribitsch, D., Dellacher, A., et al. (2013). Surface engineering of a cutinase from *Thermobifida cellulosilytica* for improved polyester hydrolysis. *Biotechnology and Bioengineering, 110*(10), 2581–2590. https://doi.org/10.1002/bit.24930
Huang, J., Meng, H., Luo, X., et al. (2022). Insights into the thermal degradation mechanisms of polyethylene terephthalate dimer using DFT method. *Chemosphere, 291*, 133112.
Huda, M. K., & Widiastuti, I. (2021). Natural fiber reinforced polymer in automotive application: A systematic literature review. *Journal of Physics: Conference Series, 1808*, 012015.
Inderthal, H., Tai, S. L., & Harrison, S. T. (2021). Non-hydrolyzable plastics – An interdisciplinary look at plastic bio-oxidation. *Trends in Biotechnology, 39*(1), 12–23.
Ishino, Y., Krupovic, M., & Forterre, P. (2018). History of CRISPR-Cas from encounter with a mysterious repeated sequence to genome editing technology. *Journal of Bacteriology, 200*(7), 10–1128.

Islam, S., Apitius, L., Jakob, F., et al. (2019). Targeting microplastic particles in the void of diluted suspensions. *Environment International, 123*, 428–435. https://doi.org/10.1016/j.envint.2018.12.029

Jaiswal, S., Sharma, B., & Shukla, P. (2020). Integrated approaches in microbial degradation of plastics. *Environmental Technology and Innovation, 17*, 100567.

Jeon, H. J., & Kim, M. N. (2015). Functional analysis of alkane hydroxylase system derived from *Pseudomonas aeruginosa* E7 for low molecular weight polyethylene biodegradation. *International Biodeterioration and Biodegradation, 103*, 141–146. https://doi.org/10.1016/j.ibiod.2015.04.024

Jiang, F., & Doudna, J. A. (2017). CRISPR–Cas9 structures and mechanisms. *Annual Review of Biophysics, 46*, 505–529.

Joo, S., Cho, I. J., Seo, H., et al. (2018). Structural insight into the molecular mechanism of poly (ethylene terephthalate) degradation. *Nature Communications, 9*(1), 382. https://doi.org/10.1038/s41467-018-02881-1

Joung, J. K., & Sander, J. D. (2013). TALENs: A widely applicable technology for targeted genome editing. *Nature Reviews. Molecular Cell Biology, 14*(1), 49–55.

Kale, S. K., Deshmukh, A. G., et al. (2015). Microbial degradation of plastic: A review. *Journal of Biochemical Technology, 6*(2), 952–961.

Kan, Y., He, L., Luo, Y., et al. (2021). IsPETase is a novel biocatalyst for poly (ethylene terephthalate) (PET) hydrolysis. *ChemBioChem, 22*(10), 1706–1716.

Kaur, K., Sharma, S., Shree, N., et al. (2023). Recent advancements and mechanism of plastics biodegradation promoted by bacteria: A key for sustainable remediation for plastic wastes. *Biosciences, Biotechnology Research Asia, 20*(1), 1–12.

Kaushal, J., Khatri, M., & Arya, S. K. (2021). Recent insight into enzymatic degradation of plastics prevalent in the environment: A mini-review. *Cleaner Engineering and Technology, 2*, 100083.

Kawai, F., Oda, M., Tamashiro, T., et al. (2014). A novel Ca $^{2+-}$activated, thermo stabilized polyesterase capable of hydrolyzing polyethylene terephthalate from *Saccharomonospora viridis* AHK190. *Applied Microbiology and Biotechnology, 98*, 10053–10064. https://doi.org/10.1007/s00253-014-5860-y

Knott, B. C., Erickson, E., Allen, M. D., et al. (2020). Characterization and engineering of a two-enzyme system for plastics depolymerization. *Proceedings of the National Academy of Sciences of the United States of America, 117*(41), 25476–25485. https://doi.org/10.1073/pnas.2006753117

Koller, M., Matsumoto, K. I., Wang, Z., et al. (2023). Biodegradation of plastics. *Frontiers in Bioengineering and Biotechnology, 11*, 1150078.

Koonin, E. V., Makarova, K. S., et al. (2017). Diversity, classification and evolution of CRISPR-Cas systems. *Current Opinion in Microbiology, 37*, 67–78.

Kopke, K., Hoff, B., & Kück, U. (2010). Application of the Saccharomyces cerevisiae FLP/FRT recombination system in filamentous fungi for marker recycling and construction of knockout strains devoid of heterologous genes. *Applied and Environmental Microbiology, 76*(14), 4664–4674.

Kumar, A., & Chandra, R. (2020). Ligninolytic enzymes and its mechanisms for degradation of lignocellulosic waste in environment. *Heliyon, 6*(2), e03170.

Kumar, M., Xiong, X., et al. (2020). Microplastics as pollutants in agricultural soils. *Environmental Pollution, 265*, 114980.

Li, P., Zou, X., Wang, X., et al. (2020). A preliminary study of the interactions between microplastics and citrate-coated silver nanoparticles in aquatic environments. *Journal of Hazardous Materials, 385*, 121601. https://doi.org/10.1016/j.jhazmat.2019.121601

Liang, J. L., JiangYang, J. H., Nie, Y., et al. (2016). Regulation of the alkane hydroxylase CYP153 gene in a Gram-positive alkane-degrading bacterium, *Dietzia sp.* strain DQ12-45-1b. *Applied and Environmental Microbiology, 82*(2), 608–619.

Liu, B., He, L., Wang, L., et al. (2018). Protein crystallography and site-direct mutagenesis analysis of the poly (ethylene terephthalate) hydrolase PETase from *Ideonella sakaiensis*. *ChemBioChem, 19*(14), 1471–1475. https://doi.org/10.1002/cbic.201800097

Liu, C., Shi, C., Zhu, S., et al. (2019). Structural and functional characterization of polyethylene terephthalate hydrolase from *Ideonella sakaiensis*. *Biochemical and Biophysical Research Communications, 508*(1), 289–294. https://doi.org/10.1016/j.bbrc.2018.11.148

Ma, Y., Yao, M., Li, B., et al. (2018). Enhanced poly (ethylene terephthalate) hydrolase activity by protein engineering. *Engineering, 4*(6), 888–893.

Ma, J., Gong, Z., Wang, Z., et al. (2022). Elucidating degradation properties, microbial community, and mechanism of microplastics in sewage sludge under different terminal electron acceptors conditions. *Bioresource Technology, 346*, 126624.

Matar, R. (2020). *Development of polyethylene terephthalate plastic eating pseudomonas putida* (Doctoral dissertation, Notre Dame University-Louaize).

Mathew, J. J., Mathews, A. I., Sajeshkumar, N. K., et al. (2021). Microbial degradation of polyethylene terephthalate (PET): An outlook study. *Journal of Medicinal Plants, 9*(5), 31–40.

Maurya, A., Bhattacharya, A., & Khare, S. K. (2020). Enzymatic remediation of polyethylene terephthalate (PET)–based polymers for effective management of plastic wastes: An overview. *Frontiers in Bioengineering and Biotechnology, 8*, 602325.

Meng, X., Yang, L., Liu, H., et al. (2021). Protein engineering of stable IsPETase for PET plastic degradation by Premuse. *International Journal of Biological Macromolecules, 180*, 667–676.

Mizutani, O., Arazoe, T., Toshida, K., et al. (2017). Detailed analysis of targeted gene mutations caused by the Platinum-Fungal TALENs in Aspergillus oryzae RIB40 strain and a ligDdisruptant. *Journal of Bioscience and Bioengineering, 123*(3), 287–293.

Mohanan, N., Montazer, Z., Sharma, P. K., et al. (2020). Microbial and enzymatic degradation of synthetic plastics. *Frontiers in Microbiology, 11*, 580709.

Mohanan, N., Wong, C. H., Budisa, N., et al. (2022). Characterization of polymer degrading lipases, LIP1 and LIP2 from *Pseudomonas chlororaphis* PA23. *Frontiers in Bioengineering and Biotechnology, 10*, 854298. https://doi.org/10.3389/fbioe.2022.854298

Moog, D., Schmitt, J., Senger, J., et al. (2019). Using a marine microalga as a chassis for polyethylene terephthalate (PET) degradation. *Microbial Cell Factories, 18*(1), 1–15. https://doi.org/10.1186/s12934-019-1220-z

Moon, H. Y., Sim, G. H., Kim, H. J., et al. (2022). Assessment of Cre-lox and CRISPR-Cas9 as tools for recycling of multiple-integrated selection markers in *Saccharomyces cerevisiae*. *Journal of Microbiology, 60*, 18–30. https://doi.org/10.1007/s12275-022-1580-7

Morgana, S., Casentini, B., et al. (2021). Uncovering the release of micro/nanoplastics from disposable face masks at times of COVID-19. *Journal of Hazardous Materials, 419*, 126507.

Nikolaivits, E., Taxeidis, G., Gkountela, C., et al. (2022). A polyesterase from the Antarctic bacterium *Moraxella* sp. degrades highly crystalline synthetic polymers. *Journal of Hazardous Materials, 434*, 128900. https://doi.org/10.1016/j.jhazmat.2022.128900

Nyyssölä, A., & Ahlgren, J. (2019). Microbial degradation of polyacrylamide and the deamination product polyacrylate. *International Biodeterioration and Biodegradation, 139*, 24–33.

Palm, G. J., Reisky, L., Böttcher, D., et al. (2019). Structure of the plastic-degrading Ideonellasakaiensis MHETase bound to a substrate. *Nature Communications, 10*(1), 1717.

Palma, C., Martınez, A. T., Lema, J. M., et al. (2000). Different fungal manganese-oxidizing peroxidases: a comparison between *Bjerkandera sp.* and *Phanerochaete chrysosporium*. *Journal of Biotechnology, 77*(2–3), 235–245.

Pandey, P., Swati, P., Harshita, M., et al. (2015). Nanoparticles accelerated in vitro biodegradation of LDPE: A review. *Advances in Applied Science Research, 6*(4), 17–22.

Pant, S., Nag, P., Ghati, A., et al. (2022). Employment of the CRISPR/Cas9 system to improve cellulase production in *Trichoderma reesei*. *Biotechnology Advances, 60*, 108022. https://doi.org/10.1016/j.biotechadv.2022.108022

Park, C., & Park, W. (2018). Survival and energy producing strategies of alkane degraders under extreme conditions and their biotechnological potential. *Frontiers in Microbiology, 9*, 1081.

Perez-Garcia, P., Chow, J., Costanzi, E., et al. (2023). An archaeal lid-containing feruloyl esterase degrades polyethylene terephthalate. *Communications Chemistry, 6*(1), 193. https://doi.org/10.1038/s42004-023-00998-z

Ping, L. F., Chen, X. Y., Yuan, X. L., et al. (2017). Application and comparison in biosynthesis and biodegradation by *Fusarium solani* and *Aspergillus fumigatus* cutinases. *International Journal of Biological Macromolecules, 104*, 1238–1245.

Pinto, M., Zhao, Z., Klun, K., et al. (2022). Microbial consortiums of putative degraders of low-density polyethylene-associated compounds in the ocean. *Msystems, 7*(2), 01415–01421.

Plastics Europe. (2022). Plastics – The facts 2022. https://plasticseurope.org/knowledge-hub/plastics-the-facts-2022/

Plastics Europe. (2019). https://www.plasticseurope.org/en/newsroom/events/2019

Ru, J., Huo, Y., & Yang, Y. (2020). Microbial degradation and valorization of plastic wastes. *Frontiers in Microbiology, 11*, 442.

Rujnić-Sokele, M., & Pilipović, A. (2017). Challenges and opportunities of biodegradable plastics: A mini review. *Waste Management & Research, 35*(2), 132–140. https://doi.org/10.1177/0734242X16683272

Rüthi, J., Rast, B. M., Qi, W., et al. (2023). The plastisphere microbiome in alpine soils alters the microbial genetic potential for plastic degradation and biogeochemical cycling. *Journal of Hazardous Materials, 441*, 129941.

Sadler, J. C., & Wallace, S. (2021). Microbial synthesis of vanillin from waste poly (ethylene terephthalate). *Green Chemistry, 23*(13), 4665–4672.

Sadowski, P. D. (1995). The Flp recombinase of th 2-μm plasmid of *Saccharomyces cerevisiae*. *Progress in Nucleic Acid Research and Molecular Biology, 51*, 53–91.

Sagong, H. Y., Seo, H., Kim, T., et al. (2020). Decomposition of the PET film by MHETase using Exo-PETase function. *ACS Catalysis, 10*(8), 4805–4812. https://doi.org/10.1021/acscatal.9b05604

Sanghvi, G., Thanki, A., Pandey, S., et al. (2020). Engineered bacteria for bioremediation. In *Bioremediation of pollutants* (pp. 359–374). Elsevier.

Sanniyasi, E., Gopal, R. K., et al. (2021). Biodegradation of low-density polyethylene (LDPE) sheet by microalga, *Uronema africanum* Borge. *Scientific Reports, 11*(1), 17233.

Santos-Beneit, F., Chen, L. M., Bordel, S., et al. (2023). Screening enzymes that can depolymerize commercial biodegradable polymers: Heterologous expression of *Fusarium solani* cutinase in *Escherichia coli*. *Microorganisms, 11*(2), 328. https://doi.org/10.3390/microorganisms11020328

Sharma, S. R. (2018). Bioremediation of polythenes and plastics: A microbial approach. In *Approaches in bioremediation: The new era of environmental microbiology and nanobiotechnology* (Vol. 97-114).

Shilpa, B. N., & Meena, S. S. (2022). Microbial biodegradation of plastics: Challenges, opportunities, and a critical perspective. *Frontiers of Environmental Science & Engineering, 16*(12), 161.

Singh, D., & Gupta, N. (2020). Microbial laccase: A robust enzyme and its industrial applications. *Biologia, 75*, 1183–1193.

Sivan, A., Szanto, M., et al. (2006). Biofilm development of the polyethylene-degrading bacterium *Rhodococcus ruber*. *Applied Microbiology and Biotechnology, 72*, 346–352.

Son, H. F., Cho, I. J., Joo, S., et al. (2019). Rational protein engineering of thermo-stable PETase from *Ideonella sakaiensis* for highly efficient PET degradation. *ACS Catalysis, 9*(4), 3519–3526.

Statista Research Department. (2023). *Global plastic production*. https://www.statista.com/statistics/282732/global-production-of-plasticssince1950/#:~:text=Global%20plastic%20production%201950%2D2021&text=Global%20plastics%20production%20was%20estimated,production%20has%20soared%20since%201950s

Tan, Y., Henehan, G. T., Kinsella, G. K., et al. (2021). An extracellular lipase from *Amycolatopsis mediterranei* is a cutinase with plastic degrading activity. *Computational and Structural Biotechnology Journal, 19*, 869–879.

Tarasava, K., Oh, E. J., Eckert, C. A., et al. (2018). CRISPR-enabled tools for engineering microbial genomes and phenotypes. *Biotechnology Journal, 13*(9), 1700586.

Then, J., Wei, R., Oeser, T., et al. (2015). Ca^{2+} and Mg^{2+} binding site engineering increases the degradation of polyethylene terephthalate films by polyester hydrolases from *Thermobifida fusca*. *Biotechnology Journal, 10*(4), 592–598. https://doi.org/10.1002/biot.201400620

Tiso, T., Narancic, T., Wei, R., et al. (2020). Bio-upcycling of polyethylene terephthalate. BioRxiv 03.

Tokiwa, Y., Calabia, B. P., Ugwu, C. U., et al. (2009). Biodegradability of plastics. *International Journal of Molecular Sciences, 10*(9), 3722–3742.

Tourova, T. P., Sokolova, D. S., Nazina, T. N., et al. (2020). Phylogenetic diversity of microbial communities from the surface of polyethylene terephthalate materials exposed to different water environments. *Microbiology, 89*, 96–106.

Urbanek, A. K., Mirończuk, A. M., García-Martín, A., et al. (2020). Biochemical properties and biotechnological applications of microbial enzymes involved in the degradation of polyester-type plastics. *Biochimica et Biophysica Acta-Proteins Proteom, 1868*(2), 140315.

Urbanek, A. K., Rymowicz, W., & Mirończuk, A. M. (2018). Degradation of plastics and plastic-degrading bacteria in cold marine habitats. *Applied Microbiology and Biotechnology, 102*, 7669–7678.

Voběrková, S., Solčány, V., Vršanská, M., et al. (2018). Immobilization of ligninolytic enzymes from white-rot fungi in cross-linked aggregates. *Chemosphere, 202*, 694–707.

Wei, R., Oeser, T., Schmidt, J., et al. (2016). Engineered bacterial polyester hydrolases efficiently degrade polyethylene terephthalate due to relieved product inhibition. *Biotechnology and Bioengineering, 113*(8), 1658–1665.

Wilkes, R. A., & Aristilde, L. (2017). Degradation and metabolism of synthetic plastics and associated products by *pseudomonas* sp.: Capabilities and challenges. *Journal of Applied Microbiology, 123*(3), 582–593. https://doi.org/10.1111/jam.13472

Xia, B. (2022). Potential solution for plastic degradation – Genetical modification. In *2021 international conference on Social Development and Media Communication (SDMC 2021)* (pp. 1076–1079). Atlantis Press.

Xu, Y., Shan, L., Zhou, Y., et al. (2019). Development of a Cre-lox P-based genetic system in *Aspergillus Niger* ATCC1015 and its application to construction of efficient organic acid-producing cell factories. *Applied Microbiology and Biotechnology, 103*, 8105–8114.

Yan, F., Wei, R., Cui, Q., et al. (2021). Thermophilic whole-cell degradation of polyethylene terephthalate using engineered *Clostridium thermocellum*. *Microbial Biotechnology, 14*(2), 374–385. https://doi.org/10.1111/1751-7915.13580

Yang, S., Cheng, Y., et al. (2022). Impact of waste of COVID-19 protective equipment on the environment, animals and human health: A review. *Environmental Chemistry Letters, 20*(5), 2951–2970.

Yeom, S. J., Le, T. K., & Yun, C. H. (2021). P450-driven plastic-degrading synthetic bacteria. *Trends in Biotechnology, 40*(2), 166–179.

Yoshida, S., Hiraga, K., Takehana, T., et al. (2016). A bacterium that degrades and assimilates poly (ethylene terephthalate). *Science, 351*(6278), 1196–1199.

Zacchi, L., Burla, G., Zuolong, D., et al. (2000). Metabolism of cellulose by *Phanerochaete chrysosporium* in continuously agitated culture is associated with enhanced production of lignin peroxidase. *Journal of Biotechnology, 78*(2), 185–192.

Zadjelovic, V., Chhun, A., Quareshy, M., et al. (2020). Beyond oil degradation: Enzymatic potential of *Alcanivorax* to degrade natural and synthetic polyesters. *Environmental Microbiology, 22*, 1356–1369.

Zeng, Y., Hong, Y., Azi, F., et al. (2022). Advanced genome-editing technologies enable rapid and large-scale generation of genetic variants for strain engineering and synthetic biology. *Current Opinion in Microbiology, 69*, 102175. https://doi.org/10.1016/j.mib.2022.102175

Zhang, J., Gao, D., Li, Q., et al. (2020). Biodegradation of polyethylene microplastic particles by the fungus *Aspergillus flavus* from the guts of wax moth *Galleria mellonella. Science of the Total Environment, 704*, 135931.

Zhu, B., Ye, Q., Seo, Y., et al. (2022). Enzymatic degradation of polyethylene terephthalate plastics by bacterial *Curli* display PETase. *Environmental Science & Technology Letters, 9*(7), 650–657.

Chapter 9
Plastic Waste and Its Eco-Friendly Management

Lázaro Adrián González-Fernández and Nahum Andrés Medellín-Castillo

Abstract The impact of plastic waste on the environment is significant. Plastic waste can take hundreds of years to decompose, and it can cause harm to wildlife, marine ecosystems, and human health. From the Mariana Trench to Mt. Everest, there is virtually no place on Earth which is left untouched by plastic pollution. Plastic waste is now so ubiquitous in the natural environment that scientists have even suggested it could serve as a geological indicator of the Anthropocene era. The world is facing a global plastics crisis, and addressing the end of life of plastic products will not be enough to solve this global crisis. Plastic waste is a complex issue that requires a multifaceted approach. Governments, municipalities, businesses, and individuals all have a role to play in reducing plastic waste and managing it in an eco-friendly manner. The current trends and future perspectives of plastic waste management involve circular economy, 3R perspective, and innovative technologies. These approaches can help reduce the burden of plastic waste on the environment and pave the way for a more sustainable future.

Keywords Sustainability · Plastics · Polystyrene · Waste

9.1 Introduction

Plastic waste is a growing concern due to the increasing issues related to its generation. However, there are several sustainable and promising approaches to reduce the environmental impacts of plastic waste. Recycling is one of the most common

L. A. González-Fernández (✉)
Multidisciplinary Graduate Program in Environmental Sciences, Autonomous University of San Luis Potosí, San Luis Potosí, Mexico

N. A. Medellín-Castillo
Multidisciplinary Graduate Program in Environmental Sciences, Autonomous University of San Luis Potosí, San Luis Potosí, Mexico

Center for Research and Graduate Studies, Faculty of Engineering, San Luis Potosí, Mexico

R. Soni et al. (eds.), *Advanced Strategies for Biodegradation of Plastic Polymers*, https://doi.org/10.1007/978-3-031-55661-6_9

methods of managing plastic waste. It not only protects the environment and resources for future generations but also reduces energy consumption and greenhouse gas emissions. Bioplastics are an alternative to traditional plastics that are made from renewable resources such as cornstarch, sugarcane, or potato starch. They are biodegradable and compostable, making them an eco-friendlier option than traditional plastics. Proper waste management is essential to reduce the negative effects of plastic waste. This includes sorting and separation of waste, quantification of waste generation, and physical composition of waste.

According to a recent study (Kibria et al., 2023), the world is facing the challenge of proper management and resource recovery of the enormous amount of plastic waste. Plastic waste exposed to the environment creates problems and is of significant concern for all life forms (Kumar et al., 2021). Out of the 8.3 billion tons of plastics produced since the 1950s, 79% ended in landfills or leaked into the environment (Chamas et al., 2020). Plastic pollution is growing relentlessly as waste management and recycling fall short (OECD, 2022). In the process of plastic production and use, major disposal problems arise, which leads to poor aesthetics, fire hazards, health hazards, and energy shortages (Kehinde et al., 2020).

Additionally, approximately 20 million informal workers already recover plastic waste from the garbage in the Global South, usually working under precarious, risky, and poorly paid conditions. The literature claims that they represent a workforce that if recognized, integrated, and valued and under decent work conditions and fair remuneration could potentially significantly increase the capturing of plastic waste and reduce the amount of fugitive plastics. In conclusion, there are several sustainable and promising approaches to reduce the environmental impacts of plastic waste, and it is essential to adopt them to protect the environment and resources for future generations.

The widespread use of plastic products in our daily lives has led to an increase in plastic waste, which poses a significant threat to the environment. Plastic waste can take hundreds of years to decompose, and it can cause harm to wildlife, marine ecosystems, and human health. As a result, there is a growing need for eco-friendly management of plastic waste. In this chapter, we will explore the impact of plastic waste on the environment, the challenges of managing plastic waste, and the eco-friendly solutions that are being developed to address this issue. We will also discuss the role of individuals, businesses, and governments in managing plastic waste and promoting sustainable practices.

9.2 Plastic Polymers

Plastics are composed of molecules that are usually very large and are known as polymers, to which additives are added to achieve or enhance a certain property in the final product. Polymers are built from relatively small molecules known as monomers. These molecules can have different shapes and sizes, and their physical and chemical properties depend on the molecular structure and the type of monomers of which they are composed and can reach a molecular weight (or chain length) of up to 103 g/mol or even higher (Asiandu et al., 2021).

Polymerization is the result of the chemical process in which monomers are chemically grouped together by addition or condensation to form the polymer, a linear chain, or a three-dimensional macromolecule. Cotton, wool, wood, silk, and leather are examples of natural polymers widely used for decades. In addition, there are biopolymers (proteins, carbohydrates, etc.) which are part of the structure of all living beings (Bahl et al., 2021).

Another group of polymers are synthetic polymers, among which we find plastics (which are nothing more than polymers with additives). This type of synthetic polymers attracted the attention of the international community in the twentieth century due to their great applicability and versatility. Since then, engineering has been devoted to carefully designing and improving them to produce materials with the desired properties (heat resistance, density, strength, stiffness, etc.) (Hacker et al., 2019). The main classifications of plastic polymers are shown in Fig. 9.1.

Plastic polymers are widely used in a variety of applications, from plastic packaging to building materials and electronic devices. Among the main characteristics of polymers are their high strength, durability, flexibility, and ease of processing. In addition, polymers can be engineered to have specific properties, such as temperature resistance, chemical resistance, and electrical conductivity (Geyer, 2020b).

Within agricultural packaging, polymers used in plastic packaging include polyethylene, polypropylene, polycarbonate, and PET. Plastic containers made from polymers offer excellent resistance to moisture, corrosion, light, and heat, making them ideal for protecting a wide range of products. In addition to being used in the manufacture of plastic containers, such as bottles, bags, boxes, and other types of packaging, polymers are also used in other types of packaging such as plastic films, which are used to package food and other products, making them light, resistant, and economical (Kan & Miller, 2022).

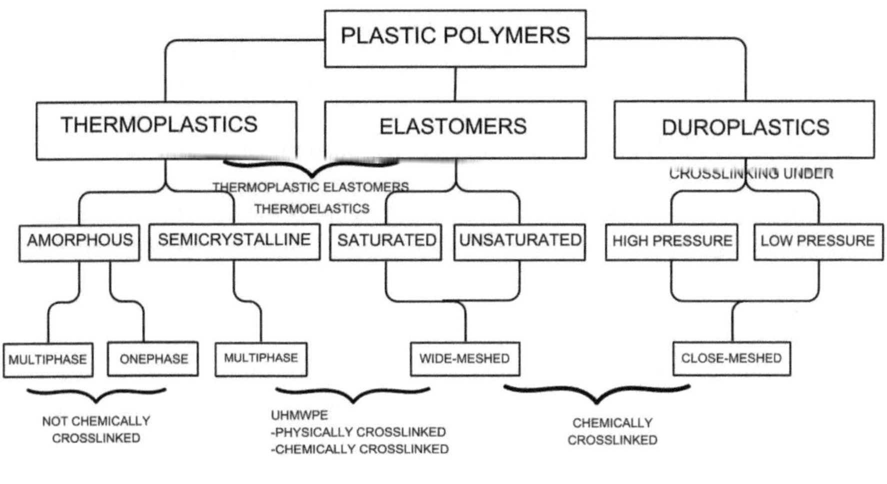

Fig. 9.1 Main classifications of plastic polymers

Like natural polymers, synthetic polymers are formed by a structure that in turn is made up of carbon chains composed of monomers that can be the same or different and are called homopolymers and copolymers. Synthetic homopolymers are macromolecules that are formed by the repetition of a single identical monomer unit. As for copolymers, they require two different monomers to form their structure and at least two different types of repeating patterns (Hacker et al., 2019).

The best known and most commonly used synthetic polymers are nylon, polyethylene (PE), polystyrene (PS), polyvinyl chloride (PVC), polymethyl methacrylate (acrylic), polypropylene, polyester, etc. Both synthetic and natural polymers can be recycled, and both are classified as thermoplastic polymers, precisely because they do not lose their properties when remolded (Khalid et al., 2022).

9.3 Bio-based Plastics

By definition, bio-based plastics or bioplastics are those plastics that come from plant products and renewable biological sources, i.e., raw materials such as corn, potato starch, agricultural by-products, and agricultural residues and even those from microorganisms such as algae, bacteria, or fungi. Bioplastics are created by transforming these biological products into raw materials that will be used to make a type of plastic. For example, sugarcane can be synthesized into bio-based ethylene which can then be transformed into polyethylene, a commonly used plastic (Abrha et al., 2022).

The biomass used for the manufacture of polymers of biological origin is mostly renewable and comes from different activities such as agriculture or the agri-food industry. In the different stages of biomass treatment, the key moment is the bacterial fermentation of sugars of different origins. Through these processes, green chemistry platform molecules are obtained, which can be used as monomers for the production of polymers of biological origin. Bio-based polymers are nowadays mostly made from renewable resources and can compete with other applications, especially human or animal food. However, this proportion attributed to the production of bioplastics is still very marginal (approximately 0.02% of the total agricultural area) (Mujtaba et al., 2023).

It is important to know that although they are made from bio-based raw materials, they are no different from other plastics nor are they necessarily biodegradable, so it is important to know where to dispose of them. If these materials are used to make packaging, you should dispose of them in the yellow bin, like any other traditional plastic. Examples of everyday bioplastics include fresh food packaging and many soft drink bottles (Sun et al., 2022).

Biodegradable bioplastics are made from plant materials such as corn, sugarcane, or cellulose. They are relatively new in the industry, although they are already being used in packaging. Examples include polylactic acid (PLA), thermoplastic starch (TPS), polyhydroxyalkanoates (PhA), and regenerated cellulose (Falua et al., 2022).

As mentioned above, the fact that a plastic is made from biomass does not necessarily mean that it can biodegrade. This is the case for materials whose chemical structure cannot be broken down by microorganisms but which are mechanically recyclable. Examples include bio-polyethylene (BioPE), bio-polypropylene (BIO-PP), bio-polyamide (Nylon 11), and bio-polyethylene terephthalate (BIO-PET) (Liwarska-Bizukojc, 2022).

9.4 Fossil-Based Plastics

Fossil-based plastics are synthetic materials widely used in our daily lives. These plastics are obtained through a complex process that begins with the extraction of nonrenewable resources, such as oil and natural gas, through drilling in subway reservoirs. The crude oil is then transported to refineries where it is separated into different chemical compounds, including hydrocarbons that are used to manufacture plastics. Through polymerization, these hydrocarbons are transformed into long, strong polymers, which are molded and shaped through various industrial processes to become final plastic products (Kawashima et al., 2019).

These fossil-based plastics have a wide range of applications in our modern society, from food packaging and cleaning products to automotive components and electronic devices (Walker & Rothman, 2020). One of the most common fossil-based plastics is polyethylene, which is used in the production of bags, bottles, and packaging materials. Another widely used plastic is polypropylene, which is found in food containers, automotive parts, and household appliances. These plastics have revolutionized many industries and have provided convenience and ease in our daily lives.

It is important to note that while fossil-based plastics have been beneficial in many ways, they also present significant environmental challenges. One of the main concerns is their slow degradation (Qin et al., 2021). Conventional plastics can take hundreds of years to decompose, which means that plastic waste accumulated in landfills or dispersed in the environment can persist for generations, causing damage to ecosystems, threatening marine life, and affecting soil quality (Nielsen et al., 2020).

In addition, the production and use of fossil-based plastics also contribute to climate change. During the extraction, transport, and processing of natural resources, greenhouse gases such as carbon dioxide and methane are emitted, which increase global warming and environmental deterioration. These negative impacts drive the search for more sustainable alternatives, such as biodegradable and compostable plastics, as well as the promotion of recycling practices and the reduction of plastic consumption in general, to mitigate the negative effects of fossil plastics on our planet (Geyer, 2020a, b).

9.5 Polymer Blends

Polymeric blends are combinations of two or more polymers that are joined together to form a new material with unique properties. These blends make it possible to combine the characteristics of different polymers, such as mechanical strength, heat resistance, flexibility, or transparency, to obtain a material with improved properties tailored to specific needs. For example, by blending a rigid polymer with a flexible polymer, it is possible to obtain a material that combines strength and flexibility, ideal for applications where impact resistance is required (Simon, 2019).

Polymer blends also offer advantages in terms of processing. By combining polymers with different melting temperatures, a blend with a wider window of processability can be obtained, making it easier to mold or extrude. In addition, polymer blends can improve the thermal stability of polymers, preventing their degradation during processing at high temperatures. This is especially useful in the plastics industry, where thermal stability is crucial to obtain high-quality and durable products (Muthuraj et al., 2018).

Polymer blends are also used in the field of materials engineering to develop advanced composite materials. These materials combine polymers with other reinforcements, such as glass, carbon, or ceramic fibers, to further improve their mechanical properties. For example, blends of polymers with carbon fibers can result in composite materials with high strength and stiffness, used in aerospace or automotive applications. These composite materials offer an optimal weight-to-strength ratio, which makes them very attractive in the industry (Gebrekrstos & Ray, 2023; Yang et al., 2020).

The degradation of polymer blends is a very broad and diverse topic. Degradation can occur at the molecular level, where the bonds between polymer chains break, affecting the structure and properties of the material. There can also be physical degradation, such as fragmentation of the polymer chains or wear due to friction. Some factors that accelerate degradation include exposure to high temperatures, ultraviolet radiation, action of microorganisms, oxidation, and humidity. These factors can cause chemical reactions in polymers, leading to degraded products and changes in the material's appearance, strength, and flexibility (Dorigato, 2021).

Fortunately, scientists are working to develop techniques to minimize degradation. Stabilizing additives have been created that protect polymers from sunlight and heat, thus reducing degradation. In addition, research is being done on modifying the molecular structure of polymers to improve their resistance to degradation. The degradation of polymer blends is not only a challenge but also an opportunity to continue innovating and improving these versatile materials.

9.6 Most Commonly Used Plastic Polymers

9.6.1 Nylon

Nylon is a synthetic polymer that is classified as a polyamide. In simpler terms, it is a human-made material using specific chemical compounds. Nylon was invented by Wallace Carothers in the 1930s at DuPont, an American chemical company. The molecular structure of nylon is very interesting (Fig. 9.2). It is composed of repeating units of carboxylic acids with functional groups that react with amines to form amide bonds (Shakiba et al., 2021).

The nylon production process involves a chemical reaction called polymerization. During this reaction, the monomers combine under controlled conditions of temperature and pressure, allowing the formation of long nylon chains. Nylon is known for its exceptional properties. It is an extremely strong, durable, and elastic material, which makes it suitable for a wide range of applications (Varghese & Grinstaff, 2022). In addition, it has a high resistance to abrasion, chemicals, and ultraviolet light (Lay et al., 2019).

Nylon is a synthetic material with high tenacity and tensile strength. It has a low density, which makes it light and easy to transport, is resistant to moisture, and has good dimensional stability. It is used in a wide variety of applications, such as hosiery, ropes, sportswear, backpacks, tents, and boat sails. It is also used in the manufacture of automotive parts and personal protective equipment (Winnacker, 2017). Thanks to its elasticity, nylon is used in the manufacture of hosiery, ropes, fabrics, tires, and various industrial applications. It is also used in the manufacture of machinery parts, such as gears and bearings, due to its high resistance to wear (Calignano et al., 2020).

Although nylon is not biodegradable, work is underway to produce recycled nylon to reduce its environmental impact. Research and development continue in this field, intending to develop nylon with improved properties, such as greater resistance to heat and chemicals, as well as greater elasticity. These innovations could open up new applications and improve the quality of nylon in the future (Tonsi et al., 2023).

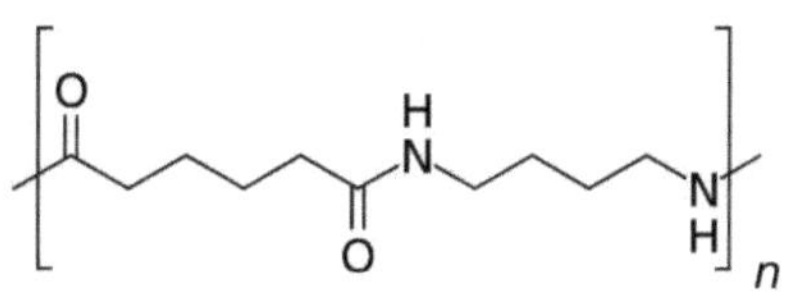

Fig. 9.2 Chemical structure of nylon

Nylon waste can present environmental challenges due to its low biodegradability. This means that nylon waste can accumulate in landfills or in the environment, negatively affecting ecosystems and wildlife. However, efforts are underway to address this problem. Recycled nylon is produced by collecting and processing nylon waste, such as discarded products or waste materials from manufacturing. These recycled materials can be used to create new nylon products, which helps reduce the amount of waste generated and the environmental impact of nylon (Singh et al., 2017).

In addition, techniques are being researched and developed to improve the biodegradability of nylon. Some scientists are exploring the possibility of using enzymes or microorganisms to accelerate the decomposition of nylon under controlled conditions (Zaaba & Jaafar, 2020). It is important to note that proper management of nylon waste is essential. This involves promoting recycling practices, such as separating and collecting nylon products for proper processing. It is also important to raise awareness of the environmental impact of nylon and promote responsible consumption.

9.6.2 Polyethylene

Polyethylene is one of the most widely used plastics in the world due to its diverse applications and properties. It is a thermoplastic polymer obtained from ethylene, a chemical compound derived from petroleum or natural gas. One of the most outstanding characteristics of PE is its versatility.

There are different types of PE, each with specific properties that give it different applications. For example, high-density polyethylene (HDPE) is strong and rigid and is used in the manufacture of pipes, food packaging, and toys (Dabees et al., 2021). On the other hand, low-density polyethylene (LDPE) is more flexible and is used in plastic bags, food wrapping film, and textiles (Rong et al., 2021). In addition to these two main categories, there are other types of polyethylene, such as linear low-density polyethylene (LLDPE) (dos Anjos et al., 2019), ultrahigh-density polyethylene (UHDPE) (Calderón et al., 2022), and cross-linked polyethylene (CLPE) (Li et al., 2019b). Each of these types has specific properties and applications.

In addition to its versatility, PE also stands out for its strength. It is a durable and impact-resistant material, making it ideal for applications requiring high wear resistance, such as chemical packaging and shipping containers. It is also resistant to moisture and most chemicals, making it a popular choice for containers and packaging (Li et al., 2019a).

Another advantage of PE is its low production cost. It is an inexpensive material and easy to manufacture in large quantities, making it affordable and accessible to many industries (Tan & Rodrigue, 2019). As for its environmental impact, it is important to note that PE is recyclable. However, it is fundamental to encourage proper recycling and the reduction of its consumption in order to minimize its impact on the environment (Raheem et al., 2019).

Fig. 9.3 Chemical structure of PE and its monomer

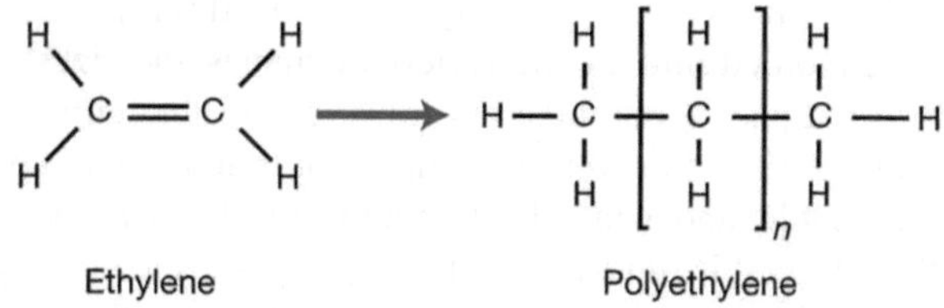

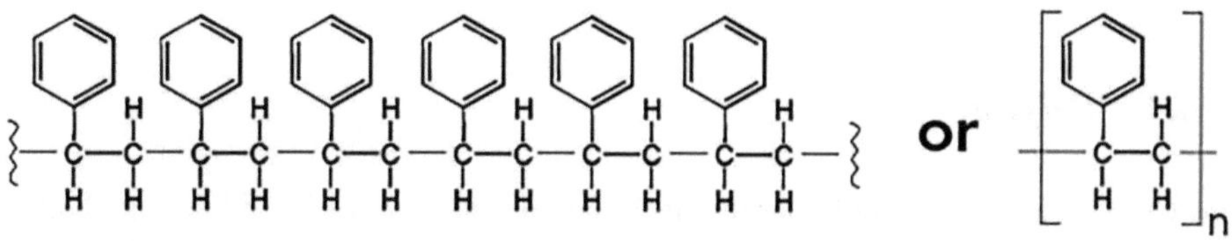

Fig. 9.4 Chemical structure of polystyrene

It is important to mention that PE is a safe material for use in contact with food and pharmaceuticals, as it is nontoxic and resistant to chemical migration. In terms of its manufacture, PE is produced by a polymerization process, in which ethylene molecules are joined together to form long polymer chains (Fig. 9.3). Depending on the polymerization conditions and the additives used, different grades of PE with specific properties can be obtained.

9.6.3 *Polystyrene*

Polystyrene, whose chemical name is polybenzene polystyrene, is a synthetic polymer obtained from styrene, a liquid compound derived from petroleum. Its chemical structure (Fig. 9.4) gives polystyrene particular properties such as lightness, chemical stability, and moisture resistance. The PS manufacturing process involves the polymerization of styrene, where styrene molecules are bonded together to form polymer chains. Depending on the type of PS desired, other compounds may be added during the process to modify its physical and chemical properties (Lerman et al., 2018).

PS is a plastic material widely used in our daily lives. It is known for its lightweight, durability, and thermal insulation capabilities, making it a popular choice in a wide range of applications. Polystyrene can be found in two main forms: expanded (EPS) and extruded (XPS). EPS is commonly known as "white cork" or "polystyrene foam" and is used in food packaging, building insulation, protective packaging, and decorative products (Ramli Sulong et al., 2019). On the other hand, XPS is denser and is used in more demanding applications, such as building thermal insulation, green roofs, and insulating board construction (Cai et al., 2017).

In addition to its versatility, PS is prized for its low production cost and its ability to be molded into a wide variety of shapes and sizes. This makes it an ideal choice for food packaging, take-out trays, disposable cups, electronics, and toys (Gounni et al., 2019). However, it is important to note that expanded polystyrene is not biodegradable and can take hundreds of years to decompose in the environment. Therefore, it is fundamental to properly recycle polystyrene products and seek more sustainable alternatives where possible (Ho et al., 2018).

9.6.4 Polyester

Polyester is a synthetic polymer that is widely used in the textile industry and other sectors. It is a strong, durable, and versatile material that has revolutionized the way various garments and products are manufactured. Polyester is produced from the chemical reaction between acids and polyhydric alcohols. This results in a chain of long, repeating molecules that form the polyester polymer (Fig. 9.5). One of the key characteristics of this material is its high tensile strength and wear resistance, making it an ideal choice for garments that are worn frequently and washed regularly (Jaffe et al., 2020).

In addition to its strength, polyester also stands out for its ability to retain color and resist wrinkling. Unlike other natural fibers such as cotton, polyester does not fade easily with wear and does not require frequent ironing. This makes it a popular choice for sportswear, bedding, and even furniture upholstery. Polyester also has waterproof and water-repellent properties. This is especially useful in the manufacture of outerwear and sports equipment that must protect the wearer from moisture and inclement weather (Darie-Niță et al., 2022).

In terms of care, polyester is easy to wash and dry, making it a practical choice for people looking for low-maintenance garments. It can be both hand- and machine-washed and dries quickly due to its ability to not retain much moisture. However, while polyester has many advantages, it also has some disadvantages. One of these is its low breathability, which can make polyester clothing uncomfortable in hot climates. Although there are technologies and finishes that help improve its breathability, it is still less breathable compared to natural fibers such as cotton or silk (Memon et al., 2016).

Fig. 9.5 Chemical structure of polyester

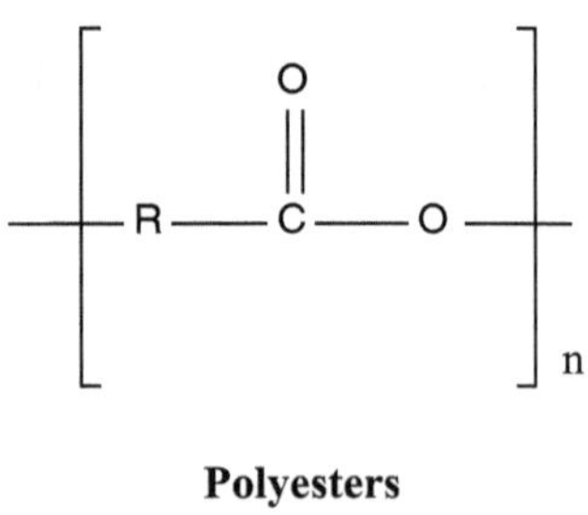

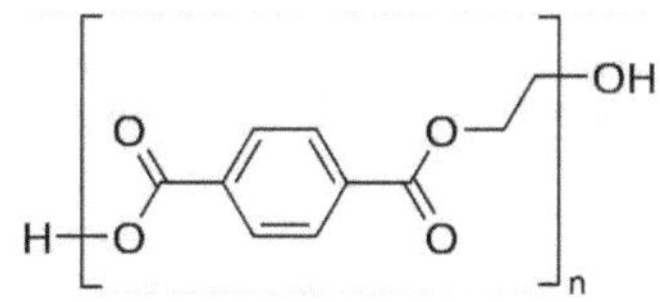

Fig. 9.6 Chemical structure of PET

9.6.5 Polyethylene Terephthalate

Polyethylene terephthalate (PET) is a type of polyester that is widely used in the manufacture of packaging, especially plastic bottles. It is a very popular material due to its strength, transparency, and barrier properties. PET (Fig. 9.6) is obtained from the polymerization of a mixture of terephthalic acid and ethylene glycol. The resulting molecules form a long, repeating chain that gives the material its unique characteristics (Dhaka et al., 2022).

One of the main advantages of PET is its physical and chemical resistance. It is a rigid and durable material that can withstand the transport and storage of liquids without deformation or loss of integrity. In addition, PET is resistant to most chemicals, making it suitable for containing a wide range of products, from food and beverages to chemicals and personal care products. Another outstanding feature of PET is its transparency. Unlike other plastics, PET is crystal clear, allowing excellent visualization of the contents of bottles and containers. This is especially important in the beverage industry, where consumers are looking to clearly see the product they are purchasing (Ge-Zhang et al., 2022).

In addition to its strength and transparency, PET also stands out for its barrier properties. It is a material that prevents the entry of oxygen, which helps to maintain the freshness and quality of packaged products. It is also an excellent thermal insulator, which makes it an ideal choice for keeping the contents of bottles and containers at the right temperature (Benyathiar et al., 2022).

One of the advantages of PET is that it is completely recyclable. It can be recovered and recycled to make new containers or used in other products, helping to reduce the amount of plastic waste in the environment. Recycled PET is commonly used for the production of textile fibers, such as recycled polyester, which is used to make clothing and other textile products (Antolinc & Filipič, 2021).

9.6.6 Polyurethane

Polyurethane is a very versatile material widely used in different industries. It is obtained by reacting a polyol with an isocyanate, which results in a resistant and highly durable polymer (Fig. 9.7).

One of the most common applications of polyurethane is in the manufacture of foams. These foams are used in mattresses, furniture, cushions, carpets, and many other products. Thanks to its properties of comfort, elasticity, and strength,

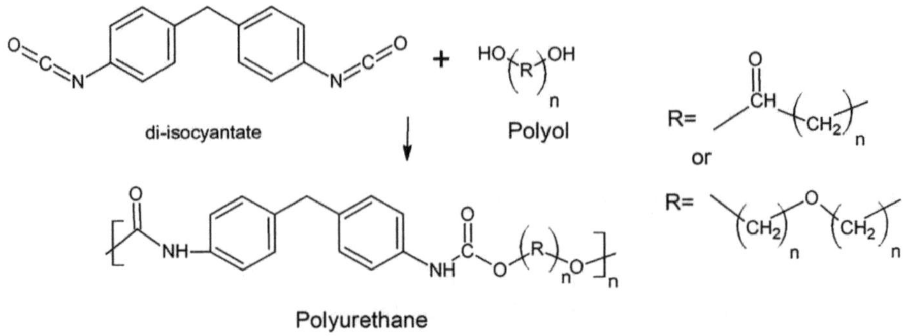

Fig. 9.7 Polyurethane obtention via di-isocyanate and a polyol reaction

polyurethane offers excellent support and comfort. In addition to foams (Gama et al., 2018), polyurethane is also used in coatings and adhesives. Polyurethane coatings are known for their resistance to abrasion, chemicals, and weathering, making them ideal for protecting surfaces such as floors, wood, and metals. On the other hand, polyurethane adhesives are very strong and are used in the automotive, construction, and furniture manufacturing industries, among others (Das & Mahanwar, 2020).

Another interesting application of polyurethane is in the manufacture of elastomers, which are materials with elastic properties and high strength. These polyurethane elastomers are used in the automotive industry to manufacture parts such as gaskets, seals, and shock absorbers, as they offer excellent resistance to wear and fatigue (Groenewolt, 2019).

Polyurethane is also used in the manufacture of insulation products. Thanks to its low thermal conductivity, polyurethane is used in the production of insulating panels for construction, food refrigeration, and freezing, as well as in air-conditioning systems. This helps to maintain stable temperatures and reduce energy consumption. In addition, polyurethane is used in the manufacture of coatings and paints. These coatings offer protection against corrosion and wear and are used in marine, industrial, and architectural applications. They are also used in the fashion and footwear industry, where they are used in primers and topcoats to improve the appearance and durability of products (Huang et al., 2018).

Another interesting application is in the automotive industry. Polyurethane is used to manufacture components such as seats, instrument panels, bumpers, and interior trim. This is due to its lightness, strength, and ability to absorb impacts, which contributes to improving the safety and comfort of vehicles. In addition, polyurethane is used in the manufacture of industrial wheels and rollers, as it offers excellent wear resistance and a long service life. It is also used in the manufacture of pipes and ducts, as it is resistant to corrosion and high temperatures (Tong et al., 2017).

In short, polyurethane is an incredibly versatile and useful material in a wide range of applications. From comfortable foams to tough coatings and durable elastomers, polyurethane has become an indispensable material in many industries.

Recycling and proper disposal of polyurethane are important issues to ensure sustainability and reduce environmental impact. Although polyurethane is a durable and versatile material, its recycling can present certain challenges due to its chemical composition and the different ways in which it is used.

In terms of recycling, there are different methods used to process polyurethane and reuse it in new products. One of the most common methods is mechanical recycling, which involves crushing and grinding polyurethane products to obtain smaller particles, which can then be used as raw material in the manufacture of new polyurethane products. There are also chemical recycling processes, where polyurethane is broken down into its basic components, such as polyols and isocyanates, which can then be used to produce new polymers (Xie et al., 2019).

In terms of polyurethane disposal, it is important to follow local regulations and practices to ensure that it is done in a safe and environmentally friendly manner. Depending on the amount and type of polyurethane, it may be possible to take it to specialized recycling facilities that can process it properly. In some cases, controlled incineration may be a viable option for polyurethane disposal, although the resulting environmental impact and emissions should always be considered.

Importantly, as consumers, we can contribute to the sustainability of polyurethane by opting for products that use recycled polyurethane or are designed to be recyclable at the end of their useful life. This promotes the circular economy and reduces the amount of polyurethane that ends up in landfills or unsustainable disposal processes.

In general, recycling and proper disposal of polyurethane are important aspects of promoting sustainability. Mechanical and chemical recycling are common methods used in polyurethane recycling, while disposal should follow local regulations and practices. By opting for recycled polyurethane products and promoting the circular economy, we can contribute to reducing the environmental impact of this material.

9.6.7 Polyvinyl Chloride

PVC, which is short for polyvinyl chloride, is a thermoplastic polymer derived from vinyl chloride monomer (Fig. 9.8). It is a material composed of carbon, hydrogen, and chlorine atoms, arranged in a linear molecular structure. The PVC manufacturing process involves the polymerization of vinyl chloride through a chemical chain

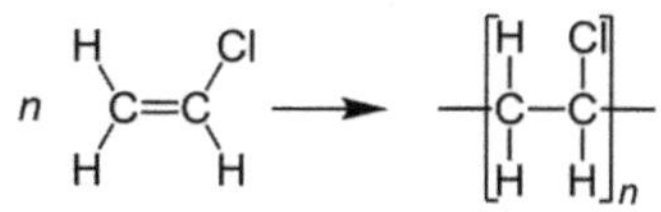

Fig. 9.8 Chemical structure of PCV and its monomer

reaction. During this polymerization, the vinyl chloride molecules bond together, forming long, flexible chains. This gives PVC its thermoplastic nature, which means that it can be softened and molded repeatedly through the application of heat.

PVC is characterized by its strength and durability, as well as its low production cost. In addition, its chemical structure gives it properties of resistance to fire, impact, and weathering. These characteristics have made PVC a material widely used in different applications, such as pipes, coatings, electrical cables, construction material, packaging, and medical products, among others (Ling et al., 2022).

In addition to its properties mentioned above, PVC is also known for being a versatile and easy to work with material. It can be molded into a wide variety of shapes and sizes, making it a popular choice in industries such as construction and product manufacturing (Abdulla, 2017). PVC can be found in different forms, such as sheets, profiles, pipes, and flexible compounds. PVC pipes are especially popular in plumbing and sewage systems due to their resistance to corrosion and their ability to withstand high pressures and temperatures (Makris et al., 2020).

In addition, PVC is a relatively lightweight material, making it easy to transport and install. It is also resistant to many chemicals, making it suitable for use in industrial applications where chemical resistance is required (Abdel-Gawad et al., 2019). In terms of sustainability, PVC is known to be durable and recyclable. When properly recycled, it can be transformed into new PVC products, thus reducing the need for virgin materials and reducing environmental impact (Ahmadian Hoseini et al., 2021).

9.7 Hazardous Effects of Plastic Waste

Plastic waste poses a serious threat to our environment. Their slow degradation means that they persist for decades and even centuries in our oceans, rivers, and land. This has devastating consequences for marine and terrestrial life. Aquatic animals often mistake plastic fragments for food and end up ingesting them, which can cause suffocation, blockage of the digestive tract, and even death. In addition, when plastic breaks down into microplastics, they can contaminate aquatic systems and negatively affect ecosystems and the balance of marine life (Avio et al., 2017).

The negative effects of plastics are widely known and have a significant impact on the environment and human health. One of the main problems is the pollution of the oceans and aquatic ecosystems. Plastics, especially single-use plastics such as bags and bottles, are indiscriminately dumped into the environment and end up in water bodies. This causes death and harm to marine wildlife, which can ingest or become trapped in plastic debris.

In addition, most single-use plastics are not biodegradable, which means they remain in the environment for hundreds of years. Over time, they break down into microplastics, which have been found in high concentrations in oceans, rivers, and soils. These microplastics can be ingested by animals and, through the food chain, end up on our plates, posing a risk to human health (Yeganeh et al., 2020).

The production of plastics also has a negative impact in terms of greenhouse gas emissions and energy consumption. The extraction of oil and natural gas, which are key components in the manufacture of plastics, contributes to deforestation and the emission of polluting gases. In addition, the production of plastics requires large amounts of energy, which in turn leads to increased consumption of natural resources and carbon emissions.

Plastic waste is also a serious threat to human health. When plastic decomposes or burns, it can release toxic chemicals, such as bisphenol A (BPA) and phthalates, which are endocrine disruptors and can negatively affect our hormonal system (Yi et al., 2020). These chemicals can leach into our water and food and eventually reach our bodies. Scientific studies have linked long-term exposure to these chemicals to developmental problems, hormone disruption, cancer, and other health problems (Rudel et al., 2003).

In addition to the environmental and health impacts, plastic waste also has a significant economic cost. The cleanup and management of this waste require a large investment of resources and time. Furthermore, the plastics that accumulate in our oceans and beaches affect the tourism industry, which is an important source of income in many regions. It is also important to mention that plastic production requires large amounts of energy and natural resources, contributing to the depletion of our resources and climate change (Geyer, 2020b).

In short, plastic waste represents a serious threat to our environment, marine life, our health, and our economy. It is essential to take action to reduce, reuse, and recycle plastic and promote sustainable alternatives. Together, we can work toward a cleaner and healthier future for all. Improper management of plastics and their final disposal in landfills or incinerators can release toxic substances and contaminate soil and groundwater. The burning of plastics also produces air pollutants harmful to health, such as dioxins and furans, which have neurotoxic and carcinogenic effects.

In summary, plastics have significant negative effects on the environment and human health, such as pollution of the oceans, presence of microplastics in the food chain, and negative effects of the production and disposal of plastics in terms of greenhouse gas emissions and air pollutants. It is essential to take actions to reduce the use of single-use plastics and promote appropriate management practices, such as recycling and reuse. In addition, more sustainable and biodegradable alternatives should be sought to replace plastics in various sectors, such as packaging and the food industry. Awareness and education on the importance of reducing plastic consumption and promoting a circular economy are also key to mitigating the negative effects of plastics on the environment.

9.8 Traditional Methods for Plastic Degradation

9.8.1 Photodegradation

Light-induced degradation in polymers is related to changes produced by irradiation with visible or ultraviolet (UV) light. Most commercial polymers undergo chemical reactions when irradiated with UV light because they have constituents or impurities capable of absorbing UV light (chromophore groups). This is fundamental because sunlight reaching the Earth has a portion of its spectrum that includes the UV region. Therefore, when organic polymers are exposed to the elements, they undergo photoreactions that are generally harmful because they make them fragile and become discolored (Zhang et al., 2021).

The essential photochemical reactions in polymers include scission of the main chains, cross-linking, generation of unsaturated groups, as well as cleavage of side groups, which result in the formation of volatile products. Although there are many lamps and devices that radiate light, as far as practical applications of polymers are concerned, the sun is the most important source. The spectrum of sunlight penetrating to the Earth's surface ranges from 290 to 3000 nm. About 10% of this light is UV, 50% visible, and 40% infrared (Kotilainen et al., 2020).

One of the prerequisites for photoreaction to occur is light absorption. Saturated bonds, such as C-C, C-H, O-H, C-Cl, etc., absorb light at wavelengths below 200 nm. Carbonyl groups and C=C double bonds have absorption maxima between 200 and 300 nm. This implies that only a small number of polymers are capable of absorbing solar radiation, whose spectral range is limited to about 300 nm. However, it is common for commercial polymers to have chromophore groups that absorb sunlight. These include carbonyl groups and polymerization catalysts (Kosloski-Oh et al., 2021).

Another prerequisite for the reaction is the photon energy. The bond dissociation energies range from 100 kcal/mol (C-C) to 40 kcal/mol (N-O), while for a photon with a wavelength of 200 nm the energy is 143 kcal/mol. Thus, the cleavage of chemical bonds is energetically possible, in principle (Wang et al., 2019).

The physical process of photon absorption consists in the transition of a molecule from one state to another of higher energy. The photochemical reaction can originate with radical ions or with excited states. The result is the photolysis of the polymer, in other words, its decomposition. The degradation generated in these cases can be oxidative, or in the absence of oxygen, the percentage of scission of the main chain is low and is mainly due to the presence of chromophore groups, especially of the carbonyl type.

Photooxidation is a powerful degradation mechanism because in the photolytic process free radicals are generated that react easily with oxygen, giving rise to autooxidation reactions. Since no polymer can withstand prolonged exposure to solar radiation, the protection of commercial polymers against the action of visible or UV light is of great importance. Since light-induced damage is largely due to oxidation chain reactions triggered by free radicals, antioxidants are effective in stabilizing polymers against light (Liu & Mabury, 2020).

9.8.2 Thermooxidative Degradation

This type of degradation technique is a deterioration process that affects plastics due to the combination of high temperatures and exposure to oxygen. This degradation can occur during different stages of the plastic life cycle, such as production, storage, use, and disposal (Gijsman, 2008). During thermooxidative degradation, plastics change their molecular and physical structure. The most noticeable effect is the loss of mechanical properties, such as strength, hardness, and flexibility. In addition, there may be a decrease in the aesthetic appearance of the plastic, turning it opaque or yellowish (Gijsman & Fiorio, 2023).

This procedure is accelerated at high temperatures, as the increased thermal energy promotes the chemical reactions that lead to degradation. In addition, exposure to oxygen facilitates the oxidation of the plastic, which can lead to the formation of free radicals and the breakdown of polymer chains (Dong & Gijsman, 2010).

This degradation process can be harmful to the environment, as it results in the release of hazardous chemicals and polluting gases. In addition, degraded plastics can become microplastics, small particles that have negative impacts on aquatic ecosystems and human health (Gama et al., 2023). There are different strategies to prevent or mitigate the thermooxidative degradation of plastics. One is to incorporate antioxidant additives into the polymer matrix, which slow degradation by acting as a barrier to free radicals. Suitable processing techniques, such as reducing the processing temperature or using controlled atmospheres, can also be used (Gensler et al., 2000).

Even with the addition of antioxidants, thermooxidative degradation is inevitable in the long term, especially in plastics exposed to extreme conditions such as high temperatures or ultraviolet radiation. Therefore, it is important to consider the shelf life of plastic products and their environmental impact prior to manufacture or use. Thermooxidative degradation can also be accelerated by other factors, such as contact with aggressive chemicals, the action of microorganisms, or the presence of impurities in the plastic. These factors can increase the degradation rate and lead to a more rapid and significant loss of plastic properties (Fa et al., 2020).

It is important to note that thermooxidative degradation of plastics not only affects the final products but also the manufacturing and recycling processes. Material deterioration can decrease the quality and efficiency of production processes, resulting in wasted time and resources. To mitigate the effects of thermooxidative degradation, proper management of plastics at the end of their life cycle is essential. This implies implementing recycling and material recovery practices, thus reducing the environmental impact of degraded plastics and promoting the circular economy.

In conclusion, thermooxidative degradation of plastics is a problem that affects both the final products and the production and recycling processes. It is important to consider this process during the design and manufacture of plastics, as well as to implement management strategies at the end of their life cycle to mitigate their environmental impact.

9.9 Eco-Friendly Management of Plastic Polymers

9.9.1 Biodegradation

The concept of biodegradability is usually combined with that of ecology and also with the adverse effects caused by the indiscriminate use of plastic products. Biodegradation is a type of decomposition by which some products and substances disintegrate due to the action of living beings such as insects, algae, bacteria, fungi, etc. (Amobonye et al., 2021).

These organisms use these substances as substrates for energy production and to create new, simpler compounds. In addition, the conditions of temperature, light, moisture, oxygen, etc. must be favorable for this to occur in a relatively short period of time (Zhang et al., 2022). The duration or time of decomposition is a very important aspect to consider in determining the degree of biodegradability (Wu et al., 2021).

Therefore, a product is said to be biodegradable when it can be broken down by environmental conditions and living beings, marking two types of biodegradation: aerobic biodegradation and anaerobic biodegradation (in the presence or absence of oxygen, respectively). During anaerobic biodegradation, the so-called biogas is produced, which is essentially methane and is one of the main greenhouse gases (Kabaivanova et al., 2022).

Organic matter can degrade aerobically or anaerobically. Matter that is biodegradable, such as substances of plant or animal origin, is bioassimilated by catabolic microorganisms that degrade, accumulate, or transform them (Chan et al., 2022). Significant methodological advances in microbial biodegradation have enabled detailed high-throughput analyses by understanding the organisms required to provide key biodegradation pathways.

In nature, materials are biodegradable at different rates. To perform the metabolic processes that lead to biodegradation, the vast majority of microorganisms need to have adequate conditions of water, oxygen, and light, among other factors such as temperature (generally warmer conditions favor biodegradation). The care and optimization of these conditions lead to a faster reproduction of the degrading microorganisms (Amobonye et al., 2021).

Biodegradation is a form of natural recycling of organic matter which means it is an important part of natural processes in the circular economy and long-term sustainability. This process has several benefits for the environment and society in general, among them:

- Waste reduction: Biodegradation helps reduce the amount of waste in the environment by breaking down organic materials and converting them into simpler substances. This means less materials taking up landfill space and less environmental pollution (Ali et al., 2021).
- Nutrient recycling: Biodegradation recycles nutrients contained in organic materials, such as carbon, nitrogen, and phosphorus. These nutrients are essential for the life of plants and other living beings, which means that biodegradation is an

important process for maintaining ecological balance and soil fertility (Kliopova et al., 2019).

- Renewable energy production: Biodegradation of organic waste can also produce biogas, a renewable energy source that can be used to generate electricity and heat. This means that biodegradation processes not only help reduce waste but can also be a clean and sustainable energy source (Raj et al., 2022).
- Pollution reduction: Biodegradation helps reduce environmental pollution by breaking down organic materials that might otherwise contaminate soil and water. This means that manufacturing biodegradable products is an important tool for reducing air, water, and soil pollution (Chia et al., 2020).
- Circular economy: Biodegradation is a key component of the circular economy, an approach to resource management that focuses on reducing, reusing, recycling, and recovering materials and resources (von Vacano et al., 2023).

As can be seen, biodegradation has an overall positive environmental impact, but it is important to consider the potential negative impacts and limitations associated with this natural process. For example, if large quantities of organic materials are biodegraded in a confined space, such as a landfill, it can become a source of carbon and expel large amounts of greenhouse gases. To maximize the benefits of biodegradation, it is important to properly manage waste and organic materials, avoid contamination, and use appropriate techniques for the management of biodegradable materials (Moshood et al., 2022).

9.9.2 Microbial Degradation of Plastics

Microbial degradation of plastics is an environmentally friendly way to degrade plastic waste. Microbial degradation offers a promising solution to this problem. Microbes can break down plastics into simpler compounds, which can be further degraded by other microorganisms (Liu et al., 2021). There are several potential microbes and enzymes that can be used to degrade plastics. For example, some bacteria can produce enzymes that break down PET, which is commonly used in plastic bottles. Other microbes can degrade polyethylene, PVC, polypropylene, PS, and polyurethane. The underpinning mechanisms for the degradation of these plastics are still being studied, but researchers have identified several enzymes and pathways involved in the process (Sebastian et al., 2023).

One of the challenges of microbial degradation of plastics is the slow rate of degradation. Plastics can take hundreds of years to degrade in the environment, but microbial degradation can take even longer (Rambabu et al., 2023). Researchers are working to optimize the conditions for microbial degradation, such as temperature, pH, and nutrient availability, to improve the rate of degradation. Another challenge is the potential release of harmful compounds during the degradation process. Researchers are studying the safety and environmental impact of microbial degradation of plastics to ensure that it is a safe and effective solution.

Despite the challenges, microbial degradation of plastics has promising future perspectives. It offers a sustainable solution to the problem of plastic waste, which is a major environmental issue. By using microbes to break down plastics, we can reduce the amount of plastic waste in landfills and oceans and potentially even recover valuable resources from the waste. Further research is needed to optimize the process and ensure its safety, but microbial degradation of plastics has the potential to be a key solution to the problem of plastic waste.

9.9.2.1 Microbial Mechanisms for Plastic Biodegradation

The mechanisms involved in the microorganism-mediated degradation of plastics are diverse. First, it is important to note that the physicochemical characteristics of the plastics, such as their chemical composition, size, shape, and structure, can influence the ability of microorganisms to degrade them. Some microorganisms can secrete extracellular enzymes, such as hydrolases, which are capable of breaking the chemical bonds of plastics. These enzymes can act as biocatalysts to break down the plastics into smaller fragments that are more easily assimilated by microorganisms (Sebastian et al., 2023).

Enzymes play a key role in the degradation process of plastics. For example, esterases are enzymes that have the ability to hydrolyze the esters present in some plastics, such as PET, leading to depolymerization and subsequent degradation of these materials. Another example is lipases, which can break down lipids and facilitate their decomposition (Liu et al., 2021).

In addition to enzymes, microorganisms can also use other strategies to degrade plastics. Some microorganisms can produce biosurfactants, which are molecules that can interact with the surface of the plastics and facilitate their breakdown. Other microorganisms can secrete metabolic products, such as organic acids, which can alter the properties of the plastic material and promote its degradation (Sebastian et al., 2023).

It is important to note that different microorganisms may have different degradation mechanisms and use different enzymes and metabolites to carry out this process. It is also possible that there is a synergistic interaction between different microorganisms, where some microorganisms may play a key role in initializing the degradation of plastics and others may degrade the intermediate products generated by the former.

The mechanisms involved in the microbial-mediated degradation under aerobic conditions of plastics include physical fragmentation, enzymatic degradation, and biofilm formation (Yuan et al., 2020). Physical fragmentation of plastics can occur due to various factors such as exposure to ultraviolet radiation, movement, and abrasion in the aquatic environment, among others. These processes can cause the fragmentation of plastics into smaller particles, which increases their availability and facilitates their transport and dispersion in the environment. However, it is important to note that the physical fragmentation of plastics is not the only mechanism involved in the degradation of these materials, as there may also be other processes such as chemical degradation and biodegradation by microorganisms (A L Andrady et al., 2022).

Enzymatic degradation plays a crucial role in the microbial degradation of plastics. Microorganisms produce various enzymes, such as lipases, esterases, and proteases, which can break down the chemical bonds in plastics. These enzymes target the polymer chains of plastics and cleave them into smaller fragments, facilitating further degradation by microorganisms (Yuan et al., 2020). For example, studies have shown that bacteria isolated from marine environments produce enzymes that can degrade PE and PP, two common types of plastics. These enzymes can hydrolyze the ester bonds in the polymer chains, leading to the breakdown of plastics into smaller molecules that can be utilized as carbon sources by microorganisms (Yoon et al., 1996).

Enzymatic degradation of MPs can occur extracellularly, where enzymes are secreted into the surrounding environment, or intracellularly, where enzymes are produced and utilized within the microbial cells. Extracellular enzymes can act on plastics that are in close proximity to the microorganisms, while intracellular enzymes can degrade plastics that have been internalized by the microorganisms (Yuan et al., 2020).

In summary, the mechanisms involved in microorganism-mediated degradation of plastics may include the secretion of enzymes (such as hydrolases, esterases, and lipases), the production of biosurfactants, and the generation of metabolic products that promote degradation. These mechanisms may vary depending on the type of microorganism and the type of plastic being degraded.

9.9.3 Biodegradation of Plastic Polymers by Fungi

Microbial degradation of plastic polymers by fungi is a promising method for dealing with plastic waste. Fungi such as *Aspergillus* spp. and *Penicillium* spp. are used as biodegradation agents to break down plastic polymers into simpler and smaller fragments including oligomer, dimer, and monomer through several steps including biodeterioration, depolymerization, assimilation, and mineralization. The fragmented plastic particles are absorbed and used by plastic-degrading fungi as their energy and carbon sources (Srikanth et al., 2022).

One example of a plastic polymer that can be degraded by fungi is PET. During the biodegradation process, enzymes hydrolyze the ester bonds present in PET, resulting in the formation of smaller molecules. The smaller molecules can then be used as a carbon and energy source by the fungi. The process may also involve the production of extracellular structures, such as mycelia, which physically break down the plastic polymer.

It is worth noting that not all fungi have the ability to degrade plastic polymers. Different species of fungi exhibit varying levels of efficiency in degrading different types of plastics. Additionally, the biodegradation process can be influenced by environmental factors such as temperature, pH, and nutrient availability. The biodegradation of plastic polymers by fungi holds potential for the development of eco-friendly solutions for plastic waste management. However, further research is still needed to optimize the process and make it practical on a larger scale.

9.9.4 Algal Degradation of Plastic Polymers

Algae, in their photosynthetic capacity, possess the ability to degrade plastics. The degradation of plastics by algae takes place through biodegradation. This process involves the breakdown of plastics into simpler compounds, such as water and carbon dioxide, through the action of enzymes secreted by algae. This phenomenon occurs mainly in aquatic environments, where algae can colonize the surfaces of plastics (Liu et al., 2022).

The degradation of plastics by algae is facilitated by factors such as the presence of oxygen in the environment and the availability of nutrients. Algae, by colonizing plastics, can release enzymes that degrade the structure of the polymers present in the plastics, thus breaking down the chains of molecules and facilitating their decomposition. The degradation process can also be influenced by the chemical composition of the plastic, as some polymers are more susceptible than others to the action of enzymes (Priya et al., 2022).

It is important to note that the degradation of plastics by algae is a relatively slow process and is dependent on several variables. The time required for the complete degradation of a plastic can vary according to factors such as the amount of algae present, the exposure to sunlight, and the temperature of the environment. In addition, it is necessary to consider that the degradation of plastics by algae can generate toxic by-products, such as microplastics, which can negatively affect aquatic ecosystems (Nabi et al., 2023).

The degradation of plastics by algae can have both positive and negative consequences. On the one hand, this capacity of algae can be harnessed for the removal of plastic waste in aquatic environments, thus contributing to the mitigation of plastic pollution. On the other hand, the release of toxic by-products during the process of plastic degradation by algae can have negative impacts on the health of aquatic organisms and water quality (Karalija et al., 2022).

In addition, it is important to consider that the degradation of plastics by algae is not a complete and efficient process. Although algae can decompose part of the plastics, they do not manage to completely get rid of all the components of the material, which can lead to the accumulation of partial residues and microplastics that persist in the environment for long periods of time.

Further studies are needed to develop more effective strategies to harness the ability of algae to degrade plastics. This includes research on the specific enzymes secreted by algae and how to improve their activity in the decomposition of different types of plastics. The possible long-term consequences of the release of toxic by-products during the degradation process should also be evaluated and solutions sought to mitigate their impact.

In summary, the degradation of plastics by algae is an interesting phenomenon from a scientific point of view and may have potential as a disposal strategy for plastic waste in aquatic environments. However, it also poses challenges and concerns in terms of its efficacy, environmental impact, and the generation of toxic by-products. Research in this field is needed to advance the development of more effective and sustainable solutions to address the plastic pollution problem.

9.10 Factors Affecting Plastic Biodegradation

9.10.1 Moisture

The presence of moisture can have a significant impact on the biodegradation of plastics. Plastics are known for their moisture resistance, which helps protect them from degrading when exposed to water or high humidity. However, when it comes to biodegradation, moisture can play a crucial role. Fungi, for example, are major organisms that have the ability to degrade plastic, and moisture provides a suitable environment for their growth and activity, enhancing the biodegradation process (Hari, 2019).

In the case of compostable plastics, moisture content is an important factor. Studies on backyard composting have shown that higher moisture levels can accelerate the degradation rate of compostable and oxodegradable plastic films (Quecholac-Piña et al., 2017). Similarly, in the case of bioplastics, moisture can impact the biodegradation process. Research on bioplastics made from cassava starch found that higher moisture levels increased the degradation rate (Wahyuningtiyas & Suryanto, 2017). This is because moisture facilitates the microbial degradation process and contributes to the breakdown of the bioplastic material (Ungprasoot et al., 2021). Therefore, moisture can either accelerate or impede the biodegradation of plastics, depending on the type of plastic and the specific environmental conditions.

9.10.2 pH and Temperature

The pH and temperature of the environment play crucial roles in the biodegradation of plastics. The pH level can have a significant impact on the activity of microorganisms involved in the biodegradation process. Different microorganisms thrive in different pH ranges, and the pH of the environment can affect their growth and metabolic activity (Quecholac-Piña et al., 2017). Acidic conditions, for example, have been found to enhance the biodegradation of certain plastics. Studies have shown that the degradation of polylactic acid (PLA) is more rapid under acidic conditions compared to neutral or alkaline conditions (Baldera-Moreno et al., 2023). However, extreme pH levels, such as highly acidic or highly alkaline conditions, can inhibit microbial activity and slow down the biodegradation process (Quecholac-Piña et al., 2017).

Temperature also plays a crucial role in the biodegradation of plastics. Higher temperatures generally accelerate the biodegradation process as they increase the metabolic activity of microorganisms. The degradation rate of plastics, such as compostable and oxodegradable plastic films, can be influenced by temperature. Higher temperatures can lead to faster degradation rates. However, it is important to note that there is an optimal temperature range for biodegradation and extreme temperatures can have negative effects. Very high temperatures can denature enzymes and inhibit microbial activity, while very low temperatures can slow down metabolic processes (Quecholac-Piña et al., 2017). The specific temperature range for optimal biodegradation depends on the type of plastic and the microorganisms involved. For example, the biodegradation of polylactic acid (PLA) is most efficient at temperatures around 50–60 °C (Baldera-Moreno et al., 2023).

In summary, pH and temperature are important factors that can influence the biodegradation of plastics. pH levels can affect the activity of microorganisms involved in the process, while temperature can impact the metabolic activity and degradation rate. It is crucial to consider the optimal pH and temperature conditions for specific plastics and microorganisms to promote efficient biodegradation.

9.10.3 Enzyme Characteristics

Enzymes play a crucial role in the biodegradation of plastics, and their characteristics can have a significant impact on the process. Enzymes are biological catalysts that facilitate chemical reactions in living organisms. In the context of plastic biodegradation, enzymes play a vital role in breaking down the complex polymer structure of plastics into smaller, more manageable compounds that can be utilized by microorganisms as a source of energy and nutrients.

Fungal enzymes have shown significant potential in the biodegradation of plastics. Fungi are major organisms that can degrade plastic, and enzymes from fungi are exploited for their ability to degrade plastics (Hari, 2019; Temporiti et al., 2022). Fungi have been found to have a remarkable efficiency in plastic biodegradation under laboratory conditions. Different types of enzymes isolated from many different fungi, from both basidiomycetes and ascomycetes, have shown good results in degrading different types of plastics. Here are some of the enzymes and the plastics they can degrade:

- Fungal laccases and peroxidases: These enzymes, commonly used by fungi to reduce lignin, show good results in degrading PE and PVC (Temporiti et al., 2022).
- Esterases such as cutinases and lipases: These enzymes were successfully used to degrade PET and polyurethane (Temporiti et al., 2022).
- Fungal proteases and ureases: These enzymes were found to be effective in degrading polyurethane (Temporiti et al., 2022).

A marine fungus called *Alternaria* sp. FB1 was found to efficiently degrade PE (Gao et al., 2021), and in another study, 30 fungi from Thailand were screened for their biodegradation potential of LDPE films, and 4 fungi were found to grow significantly by culturing with LDPE film as the only sole carbon source. The degradation of LDPE films by culturing with these fungi was further determined, and weight loss of up to 48.78% was observed (Khruengsai et al., 2021).

In recent years, enzymes capable of degrading PET have received significant attention. PET is one of the most produced plastics, and many enzymes with PET degradation activity have been discovered, characterized, and engineered in the last few years. However, classification and integrated knowledge of these enzymes are not trivial. Several high potential enzymes have been discovered, but further efforts to improve activity and thermal stability are necessary (Magalhães et al., 2021).

Ideonella sakaiensis is a bacterium that plays a notable role in plastic biodegradation, specifically in the biodegradation of PET (Walter et al., 2022; Yoshida et al., 2016). This bacterium produces PETase enzymes that break down PET, which has the potential to degrade PET faster with the PETase enzyme (Walter et al., 2022). The biodegradation mechanism involves the process of transforming the PETase coding gene from *I. sakaiensis* to *Escherichia coli* bacteria influenced by pH and the effectiveness of the PETase enzyme work until the use of harvested products interacts with PET waste, which decomposes plastic waste (Walter et al., 2022). Enzymes from fungi and bacteria, including *Ideonella sakaiensis*, play a crucial role in the biodegradation of plastics. Fungal enzymes such as laccases, peroxidases, esterases, proteases, and ureases have shown remarkable efficiency in plastic biodegradation under laboratory conditions. Many enzymes with PET degradation activity have been discovered, but further efforts to improve activity and thermal stability are necessary. Understanding the characteristics of these enzymes is crucial for the development of effective plastic biodegradation processes (Yoshida et al., 2016).

9.10.4 Molecular Weight

The molecular weight of plastics is an important factor that can affect their biodegradability. High molecular weight plastics are typically more difficult to degrade due to their closely cross-linked molecules, which make it challenging for microorganisms to break them down. The molecular weight of plastics can also affect the effectiveness of enzymes released by microorganisms during biodegradation. Enzymes work on the polymer backbone of plastics, and the effectiveness of these enzymes can be influenced by the molecular weight of the plastic (Singh, 2021).

However, there are ways to enhance the biodegradation of plastics with high molecular weight. One approach is the use of nanoparticles. Nanoparticles can increase the surface area of the plastic, making it more accessible to microorganisms (Bonhomme et al., 2003). Additionally, nanoparticles can act as carriers for enzymes, protecting them from degradation and increasing their stability. A review article on the biodegradation of plastic using microorganisms with the help of nanoparticles found that nanoparticles can enhance the biodegradation of plastics, particularly those with high molecular weight (Singh, 2021).

Another approach to enhancing the biodegradation of plastics is through preheating. A study on the biodegradation of a commercial environmentally degradable polyethylene found that molecular enlargement and broadening of molecular weight distribution occurred after preheating in air at 60 °C but not at ambient temperatures. This suggests that preheating can alter the molecular weight of plastics and make them more accessible to microorganisms during biodegradation (Juliana et al., 2022).

As can be seen, the molecular weight of plastics can affect their biodegradability, with high molecular weight plastics being more difficult to degrade. However, microorganisms and nanoparticles can play a role in enhancing the biodegradation process. Microorganisms can release enzymes that work on the polymer backbone, while nanoparticles can increase the surface area of the plastic and act as carriers for enzymes. Further research is needed to explore the potential of these approaches for large-scale plastic waste management.

9.10.5 Shape and Size

The shape and size of plastics are important factors that can affect their biodegradation. The surface area of plastics is a key factor that determines their accessibility to microorganisms and enzymes. Smaller plastic particles have a larger surface area-to-volume ratio, which makes them more accessible to microorganisms and enzymes, and can enhance the biodegradation process.

The shape of plastics can also influence their biodegradation rate. For example, a study on the biodegradation of polystyrene (PS) found that spherical PS particles degraded faster than cubic or cylindrical particles. This suggests that the shape of plastics can affect the accessibility of microorganisms and enzymes to the plastic surface and thus influence the biodegradation rate (Swati et al., 2015).

In the case of compostable plastics, the size and shape of the plastic can also affect the biodegradation process. Composting is a process that involves the degradation of organic matter by microorganisms under controlled conditions. A study on the biodegradation of compostable and oxodegradable plastic films found that the size and shape of the plastic particles affected the degradation rate. Smaller particles degraded faster than larger particles, and films with a higher surface area-to-volume ratio degraded faster than films with a lower ratio. This suggests that the size and shape of compostable plastics can influence their accessibility to microorganisms and enzymes during the composting process (Quecholac-Piña et al., 2017).

In the marine environment, microplastics are a major concern due to their small size and accessibility to marine organisms. Microplastics are small plastic particles less than 5 mm in size that can be ingested by marine organisms, causing harm to the ecosystem. The shape and size of microplastics can affect their distribution and transport in the marine environment. For example, smaller microplastics are more likely to be transported by ocean currents, while larger microplastics tend to accumulate in coastal areas. The shape of microplastics can also affect their ingestion and toxicity to marine organisms. For example, microplastics with sharp edges or irregular shapes can cause physical damage to the digestive system of marine organisms (Swati et al., 2015).

Overall, the shape and size of plastics can affect their biodegradation. Smaller plastic particles with a larger surface area-to-volume ratio are more accessible to microorganisms and enzymes, which can enhance the biodegradation process. The shape of plastics can also influence their biodegradation rate, with spherical particles degrading faster than cubic or cylindrical particles. In the case of compostable plastics, smaller particles and films with a higher surface area-to-volume ratio degrade faster. In the marine environment, microplastics are a major concern due to their small size and accessibility to marine organisms.

9.11 Current Trends and Future Perspectives

The challenges of managing plastic waste are numerous. Most plastic pollution comes from inadequate collection and disposal of larger plastic debris, but the leakage of microplastics and chemicals into the environment from inadequate disposal is also a significant concern. It is important to reduce the amount of plastic debris that is deposited in marine ecosystems by limiting landfilling. Relying heavily on landfilling can lead to burning waste to make space for more waste, which can cause environmental harm. Developing nations should follow the lead of EU countries and prioritize material and energy recovery over landfilling. Proper handling of plastic waste in terms of production and consumption requires thorough international investigations, which is an important aspect that must be considered from a legal perspective.

There is a growing need for eco-friendly management of plastic waste. More than 60 countries have imposed bans and levies to limit single-use plastic waste (Kumar et al., 2021). In the current situation, it is difficult to impose a complete ban on plastic products; however, it is possible and important to ban single-use plastics and encourage alternatives to plastics. Plastic pollution is a huge problem, but it's not too late to fix it. The global campaign to gain control of plastic waste is one of the fastest-growing environmental causes ever mounted.

Current trends and future perspectives on eco-friendly management of plastic waste are crucial in addressing the global plastic pollution crisis and promoting a sustainable environment. Here are some key insights:

- Plastic Waste Recycling and Reprocessing: Conventional strategies such as landfilling, incineration, and mechanical reprocessing have been widely used for plastic waste management. However, emerging strategies like pyrolysis, gasification, and biochemical conversion are gaining attention as more sustainable alternatives. These methods aim to convert plastic waste into valuable resources, such as fuel or raw materials for manufacturing (Maitlo et al., 2022).

- Biodegradation Approach: The use of biodegradation to manage plastic waste is an area of active research and development. Future perspectives highlight the potential of biodegradation in ameliorating plastic waste pollution. This approach involves harnessing the natural ability of microorganisms to break down plastics into harmless by-products (Er et al., 2023).
- Circular Economy and Policy Interventions: To address plastic pollution, there is a need to focus on circular economy principles and policy interventions. Efforts should be made to limit landfilling and deposition of plastic debris in marine ecosystems. Many countries have already implemented bans and levies to reduce single-use plastic waste. Encouraging alternatives to plastics and promoting sustainable consumption practices are also important aspects of eco-friendly plastic waste management (Kumar et al., 2021).
- Sustainable Waste Management Techniques: Various techniques and approaches have been explored for sustainable plastic waste management. These include adsorption, coagulation, photocatalysis, microbial degradation, reduction, reuse, and recycling. By adopting these techniques, plastic waste can be effectively managed, reducing its impact on the environment (Pandey et al., 2023).
- Scientometric Analysis: A scientometric analysis of the available literature on plastic waste management revealed key trends and research focus areas. Waste management, recycling, plastics, and plastic waste were among the most frequently occurring keywords. This analysis provides insights into the current state of research and helps identify areas for further exploration and development (Huang et al., 2022).

To conclude, current trends in eco-friendly management of plastic waste focus on recycling, emerging technologies, biodegradation, circular economy principles, and policy interventions. These approaches aim to reduce plastic waste accumulation, promote resource recovery, and mitigate the environmental impact of plastic pollution. Continued research, innovation, and collaboration among governments, industries, and individuals are essential for a sustainable future in managing plastic waste.

9.12 Conclusions

Plastic waste is a significant environmental issue that has been growing over the years. According to a report by the United Nations, the world generates around 300 million tons of plastic waste every year, and this number is expected to double by 2030. The improper disposal of plastic waste can lead to various environmental problems, including soil and water pollution, harm to wildlife, and greenhouse gas emissions.

To address this issue, several sustainable approaches have been developed to manage plastic waste. Recycling is one of the most common methods of managing plastic waste. It involves the collection, sorting, and processing of plastic waste to produce new products. This not only reduces the amount of plastic waste in the environment but also conserves resources and reduces energy consumption and greenhouse gas emissions.

Another approach is the use of bioplastics, which are made from renewable resources such as cornstarch, sugarcane, or potato starch. Bioplastics are biodegradable and compostable, making them an eco-friendlier option than traditional plastics. However, the production of bioplastics requires a significant amount of resources and energy, and their disposal requires specific conditions to ensure proper decomposition.

Proper waste management is also essential to reduce the negative effects of plastic waste. This includes sorting and separation of waste, quantification of waste generation, and physical composition of waste. Additionally, informal waste pickers can play a significant role in managing plastic waste. Approximately 20 million informal workers already recover plastic waste from the garbage in the Global South, usually working under precarious, risky, and poorly paid conditions. Recognizing and integrating these workers under decent work conditions and fair remuneration could potentially significantly increase the capturing of plastic waste and reduce the amount of fugitive plastics.

In conclusion, plastic waste is a significant environmental issue that requires sustainable and promising approaches to reduce its negative impacts. Recycling, bioplastics, proper waste management, and informal waste pickers are some of the current trends and future perspectives of plastic waste and its eco-friendly management.

References

Abdel-Gawad, N. M. K., El Dein, A. Z., Mansour, D. A., Ahmed, H. M., Darwish, M. M. F., & Lehtonen, M. (2019). Development of industrial scale PVC nanocomposites with comprehensive enhancement in dielectric properties. *IET Science, Measurement & Technology, 13*(1), 90–96.

Abdulla, N. A. (2017). Concrete filled PVC tube: A review. *Construction and Building Materials, 156*, 321–329.

Abrha, H., Cabrera, J., Dai, Y., Irfan, M., Toma, A., Jiao, S., & Liu, X. (2022). Bio-based plastics production, impact and end of life: A literature review and content analysis. *Sustainability, 14*(8), 4855.

Ahmadian Hoseini, A. H., Erfanian, E., Kamkar, M., Sundararaj, U., Liu, J., & Arjmand, M. (2021). Waste to value-added product: Developing electrically conductive nanocomposites using a non-recyclable plastic waste containing vulcanized rubber. *Polymers, 13*(15), 2427.

Ali, S. S., Elsamahy, T., Al-Tohamy, R., Zhu, D., Mahmoud, Y. A.-G., Koutra, E., Metwally, M. A., Kornaros, M., & Sun, J. (2021). Plastic wastes biodegradation: Mechanisms, challenges and future prospects. *Science of the Total Environment, 780*(146), 590.

Amobonye, A., Bhagwat, P., Singh, S., & Pillai, S. (2021). Plastic biodegradation: Frontline microbes and their enzymes. *Science of the Total Environment, 759*(143), 536.

Andrady, A. L., Barnes, P. W., Bornman, J. F., Gouin, T., Madronich, S., White, C. C., Zepp, R. G., & Jansen, M. A. K. (2022). Oxidation and fragmentation of plastics in a changing environment; from UV-radiation to biological degradation. *Science of The Total Environment, 851*(158), 022.

Antolinc, D., & Filipič, K. E. (2021). Recycling of nonwoven polyethylene terephthalate textile into thermal and acoustic insulation for more sustainable buildings. *Polymers, 13*(18), 3090.
Asiandu, A. P., Wahyudi, A., & Sari, S. W. (2021). A review: Plastics waste biodegradation using plastics-degrading bacteria. *Journal of Environmental Treatment Techniques, 9*(1), 148–157.
Avio, C. G., Gorbi, S., & Regoli, F. (2017). Plastics and microplastics in the oceans: From emerging pollutants to emerged threat. *Marine Environmental Research, 128*, 2–11.
Bahl, S., Dolma, J., Singh, J. J., & Sehgal, S. (2021). Biodegradation of plastics: A state of the art review. *Materials Today: Proceedings, 39*, 31–34.
Baldera-Moreno, Y., Rojas-Palma, A., Andler, R., & Cuesta-Herrera, L. (2023). Mathematical model of polylactic acid biodegradation under controlled composting conditions. *Journal of Physics: Conference Series, 2515*(1), 12,004.
Benyathiar, P., Kumar, P., Carpenter, G., Brace, J., & Mishra, D. K. (2022). Polyethylene terephthalate (PET) bottle-to-bottle recycling for the beverage industry: A review. *Polymers, 14*(12), 2366.
Bonhomme, S., Cuer, A., Delort, A. M., Lemaire, J., Sancelme, M., & Scott, G. (2003). Environmental biodegradation of polyethylene. *Polymer Degradation and Stability, 81*(3), 441–452.
Cai, S., Zhang, B., & Cremaschi, L. (2017). Review of moisture behavior and thermal performance of polystyrene insulation in building applications. *Building and Environment, 123*, 50–65.
Calderón, S., Centella, M., Lozano, L., Sustaita, A., Collado, E., & Sáez, Y. (2022). Thermal-structural analysis of polymers for the manufacture of cases for outdoor pollutants monitoring devices. In *2022 eighth International Engineering, Sciences and Technology Conference (IESTEC)* (pp. 707–714).
Calignano, F., Lorusso, M., Roppolo, I., & Minetola, P. (2020). Investigation of the mechanical properties of a carbon fibre-reinforced nylon filament for 3D printing. *Machines, 8*(3), 52.
Chamas, A., Moon, H., Zheng, J., Qiu, Y., Tabassum, T., Jang, J. H., Abu-Omar, M., Scott, S. L., & Suh, S. (2020). Degradation rates of plastics in the environment. *ACS Sustainable Chemistry & Engineering, 8*(9), 3494–3511.
Chan, S. S., Khoo, K. S., Chew, K. W., Ling, T. C., & Show, P. L. (2022). Recent advances biodegradation and biosorption of organic compounds from wastewater: Microalgae-bacteria consortium-A review. *Bioresource Technology, 344*(126), 159.
Chia, W. Y., Tang, D. Y. Y., Khoo, K. S., Lup, A. N. K., & Chew, K. W. (2020). Nature's fight against plastic pollution: Algae for plastic biodegradation and bioplastics production. *Environmental Science and Ecotechnology, 4*(100), 065.
Dabees, S., Tirth, V., Mohamed, A., & Kamel, B. M. (2021). Wear performance and mechanical properties of MWCNT/HDPE nanocomposites for gearing applications. *Journal of Materials Research and Technology, 12*, 2476–2488.
Darie-Niţă, R. N., Râpă, M., & Frąckowiak, S. (2022). Special features of polyester-based materials for medical applications. *Polymers, 14*(5), 951.
Das, A., & Mahanwar, P. (2020). A brief discussion on advances in polyurethane applications. *Advanced Industrial and Engineering Polymer Research, 3*(3), 93–101.
Dhaka, V., Singh, S., Anil, A. G., Sunil Kumar Naik, T. S., Garg, S., Samuel, J., Kumar, M., Ramamurthy, P. C., & Singh, J. (2022). Occurrence, toxicity and remediation of polyethylene terephthalate plastics. A review. *Environmental Chemistry Letters, 91*, 1–24.
Dong, W., & Gijsman, P. (2010). Influence of temperature on the thermo-oxidative degradation of polyamide 6 films. *Polymer Degradation and Stability, 95*(6), 1054–1062.
Dorigato, A. (2021). Recycling of polymer blends. *Advanced Industrial and Engineering Polymer Research, 4*(2), 53–69.
dos Anjos, E. G. R., Backes, E. H., Marini, J., Pessan, L. A., Montagna, L. S., & Passador, F. R. (2019). Effect of LLDPE-g-MA on the rheological, thermal, mechanical properties and morphological characteristic of PA6/LLDPE blends. *Journal of Polymer Research, 26*, 1–10.

Er, C. T. X., Sen, L. Z., Srinophakun, P., & Wei, O. C. (2023). Recent advances and challenges in sustainable management of plastic waste using biodegradation approach. *Bioresource Technology, 128*, 772.

Fa, W., Wang, J., Ge, S., & Chao, C. (2020). Performance of photo-degradation and thermo-degradation of polyethylene with photo-catalysts and thermo-oxidant additives. *Polymer Bulletin, 77*, 1417–1432.

Falua, K. J., Pokharel, A., Babaei-Ghazvini, A., Ai, Y., & Acharya, B. (2022). Valorization of starch to biobased materials: A review. *Polymers, 14*(11), 2215.

Gama, N. V., Ferreira, A., & Barros-Timmons, A. (2018). Polyurethane foams: Past, present, and future. *Materials, 11*(10), 1841.

Gama, N., Evtyugin, D. D., Lourenço, A., Lopes, C., & Evtuguin, D. V. (2023). Ellagic acid as stabilizer in the thermo-oxidative degradation of thermoplastic polyurethane. *Polymer Degradation and Stability, 110*, 456.

Gao, R., Liu, R., & Sun, C. (2021). A marine fungus efficiently degrades polyethylene. *BioRxiv*, 2011–2021.

Gebrekrstos, A., & Ray, S. S. (2023). Superior electrical conductivity and mechanical properties of phase-separated polymer blend composites by tuning the localization of nanoparticles for electromagnetic interference shielding applications. *Journal of Polymer Science, 61*, 2567–2584.

Gensler, R., Plummer, C. J. G., Kausch, H.-H., Kramer, E., Pauquet, J.-R., & Zweifel, H. (2000). Thermo-oxidative degradation of isotactic polypropylene at high temperatures: Phenolic antioxidants versus HAS. *Polymer Degradation and Stability, 67*(2), 195–208.

Geyer, R. (2020a). A brief history of plastics. In *Mare Plasticum-the plastic sea: Combatting plastic pollution through science and art* (pp. 31–47).

Geyer, R. (2020b). Production, use, and fate of synthetic polymers. In *Plastic waste and recycling* (pp. 13–32). Elsevier.

Ge-Zhang, S., Liu, H., Song, M., Wang, Y., Yang, H., Fan, H., Ding, Y., & Mu, L. (2022). Advances in polyethylene terephthalate beverage bottle optimization: A mini review. *Polymers, 14*(16), 3364.

Gijsman, P. (2008). Review on the thermo-oxidative degradation of polymers during processing and in service. *E-Polymers, 8*(1), 65.

Gijsman, P., & Fiorio, R. (2023). Long term thermo-oxidative degradation and stabilization of polypropylene (PP) and the implications for its recyclability. *Polymer Degradation and Stability, 110*, 260.

Gounni, A., Mabrouk, M. T., El Wazna, M., Kheiri, A., El Alami, M., El Bouari, A., & Cherkaoui, O. (2019). Thermal and economic evaluation of new insulation materials for building envelope based on textile waste. *Applied Thermal Engineering, 149*, 475–483.

Groenewolt, M. (2019). Polyurethane coatings: a perfect product class for the design of modern automotive clearcoats. *Polymer International, 68*(5), 843–847.

Hacker, M. C., Krieghoff, J., & Mikos, A. G. (2019). Synthetic polymers. In *Principles of regenerative medicine* (pp. 559–590). Elsevier.

Hari, S. (2019). Review on effect of fungi on plastic degradation. *JRAR, 6*(1), 261–265.

Ho, B. T., Roberts, T. K., & Lucas, S. (2018). An overview on biodegradation of polystyrene and modified polystyrene: the microbial approach. *Critical Reviews in Biotechnology, 38*(2), 308–320.

Huang, Y., Tang, Z., Liu, Z., Wei, J., Hu, H., & Zhi, C. (2018). Toward enhancing wearability and fashion of wearable supercapacitor with modified polyurethane artificial leather electrolyte. *Nano-Micro Letters, 10*, 1–8.

Huang, S., Wang, H., Ahmad, W., Ahmad, A., Ivanovich Vatin, N., Mohamed, A. M., Deifalla, A. F., & Mehmood, I. (2022). Plastic waste management strategies and their environmental aspects: A scientometric analysis and comprehensive review. *International Journal of Environmental Research and Public Health, 19*(8), 4556.

Jaffe, M., Easts, A. J., & Feng, X. (2020). Polyester fibers. In *Thermal analysis of textiles and fibers* (pp. 133–149). Elsevier.

Juliana, S., Parhusip, M., Simanullang, A., Tita, E., & Irawati, W. (2022). Potential of Ideonella sakaiensis bacteria in degrading plastic waste type polyethylene terephthalate. *Jurnal Biologi Tropis, 22*(2), 381–389.

Kabaivanova, L., Petrova, P., Hubenov, V., & Simeonov, I. (2022). Biogas production potential of thermophilic anaerobic biodegradation of organic waste by a microbial consortium identified with metagenomics. *Life, 12*(5), 702.

Kan, M., & Miller, S. A. (2022). Environmental impacts of plastic packaging of food products. *Resources, Conservation and Recycling, 180*(106), 156.

Karalija, E., Carbó, M., Coppi, A., Colzi, I., Dainelli, M., Gašparović, M., Grebenc, T., Gonnelli, C., Papadakis, V., & Pilić, S. (2022). Interplay of plastic pollution with algae and plants: Hidden danger or a blessing? *Journal of Hazardous Materials, 438*(129), 450.

Kawashima, N., Yagi, T., & Kojima, K. (2019). How do bioplastics and fossil-based plastics play in a circular economy? *Macromolecular Materials and Engineering, 304*(9), 1,900,383.

Kehinde, O., Ramonu, O. J., Babaremu, K. O., & Justin, L. D. (2020). Plastic wastes: Environmental hazard and instrument for wealth creation in Nigeria. *Heliyon, 6*(10), e05131.

Khalid, M. Y., Arif, Z. U., Ahmed, W., & Arshad, H. (2022). Recent trends in recycling and reusing techniques of different plastic polymers and their composite materials. *Sustainable Materials and Technologies, 31*, e00382.

Khruengsai, S., Sripahco, T., & Pripdeevech, P. (2021). Low-density polyethylene film biodegradation potential by fungal species from Thailand. *Journal of Fungi, 7*(8), 594.

Kibria, M. G., Masuk, N. I., Safayet, R., Nguyen, H. Q., & Mourshed, M. (2023). Plastic waste: Challenges and opportunities to mitigate pollution and effective management. *International Journal of Environmental Research, 17*(1), 20. https://doi.org/10.1007/s41742-023-00507-z

Kliopova, I., Staniškis, J. K., Stunžėnas, E., & Jurovickaja, E. (2019). Bio-nutrient recycling with a novel integrated biodegradable waste management system for catering companies. *Journal of Cleaner Production, 209*, 116–125.

Kosloski-Oh, S. C., Wood, Z. A., Manjarrez, Y., de Los Rios, J. P., & Fieser, M. E. (2021). Catalytic methods for chemical recycling or upcycling of commercial polymers. *Materials Horizons, 8*(4), 1084–1129.

Kotilainen, T., Aphalo, P. J., Brelsford, C. C., Böök, H., Devraj, S., Heikkilä, A., Hernández, R., Kylling, A., Lindfors, A. V., & Robson, T. M. (2020). Patterns in the spectral composition of sunlight and biologically meaningful spectral photon ratios as affected by atmospheric factors. *Agricultural and Forest Meteorology, 291*(108), 041.

Kumar, R., Verma, A., Shome, A., Sinha, R., Sinha, S., Jha, P. K., Kumar, R., Kumar, P., & Shubham, & Das, S. (2021). Impacts of plastic pollution on ecosystem services, sustainable development goals, and need to focus on circular economy and policy interventions. *Sustainability, 13*(17), 9963.

Lay, M., Thajudin, N. L. N., Hamid, Z. A. A., Rusli, A., Abdullah, M. K., & Shuib, R. K. (2019). Comparison of physical and mechanical properties of PLA, ABS and nylon 6 fabricated using fused deposition modeling and injection molding. *Composites Part B: Engineering, 176*(107), 341.

Lerman, M. J., Lembong, J., Muramoto, S., Gillen, G., & Fisher, J. P. (2018). The evolution of polystyrene as a cell culture material. *Tissue Engineering Part B: Reviews, 24*(5), 359–372.

Li, D., Zhou, L., Wang, X., He, L., & Yang, X. (2019a). Effect of crystallinity of polyethylene with different densities on breakdown strength and conductance property. *Materials, 12*(11), 1746.

Li, S., Yang, P., Liu, X., Zhang, J., Xie, W., Wang, C., Liu, C., & Guo, Z. (2019b). Graphene oxide based dopamine mussel-like cross-linked polyethylene imine nanocomposite coating with enhanced hexavalent uranium adsorption. *Journal of Materials Chemistry A, 7*(28), 16,902–16,911.

Ling, M., Ma, D., Hu, X., Liu, Z., Wang, D., & Feng, Q. (2022). Hydrothermal treatment of polyvinyl chloride: Reactors, dechlorination chemistry, application, and challenges. *Chemosphere, 137*, 718.

Liu, R., & Mabury, S. A. (2020). Synthetic phenolic antioxidants: A review of environmental occurrence, fate, human exposure, and toxicity. *Environmental Science & Technology, 54*(19), 11,706–11,719.

Liu, T., Xin, Y., Liu, X., Wu, B., & Xiang, M. (2021). Advances in microbial degradation of plastics. *Sheng Wu Gong Cheng Xue Bao = Chinese Journal of Biotechnology, 37*(8), 2688–2702.

Liu, L., Xu, M., Ye, Y., & Zhang, B. (2022). On the degradation of (micro) plastics: Degradation methods, influencing factors, environmental impacts. *Science of the Total Environment, 806*(151), 312.

Liwarska-Bizukojc, E. (2022). Phytotoxicity assessment of biodegradable and non-biodegradable plastics using seed germination and early growth tests. *Chemosphere, 289*(133), 132.

Magalhães, R. P., Cunha, J. M., & Sousa, S. F. (2021). Perspectives on the role of enzymatic biocatalysis for the degradation of plastic PET. *International Journal of Molecular Sciences, 22*(20), 11,257.

Maitlo, G., Ali, I., Maitlo, H. A., Ali, S., Unar, I. N., Ahmad, M. B., Bhutto, D. K., Karmani, R. K., Naich, S., & ur R., & Sajjad, R. U. (2022). Plastic waste recycling, applications, and future prospects for a sustainable environment. *Sustainability, 14*(18), 11,637.

Makris, K. F., Langeveld, J., & Clemens, F. H. L. R. (2020). A review on the durability of PVC sewer pipes: Research vs. practice. *Structure and Infrastructure Engineering, 16*(6), 880–897.

Memon, H., Yasin, S., Khoso, N. A., & Memon, S. (2016). Study of wrinkle resistant, breathable, anti-UV nanocoated woven polyester fabric. *Surface Review and Letters, 23*(03), 1,650,003.

Moshood, T. D., Nawanir, G., Mahmud, F., Mohamad, F., Ahmad, M. H., & AbdulGhani, A. (2022). Sustainability of biodegradable plastics: New problem or solution to solve the global plastic pollution? *Current Research in Green and Sustainable Chemistry, 100*, 273.

Mujtaba, M., Fraceto, L., Fazeli, M., Mukherjee, S., Savassa, S. M., de Medeiros, G. A., Santo Pereira, A., & do E., Mancini, S. D., Lipponen, J., & Vilaplana, F. (2023). Lignocellulosic biomass from agricultural waste to the circular economy: A review with focus on biofuels, biocomposites and bioplastics. *Journal of Cleaner Production, 136*, 815.

Muthuraj, R., Misra, M., & Mohanty, A. K. (2018). Biodegradable compatibilized polymer blends for packaging applications: A literature review. *Journal of Applied Polymer Science, 135*(24), 45,726.

Nabi, I., Zaheer, M., Jin, W., & Yang, L. (2023). Biodegradation of macro-and micro-plastics in environment: A review on mechanism, toxicity, and future perspectives. *Science of The Total Environment, 858*(160), 108.

Nielsen, T. D., Hasselbalch, J., Holmberg, K., & Stripple, J. (2020). Politics and the plastic crisis: A review throughout the plastic life cycle. *Wiley Interdisciplinary Reviews: Energy and Environment, 9*(1), e360.

OECD. (2022). *Plastic pollution is growing relentlessly as waste management and recycling fall short, says OECD*. Newsroom.

Pandey, P., Dhiman, M., Kansal, A., & Subudhi, S. P. (2023). Plastic waste management for sustainable environment: Techniques and approaches. *Waste Disposal & Sustainable Energy, 5*, 205–222.

Priya, A. K., Jalil, A. A., Dutta, K., Rajendran, S., Vasseghian, Y., Karimi-Maleh, H., & Soto-Moscoso, M. (2022). Algal degradation of microplastic from the environment: Mechanism, challenges, and future prospects. *Algal Research, 102*, 848.

Qin, Z., Mou, J., Chao, C. Y. H., Chopra, S. S., Daoud, W., Leu, S., Ning, Z., Tso, C. Y., Chan, C. K., & Tang, S. (2021). Biotechnology of plastic waste degradation, recycling, and valorization: Current advances and future perspectives. *ChemSusChem, 14*(19), 4103–4114.

Quecholac-Piña, X., García-Rivera, M. A., Espinosa-Valdemar, R. M., Vázquez-Morillas, A., Beltrán-Villavicencio, M., de la Cisneros-Ramos, A., & L. (2017). Biodegradation of compostable and oxodegradable plastic films by backyard composting and bioaugmentation. *Environmental Science and Pollution Research, 24*, 25,725–25,730.

Raheem, A. B., Noor, Z. Z., Hassan, A., Abd Hamid, M. K., Samsudin, S. A., & Sabeen, A. H. (2019). Current developments in chemical recycling of post-consumer polyethylene terephthalate wastes for new materials production: A review. *Journal of Cleaner Production, 225*, 1052–1064.

Raj, T., Chandrasekhar, K., Kumar, A. N., & Kim, S.-H. (2022). Lignocellulosic biomass as renewable feedstock for biodegradable and recyclable plastics production: A sustainable approach. *Renewable and Sustainable Energy Reviews, 158*(112), 130.

Rambabu, K., Bharath, G., Govarthanan, M., Kumar, P. S., Show, P. L., & Banat, F. (2023). Bioprocessing of plastics for sustainable environment: Progress, challenges, and prospects. *TrAC Trends in Analytical Chemistry, 117*, 189.

Ramli Sulong, N. H., Mustapa, S. A. S., & Abdul Rashid, M. K. (2019). Application of expanded polystyrene (EPS) in buildings and constructions: A review. *Journal of Applied Polymer Science, 136*(20), 47,529.

Rong, L., Zhao, L., Zhao, L., Cheng, Z., Yao, Y., Yuan, C., Wang, L., & Sun, H. (2021). LDPE microplastics affect soil microbial communities and nitrogen cycling. *Science of the Total Environment, 773*(145), 640.

Rudel, R. A., Camann, D. E., Spengler, J. D., Korn, L. R., & Brody, J. G. (2003). Phthalates, alkylphenols, pesticides, polybrominated diphenyl ethers, and other endocrine-disrupting compounds in indoor air and dust. *Environmental Science & Technology, 37*(20), 4543–4553.

Sebastian, A., Paul, A. M., Dominic, D., Shaji, M., Jose, P., Sasi, S., & Prasad, M. N. V. (2023). Microbial degradation of plastics. In *Microplastics in the Ecosphere: Air, Water, Soil, and Food* (pp. 305–320).

Shakiba, M., Rezvani Ghomi, E., Khosravi, F., Jouybar, S., Bigham, A., Zare, M., Abdouss, M., Moaref, R., & Ramakrishna, S. (2021). Nylon – A material introduction and overview for biomedical applications. *Polymers for Advanced Technologies, 32*(9), 3368–3383.

Simon, G. P. (2019). *Polymer blends and alloys*. Routledge.

Singh, D. P. (2021). Biodegradation of plastic using micro-organisms. *International Journal of Environmental Chemistry, 7*(2), 1–6.

Singh, N., Hui, D., Singh, R., Ahuja, I. P. S., Feo, L., & Fraternali, F. (2017). Recycling of plastic solid waste: A state of art review and future applications. *Composites Part B: Engineering, 115*, 409–422.

Srikanth, M., Sandeep, T., Sucharitha, K., & Godi, S. (2022). Biodegradation of plastic polymers by fungi: a brief review. *Bioresources and Bioprocessing, 9*(1), 42.

Sun, X., Xie, M., Mai, L., & Zeng, E. Y. (2022). Biobased plastic: A plausible solution toward carbon neutrality in plastic industry? *Journal of Hazardous Materials, 435*(129), 037.

Swati, P., Poornima, P., Ey, H., Manimita, S., Yadav, M., & Tiwari, A. (2015). Biodegradation of plastic using microorganisms with help of nanoparticles: A review. *Advances in Applied Science Research, 6*, 1–18.

Tan, X., & Rodrigue, D. (2019). A review on porous polymeric membrane preparation. Part II: Production techniques with polyethylene, polydimethylsiloxane, polypropylene, polyimide, and polytetrafluoroethylene. *Polymers, 11*(8), 1310.

Temporiti, M. E. E., Nicola, L., Nielsen, E., & Tosi, S. (2022). Fungal enzymes involved in plastics biodegradation. *Microorganisms, 10*(6), 1180.

Tong, Y., Bohm, S., & Song, M. (2017). The capability of graphene on improving the electrical conductivity and anti-corrosion properties of Polyurethane coatings. *Applied Surface Science, 424*, 72–81.

Tonsi, G., Maesani, C., Alini, S., Ortenzi, M. A., & Pirola, C. (2023). Nylon recycling processes: a brief overview. *Chemical Engineering Transactions, 100*, 727–732.

Ungprasoot, P., Muanruksa, P., Tanamool, V., Winterburn, J., & Kaewkannetra, P. (2021). Valorization of aquatic weed and agricultural residues for innovative biopolymer production and their biodegradation. *Polymers, 13*(17), 2838.

Varghese, M., & Grinstaff, M. W. (2022). Beyond nylon 6: polyamides via ring opening polymerization of designer lactam monomers for biomedical applications. *Chemical Society Reviews, 51*, 8258–8275.

von Vacano, B., Mangold, H., Vandermeulen, G. W. M., Battagliarin, G., Hofmann, M., Bean, J., & Künkel, A. (2023). Sustainable design of structural and functional polymers for a circular economy. *Angewandte Chemie International Edition, 62*(12), e202210823.

Wahyuningtiyas, N. E., & Suryanto, H. (2017). Analysis of biodegradation of bioplastics made of cassava starch. *Journal of Mechanical Engineering Science and Technology (JMEST), 1*(1), 24–31.

Walker, S., & Rothman, R. (2020). Life cycle assessment of bio-based and fossil-based plastic: A review. *Journal of Cleaner Production, 261*(121), 158.

Walter, A., Sopracolle, L., Mutschlechner, M., Spruck, M., & Griesbeck, C. (2022). Biodegradation of different PET variants from food containers by Ideonella sakaiensis. *Archives of Microbiology, 204*(12), 711.

Wang, Z., Zhu, C., Mo, J., Fu, P., Zhao, Y., Yin, S., Jiang, J., Pan, M., & Su, C. (2019). White-light emission from dual-way photon energy conversion in a dye-encapsulated metal–organic framework. *Angewandte Chemie International Edition, 58*(29), 9752–9757.

Winnacker, M. (2017). Polyamides and their functionalization: recent concepts for their applications as biomaterials. *Biomaterials Science, 5*(7), 1230–1235.

Wu, F., Misra, M., & Mohanty, A. K. (2021). Challenges and new opportunities on barrier performance of biodegradable polymers for sustainable packaging. *Progress in Polymer Science, 117*(101), 395.

Xie, F., Zhang, T., Bryant, P., Kurusingal, V., Colwell, J. M., & Laycock, B. (2019). Degradation and stabilization of polyurethane elastomers. *Progress in Polymer Science, 90*, 211–268.

Yang, Y., Deng, H., & Fu, Q. (2020). Recent progress on PEDOT: PSS based polymer blends and composites for flexible electronics and thermoelectric devices. *Materials Chemistry Frontiers, 4*(11), 3130–3152.

Yeganeh, F. M., SHAKERI, A. T. A., Rastegari, M. M., & Lahijani, O. (2020). *Investigating abundance and characteristics of microplastics as emerging pollutants in sediments of Taleqan dam and upstream river in Alborz province*.

Yi, L., Zuo, L., Wei, C., Fu, H., Qu, X., Zheng, S., Xu, Z., Guo, Y., Li, H., & Zhu, D. (2020). Enhanced adsorption of bisphenol A, tylosin, and tetracycline from aqueous solution to nitrogen-doped multiwall carbon nanotubes via cation-π and π-π electron-donor-acceptor (EDA) interactions. *Science of The Total Environment, 719*(137), 389.

Yoon, J.-S., Chin, I.-J., Kim, M.-N., & Kim, C. (1996). Degradation of microbial polyesters: A theoretical prediction of molecular weight and polydispersity. *Macromolecules, 29*(9), 3303–3307.

Yoshida, S., Hiraga, K., Takehana, T., Taniguchi, I., Yamaji, H., Maeda, Y., Toyohara, K., Miyamoto, K., Kimura, Y., & Oda, K. (2016). A bacterium that degrades and assimilates poly (ethylene terephthalate). *Science, 351*(6278), 1196–1199.

Yuan, J., Ma, J., Sun, Y., Zhou, T., Zhao, Y., & Yu, F. (2020). Microbial degradation and other environmental aspects of microplastics/plastics. *Science of the Total Environment, 715*(136), 968.

Zaaba, N. F., & Jaafar, M. (2020). A review on degradation mechanisms of polylactic acid: Hydrolytic, photodegradative, microbial, and enzymatic degradation. *Polymer Engineering & Science, 60*(9), 2061–2075.

Zhang, F., Zhao, Y., Wang, D., Yan, M., Zhang, J., Zhang, P., Ding, T., Chen, L., & Chen, C. (2021). Current technologies for plastic waste treatment: A review. *Journal of Cleaner Production, 282*(124), 523.

Zhang, Z., Sarkar, D., Biswas, J. K., & Datta, R. (2022). Biodegradation of per-and polyfluoroalkyl substances (PFAS): A review. *Bioresource Technology, 344*(126), 223.

Chapter 10
Microplastics Waste and Its Eco-Friendly Management

Sanjay Chandravanshi, Adyasha Sahu, Jham Lal, Narsingh Kashyap, and Sudhan Chandran

Abstract The microplastics waste has a major adverse impact on oceans, lakes, seas, rivers, and coastal region, the environment seriously threatened by them. Microplastics (MPs) are particles of plastic with a diameter of <5 mm that can be found in both primary and secondary sources. Global research is being done on the growing issue of MP contamination in every aspect of our environment. A wide range of marine organisms are negatively impacted by microplastics, and studies show that these pollutants are known to move into the marine ecosystem through land-based activities like runoffs. Microplastic particles have been found dispersed extensively across Africa, South Asia, India, North America, and Europe. In this work, we discuss the sources, worldwide distribution, fate, and organisms' life, especially in their food web and trophic levels. Additionally, international and national environmental association or organizations have advocated the mitigation techniques offered to decrease the impact of microplastics. Microplastics pollution in the environment can be greatly reduced by identifying the primary causes of the problem and raising public, private, and governmental knowledge of it through education. Additionally, being aware of the underlying behavioral mechanisms will help one comprehend the impacts on the marine ecosystems.

Keywords Microplastics · Sources · Ecosystem effects · Freshwater and marine environment · Removal techniques

---

S. Chandravanshi (✉) · A. Sahu · S. Chandran
Fisheries College and Research Institute, Thoothukudi, Tamil Nadu, India

J. Lal
College of Fisheries, Lembucherra, Tripura, India

N. Kashyap
Institute of Fisheries, Post Graduate Studies, Kanchipuram, Tamil Nadu, India

R. Soni et al. (eds.), *Advanced Strategies for Biodegradation of Plastic Polymers*, https://doi.org/10.1007/978-3-031-55661-6_10

10.1 Introduction

Plastic wastes pollute more and more the environment. Because they may result in a variety of environmental issues for creatures and the assemblages in which they exist, small plastic particles, or microplastics, are a growing source of worry. Due to its low cost and straightforward manufacturing, plastic has emerged as a crucial material in the modern era. The consistent use of plastic is a big barrier in achieving the UN Sustainable Development Goals because of poor management and open disposal, nevertheless.

10.1.1 What Is Microplastic?

The majority of authors use the definition of "microplastics," which generally refers to particles of plastics with a diameter of b < 5 mm. According to others, the term "microplastics" should be changed to "items 1 mm" in order to exclusively refer to particles that are created when several plastic-based products are exfoliated and broken down in ecosystems (Sommer et al., 2018). Microplastics have been found in rural and urban areas, seawaters, freshwaters, and marine sediments (Romeo et al., 2015). Majority of research findings show that living things are exposed to microplastics and the breakdown products they produce more frequently when there is a significant amount of microplastics waste in aquatic ecosystems. In general, the two types of microplastics are primary microplastics, which are used as raw materials to make household and personal care items, and secondary microplastics, which are materials left over from production that degrade chemically, physically, and biologically in the environment (Vaid et al., 2021). Environmental experts are concerned about microplastics because of their durability over time and their ease of movement between various habitats (Scheurer & Bigalke, 2018). The popular raw polymer materials are polyvinyl chloride (PVC), polystyrene (PS), polyethylene, (PE) polypropylene (PP), polyamides, and polyesters.

10.1.2 Microplastics Waste: A Global Environmental Issue

Sustainable waste management of waste is a key method to environmental conservation, and plastic is one of the more difficult types of waste to manage. It is still being produced in enormous quantities all around the world. Reason behind is the recycling and other ways of processing plastic waste have fallen behind, the leftover waste of plastic has built up in the environment, especially at unofficial recycle locations such as backyards in developing countries. The percentage-wise distribution of plastic waste is given in Table 10.1. As a result, a large amount of this

Table 10.1 Distribution of plastic waste production in 2022, by country wise

Country	Contribution of global plastic waste (%)
China	32
North America	18
Rest of Asia	17
Europe	15
Middle East and Africa	8
Latin America	4
Japan	3

Source: Statista (2023)

debris is dumped close to water resources, in urban drainage systems, where it eventually flows into rivers and ends up in the oceans as microplastics. According to the Global Chemicals Outlook report, microplastics are dispersed in sludge and wastewater by municipal sewage systems (Chen et al., 2020). Water is frequently used for irrigation and soil additives, while sewage sludge has been applied to agricultural soil (de Souza Machado et al., 2018).

MPs are easy consumed by marine species and can enter the food chain. According to a 2022 UNEP (United Nations Environment Programme) research, 99% of seabirds have consumed MPs, and by 2050, more than 600 (15%) marine species are anticipated to be endangered by ingesting microplastics or becoming entangled in marine debris made of the microplastics. According to the findings, microplastics may have an influence on human health. In addition to its negative effects on the environment and human health, microplastics pollution has an economic cost, especially in small island developing nations where tourism is a major source of income (Lwanga et al., 2017).

10.2 Sources of Microplastics Waste

Microplastics are present everywhere in marine environments and coastal areas due to waves and currents of ocean. The aquatic ecosystem contains MPs particles of different sizes, densities, chemical compositions, and shape pattern (Wagner et al., 2014).

10.2.1 Primary Microplastics

The primary MPs are tiny, microscopic particle especially made for specific commercial industry or home uses. The plastics particle are found in face washes, toothpaste, and beauty and cosmetics products such as bath soaps, face or body scrubs, deodorant, eye shadow, shaving creams, mascara, body lotions, and nail polish

(Fendall & Sewell, 2009). Another includes air-blasting media, drilling liquids, artificial clothes, and abrasives used in cleaning supplies (Alomar et al., 2016). More and more reports have been made about the use of microplastics as ways to deliver drugs in medicine (Patel et al., 2009). Microplastics "scrubbers" have taken the role of commonly used natural components like oats, powdered almonds, and pumice in exfoliating hand cleaners and scrubs for faces (Fendall & Sewell, 2009).

Consider the inclusion of polystyrene spheres, polypropylene, and polyethylene in a cosmetics product. Recently, Fendall and Sewell (2009) observed a distribution and abundance of MPs with irregularity in shapes in other commercial cosmetic products, and the diameter of that plastics is $b < 0.5$ mm and the mode size $b < 0.1$ mm. Primary MPs have been manufactured by air-blasting method. In this microplastic production process, machines and engines are shot with acrylic resin, polyurethane microplastic scrubbing engine or machines to eliminate rust and paint. Such scrubber are regularly utilized until they get smaller and lose their ability to cut, at which point they are frequently affected with toxic heavy metals (including Cd, Cr, and Pb) (Cole et al., 2016).

10.2.2 Secondary Microplastics

Larger pieces of plastic waste on lands and in the ocean are gradually broken down into small fragments and produce microplastics when left out in the elements for an extended period of time. They come under secondary microplastics category. The structural stability of microplastics waste is reduced as a result of a confluence of chemical, physical, and biological process, which fragments the debris (Cole et al., 2013). The breakdown of macroplastics trash is influenced by a variety of environmental conditions (including temperature, sunlight) in addition to the polymer characteristics (like densities and sizes). Larger waste of plastics that gets exposed to the ultraviolet (UV) rays suffers from photodegradation (Brandon et al., 2016). Beaches are the perfect environments for the formation of microplastics by fragmentation into smaller sizes because of the intense ultraviolet lights, physical friction by waves, and availability of oxygen and water turbulent (Bianco & Passananti, 2020). Cooler temperatures and less UV light make the breakdown take slower once these fragments are submerged in surface waters or deep habitats (GESAMP, 2014).

The marine environment contaminated by microplastics increases, when larger plastic items break down into smaller particles, possibly magnifying their potential effects on aquatic species. As the size of particles reduces, a wider range of organisms have the capacity of ingesting waste. Because of their small size and easy consumption, the microplastics are more susceptible and have improved leaching or the process of desorption and adsorption possibilities (Law & Thompson, 2014; Shim & Thompson, 2015). The amount of microplastics created annually is thought to be around 245 tons, and they wind up in a body of water where aquatic species ingest and absorb them (Morris, 2015).

10.3 Impact of Microplastics Waste on Agriculture Sector

Large amounts of microplastics have been discovered in soils used for farming. The stated abundance values for each dry kilogram of soil usually vary from tens to thousands of particles. One possible method for introducing microplastics into agricultural soils is the use of contaminated organic matter on lands or in fields (Golwala et al., 2021). The breakdown of macroplastics produces secondary microplastics, which can be mistakenly added to soils as contaminated soil additions (e.g., biosolids, composts, digestates) (Corradini et al., 2021; Zhu et al., 2021) or by the breakdown of plastic mulching (Holmes et al., 2012; Stubenrauch & Ekardt, 2020).

Plastic mulching made from LDPE (low-density polyethylene) or biodegradable-type polymer materials is frequently used in agriculture to improve yield of crops, suppress unwanted weeds, and minimize the need of fertilizer (Brodhagen et al., 2017; Serrano-Ruiz et al., 2021; Feng et al., 2021; Zhou et al., 2020). Not all probable sources will necessarily have an impact on microplastics abundance at a specific location (Corradini et al., 2021; Yu et al., 2021). To ascertain the relative effects of several microplastics introduction pathways into crops, including the use of soil additives produced from food waste, more research is required. Understanding the current levels of microplastics intakes from all potential sources and using reference soils to distinguish between microplastics inputs from various sources are both necessary for effective mitigation (Harms et al., 2021; Kumar & Sheela, 2020).

Recent research summarizes the effects of microplastics on physical characteristics, biota, and crops (Iqbal et al., 2020; Qi et al., 2020; Xu et al., 2020). According to their investigation, there is currently a lack of knowledge regarding the lasting consequences of microplastics in agricultural soils. Multiple crop types have encountered delayed or decreased seed germination, reduced plant growth, and absorption into the tissues of plants (e.g., Boots et al., 2019; Pflugmacher et al., 2020; Tympa et al., 2021; Beriot et al., 2021; Yang et al., 2021). These effects are likely to differ depending on shape, size, and type of polymer and soil parameters (Lozano et al., 2021; Zhao et al., 2021).

10.4 Impact of Microplastics Waste into Marine Waters

Microplastics reach the marine ecosystems through a variety of distinct channels (Alomar et al., 2016). Through domestic and commercial drainage systems, the microplastics beads found in cosmetics like scrubs, toothpaste, air-blasting media, and clothing can get into the water supply. Similarly, synthetic garment fibers microplastic sheds that is flushed into wastewater treatment effluents (GESAMP, 2014; Murphy et al., 2016).

The storm, wind, and currents are further ways that microplastics enter the oceanic zone (Zalasiewicz et al., 2016). The degradation of macroplastics waste is another source that frequently travels through recycling of marine ports and landfill, where unfavorable climate conditions facilitate the dumping of macroplastics at sea coastline. Some of this trash is washed out to sea by runoff. Another possible cause of microplastics contamination is sewage sludge, which contains more of them than effluent and is dispersed into the marine ecosystem (Alomar et al., 2016). Microplastics are widely disseminated by currents over greater distances and are widely transported due to their small size and low density. In all aquatic environments around the world, these marine plastic particles are common and plentiful (Cole et al., 2014; Eerkes-Medrano et al., 2015).

Microplastics are found in a broad diversity of aquatic organisms, including fishes, bivalves, mammals, crabs, sea snakes, and seabirds. They are also found on beaches, bottom sediments, sediments, and surface waters. The decaying matter of zooplankton is another way that microplastics could enter the ocean. After exposure, the organisms quickly began to feed on the microplastics that had entered the body through the gut, been encapsulated in excrement, and been ingested. The larger copepod eventually swallowed the feces after they had been digested and sunk to the bottom of the exposure vessel. According to several estimates, 92% of the 269 million tons of plastic waste of the world's 5.25 trillion particle are microplastics. These microplastics are also estimated to be 100 times less abundant on the sea surface (Auta et al., 2017; Cole et al., 2014).

10.5 Microplastics Waste Impacts into Freshwater Environment

Similar to bulk plastics, microplastics can infiltrate the environment in a variety of ways, and a particular route that is significant in one region may not be as significant in another. MPs may leak substances into the environment in different ways: (1) the discharge of MP into surface waterways through wastewater treatment plants (WWTPs), either through the use of MP in personal care items or from the release of textile fibers during cloth washing, (2) storm water overflow incidents, (3) the application of WWTP biosolids to agricultural areas, (4) release from industrial goods or processes, and (5) fiber deposition in the atmosphere. Another most important contributor to the microplastics pollution of agricultural soils and a substantial agricultural emission is the use of disposable plastic sheets for crop cultivation (Klein et al., 2018; Wagner & Lambert, 2018; Ererkes-Medrano et al., 2015).

There are several advantages, such as moisture conservation, protection from inclement weather, and reduction of weed growth and soil temperature. It can be quite difficult to isolate MPs in environmental matrices, especially when working with samples that contain a lot of organic matter, including sediments and soil (Liu et al., 2021).

10.6 Fate and Transportation in Freshwater Ecosystems

MPs will undergo mechanisms that define their fate and mode of transit once they are released or generated in freshwater. The section that follows provides an explanation of these processes.

10.6.1 Environmental Transportation

The MPs can then scatter into overland runoff or roadside ditches to move from land to water. The movement of bulk plastics and MPs within the riverine system will be significantly impacted by the hydrology (daily discharge and flow condition of water), as well as the morphology (vegetation pattern), at a specific flow site of river. The development of biofilm-associated microplastics for LDPE in a maritime environment has been researched. Both the biofilm formation and the sediment transportation on MP surfaces, which are exposed to light, may decrease the rate of deterioration.

10.6.2 Environmental Sustainability and Deterioration

The majority of the knowledge we currently have about plastic degradation processes comes from laboratory experiments, many of which focus on a specific mechanism like photothermal or biodegradation. There is a paucity of information on the degradation of plastics in environmental settings where a variety of degradation mechanisms coexist. This is significant since polymers like PE typically disintegrate into smaller fragments and do not readily depolymerize. These pieces eventually split out into smaller pieces, finally generating nanoplastics.

10.6.3 Interaction of Other Substances

Since this could affect the pollutants' mobility and bioavailability, the hydrophobic sorption contaminant to microplastics is regarded as a crucial environmental activity. It is common knowledge that POPs like DDT, PCBs, and dioxins accumulate in MPs in marine environment. In order to collect metals from the surrounding saltwater, fresh pellet polypropylene was floated in a harbor for 8 weeks. The metal concentrations ranged from 0.25 g g^{-1} for zinc to 17.98 g g^{-1} for iron. This was done to see if the metals were, in fact, linked to the pellet sample's nonremovable fine organic debris. As a result, the microplastics are likely to co-occur in freshwater ecosystems with other emerging pollutants such as drugs, cosmetics, and chemical industries.

10.7 Microplastics Waste in Environment

10.7.1 Microplastics Waste in Marine Ecosystem

Microplastics with density higher than those of seawater accumulate in sediments, while those with density lower than those of seawater drift on the surface of the water (Suaria & Aliani, 2014). Microplastics may sink as their density increases due to biofouling by marine organisms. As biofouling progresses, plastic material sinks to the ocean floor when its density surpasses that of seawater (Jorissen, 2014). According to Cozar et al. (2014), marine sediments have the capacity to accumulate the microplastics in long-term period (Nuelle et al., 2014). Marine sediments now include extremely high levels of microplastics; on significantly impacted beaches, these plastics can account for 3.3% of the sediment weight (Van Cauwenberghe et al., 2015a, b; Boucher et al., 2016; Alomar et al., 2016).

10.7.2 Microplastics Waste in Mangrove Debris

Mangrove is sometimes called as an "enhancer of sedimentation" since it accumulates carbon, nutrients, and sediments. There are two types of sediments that are deposited into mangrove debris, i.e., allochthonous sediments, which are sediments that originate from outside sources like terrestrial sources, and autochthonous sediments, which are sediments that originate from resuspended sources (Adame et al., 2010). Similar to other aquatic environments, marsh sediment accumulates microplastics. Polypropylene, polyethylene, polyvinyl chloride, and nylon were all present in the plastic particles that were extracted, which were smaller than 20 micrometers. Microplastics were present in dry sediment in concentrations of 12.0–62.7% particles. These various microplastics polymers may be the result of the decomposition of marine macroplastics waste that may have collected in the mangroves.

10.8 Effects of Microplastics Waste in the Marine Ecosystem

10.8.1 Interaction with Marine Fauna

The bioavailability of microplastics to marine life is rising along with their prevalence. The potential bioavailability of these plastic particles to marine animals is influenced by their color, shape, size, and abundance (Wright et al., 2013; Van Cauwenberghe et al., 2015a, b; Clark et al., 2016). In oceanic regions throughout the world, a variety of aquatic organisms that consume microplastics particles have been noticed (Ferreira et al., 2016; Setala et al., 2016; Devriese et al., 2015; Green, 2016). Although some microplastics can be selectively targeted by some marine organism, the majority of microplastics consumption by marine organisms is inadvertent as the particle is commonly misidentified as food (Lonnstedt & Eklov, 2016; Fossi et al., 2016).

Microplastics cause chemical and physical harm when ingested by marine organism. The ingestion of MPs by marine organisms may have chemical special effects such as hepatic stress, decreased growth rate, and inflammation, or it may have some other mechanical effects such as obstructing mobility, blocking, or clogging of the digestive tract. Numerous marine organisms from various trophic levels, such as invertebrates, particularly lugworms (Green, 2016), mussels (Avio et al., 2016), barnacles (Rehse et al., 2016; Watts et al., 2014), sea cucumbers, amphipods, zooplankton, as well as fishes, turtles, and mammals that eats indirectly or directly to microplastics (Nobre et al., 2015; Fossi et al., 2016). Microplastics ingested by zooplankton and copepods, as well as other species at the lower trophic level, may migrate up the food chain when absorbed by organisms at higher trophic level, potentially disrupting the food chain (Brennecke et al., 2016; Hollman et al., 2013).

10.8.2 Microplastics in Fishes

Based on studies, fish tissues contain the same compounds that go into making plastic. Interactions between predators and their victims make it easier for harmful compounds to be passed in higher amounts since they can accumulate in organisms from numerous sources (Zarfl & Matthies, 2010; Wang et al., 2016). In response to concern about the movement of dangerous chemicals and microplastics between trophic levels, laboratory experiments have been conducted to demonstrate the effect of microplastics on marine biota (Ren et al., 2020). *Artemia* sp. *nauplii* were discovered to have swallowed and accumulated microplastics particles in high concentrations in the study. These particles ranged in size from 1 to 20 m and were exposed to high concentrations of microplastics. These particles were then transferred to zebra fish that consumed the *nauplii*. Despite the fact that several of the collected microplastics particles were ingested by the organisms, others were retained in the intestinal villi and epithelial cells. The research also showed that the related persistent organic pollutant benzo[a]pyrene (BaP), which was retained in the intestinal system of the zebra fish, was transported from the *nauplii* to the fish via the microplastics particles (Cole et al., 2013).

However, neither *nauplii* nor zebra fish showed signs of physical injury due to the presence of the hazardous chemicals and microplastics in the aquatic ecosystem. The study unequivocally demonstrated that hazardous chemicals and microplastics can go up food chains at different trophic levels. The impact of toxic compounds that are present in microplastics that were reported in marine fish (*Oryzias latipes*) (Do Sul et al., 2013; Rochman et al., 2013). Other studies have been done on the consumption of microplastics by several fish species. 1–15 bits of plastics were present on average of every fish. FT-IR spectroscopy was used to identify 351 different plastic particle fragments. Microplastics were identified in 63.5% of demersal fish and 36.5% of pelagic fish. The stomach contents of the fish were found to contain a total of 73 microplastics particle (Neves et al. (2015). In several investigation, the ingestion and adverse impacts of microplastics on zebra fish was found that most of plastic fragments (diameter less than 5 mm) had accumulated in the gills, liver, and stomach of fishes, but some plastic fragments with a diameter of more than 20 mm could only accumulate in the gills and gut of fishes. The size and shape of the polystyrene microplastics may have rendered them suitable for ingestion, which may explain why they preferred them to natural food particles. Ingesting microplastics is a problem affecting not only fish but also zooplankton and marine turtles (Besseling et al., 2012; Von Moos et al., 2012).

10.9 Microplastics' Impacts on Freshwater Environment

10.9.1 Take-Up and Biological Impacts

A variety of species are capable of absorbing MPs from the sediment. The most important way for this to happen is directly through the gills or by cutaneous uptake. *Bosmina coregoni*, a freshwater zooplankton species studied in the past, failed to distinguish between particle size (2–6 m) beads when exposed to a combination of both freshwater zooplankton and algae. Particle size determines the amount of physical stress organisms are exposed to as a result of microplastics absorption since the particle size of microplastics is larger than soil or food particles may be more difficult to digest. Physical irritability may be one of the indirect effects of MPs, depending on the shape and size of the microplastics.

10.9.2 *Impact of Sub-micrometer Plastics on Biological Systems*

Plastic materials may contain elements like antimicrobial agents and nanoparticles, depending on how they are meant to be used, which could be harmful to microbial and fungal species vital to the health of ecosystems. It is potential that the mixture of microscopic particles, dissolved chemicals, and other degraded products could have modest effects on aquatic and terrestrial organisms that are difficult to detect in the testing techniques currently in use. It is very likely that during degradation it may convert into submicron and nanoscale.

10.10 Eco-Friendly Sustainable Management of Microplastics Waste

Although larger plastic particles are successfully removed during the treatment process, but microplastics frequently bypass in this process finally accumulate in aquatic habitats. Despite the consolation of being separated from bigger plastic particles during wastewater treatment, microplastics particles nevertheless accumulate in aquatic habitats. There are a lot of water treatment facilities close to oceans and seawater, which result in the discharge of microplastics into the environment. In order to address this issue, numerous studies are being done on the fate, occurrence, and removal of microplastics from water treatment plants (Alrbaihat & Abu-Afifeh, 2023).

10.10.1 *Physical Approaches for Removing Microplastics*

This study examined many physical techniques that are effective at removing microplastics (MPs) from treated water.

10.10.1.1 Modern Filtration Techniques to Remove Microplastics

The flowchart illustrates the various filtration procedures used to remove MPs that are classed as physical approaches (Fig. 10.1). Membrane bioreactors considerably outperformed traditional initiate sludge treatment methods (98.3%) in the removal of microplastics (99.4%). The authors of the study also noted that the small range of final microplastics contents may have been a result of their study's use of wastewater samples and somewhat varied processing procedures. Over 90% of the microplastics in the influents were removed from almost all plant utilizing a combination

Fig. 10.1 Flowchart of filtration process for microplastics removal

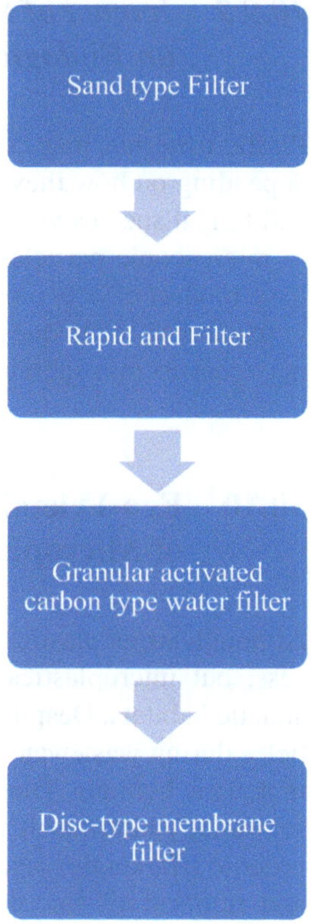

of floating and sedimentation tanks. 97.15% of the microplastics were successfully removed. The removal efficiency could differ significantly depending on the daily processing volume, the kind of raw water, and the procedure treatment. The high degree of clearance further demonstrated that these microplastics were primarily made of polyethylene terephthalate and rayon fiber. Municipal sewage treatment facilities were cleaned of microplastics. The influents were initially treated using anoxic, aerobic, and anaerobic A_2O treatments. This comprised a primary and secondary sedimentation tank, as well as aerated grit chamber, and other components. Using the processes of denitrification, ultrafiltration, ozonation, and ultrasound, microplastics are removed from wastewater.

According to Fourier transform infrared spectra (FTIR) analysis, polyester and polyethylene terephthalate are the two substances that are present in the effluent in the highest amounts (42.26% and 19.1%, respectively). Following the first treatment utilizing aerated grit, microplastics were eliminated at an efficiency of 58.84% and 71.67% after the advanced treatment techniques. Microplastics cannot be eliminated from sewage treatment facilities using the present treatment methods.

10.10.1.2 Membrane-Based Technology for the Removal of Microplastics

Microplastics wastes were successfully removed from artificial wastewaters using the dynamic flexible membranes (Li et al., 2018). The dynamic flexible membranes were used to successfully remove microplastics. Effects of flow and particle density on the performance of dynamic flexible membranes constructed on the diatomite platforms with 90 meters of supporting mesh when filtering wastewater (Fig. 10.1).

The effects of the influent flux and concentration of particles were investigated when artificial waste had been passed or filtered via a diatomite stage with a supportive mesh of 90μm. The MPs were filtered to practically low or zero turbidity in just 20 min by decreasing the influent turbidity from 195 NTU (nephelometric turbidity unit) to a value of <1 for the effluent. Both microplastics concentrations and input fluxes aid in the development of membranes. Researchers have developed an effective microplastics cleanup method using mesh screens and elongated polymer coatings. He claims that the tool can be quickly and easily made from basic components and has good endurance. These tools are also devoid of any mechanical or electrical components (Ng et al., 2018; Wang et al., 2019).

However, compared to basic dynamic membranes, membrane bioreactors have a stronger ability to remove micro-sized plastics. The potential of purifying combined systems utilize both porous membranes and biological processes for filtration. The ability of bioreactor covering membrane to effectively remove highly contaminated substances including polymeric waste and MPs shows that this technology is suitable for handling challenging industrial effluent. The elimination of microplastics from wastewater treatment plant effluents was investigated in a study by Talvitie et al. (2017) using end-stage technologies such as air-dissolved flotation process, quick sand filters, disc filters, and membrane bioreactors. According to their analyses, 99.9% of the particles of microplastics from 6.9 to 0.005/L water have been eliminated by the membrane bioreactor (Lusher et al., 2013). Additionally, bioreactor membrane, rapid filter, dissolved air flotation, and sand filtration were used to remove any size of microplastics, including between 20 and 100 microns. Furthermore, regardless of the shape of the microplastics, a sizeable amount was effectively removed during influent treatment. Membrane-based technologies have been successful at removing microplastics from contaminated aquatic environment.

10.10.1.3 Adsorption of Algae

Microplastics are more important than other pollutants because they can entangle and bioaccumulate in aquatic ecosystems. Microplastics may harm aquatic species, like reptiles, fish, mammals, and birds, among other harmful effects. Removal techniques are needed since they are difficult to remove and are persistent. Although the majority of microplastics are considered persistent materials, how rapidly they degrade depends on their natural and chemical structure. It is well known that the surfaces of microplastics can adsorb and carry many kinds of pollutants from water, bringing them into surrounding habitats and desorbing them.

The edible marine microalga *Fucus vesiculosus*, which is a type of seaweed, readily resisted fluorescent microplastics particles, due to the 20 mm diameter of the polystyrene microplastics, microchannels, and polystyrene microplastics movement into tissues is controlled by the sorbent into the plant cell. The release of compound of alginates during the cutting operations is what is responsible for the results, which demonstrated a very high sorption rate (94.5%) from microplastics to seaweeds near the cut surface. Alginates which are anionic in nature cause it to act as a gelatinous-like material that can be used to improve polystyrene's adhesion to seaweeds' surface. In this research and earlier investigations about the ability of microalgae to absorb plastic particles, surface properties of microplastics and microalgae are controlled by surface charge.

10.10.2 Chemical Treatments of Microplastics

In wastewater treatment plants, agglomeration and flotation processes are routinely used to produce large component particles that are easier for separation. These procedures bind small particles by activating uptake-complexation pathways that are started by ligand exchanges, producing strong connections between waste particles. They do this by using iron- (Fe) and aluminum (Al)-based salt conjunction with other coagulating agents. Aza-Tarazona et al. (2019) examined how pH, anion polyacrylamide, and development of Al-based coagulating agents affected the effectiveness of microplastics removal utilizing coagulants made of iron and aluminum salts and ultrafiltration process. Furthermore, pH had no apparent impact on the effectiveness of removing microplastics when a relatively low-level concentration of an Al-based coagulant (0.5 mM) was present. When pH was raised, the efficiency of removal of microplastics was also reduced. Using the enhancer coagulant polyacrylamide (PAM), the elimination efficacy of small microplastics did not increase.

The mechanisms underlying the degradation of MPs are completely not known. Brandon et al. (2016) examined the chemical modifications in composition of polyethylene, polypropylene, and other kinds of microplastics. Fourier transform infrared spectra (FTIR) analyses of several metabolite compounds, including hydroxyl, carbon-oxygen bonds, and carbonyl, revealed inadequate irregular fluctuations over time, demonstrating that microplastics require a long period of time to degrade.

10.11 Other Management Methods for Pollution Caused by Microplastics Waste

For several decades, the expansion of plastic production has been exponential, and it appears expected that the quantity of microplastics waste particles will continue to increase in the coming years. Compared to macroplastics, which are a developing pollutant of major concern, there is little to no knowledge in the private and public sectors of the potential negative hazards they represent. Additionally, raising the awareness of microplastics through education in the government, private, and public sectors will help significantly. The first comprehensive investigation of the impacts of microplastics on the marine environment and biota was conducted by Ivar do Sul et al. in 2013. They focused the scientific community's attention on the need to monitor tainted pellets in order to identify temporal trends of the numerous pollutants that might assist in planning future projects. Several organizations have developed management guidelines in response to concerns about microplastics.

10.12 Future Perspectives of Microplastics Waste

At certain amounts, microplastics may slowly break down to an extent in a suitable environment. Future polymer-degrading native microbes may be highly efficient and environmentally beneficial, for example, by enzymatic reactions. Further research investigations are still required to examine their capacity to break down microplastics debris. Microplastics waste might be addressed through biotechnological techniques, at least in tiny doses. However, it might be difficult to put such a strategy into practice on a technological development of infrastructure on an enormous scale. In any case, developing a technical mechanism to stop microplastics from entering the oceans requires organized and innovative approaches. It is certain that additional study is required to address the serious issue of microplastics pollution. This might entail eliminating microplastics from wastewater treatment facilities, minimizing waste of plastics material at the source, or using more bioplastic or readily biodegradable plastic.

With the lack of specific information addressing the effects of MPs on the environment, it is vital and necessary to monitor and stop additional wastewater pollution by microplastics. In determining its effect on human's health, it is critical to know the extent and intensity of MPs exposure. The environment has been the subject of a restricted number of researches, leaving room for novel concepts. MPs influence the ecosystem and the food chain in aquatic habitats, making it difficult to amass them. The future monitoring and surveillance of substance inventory may be made simpler if MP origins, routes, and paths in soil, sediment, and water are found and completely eradicated or if wastewater treatment plants use more effective machinery and environmentally friendly technology.

Plastic pollution still has a harmful impact on the environment. The cleaning strategies that aim to lessen the consequences of it were inadequate to keep up with the rising volume of plastics in the environment. The assessment suggests keeping an eye on how much plastic is produced and consumed, increasing consumer demand for recycled plastic, enhancing removal technology, developing sustainable removal technology turning microplastics to sources of renewable energy of information, educating consumers, and extending customer lifetime. These suggestions can lead to the implementation of a circular economy and a decline of poorly managed pollutants in the aquatic system.

10.13 Controlling Measures of Microplastics Waste

There are at present no studies or reports of any technologies that could be used to control MPs pollution. Therefore, minimizing the sources is our only choice for preventing the current scenario from getting worse. Since it can be tough to avoid using plastic in our daily life, reducing its use will undoubtedly reduce secondary sources of microplastics pollution. Furthermore, limiting the usage of plastic products and other cosmetic items labeled as key sources of microplastics will definitely help in reducing the amount of microplastics enter the environment.

Another important strategy to lessen the plastic threat is to use biodegradables plastic substances as much as feasible instead of synthetic plastic materials. The responsibility of policymakers to monitor microplastics pollution from industry and various other significant sources is highly advised given the widespread presence of MPs and their increasing level in the environment. Additionally, one of the most valuable approaches to reduce the threat of MPs pollution is to increase public awareness of the problem of microplastics pollution and its negative impact on the environment.

Overall, it is possible to draw the conclusion that the pollution caused by microplastics is getting now in alarming condition. Now, it is simple to understand its harmful effects and other environmental issues. There is an urgent need to stop microplastics pollution from spreading throughout the community before the issue gets worse.

10.14 Conclusion

Microplastics are incredibly tiny plastic fragments present in the fresh water and marine ecosystem that come from two sources: cosmetic products and weathering of large size of plastic waste into smaller fragments. Typically, this type of plastic is introduced to the marine and freshwater environment by rivers, domestic waste, drainage systems, and discharge from wastewater treatment plants and through the acts of waves, winds, and currents. Given the degree to which a variety of marine biota ingests them as well as their inclusion in sea salt, microplastics clearly represent a threat to the entire ecosystem. Because microplastics can seriously affect both humans and aquatic creatures, this is a serious concern. Microplastics degrade an environment's recreational, aesthetic, and historical value, and it seems probable that their concentration will rise in the years to come given the lack of practical solutions to eliminate their existence. The general people, socioeconomic sectors, and cosmetic industry that concentrate in waste management must all be involved in order to reduce the issue of microplastics.

References

Adame, F. M., Neil, D., Wright, S. F., & Lovelock, C. E. (2010). Sedimentation within and among mangrove forests along a gradient of geomorphological settings. *Estuarine, Coastal and Shelf Science, 86*, 21–30.

Alomar, C., Estarellas, F., & Deudero, S. (2016). Microplastics in the Mediterranean Sea: Deposition in coastal shallow sediments, spatial variation and preferential grain size. *Marine Environmental Research, 115*, 1–10.

Alrbaihat, M. R., & Abu-Afifeh, Q. (2023). Eco-friendly microplastic removal through physical and chemical techniques: A review. *Annals of Advances in Chemistry, 7*, 14–24. https://doi.org/10.29328/journal.aac.1001038

Ariza-Tarazona, M. C., Villarreal-Chiu, J. F., Barbieri, V., Siligardi, C., & Cedillo-González, E. I. (2019). New strategy for microplastic degradation: Green photocatalysis using a protein-based porous N-TiO_2 semiconductor. *Ceramics International, 45*, 9618–9624. https://doi.org/10.1016/j.ceramint.2018.10.208

Auta, H. S., Emenike, C. U., & Fauziah, S. H. (2017). Distribution and importance of microplastics in the marine environment: A review of the sources, fate, effects, and potential solutions. *Environment International, 102*, 165–176.

Avio, C. G., Gorbi, S., Milan, M., Benedetti, M., Fattorini, D., d'Errico, G., Pauletto, M., Bargelloni, L., & Regoli, F. (2016). Pollutants bioavailability and toxicological risk from microplastics to mussels. *Environmental Pollution, 198*, 211–222.

Beriot, N., Peek, J., Zornoza, R., Geissen, V., & Lwanga, E. H. (2021). Low density-microplastics detected in sheep faeces and soil: A case study from the intensive vegetable farming in Southeast Spain. *Science of the Total Environment, 755*, 142653.

Besseling, E., Wegner, A., Foekema, E. M., van den Heuvel-Greve, M. J., & Koelmans, A. A. (2012). Effects of microplastics on fitness and PCB bioaccumulation by the *lugworm Arenicola marina* (L). *Environmental Science & Technology, 47*(1), 593–600. https://doi.org/10.1021/es302763x

Bianco, A., & Passananti, M. (2020). Atmospheric micro and nanoplastics: An enormous microscopic problem. *Sustainability, 12*(18), 7327.

Boots, B., Russell, C. W., & Green, D. S. (2019). Effects of microplastics in soil ecosystems: Above and below ground. *Environmental Science & Technology, 53*(19), 11496–11506.

Boucher, C., Morin, M., & Bendell, L. I. (2016). The influence of cosmetic microbeads on the sorptive behavior of cadmium and lead within intertidal sediments. A laboratory study. *Regional Studies in Marine Science, 3*, 1–7.

Brandon, J., Goldstein, M., & Ohman, M. D. (2016). Long-term aging and degradation of microplastic particles: Comparing in situ oceanic and experimental weathering patterns. *Marine Pollution Bulletin, 110*(1), 299–308. https://doi.org/10.1016/j.marpolbul.2016.06.048

Brennecke, D., Duarte, B., Paiva, F., Cacador, I., & Canning-Clode, J. (2016). Microplastics as vectors for heavy metal contamination from the marine environment. *Estuarine, Coastal and Shelf Science*, 1–7. https://doi.org/10.1016/j.ecss.2015.12.003

Brodhagen, M., Goldberger, J. R., Hayes, D. G., Inglis, D. A., Marsh, T. L., & Miles, C. (2017). Policy considerations for limiting unintended residual plastic in agricultural soils. *Environmental Science & Policy, 69*, 81–84.

Chen, G., Feng, Q., & Wang, J. (2020). Mini-review of microplastics in the atmosphere and their risks to humans. *Science of the Total Environment, 703*, 135504.

Clark, J. R., Cole, M., Lindeque, P. K., Fileman, E., Blackford, J., Lewis, C., Lenton, T. M., & Galloway, T. S. (2016). Marine microplastic debris: A targeted plan for understanding and quantifying interactions with marine life. *Frontiers in Ecology and the Environment, 14*(6), 317–324. https://doi.org/10.1002/fee.1297

Cole, M., Lindeque, P., Fileman, E., Halsband, C., Goodhead, R., Moge, R. J., & Galloway, T. S. (2013). Microplastic ingestion by zooplankton. *Environmental Science & Technology, 47*, 6646–6655. https://doi.org/10.1021/es400663f

Cole, M., Webb, H., Lindeque, P. K., Fileman, E. S., Halsband, C., & Galloway, T. S. (2014). Isolation of microplastics in biota-rich seawater samples and marine organisms. *Scientific Reports, 4*, 4528. https://doi.org/10.1038/srep04528

Cole, M., Lindeque, P. K., Fileman, E., Clark, J., Lewis, C., Halsband, C., & Galloway, T. S. (2016). Microplastics alter the properties and sinking rates of zooplankton Faecal pellets. *Environmental Science & Technology, 50*, 3239–3246. https://doi.org/10.1021/acs.est.5b05905

Corradini, F., Casado, F., Leiva, V., Huerta-Lwanga, E., & Geissen, V. (2021). Microplastics occurrence and frequency in soils under different land uses on a regional scale. *Science of the Total Environment, 752*, 141917.

Cozar, A., Echevarria, F., Gonzalez-Gordillo, J. I., Irigoien, X., Úbeda, B., Hernàndez-Leon, S., Palma, A. T., Navarro, S., Garcia-de-Lomas, J., Ruiz, A., Fernandez-de-Puelles, M. L., & Duarte, C. M. (2014). Plastic debris in the open ocean. *PNAS, 111*, 10239–10244.

de Souza Machado, A. A., Lau, C. W., Till, J., Kloas, W., Lehmann, A., Becker, R., & Rillig, M. C. (2018). Impacts of microplastics on the soil biophysical environment. *Environmental Science & Technology, 52*(17), 9656–9665.

Devriese, L. I., van der Meulen, M. D., Maes, T., Bekaert, K., Paul-Pont, I., Frére, L., Robbens, J., & Vethaak, A. D. (2015). Microplastic contamination in brown shrimp (*Crangon crangon*, Linnaeus 1758) from coastal waters of the southern North Sea and channel area. *Marine Pollution Bulletin, 98*, 179–187.

Do Sul, J. A. I., Costa, M. F., Barletta, M., & Cysneiros, F. J. A. (2013). Pelagic microplastics around an archipelago of the equatorial Atlantic. *Marine Pollution Bulletin, 75*(1–2), 305–309.

Eerkes-Medrano, D., Thompson, R. C., & Aldridge, D. C. (2015). Microplastics in freshwater systems: A review of the emerging threats, identification of knowledge gaps and prioritization of research needs. *Water Research, 75*, 63–82.

Fendall, L. S., & Sewell, M. A. (2009). Contributing to marine pollution by washing your face: Microplastics in facial cleansers. *Marine Pollution Bulletin, 58*(8), 1225–1228.

Feng, S., Lu, H., & Liu, Y. (2021). The occurrence of microplastics in farmland and grassland soils in the Qinghai-Tibet plateau: Different land use and mulching time in facility agriculture. *Environmental Pollution, 279*, 116939.

Ferreira, P., Fonte, E., Soares, M. E., Carvalho, F., & Guilhermino, L. (2016). Effects of multi-stressors on juveniles of the marine fish *Pomatoschistus microps*: Gold nanoparticles, microplastics and temperature. *Aquatic Toxicology, 170*, 89–103.

Fossi, M. C., Marsili, L., Baini, M., Giannetti, M., Coppola, D., Guarranti, C., Caliani, I., Minutoli, R., Lauriano, G., Finoia, M. G., Rubegni, F., Panigada, S., Bérubé, M., Ramirez, J. U., & Panti, C. (2016). Fin whales and microplastics: The Mediterranean Sea and the sea of Cortez scenarios. *Environmental Pollution, 209*, 68–78.

GESAMP. (2014). *Microplastics in the ocean – A global assessment*. GESAMP -IOC, IMO/FAO/IOC/WMO/UNIDO/IAEA/UN/UNEP. Joint Group of Experts on the Scientific Aspects of Marine Environmental Protection.

Golwala, H., Zhang, X., Iskander, S. M., & Smith, A. L. (2021). Solid waste: An overlooked source of microplastics to the environment. *Science of the Total Environment, 769*, 144581.

Green, D. S. (2016). Effects of microplastics on European flat oysters, *Ostrea edulis* and their associated benthic communities. *Environmental Pollution, 216*, 95–103.

Harms, I. K., Diekötter, T., Troegel, S., & Lenz, M. (2021). Amount, distribution and composition of large microplastics in typical agricultural soils in northern Germany. *Science of the Total Environment, 758*, 143615.

Hollman, P. C. H., Bouwmeester, H., & Peters, R. J. B. (2013). *Microplastics in the aquatic food chain: Sources, measurements, occurrence and potential health risks*. RIKILT report. RIKILT Wageningen UR (University & Research Centre).

Holmes, L. A., Turner, A., & Thompson, R. C. (2012). Adsorption of trace metals to plastic resin pellets in the marine environment. *Environmental Pollution, 160*, 42–48. https://doi.org/10.1016/j.envpol.2011.08.052

Iqbal, S., Xu, J., Allen, S. D., Khan, S., Nadir, S., Arif, M. S., & Yasmeen, T. (2020). Unraveling consequences of soil micro-and nano-plastic pollution on soil-plant system: Implications for nitrogen (N) cycling and soil microbial activity. *Chemosphere, 260*, 127578.

Jorissen, F. J. (2014). Colonization by the benthic foraminifer *Rosalina (Trettomphalus) concinna* of Mediterranean drifting plastics in CIESM2014. In: F. Briand (Ed.), *Marine Litter in the Mediterranean and Black Seas* (CIESM Workshop Monograph no. 46., p. 180). CIESM.

Klein, S., Dimzon, I. K., Eubeler, J., & Knepper, T. P. (2018). Analysis, occurrence, and degradation of microplastics in the aqueous environment. In *Freshwater microplastics* (pp. 51–67). Springer.

Kumar, M. V., & Sheela, A. M. (2020). Effect of plastic film mulching on the distribution of plastic residues in agricultural fields. *Chemosphere, 273*, 128590–128590.

Law, K. L., & Thompson, R. C. (2014). Microplastics in the seas. *Science, 345*(6193), 144–145. https://doi.org/10.1126/science.1254065

Li, J., Liu, H., & Chen, J. P. (2018). Microplastics in freshwater systems: A review on occurrence, environmental effects, and methods for microplastics detection. *Water Research, 137*, 362–374.

Liu, Y., Liu, W., Yang, X., Wang, J., Lin, H., & Yang, Y. (2021). Microplastics are a hotspot for antibiotic resistance genes: Progress and perspective. *Science of the Total Environment, 773*, 145643.

Lönnstedt, O. M., & Eklöv, P. (2016). Environmentally relevant concentrations of microplastic particles influence larval fish ecology, ecotoxicology. *Science, 352*, 6290.

Lozano, Y. M., Lehnert, T., Linck, L. T., Lehmann, A., & Rillig, M. C. (2021). Microplastic shape, polymer type, and concentration affect soil properties and plant biomass. *Frontiers in Plant Science, 12*, 616645.

Lusher, A. L., McHugh, M., & Thompson, R. C. (2013). Occurrence of microplastics in the gastrointestinal tract of pelagic and demersal fish from the English Channel. *Marine Pollution Bulletin, 67*(1), 94–99.

Lwanga, E. H., Gertsen, H., Gooren, H., Peters, P., Salánki, T., van der Ploeg, M., & Geissen, V. (2017). Incorporation of microplastics from litter into burrows of *Lumbricus terrestris*. *Environmental Pollution, 220*, 523–531.

Morris, C. (2015). *Marine litter, green facts: facts on health and the environment, 2001–2015*.

Murphy, F., Ewins, C., Carbonnier, F., & Quinn, B. (2016). Wastewater treatment works (WwTW) as a source of microplastics in the aquatic environment. *Environmental Science & Technology*. https://doi.org/10.1021/acs.est.5b05416

Neves, D., Sobral, P., Ferreira, J. L., & Pereira, T. (2015). Ingestion of microplastics by commercial fish off the Portuguese coast. *Marine Pollution Bulletin, 101*, 119–126.

Ng, E. L., Lwanga, E. H., Eldridge, S. M., Johnston, P., Hu, H. W., Geissen, V., & Chen, D. (2018). An overview of microplastic and nanoplastic pollution in agroecosystems. *Science of the Total Environment, 627*, 1377–1388.

Nobre, C. R., Santana, M. F., Maluf, A., Cortez, F. S., Cesar, A., Pereira, C. D. S., & Turra, A. (2015). Assessment of microplastic toxicity to embryonic development of the sea urchin *Lytechinus variegatus* (Echinodermata: Echinoidea). *Marine Pollution Bulletin, 92*(1–2), 99–104.

Nuelle, M. T., Dekiff, J. H., Remy, D., & Fries, E. (2014). A new analytical approach for monitoring microplastics in marine sediments. *Environmental Pollution, 184*, 161–169.
Patel, M. M., Goyal, B. R., Bhadada, S. V., Bhatt, J. S., & Amin, A. F. (2009). Getting into the brain: Approaches to enhance brain drug delivery. *CNS Drugs, 23*, 35–58. https://doi.org/10.2165/0023210-200923010-00003
Pflugmacher, S., Sulek, A., Mader, H., Heo, J., Noh, J. H., Penttinen, O. P., & Esterhuizen, M. (2020). The influence of new and artificial aged microplastic and leachates on the germination of Lepidium sativum L. *Plants, 9*(3), 339.
Qi, R., Jones, D. L., Li, Z., Liu, Q., & Yan, C. (2020). Behavior of microplastics and plastic film residues in the soil environment: A critical review. *Science of the Total Environment, 703*, 134722.
Rehse, S., Kloas, W., & Zarfi, C. (2016). Short-term exposure with high concentrations of pristine microplastic particles leads to immobilization of *Daphnia magna. Chemosphere, 153*, 91–99.
Ren, X., Tang, J., Liu, X., & Liu, Q. (2020). Effects of microplastics on greenhouse gas emissions and the microbial community in fertilized soil. *Environmental Pollution, 256*, 113347.
Rochman, C. M., Hoh, E., Kurobe, T., & The, S. J. (2013). Ingested plastic transfers hazardous chemicals to fish and induces hepatic stress. *Scientific Reports, 3*, 3263. https://doi.org/10.1038/srep03263
Romeo, T., Pietro, B., Pedà, C., Consoli, P., Andaloro, F., & Fossi, M. (2015). First evidence of presence of plastic debris in stomach of large pelagic fish in the Mediterranean Sea Mar. *Pollution Bulletin, 95*(1), 358–361. https://doi.org/10.1016/j.marpolbul.2015.04.048
Scheurer, M., & Bigalke, M. (2018). Microplastics in Swiss floodplain soils. *Environmental Science & Technology, 52*(6), 3591–3598.
Serrano-Ruiz, H., Martin-Closas, L., & Pelacho, A. M. (2021). Biodegradable plastic mulches: Impact on the agricultural biotic environment. *Science of the Total Environment, 750*, 141228.
Setala, O., Norkko, J., & Lehtiniemi, M. (2016). Feeding type affects microplastic ingestion in a coastal Invertebrate community. *Marine Pollution Bulletin, 102*, 95–101.
Shim, W. J., & Thompson, R. C. (2015). Microplastics in the ocean. *Archives of Environmental Contamination and Toxicology, 69*, 265–268.
Sommer, F., Dietze, V., Baum, A., Sauer, J., Gilge, S., Maschowski, C., & Gieré, R. (2018). Tire abrasion as a major source of microplastics in the environment. *Aerosol and Air Quality Research, 18*(8), 2014–2028.
Statista (2023). Plastic waste worldwide - statistics & facts. https://www.statista.com/topics/5401/global-plasticwaste/#topicOverview
Stubenrauch, J., & Ekardt, F. (2020). Plastic pollution in soils: Governance approaches to foster soil health and closed nutrient cycles. *Environments, 7*(5), 38.
Suaria, G., & Aliani, S. (2014). Floating debris in the Mediterranean Sea. *Marine Pollution Bulletin, 86*, 494–504.
Talvitie, J., Mikola, A., Koistinen, A., & Setälä, O. (2017). Solutions to microplastic pollution–removal of microplastics from wastewater effluent with advanced wastewater treatment technologies. *Water Research, 123*, 401–407.
Tympa, L. E., Katsara, K., Moschou, P. N., Kenanakis, G., & Papadakis, V. M. (2021). Do microplastics enter our food chain via root vegetables? A Raman based spectroscopic study on Raphanus sativus. *Materials (Basels), 14*(9), 2329.
Vaid, M., Mehra, K., & Gupta, A. (2021). Microplastics as contaminants in Indian environment: A review. *Environmental Science and Pollution Research*, 1–28.
Van Cauwenberghe, L., Claessens, M., Vandegehuchte, M. B., & Janssen, C. R. (2015a). Microplastics are taken up by mussels (*Mytilus edulis*) and lugworms (*Arenicola marina*)living in natural habitats. *Environmental Pollution, 199*, 10–17.
Van Cauwenberghe, L., Devriese, L., Galgani, F., Robbens, J., & Janssen, C. R. (2015b). Microplastics in sediments: A review of techniques, occurrence and effects. *Marine Environmental Research, 111*, 5–17.

von Moos, N., Burkhardt-Holm, P., & Köhler, A. (2012). Uptake and effects of microplastics on cells and tissues of the blue mussel *Mytilus edulis* L. after experimental exposure. *Environmental Science & Technology, 46*, 11327–11335.

Wagner, M., & Lambert, S. (2018). *Freshwater microplastics: Emerging environmental contaminants?* (p. 303). Springer.

Wagner, M., Scherer, C., Alvarez-Muñoz, D., Brennholt, N., Bourrain, X., Buchinger, S., Fries, E., Grosbois, C., Klasmeier, J., Marti, T., Rodriguez-Mozaz, S., Urbatzka, R., Vethaak, A. D., Winther-Nielsen, M., & Reifferscheid, G. (2014). Microplastics in freshwater ecosystems: What we know and what we need to know. *Environmental Sciences Europe, 26*(12), 1–9.

Wang, J., Tan, Z., Peng, J., Qiu, Q., & Li, M. (2016). The behaviors of microplastics in the marine environment. *Marine Environmental Research, 133*, 7–17.

Wang, J., Liu, X., Li, Y., Powell, T., Wang, X., Wang, G., & Zhang, P. (2019). Microplastics as contaminants in the soil environment: A mini-review. *Science of the Total Environment, 691*, 848–857.

Watts, A. J. R., Lewis, C., Goodhead, R. M., Beckett, S., Moger, J., Tyler, C. R., & Galloway, T. S. (2014). Uptake and retention of microplastics by the shore crab *Carcinus maenas*. *Environmental Science & Technology, 24*, 8823–8830.

Wright, S. L., Thompson, R. C., & Galloway, T. S. (2013). The physical impacts of microplastics on marine organisms: A review. *Environmental Pollution, 178*, 483–492.

Xu, B., Liu, F., Cryder, Z., Huang, D., Lu, Z., He, Y., & Xu, J. (2020). Microplastics in the soil environment: Occurrence, risks, interactions and fate–A review. *Critical Reviews in Environmental Science and Technology, 50*(21), 2175–2222.

Yang, M., Huang, D. Y., Tian, Y. B., Zhu, Q. H., Zhang, Q., Zhu, H. H., & Xu, C. (2021). Influences of different source microplastics with different particle sizes and application rates on soil properties and growth of Chinese cabbage (*Brassica chinensis L.*). *Ecotoxicology and Environmental Safety, 222*, 112480.

Yu, L., Zhang, J., Liu, Y., Chen, L., Tao, S., & Liu, W. (2021). Distribution characteristics of microplastics in agricultural soils from the largest vegetable production base in China. *Science of the Total Environment, 756*, 143860.

Zalasiewicz, J., Waters, C. N., Ivar do Sul, J. A., Corcoran, P. L., Barnosky, A. D., Cearreta, A., Edgeworth, M., Galuszka, A., Jeandel, C., Leinfelder, R., McNeill, J. R., Steffen, W., Summerhayes, C., Wagreich, M., Williams, M., Wolfe, A. P., & Yonan, Y. (2016). The geological cycle of plastics and their use as a stratigraphic indicator of the Anthropocene. *Anthropocene*. https://doi.org/10.1016/j.ancene.2016.01.002

Zarfl, C., & Matthies, M. (2010). Are marine plastic particles transport vectors for organic pollutants to the Arctic? *Marine Pollution Bulletin, 60*, 1810–1814.

Zhao, T., Lozano, Y. M., & Rillig, M. C. (2021). Microplastics increase soil pH and decrease microbial activities as a function of microplastic shape, polymer type, and exposure time. *Frontiers in Environmental Science, 9*, 675803.

Zhou, B., Wang, J., Zhang, H., Shi, H., Fei, Y., Huang, S., & Barceló, D. (2020). Microplastics in agricultural soils on the coastal plain of Hangzhou Bay, East China: Multiple sources other than plastic mulching film. *Journal of Hazardous Materials, 388*, 121814.

Zhu, F., Zhu, C., Wang, C., & Gu, C. (2021). Occurrence and ecological impacts of microplastics in soil systems: A review. *Bulletin of Environmental Contamination and Toxicology, 102*, 741–749.

Chapter 11
Electronic Plastic (E-plastic) and Their Biodegradation

Jham Lal, Soibam Khogen Singh, Pradyut Biswas, Anand Vaishnav, Nitesh Kumar Yadav, and Sanjay Chandravanshi

Abstract Electronic waste (e-waste) has become a significant global concern due to the rapid growth of consumer electronics and technological advancements. Among e-waste components, electronic plastic (e-plastic) is a major contributor, often composed of complex polymer blends with hazardous substances. The improper disposal of e-plastic poses severe environmental and health risks. This study aims to explore the current state of e-plastic waste management, focusing on its biodegradation potential as a sustainable solution. Several biodegradation methods, including natural and synthetic approaches, have been investigated. The role of microorganisms, enzymes, and bioactive compounds in breaking down e-plastic has been studied to understand the mechanisms behind biodegradation. Furthermore, the challenges associated with e-plastic biodegradation, such as the slow degradation rate, limited understanding of microbial communities involved, and potential toxic by-products during degradation, are also discussed. Additionally, the importance of eco-friendly design and sustainable materials selection in electronic manufacturing is emphasized to reduce the environmental impact of e-plastic waste. This research contributes to the growing body of knowledge on e-waste management and the need for sustainable disposal options for e-plastic. By understanding the biodegradation potential of e-plastic, policymakers, industry stakeholders, and researchers can work together to develop effective waste management strategies that prioritize environmental protection and resource conservation.

Keywords Electronic plastic · E-plastic · Biodegradation · Eco-friendly electronics · Sustainable waste management

---

J. Lal (✉) · S. K. Singh · P. Biswas · A. Vaishnav · N. K. Yadav
College of Fisheries, Central Agricultural University, Lembucherra, Tripura, India

S. Chandravanshi
College of Fisheries, Dr. J. Jayalalithaa Fisheries University (TNJFU), Thoothukudi, Tamil Nadu, India

R. Soni et al. (eds.), *Advanced Strategies for Biodegradation of Plastic Polymers*, https://doi.org/10.1007/978-3-031-55661-6_11

11.1 Introduction

In the modern digital era, electronics have become an integral part of our daily lives, revolutionizing communication, entertainment, and various industries. However, the widespread use and disposal of electronic devices have led to a significant environmental concern—electronic plastic waste (e-plastic). Electronic plastic, also known as e-plastic or e-waste, refers to the nonbiodegradable plastic materials used in electronic devices, such as smartphones, computers, TVs, and other electronic appliances (Bhutta et al., 2011). The alarming growth of e-plastic waste has raised concerns about its impact on the environment and human health. Traditional plastics used in electronics, such as polyvinyl chloride (PVC), polystyrene (PS), and polyethylene terephthalate (PET), are not easily biodegradable, leading to long-lasting environmental pollution. Improper disposal of e-plastic waste, including burning or landfilling, further exacerbates the problem, releasing toxic chemicals and hazardous substances into the environment (Evode et al., 2021).

Global electronic trash has accumulated as a result of the tremendous advancement of technology. The term "waste electrical and electronic equipment" (WEEE), also referred to as "e-waste," refers to outdated or abandoned electronic devices. According to Sekhar et al. (2016), e-waste is an environmental issue that is exponentially growing and comprises both useful and harmful components that call for specific management and recycling techniques. In the manufacture of electronic products, high-impact polystyrene (HIPS) is frequently utilized as a plastic resin (Othman et al., 2009). With polybutadiene distributed in a rigid polystyrene (PS) matrix, HIPS polymer is a multiphase system with enhanced breakage resistance, decreased transparency, modulus, as well as tensile strength. Approximately one million tons of HIPS are being produced annually in the entire world (Prasad & Singh, 1997). Decabromodiphenyl ether and antimony trioxide (Sb_2O_3), two polybrominated substances that work well together as flame retardants, are routinely added to HIPS (Alaee et al., 2003; Sekhar et al., 2016).

A novel category of semiconductor technology known as "transient electronics" is developing, and its key feature is its capability to completely dissolve, in controlled conditions, upon engagement in groundwater or biofluids. The biodegradable substrates as well as encapsulating substances play crucial roles in this progression. The management of electronic trash (e-waste), sustainable development, and potential implantable technologies are all thought to benefit from biodegradable electronics. For human-related applications, elastic electronics that better mimic the mechanical properties of human tissues have become essential. In particular for wearable and implantable electronics, the convergence of biodegradability and elasticity has given rise to an entirely novel approach of next-generation electronics (Chen et al., 2022).

11.2 E-plastic Waste Definition

E-waste is described as outdated, carelessly eliminated electronic and electrical equipment that turns into waste after its useful life. Discarded electrical or electronic equipment is referred to as "e-waste." E-waste also includes used electronics that are intended for reuse and recycling, disposal, salvage, or resale. Due to the lack of governmental control of e-waste processing in poor nations, informal processing of electronic trash could have detrimental effects on both health and the environment (Subramanian, 2000). Iron, aluminum, copper, gold, and other materials make up about 60% of e-waste, while plastics account for 30% of it and other hazardous materials (such as lead, palladium, beryllium, etc.) for 2.7% (Akram et al., 2015).

11.3 E-plastic Waste as a Construction Material

Electrical and electronic equipment output and consumption have increased over the past few decades as a result of very dynamic technical innovation. The cost of electronic devices decreased due to the quick changes in equipment manufacture, components, and features, and consumers began tossing out their old products in favor of new ones that incorporated the newest technological advancements. As a result, many obsolete electrical and electronic equipment have been produced. Electronic waste includes trash produced by devices like computers, mobile phones, televisions, printers, laptops, personal stereos, washing machines, air conditioners, refrigerators, and other home appliances. Around the world, the pace of growth of electronic garbage is roughly three times that of municipal waste. Electronic waste is made up of a variety of substances and chemicals, such as mercury, lead, beryllium, brominated flame retardants, cadmium, and plastics like polychlorinated biphenyls, polyvinyl chloride, and polystyrene, several of which are toxic and could have a negative impact on both the environment and human wellness if not properly disposed of. Because they can seriously harm people's health, these electronic wastes must be disposed of with extreme caution. These call for a sizable collecting network, a recycling infrastructure, and a reliable processing and disposal equipment for electronic trash. Both rich and developing countries have dangerous environmental concerns as a result of the global trade of this sophisticated and toxic electronic trash (Senthil Kumar & Baskar, 2015).

11.4 Types of Electronic Plastics

Electronic plastics are a diverse group of materials used in various electronic devices due to their specific properties, such as insulation, flexibility, and durability. Here are some common types of electronic plastics along with references for further reading:

11.4.1 Polycarbonate (PC)

Polycarbonate is a versatile and widely used thermoplastic known for its excellent impact resistance and optical clarity. It is commonly used in items like optical discs (CDs, DVDs), eyewear lenses, and electronic device casings. Polycarbonate (PC) electronic waste plastics refer to discarded electronic products or components made from polycarbonate material. These items include electronic device casings, computer components, optical discs (e.g., CDs, DVDs), and other electronic parts that are no longer in use or have reached the end of their life cycle (Fig. 11.1).

Electronic waste (e-waste) is a rapidly growing global concern due to the proliferation of electronic devices and their relatively short lifespans. When electronic

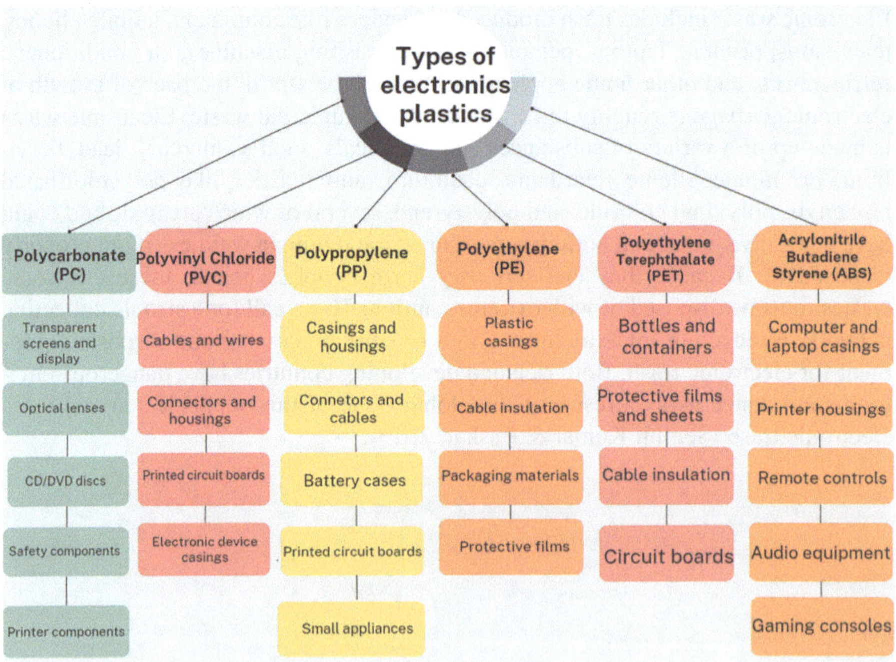

Fig. 11.1 Types of electronic plastics

devices are no longer functional or become obsolete, they often end up as e-waste. Unfortunately, e-waste is challenging to manage because it contains various materials, including plastics, metals, and toxic substances, which can be harmful to human health and the environment if not properly treated. Polycarbonate electronic waste plastics pose specific challenges in terms of recycling and disposal because of the composition's complexity of electronic devices and the presence of other materials, such as metals and other types of plastics. Proper management of e-waste, including the recycling of polycarbonate and other materials, is essential to reduce environmental pollution and conserve valuable resources (Suh et al., 2015).

11.4.2 Polyvinyl Chloride (PVC)

PVC is a widely used synthetic polymer with good chemical resistance and electrical properties. It is commonly used in cables, wires, connectors, and casings for electronic devices (Qu et al., 2018).

Polyvinyl chloride (PVC) is a commonly used plastic in various electronic applications due to its excellent electrical properties and chemical resistance. However, the recycling and management of PVC electronic waste pose challenges due to the presence of additives and potential environmental impacts. Below are some references that discuss PVC electronic waste and its recycling (Song et al., 2020). PVC electronic waste may include various items, such as the following:

1. Cables and Wires: PVC is commonly used as the insulation material for electrical cables and wires in electronic devices.
2. Connectors and Housings: Many electronic components, such as connectors and device housings, are made from PVC due to its versatility and cost-effectiveness.
3. Printed Circuit Boards (PCBs): Some PCBs and their components, such as connectors or shielding materials, may contain PVC.
4. Electronic Device Casings: PVC is used in the casings of electronic devices like smartphones, laptops, and various consumer electronics.
5. Other Electronic Components: PVC may be used in other electronic components and parts within devices. The proper management of PVC electronic waste is essential to reduce environmental pollution and health risks. Uncontrolled burning or disposal of PVC waste can release toxic substances and emit harmful gases, impacting both the environment and human health (Eze et al., 2018).

11.4.3 Polyethylene (PE)

Polyethylene is a popular plastic known for its low cost, flexibility, and resistance to moisture and chemicals. It is used in various electronic applications, including cable insulation and packaging (Covino & Smith, 2014). Polyethylene (PE) electronic plastic waste refers to discarded electronic products or components made from polyethylene material (Table 11.1). Polyethylene is a widely used plastic in various electronic applications due to its flexibility, chemical resistance, and electrical insulating properties. However, the recycling and management of polyethylene electronic waste pose challenges, especially when it is mixed with other materials commonly found in electronic devices.

Polyethylene electronic plastic waste may include various items, such as the following:

1. Plastic Casings: Many electronic devices, such as smartphones, tablets, and computers, have casings made from polyethylene (Dudnik et al., 2019).
2. Cable Insulation: Polyethylene is commonly used as the insulation material for electrical cables and wires in electronic devices (Chancerel et al., 2009).
3. Packaging Materials: Polyethylene may be used as packaging materials for electronic components or devices (Petersen et al., 2018).
4. Protective Films: Some electronic devices have protective films or coatings made from polyethylene (Velis & Bartl, 2018).

Table 11.1 Different types of e-plastic waste used for recycling

S. no.	Different types of e-plastics	Uses	References
1.	E-waste (e-plastic)	Utilizing e-plastic in concrete will lessen the need for extremely fine aggregate, which will help conserve natural resources	Gavhane et al. (2016)
2.	Biodegradable plastic witches	The determination to move to biodegradable plastics is affected by hedonic motivations, environmental reasons, and views	Moshood et al. (2022)
3.	Printed circuit board (PCB)	Possible soldering methods and thermal stability of the biodegradable PCBs	Géczy et al. (2012)
4.	Transient electronics	Utilize in transitory electronics for short-term biomedical implants, biodegradable environmental monitoring, and less wasteful consumer electronics	Kang et al. (2015)
5.	Printable biodegradable electronics	Environmentally friendly sensors, momentary electronic circuits, and implantable medical devices	Feng et al. (2019)
6.	Flexible transient electronics	Environmentally friendly gadgets and implantable medical devices	Feng et al. (2019)
7.	E-plastics	Use the polymers as sources of carbon or nitrogen	Sheth et al. (2019)

11.4.4 Polypropylene (PP)

Polypropylene is a versatile thermoplastic with excellent chemical resistance and mechanical properties. It is used in electronic components, battery casings, and automotive parts (Erlandsson & Stridh, 2012).

Polypropylene (PP) is a commonly used plastic in electronic waste (e-waste) due to its properties such as flexibility, durability, and resistance to chemical reactions. It is commonly found in various electronic components and casings of electronic devices. While polypropylene is a widely used material in electronics, it is crucial to manage e-waste responsibly. Improper disposal and recycling of electronic waste, including polypropylene components, can lead to environmental pollution and health hazards. Sustainable e-waste management practices, such as recycling and proper disposal, are essential to minimize the impact of electronic waste on the environment. Proper recycling of e-waste involves the separation and recovery of valuable materials, including polypropylene, which can be reused to manufacture new products. Additionally, recycling reduces the need for virgin plastics, conserving natural resources and lowering greenhouse gas emissions. As technology continues to evolve, the proper handling and recycling of polypropylene and other materials in electronic waste will remain critical to promote environmental sustainability and reduce the global e-waste burden (Ahmad et al., 2023).

11.4.5 Polyethylene Terephthalate (PET)

PET is a widely used plastic known for its clarity, strength, and resistance to moisture. It is commonly used in electronic packaging, such as blister packs and protective films (Rahimi & García, 2017).

Polyethylene terephthalate (PET) is another common plastic found in electronic waste (e-waste). It is widely used in various electronic components and packaging materials due to its excellent mechanical properties, transparency, and recyclability. As with other plastics in e-waste, the proper management and recycling of PET-containing electronic waste are essential to minimize environmental impact and promote resource conservation. PET is a highly recyclable plastic, and recycling it from e-waste can help reduce the demand for virgin PET production, thus saving energy and reducing greenhouse gas emissions (Smith et al., 2022).

Responsible e-waste recycling practices aim to recover valuable materials like PET and process them into reusable forms for manufacturing new products. By recycling PET and other plastics from electronic waste, we can contribute to a more sustainable and circular approach to electronic waste management. It is crucial to support and participate in e-waste recycling initiatives to ensure that PET and other materials from electronic waste are handled properly and diverted from improper disposal methods like landfilling or incineration, which can lead to environmental pollution and health hazards (Hopewell et al., 2009).

11.4.6 Acrylonitrile Butadiene Styrene (ABS)

ABS is a durable and impact-resistant plastic often used in electronic housings and 3D printing applications (Kalaitzidou et al., 2007). Acrylonitrile butadiene styrene (ABS) is another commonly used plastic in electronic waste (e-waste). It is a versatile and durable thermoplastic known for its impact resistance and mechanical strength. ABS is often used in various electronic components and casings due to its ability to withstand mechanical stress and temperature variations. As with other plastics in e-waste, responsible management and recycling of ABS-containing electronic waste are vital to mitigate environmental impacts and promote sustainability. Recycling ABS from e-waste can help reduce the need for virgin ABS production, conserving resources and reducing greenhouse gas emissions associated with plastic manufacturing (Teixeira et al., 2023).

E-waste recycling processes aim to recover valuable materials like ABS and other plastics from discarded electronic devices and transform them into reusable forms for the production of new products. By supporting e-waste recycling initiatives, we can contribute to a more sustainable electronic waste management and a circular economy approach. It is essential to dispose of electronic waste responsibly and encourage the proper handling and recycling of ABS and other materials to address the environmental challenges posed by electronic waste disposal and minimize its impact on the environment and human health (Van Yken et al., 2021).

11.5 Management of E-plastic Waste

One of the main pollutants that are dumped in the natural environment, along with other contaminants, is plastic. They are necessary for human survival because of their affordable price and adaptability. Plastics, especially used in a range of household applications, are known to include a mixture of diverse chemical compounds. Despite their numerous beneficial applications, plastics require a long time to degrade. When plastics are burned, hazardous compounds to the environment such as phosgene and dioxides are emitted. Microplastics, which are harmful waste particles generated from plastics, find their way into the food chain and water supplies. Serious health effects like congenital illnesses and malignant malignancies are caused by ingesting microplastic-contaminated meals and tiny levels of phthalates in toys. Lethally persistent organic pollutants called dioxins are emitted from plastic polymers, and they harm human neurological function and cause tumors. Water bodies now contain severe plastic pollution as a result of poor waste management techniques. Beaches frequently accumulate plastics, which reduces their aesthetic and recreational value. In this article, we've covered how to recover resources from plastic wastes, potential environmental repercussions of plastics, and current safety standards for using plastics (Rios et al., 2015). E-waste is the term used to describe electrical or electronic items that have been carelessly abandoned, are superfluous, are out of date, or are defective. There is a rapidly expanding surplus of electronic waste worldwide as a result of quick technological advancement and inexpensive beginning costs. E-garbage must be disposed of yearly in many tons. Traditional disposal methods, such as using landfills or stockpiles, are not environmentally friendly, and it is very difficult to follow EPA rules during the disposal process (Gull & Balasubramanian, 2014).

11.6 Biodegradation Process of Electronic Plastics

The biodegradation process of electronic plastics is a complex and challenging phenomenon due to the synthetic nature and composition of these materials. Electronic plastics, such as polycarbonate, polyvinyl chloride (PVC), and various flame retardants, are not easily biodegradable by natural microorganisms in the environment (Fig. 11.2). Biodegradation is the natural process by which microorganisms break down complex plastic molecules into simpler compounds, reducing their ecological impact (Singh et al., 2021). However, there are some potential mechanisms and factors that can influence the biodegradation of electronic plastics:

1. Microbial Activity: Biodegradation is primarily driven by microorganisms, such as bacteria and fungi, that possess enzymes capable of breaking down organic compounds. While some microorganisms have been found to partially degrade

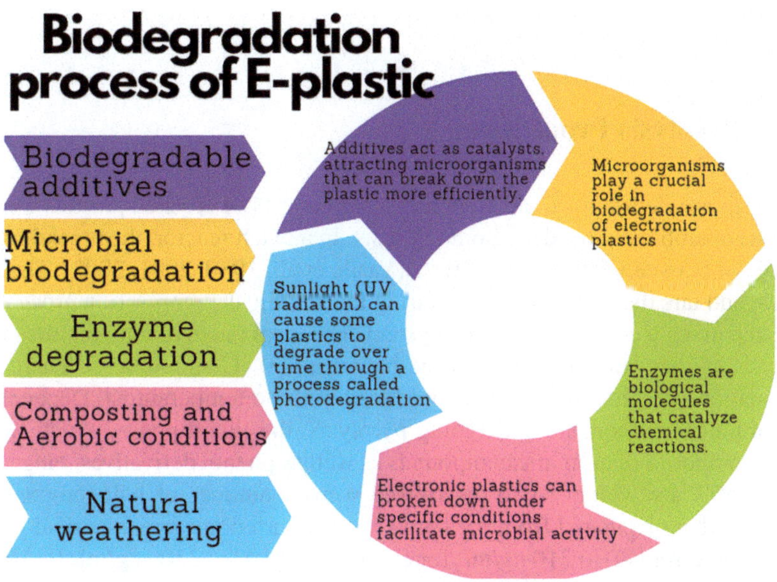

Fig. 11.2 Biodegradation process of e-plastic

certain plastics, the overall degradation of electronic plastics is slow and limited due to their complex chemical structures and additives (Jambeck et al., 2015).
2. Environmental Conditions: The rate of biodegradation depends on environmental conditions such as temperature, humidity, oxygen availability, and pH. For instance, higher temperatures and humidity can promote microbial activity, potentially enhancing the biodegradation process to some extent (Shah et al., 2008).
3. Plastic Composition: The type and composition of the electronic plastic play a significant role in its biodegradability. Some types of biodegradable plastics, such as those made from polylactic acid (PLA) or polyhydroxyalkanoates (PHA), may be more susceptible to microbial degradation compared to conventional electronic plastics (Bhattacharya & Sharma, 2004).
4. Additives and Flame Retardants: Electronic plastics often contain additives, including flame retardants, to meet specific performance requirements. These additives can hinder biodegradation and may even be toxic to microbial communities (Achilleos & Hapeshi, 2015).
5. Mechanical Processes: Mechanical stress and fragmentation caused by weathering, UV radiation, or physical abrasion can lead to the breakdown of larger plastic pieces into smaller particles. While this process is not biodegradation itself, it can increase the surface area available for microbial colonization and potentially accelerate the degradation process (Urbanek & Rymowicz, 2018).
6. Chemical and Biological Treatments: Some research has explored the use of chemical and biological treatments to improve the biodegradation of plastics. For example, enzymatic treatments or microbial-assisted degradation approaches have shown promise in breaking down certain plastic materials, including electronic plastics (Kumar et al., 2017).

11.7 Pyrolysis Procedure

The thermoset debris from old computers and batches of crushed plastics weighing 200 g were employed for the laboratory tests. In a batch reactor, the breakdown was carried out between 450 °C and 470 °C (Vasile et al., 1980, 2001, 2002). The gases, liquids, and tars that were produced during the volatile decomposition process were separated using a cooling/collection system. The ideal range of temperatures for the pyrolysis studies was discovered to be between 430 °C and 460 °C. There was a combination of light hydrocarbons produced at these temperatures. Depending on the computer components undergoing pyrolysis, the liquid phase, which was primarily composed of aromatic compounds as well as phenol derivatives, ranged from 20% to 60% by weight. Also discovered were a number of halogen-containing chemicals that originated from the flame retardant used in these electrical gadgets. They ranged from 40 to 210 ng/mL for chlorine and 100 to 3300 ng/mL for bromine in the pyrolysis oil. Comparing PCBs to casings and keyboards, PCBs produced far more halogens. Consequently, prior to utilizing the process of pyrolysis oils safely as fuels or in refinery or petrochemical industry flows, it is required to remove harmful toxic chemicals, especially those that contain bromine (Vasile et al., 2006).

11.8 Recycle of Electronic Plastic Waste

Recycling electronic waste plastics is essential to mitigate the environmental impact of e-waste and to recover valuable resources from discarded electronic devices. E-waste plastics, which are found in various electronic components, can be recycled through different methods, including mechanical recycling, chemical recycling, and energy recovery. Here are some key points on the recycling of electronic waste plastics, along with references for further reading:

1. *Mechanical Recycling*

Mechanical recycling involves the physical processing of e-waste plastics to reclaim the material for reuse. This process typically includes shredding, granulation, and washing to remove contaminants. The resulting plastic flakes or pellets can be used to manufacture new products. Alternatives to offer impact to India's PW problem include mechanical recycling, chemical recycling, recycling waste into energy methods, and bio-based polymers. The three materials that appeared to be most competitive for chemical recycling, however, are polyurethane, nylon, as well as polyethylene terephthalate. The three primary polymers with greater calorific values appropriate for energy production when compared to traditional fossil fuel energy sources are polyethylene (PE), polypropylene (PP), and polystyrene. The future of India's waste flow chain and waste management may also include multisensor-based artificial intelligence, blockchain-based technology, and digitization for PW recycling (Shanker et al., 2023).

2. *Chemical Recycling*

Chemical recycling involves breaking down e-waste plastics into monomers or other valuable chemicals through processes like depolymerization, pyrolysis, or gasification. These chemical intermediates can then be used to produce new plastics or other products; chemical recycling is an alternate method for processing plastic solid waste; its effectiveness depends on how affordable the procedures are and how effective the catalysts are. Modern methods for chemically recycling commercial plastics are also available, as are technology for sorting and turning plastic solid waste into feedstocks (Rahimi & García, 2017).

3. Energy Recovery

If e-waste plastics are not suitable for mechanical or chemical recycling due to contamination or other factors, they can be used as a source of energy through processes like incineration. This method can help reduce the environmental burden and recover energy from nonrecyclable plastic fractions. The energy utilization and climate change consequences of chemical recycling via pyrolysis are both 50% lower than those of energy restoration. If the quality of the recycled material is taken into consideration, the effects on climate change and energy consumption of pyrolysis as well as mechanical recycling of mixed plastic waste are similar. Additionally, pyrolysis-recycled mixed plastic waste has a substantially smaller climate change effect than its fossil fuel-based counterpart (−0.45 vs 1.89 t CO_2 eq./t plastic). According to sensitivity assessments, particular hypotheses, such as the expected geographical area and its energy mix, the pyrolysis' effectiveness in converting carbon to carbon dioxide, and the caliber of the recycled materials, all have significant impacts (Jeswani et al., 2021).

4. *Sustainable E-waste Management*

Sustainable e-waste management involves a combination of recycling, proper waste disposal, and circular economy principles to reduce the environmental impact of e-waste plastics and maximize resource recovery. Electronic garbage (e-waste) is one of the fastest-growing solid waste streams, expanding at an average of 3–5% each year. Techniques that may have a long-term effect on the natural environment are detailed, including physical dismantling, component recycling, metal extraction, as well as the reuse of nonmetallic elements. Regional collaboration, governmental management, technological advancement, and environmentally acceptable architecture should all assist worldwide sustainable recycling of e-waste (Liu et al., 2023).

11.9 Biodegradation of Electronic Plastics

Biodegradation is a process by which natural microorganisms break down organic materials into simpler compounds over time. However, traditional e-plastics, such as those made from polycarbonate, polyvinyl chloride (PVC), or brominated flame retardants, are not easily biodegradable (Table 11.2). These materials can persist in the environment for hundreds of years, causing pollution and harm to ecosystems (Zhu et al., 2021):

1. Biodegradable Plastics: Scientists are exploring the use of biodegradable plastics made from renewable sources, such as cornstarch or plant-based polymers. These materials can be designed to break down more readily in natural environments.
2. Biodegradable Additives: Some companies are working on incorporating biodegradable additives into conventional plastics. These additives help accelerate the biodegradation process without significantly altering the original material's properties.
3. Recycling: While not biodegradation, recycling is a crucial aspect of managing e-plastics responsibly. Proper recycling can prevent these materials from ending up in landfills or polluting the environment.

Table 11.2 The biodegradation of plastics by microorganisms

S. no.	Types of microorganisms	Utilization/biodegradation	References
1.	Bacteria	A viable choice for LDPE bioremediation, *Paenibacillus* sp. isolate has a high capacity to break down polyethylene	Bardaji et al. (2019)
2.	Fungi	Fungi have great potential for the advancement of the polyester PU waste treatment procedure	Magnin et al. (2019)
3.	Fungi, *Aspergillus* sp.	Over a 4-week period, strain S45 decreased the mass of PU films by 15–20%	Osman et al. (2018)
4.	*Paenibacillus* sp.	30% of PE biodegraded at 37 °C after pretreatment with ethanol and bleach	Bardaji et al. (2019)
5.	*Galleria mellonella*	Biodegradation of WEEE plastics by insect larvae	Zhu et al. (2022)

4. Chemical Recycling: This process involves breaking down plastics into their basic components, allowing them to be used as feedstock for new plastic production. While not true biodegradation, it can help reduce the need for new plastic production and the associated environmental impacts.
5. E-Waste Management: Implementing effective e-waste management practices, including proper collection, disposal, and recycling of electronic devices, can minimize the negative impact of e-plastics on the environment.
6. Regulations and Policy: Governments can play a crucial role by implementing regulations that encourage the use of more sustainable materials in electronic products and promoting responsible e-waste management (Chiong et al., 2021).

11.10 Current Challenges for Electronic Waste Elimination

1. *Lack of Proper Recycling Infrastructure*: The electronic waste recycling infrastructure varies widely across countries and regions. Many areas lack adequate facilities and technologies to handle e-waste properly. As a result, a significant portion of electronic waste ends up in landfills, incinerators, or informal recycling operations, leading to environmental and health hazards (Rajesh et al., 2022).
2. *Informal E-Waste Recycling*: Informal e-waste recycling, often conducted in developing countries, involves rudimentary and hazardous practices. People involved in these operations are exposed to harmful chemicals and materials while attempting to recover valuable components from electronic devices. This informal sector often lacks proper safety measures and environmentally friendly methods, exacerbating the pollution problem (Andeobu et al., 2023).
3. *Rapid Technological Advancements*: The fast pace of technological advancements leads to shorter product lifespans and a higher turnover of electronic devices. As consumers frequently upgrade their gadgets, the amount of electronic

waste generated increases, challenging waste management systems to keep up with the growing quantities (Saha et al., 2021).
4. *Global Trade and Dumping*: Some developed countries export their electronic waste to developing countries, taking advantage of lax regulations. This practice, known as e-waste dumping, not only burdens these countries with handling vast amounts of waste but also exposes their inhabitants to the hazards of improper recycling (Abalansa et al., 2021).
5. *Recovery of Rare and Valuable Metals*: Electronic devices contain valuable and scarce metals like gold, silver, palladium, and rare earth elements. Properly recovering and recycling these materials are challenging and require specialized processes and equipment (Ding et al., 2019).
6. *Consumer Awareness and Behavior*: Many consumers are unaware of the environmental impact of improper e-waste disposal. Lacking proper awareness and knowledge, they may dispose of electronic devices in regular trash bins, leading to more e-waste in landfills (Sofian et al., 2023).
7. *Design for Disassembly*: Many electronic products are not designed with easy disassembly in mind, making it challenging to extract components for recycling and reusing. Designing products with easier disassembly and modularity can enhance the recycling potential of electronic devices (Rios et al., 2015).

11.11 Future Perspectives of Electronic Plastics

Future perspectives of electronic plastics are poised to be shaped by sustainability, performance, and innovation. As the industry strives for greener solutions, the focus will be on eco-friendly materials, reducing environmental impact, and adopting a circular economy approach. Advancements in electronic plastics will also cater to miniaturization and flexible electronics, enabling cutting-edge devices with improved functionality. The integration of smart materials will further revolutionize electronics, paving the way for interactive and responsive devices. Through continuous research and technological breakthroughs, electronic plastics are set to usher in a more sustainable and advanced era for the electronic industry (Taqi Ghulam & Abushammala, 2023):

1. *Sustainable Materials*: The electronic industry has been increasingly focused on sustainability and environmental impact. In the future, there may be a significant shift toward using more eco-friendly plastics derived from renewable sources, biodegradable materials, or recycled plastics to reduce the industry's carbon footprint.
2. *Enhanced Performance*: Advancements in materials science and engineering may lead to the development of new electronic plastics with enhanced properties. These could include improved electrical conductivity, thermal stability, and mechanical strength, allowing for more efficient and durable electronic devices.

3. *Miniaturization and Flexible Electronics*: As the demand for smaller and flexible electronic devices continues to grow, there may be a need for plastics that can withstand bending and folding without compromising performance. Flexible plastics could enable the development of innovative wearable devices, foldable screens, and other portable electronics.
4. *3D Printing and Customization*: The rise of 3D printing in electronic manufacturing could create opportunities for customized plastics. Companies might be able to create unique casings and components tailored to specific device designs and functionalities, reducing waste and assembly costs.
5. *Integration of Smart Materials*: The integration of smart or functional materials into electronic plastics may open up new possibilities for interactive and responsive devices. For instance, plastics with embedded sensors, actuators, or self-healing capabilities could become more prevalent.
6. *Improved Recycling and Circular Economy*: As electronic waste becomes a growing concern, advancements in recycling technologies could lead to more efficient and effective ways to recycle electronic plastics. This would contribute to the development of a circular economy, reducing the reliance on virgin plastic materials.
7. *Advanced Packaging*: Electronic plastics play a crucial role in packaging and protecting electronic components. Future developments may focus on creating packaging materials that offer better thermal management, electromagnetic shielding, and protection against environmental factors like moisture and dust.
8. *Energy Harvesting*: In the quest for sustainable and energy-efficient electronics, there might be a push to develop plastics with energy-harvesting capabilities. These materials could convert ambient energy, such as light or vibration, into electricity to power low-energy devices or recharge batteries (Veettil et al., 2023).

11.12 Conclusion

In conclusion, electronic plastic (e-plastic) presents a significant challenge in the management of electronic waste (e-waste) due to its complex polymer composition and hazardous content. The improper disposal of e-plastic poses serious environmental and health risks, necessitating the exploration of sustainable waste management strategies. Biodegradation has emerged as a promising approach to address the e-plastic waste problem. Various natural and synthetic methods have been investigated to break down e-plastic materials. Microorganisms, enzymes, and bioactive compounds play crucial roles in the biodegradation process, offering hope for potential solutions. Despite the promising advancements, several challenges remain in e-plastic biodegradation. The slow degradation rate, limited understanding of microbial communities involved, and potential toxic by-products during degradation require further research and development. Additionally, integrating eco-friendly design and sustainable materials selection in electronic manufacturing is essential to minimize the environmental impact of e-plastic waste generation. In order to effectively address the issue of e-plastic waste, collaboration among policymakers, industry stakeholders, and researchers is crucial. By fostering research into e-plastic biodegradation and promoting responsible waste management practices, we can strive toward a more sustainable and environmentally conscious future. Embracing innovative solutions and adopting eco-friendly practices will pave the way for a cleaner and healthier planet for generations to come.

References

Abalansa, S., El Mahrad, B., Icely, J., et al. (2021). Electronic waste, an environmental problem exported to developing countries: The GOOD, the BAD and the UGLY. *Sustainability, 13*(9), 5302.

Achilleos, A., & Hapeshi, E. (2015). Biodegradation of polymers: Recent full-scale applications for polyesters and polylactic acid. *Waste Management & Research, 33*(7), 589–600.

Ahmad, Z., Alsulamy, S., Raza, A., et al. (2023). Life cycle assessment (LCA) of polypropylene fibers (PPF) on mechanical, durability, and microstructural efficiency of concrete incorporating electronic waste aggregates. *Case Studies in Construction Materials, 18*, e01979.

Akram, A., Sasidhar, C., & Pasha, K. M. (2015). E-waste management by utilization of E-plastics in concrete mixture as coarse aggregate replacement. *International Journal of Innovative Science Engineering and Technology, 4*(7), 5087–5095.

Alaee, M. P., Arias, A., & Sjödin, A. (2003). Bergman, an overview of commercially used brominated flame retardants (BFRs) and their applications, changes in their use pattern in different countries/regions over time and possible modes of release. *Environment International, 29*, 683–689.

Andeobu, L., Wibowo, S., & Grandhi, S. (2023). Informal e-waste recycling practices and environmental pollution in Africa: What is the way forward? *International Journal of Hygiene and Environmental Health, 252*, 114192.

Bardaji, D. K. R., Furlan, J. P. R., & Stehling, E. G. (2019). Isolation of a polyethylene degrading Paenibacillus sp. from a landfill in Brazil. *Archives of Microbiology, 201*(5), 699–704.

Bhattacharya, M., & Sharma, A. (2004). Polyethylene glycol (PEG) assisted biodegradation of polyvinyl chloride (PVC) in plastic waste mixtures. *Journal of Applied Polymer Science, 94*(3), 1179–1185.

Bhutta, M. K. S., Omar, A., & Yang, X. (2011). Electronic waste: A growing concern in today's environment. *Economics Research International, 2011*, 8.

Chancerel, P., Meskers, C. E., Hagelüken, C., et al. (2009). Assessment of precious metal flows during preprocessing of waste electrical and electronic equipment. *Waste Management, 29*(4), 1192–1202.

Chen, S., Wu, Z., Chu, C., et al. (2022). Biodegradable elastomers and gels for elastic electronics. *Advancement of Science, 9*(13), 2105146.

Chiong, J. A., Tran, H., Lin, Y., et al. (2021). Integrating emerging polymer chemistries for the advancement of recyclable, biodegradable, and biocompatible electronics. *Advancement of Science, 8*(14), 2101233.

Covino, B. S., & Smith, K. M. (Eds.). (2014). *Recycling of polyethylene waste to produce new plastic products*. Wiley.

Ding, Y., Zhang, S., Liu, B., et al. (2019). Recovery of precious metals from electronic waste and spent catalysts: A review. *Resources, Conservation and Recycling, 141*, 284–298.

Dudnik, V., Othman, M. R., Kattan, A., et al. (2019). A review on the current e-waste management practices in Malaysia. *Journal of Material Cycles and Waste Management, 21*(3), 602–613.

Erlandsson, B., & Stridh, G. (2012). Assessment of the environmental fate of polypropylene in life cycle assessment. *International Journal of Life Cycle Assessment, 17*(8), 1018–1027.

Evode, N., Qamar, S. A., Bilal, M., et al. (2021). Plastic waste and its management strategies for environmental sustainability. *Case Studies in Chemical and Environmental Engineering, 4*, 100142.

Eze, F. U., Forton, O. T., Mbohwa, C., et al. (2018). A review of the recycling of PVC and its composites. *Resources, Conservation and Recycling, 129*, 83–96.

Feng, S., Cao, S., Tian, Z., et al. (2019). Maskless patterning of biodegradable conductors by selective laser sintering of microparticle inks and its application in flexible transient electronics. *ACS Applied Materials & Interfaces, 11*(49), 45844–45852.

Gavhane, A., Sutar, D., Soni, S., et al. (2016). Utilisation of e-plastic waste in concrete. *International Journal of Engineering Research & Technology, 5*(2), 594–601.

Géczy, A., Kovács, M., & Hajdu, I. (2012). Conductive layer deposition and peel tests on biodegradable printed circuit boards. In *In 2012 IEEE 18th international symposium for design and technology in electronic packaging (SIITME)* (pp. 139–142). IEEE.

Gull, I., & Balasubramanian, M. (2014). A new paradigm on experimental investigation of concrete for e-plastic waste management. *International Journal of Engineering Trends and Technology (IJETT), 10*(4), 180–186.

Hopewell, J., Dvorak, R., & Kosior, E. (2009). Plastics recycling: Challenges and opportunities. *Philosophical Transactions of the Royal Society of London. Series B, Biological Sciences, 364*(1526), 2115–2126.

Jambeck, J. R., Geyer, R., Wilcox, C., et al. (2015). Plastic waste inputs from land into the ocean. *Science, 347*(6223), 768–771.

Jeswani, H., Krüger, C., Russ, M., et al. (2021). Life cycle environmental impacts of chemical recycling via pyrolysis of mixed plastic waste in comparison with mechanical recycling and energy recovery. *Science of The Total Environment, 769*, 144483.

Kalaitzidou, K., Fukushima, H., & Drzal, L. T. (2007). Mechanical properties and morphological characterization of exfoliated graphite–polypropylene nanocomposites. *Composites Part A: Applied Science and Manufacturing, 38*(7), 1675–1682.

Kang, S. K., Hwang, S. W., Yu, S., et al. (2015). Biodegradable thin metal foils and spin-on glass materials for transient electronics. *Advanced Functional Materials, 25*(12), 1789–1797.

Kumar, A., Negi, Y. S., & Choudhary, V. (2017). A review on biodegradation of polymeric materials. *International Biodeterioration and Biodegradation, 124*, 52–69.

Liu, K., Tan, Q., Yu, J., et al. (2023). A global perspective on e-waste recycling. *Circular Economy, 2*, 100028.

Magnin, A., Hoornaert, L., Pollet, E., et al. (2019). Isolation and characterization of different promising fungi for biological waste management of polyurethanes. *Microbial Biotechnology, 12*(3), 544–555.

Moshood, T. D., Nawanir, G., Mahmud, F., et al. (2022). Why do consumers purchase biodegradable plastic? The impact of hedonics and environmental motivations on switching intention from synthetic to biodegradable plastic among the young consumers. *Journal of Retailing and Consumer Services, 64*, 102807.

Osman, M., Satti, S. M., Luqman, A., et al. (2018). Degradation of polyester polyurethane by aspergillus sp. strain S45 isolated from soil. *Journal of Polymers and the Environment, 26*(1), 301–310.

Othman, N., Sidek, L. M., Basri, N. E. A., et al. (2009). Electronic plastic waste management in Malaysia: The potential of waste to energy conversion. *Proceedings of ICEE 2009 3rd international conference on energy and environment*, 7–8 December 2009, Malacca, Malaysia.

Petersen, J., Frydendal, J., & Astrup, T. F. (2018). Waste electrical and electronic equipment: Waste composition, collection and governance in the EU. *Waste Management, 79*, 105–118.

Prasad, V. A., & Singh, R. P. J. (1997). Recent developments in degradation and stabilisation of high impact polystyrene. *Macromolecular Science: Part C, Reviews in Macromolecular Chemistry & Physics, 4*, 581–598.

Qu, C., Wang, J., Zhou, Z., et al. (2018). A review of plastic waste biodegradation. *Critical Reviews in Biotechnology, 38*(3), 383–392.

Rahimi, A., & García, J. M. (2017). Chemical recycling of waste plastics for new materials production. *Nature Reviews Chemistry, 1*(7), 0046.

Rajesh, R., Kanakadhurga, D., & Prabaharan, N. (2022). Electronic waste: A critical assessment on the unimaginable growing pollutant, legislations and environmental impacts. *Environmental Challenges, 7*, 100507.

Rios, F. C., Chong, W. K., & Grau, D. (2015). Design for disassembly and deconstruction-challenges and opportunities. *Procedia Engineering, 118*, 1296–1304.

Saha, L., Kumar, V., Tiwari, J., et al. (2021). Electronic waste and their leachates impact on human health and environment: Global ecological threat and management. *Environmental Technology and Innovation, 24*, 102049.

Sekhar, V. C., Nampoothiri, K. M., Mohan, A. J., et al. (2016). Microbial degradation of high impact polystyrene (HIPS), an e-plastic with decabromodiphenyl oxide and antimony trioxide. *Journal of Hazardous Materials, 318*, 347–354.

Senthil Kumar, K., & Baskar, K. (2015). Recycling of E-plastic waste as a construction material in developing countries. *Journal of Material Cycles and Waste Management, 17*, 718–724.

Shah, A. A., Hasan, F., Hameed, A., et al. (2008). Biological degradation of plastics: A comprehensive review. *Biotechnology Advances, 26*(3), 246–265.

Shanker, R., Khan, D., Hossain, R., et al. (2023). Plastic waste recycling: Existing Indian scenario and future opportunities. *International journal of Environmental Science and Technology, 20*(5), 5895–5912.

Sheth, M. U., Kwartler, S. K., Schmaltz, E. R., et al. (2019). Bioengineering a future free of marine plastic waste. *Frontiers in Marine Science, 6*, 624.

Singh, R., Singh, D. P., & Singh, J. (2021). Biodegradation of plastic: Current status and future prospects. *Journal of Environmental Sustainability, 4*(4), 159–169.

Smith, R. L., Takkellapati, S., & Riegerix, R. C. (2022). Recycling of plastics in the United States: Plastic material flows and polyethylene terephthalate (PET) recycling processes. *ACS Sustainable Chemistry & Engineering, 10*(6), 2084–2096.

Sofian, D. D., Hanafiah, M. M., Woon, K. S., et al. (2023). Characteristics of consumer towards development of sustainable e-waste management in Malaysia. In *IOP conference series: Earth and environmental science* (Vol. 1167, No. 1, p. 012043). IOP Publishing.

Song, Q., Li, J., Liu, H., et al. (2020). A review on the recycling of PVC waste materials: Status and prospect. *Waste Management, 106*, 125–138.

Subramanian, P. M. (2000). Plastic recycling and waste management in the US. *Resources, Conservation and Recycling, 28*, 253263. on net.

Suh, D., Suh, J., Suh, Y. W., et al. (2015). Recycling of polycarbonate waste. *Journal of Material Cycles and Waste Management, 17*(3), 527–534.

Taqi Ghulam, S., & Abushammala, H. (2023). Challenges and opportunities in the management of electronic waste and its impact on human health and environment. *Sustainability, 15*(3), 1837.

Teixeira, F. D. S. M., Peres, A. C. D. C., & Pacheco, E. B. A. V. (2023). Mechanical recycling of acrylonitrile-butadiene-styrene copolymer and high impact polystyrene from waste electrical and electronic equipment to comply with the circular economy. *Frontiers in Sustainability, 4*, 1203457.

Urbanek, A. K., & Rymowicz, W. (2018). Degradation of plastics and plastic-degrading bacteria in cold marine habitats. *Applied Microbiology and Biotechnology, 102*(21), 9159–9169.

Van Yken, J., Boxall, N. J., Cheng, K. Y., et al. (2021). E-waste recycling and resource recovery: A review on technologies, barriers and enablers with a focus on Oceania. *Metals, 11*(8), 1313.

Vasile, C., Călugăru, C. M., Sabliovschi, M., et al. (1980). Romanian Patents 74577 and 78462.

Vasile, C., Pakdel, H., Brebu, M., et al. (2001). Thermal and catalytic decomposition of mixed plastics. *Journal of Analytical and Applied Pyrolysis, 57*, 287–303.

Vasile, C., Brebu, M., Darie, R., et al. (2002). Pyrolysis of complex mixtures containing biomass and its components. In *Proceedings of ISFR 2002, the 2nd international symposium on feedstock recycling of plastics and other innovative recycling techniques*, 8–11 September, Ostend, Belgium, CD-A79.

Vasile, C., Brebu, M. A., Karayildirim, T., et al. (2006). Feedstock recycling from plastic and thermoset fractions of used computers (I): Pyrolysis. *Journal of Material Cycles and Waste Management, 8*, 99–108.

Veettil, B. K., Van, D. D., Quang, N. X., & Hoai, P. N. (2023). Remote sensing of plastic-covered greenhouses and plastic-mulched farmlands: Current trends and future perspectives. *Land Degradation & Development, 34*(3), 591–609.

Velis, C. A., & Bartl, P. (2018). Current recycling approaches for polyethylene films and bags. *Resources, Conservation and Recycling, 128*, 36–51.

Zhu, P., Pan, X., Li, X., et al. (2021). Biodegradation of plastics from waste electrical and electronic equipment by greater wax moth larvae (*Galleria mellonella*). *Journal of Cleaner Production, 310*, 127346.

Zhu, P., Shen, Y., Li, X., et al. (2022). Feeding preference of insect larvae to waste electrical and electronic equipment plastics. *Science of The Total Environment, 807*, 151037.

Chapter 12
Microbial Consumption of Plastics: A Prospective Solution for Plastic Mitigation

Md. Shahbaz Anwar

Abstract In our current lifestyle, plastics place a very important role; therefore, its gathering in the environment is major alarms that harm the human health. Petroleum-derived plastics such as PE, PET, PU, PS, PP, and PVC are awfully recalcitrant to natural biodegradation pathways. Large number microorganisms have the capacity to degrade petro-polymers under natural as well as lab conditions which have been investigated by different scientists. The enzymes produced by different microbes have been genetically engineered. The polymer biodegradation rate depends on various factors like molecular weights, chemical structures, and degrees of crystallinity. These are bulky molecules having crystals as well as irregular groups that provide hardness and flexibility to the polymers. Polyethylenes (95%) are rigid, highly crystalline polymers with a limited capacity to oppose impact. Crystallinity of a polymer (30–50%) in plastic is a main cause of slow microbial degradation. The microbial degradation as well as enzymatic degradation occurs via enzyme adsorption on the surface of polymer, followed by hydro-bond peroxidation and hydrolysis. The petro-plastic disintegration or depolymerization into polymer monomers for different purposes like recycling or converting waste plastics into higher-value by-products is considered as a promising strategy. Finally, plastic degradation can be measured by applying different techniques like SEM, TEM, FTIR, TGA, UV spectroscopy, DSC, HPLC, etc. The purpose of this section is to sketch the different categories of plastic, their impact on environment, the impact of microbes, or their enzymes toward the degradability of synthetic plastics to clean up the environment.

Keywords Synthetic polymers · PE · PVC · PS · Cutinase · Elastases · FTIR · SEM

M. S. Anwar (✉)
Department of Microbiology, Dum Dum Motijheel College, WBSU, Kolkata, West Bengal, India

Department of Microbiology, Dum Dum motijheel College, Kolkata, India

R. Soni et al. (eds.), *Advanced Strategies for Biodegradation of Plastic Polymers*, https://doi.org/10.1007/978-3-031-55661-6_12

12.1 Introduction

The word plastic is derived from the Greek word *plastikos* meaning capable of being shaped or molded (Kesti & Shivasharana, 2018). Plastic is made up of different polymers which have linear as well as aliphatic carbon atoms in their structural chains (Pavani & Rajeswari, 2014). This plastic is being made from inorganic and organic raw materials, such as carbon, silicon, hydrogen, nitrogen, oxygen, and chloride. It was considered that this polymer is resistant to microbial action, due to short time presence in nature (Mueller, 2006). A lot of petroleum-based variety of synthetic plastic are produced globally (approximately 140 million tons per year), and they enter in the ecosystem as waste products (Shimao, 2001). Although various types of available polymers are reported, eight major types of plastics that build up 95% of the entire main plastics are continually made, such as polyethylene and polypropylene, encompassing 45% production globally (Geyer, 2020). Main application of plastic is for packaging (36%), followed by building and construction (16%) (Lear et al., 2021). Mainly there are three main fates of plastics: incineration, recycling, and discarding. Globally, it is approximated that only 10% are recycled plastic while incinerated plastic are 14%; the 76% remaining goes to the natural environment (Geyer, 2020). The waste material of plastic creates problem for both terrestrial and marine ecosystems, which is due to huge production of polyethylene terephthalate (Carr et al., 2020). Approximately 800 animal species in marine have overstated by the plastic pollution and through an increasing number are from the detriti-vorous like sea snails, (Li et al., 2020) to top predators of marine environment (Nelms et al., 2019; Santos et al., 2020), they have internalized plastics in their body. Rarely, harmful effects of plastics also reported in the case of *Caenorhabditis elegans* and *Daphnia magna* nematodes in which their toxicological impacts and the digestive system blockage lead to reduced feeding rates (Lear et al., 2021). In environments like terrestrial background, the presence of plastics applies physical impacts directly on the major microbial communities. For example, plastic related to agricultural farming like mulch films useful to improve productivity of crop that covers about 20 million hectares of farmland globally (Steinmetz et al., 2016) is considered as a major source of plastic contamination in terrestrial environment (Huang et al., 2020). Plastics those are buoyant in nature, such as polypropylene, polyethylene, polystyrene, are conveyed over wide distances by the velocity of oceanic currents and winds (Lacerda et al., 2019), whereas PET and PLA which are non-buoyant plastics act as a vector to transport microbes from surface to deeper water (Audrézet et al., 2020). Different organisms like toxic microalgae (Casabianca et al., 2019) and potential human and animal pathogens have been identified to be directly linked to freshwater and marine plastics (Lear et al., 2021) along with different taxa which are antibiotic-resistant (Moore et al., 2020). Microbial or enzymatic degradation of plastics is a means of gifted scheme to depolymerize waste petro-plastics into by-product of monomer for recycling or mineralize them into water, carbon dioxide, and new biomass, with the production of higher-value bioproducts (Montazer et al., 2019). A lot of fungi and bacteria have been reported to metabolize petroleum-based plastics mostly in situ condition (Shahnawaz et al., 2019) as well as in vitro condition (Mohanan et al., 2020). Biodegradation is a composite method which is based on several factors, such as surface characteristics, availability of a substrate, morphology, and molecular weight of the polymers (Harrison et al., 2018). Various groups of microorganisms like *Aureobasidium pullulans*, *Pseudomonas aeruginosa*, *Kluyveromyces* sp., and *Rhodotorula aurantiaca* have been accounted to cause in situ colonization and substratum deterioration of plasticized PVC (Anwar et al., 2013). *Pseudomonas aeruginosa*, *Aureobasidium pullulans*, *Rhodotorula aurantiaca*, and *Kluyveromyces* sp. have been accounted to cause in situ colonization and deterioration of PVC (Kawai, 2010). *Curvularia* sp., *Phanerochaete chrysosporium*, and *Trogia buccinalis* have been documented to mortify the PVC blend (Sandra et al., 2010), However, disintegration of these plastics is possible in the natural environment, through long periods of time like 50 to 100 years (Table 12.1) (Mohanan et al., 2020).

The triangle, in the recycle code of plastic, indicates that the plastic material is recyclable, and each number inside the triangle indicates a specific type of plastic as follows (Pavani & Rajeswari, 2014):

1. Polyethylene terephthalate (PET): Safe and recyclable. It is used for making plastic bottles and containers for water, juice, soft drinks, and peanut butter.
2. High-density polyethylene (HDPE): Safe and recyclable. It is used for the manufacturing of packages for shampoo, detergents, and milk, as well as plastic toys. It is safer than any other plastic material especially the transparent type.

Table 12.1 Selected properties of major synthetic thermoplastic polymer (Ojeda, 2013; Kestil & Shivasharana, 2018) with their uses and annual production (Cornwall, 2021)

Polymer	Plastic recycling symbol	Density (g/l)	Crystallinity (%)	Life span (years)	Produced annually (Mt)	Uses
PETE	1	1.35	0–50	450	33 Mt	Soft drink bottles
HDPE	2	0.94–0.97	70	> 600	116 Mt	Trays and containers
PVC	3	1.35–1.45	00	50–150	38 Mt	Cable insulation
LDPE	4	0.19–0.93	50	10–600	116 Mt	Food packaging film
PP	5	0.90–0.91	50	10–600	68 Mt	Food packaging
PS	6	1.03–1.09	00	50–80	25 Mt	Disposable cutlery

PET polyethylene terephthalate, *PS* polystyrene, *LDPE* low-density polyethylene, *PP* polypropylene, *PVC* polyvinyl chloride, *HDPE* high-density polyethylene, *Mt.*, millions of tons

3. Polyvinyl chloride (PVC): Harmful and toxic if used for a long time. It is used in the manufacturing of plumping pipes, bathroom curtains, toys, and transparent plastic wraps for meat and cheese. It is widely utilized for its cheap price, so it is considered the most dangerous type of plastic.
4. Low-density polyethylene (LDPE): Proportionally safe and recyclable. It is used for the manufacturing of CDs, bottles, and grocery bags.
5. Polypropylene (PP): One of the best and safest types of plastic. It is suitable for cold and hot liquids and other products, unharmful, and used in manufacturing of food containers, plates, medicine bottles, and all food-related products. It suggested that majority of food containers should be made of substance, especially containers related to children's foods that are used for packaging school meal.
6. Polystyrene (PS): Dangerous and unsafe. It is used for burger, hotdog, and teacup packaging. It resembles the cork and generally used in the international fast food chains, although the use of PS for 20 years has prohibited in the USA.
7. Others: Uncategorized material.

12.2 Classification of Plastics

Different types of plastics are reported, depending on the starting monomer selected, the length of chain of polymer, and the types of modifying compounds added. Each plastic has been developed for a different utility group:

Temperature Sensitivity: On the basis of behavior of plastics toward temperature, they can be categorized into two main groups, viz., thermoplastics and thermosets.

Thermoplastics: Thermoplastics are those that can be softened again and again and remade by the action of heat and pressure. On the action of pressure and heat, the molecular chain of thermoplastics undergoes change, and the polymers slide past each other, which results in the property of plasticity. Some typical thermoplastics are (www.encyclopedia.com) as follows:

- Acrylic (Perspex).
- Acrylonitrile (Orlon).
- Polyethylene (polythene).
- Polypropylene(PP).
- Polyvinyl acetate (PVAc).
- Polyvinyl chloride (PVC).
- Polystyrene and ABS.
- PTFE (Teflon).

Thermosets: The thermosetting plastics are those that cannot be softened again after exposure to heat and pressure. On the action of pressure and heat, the molecular chain of thermosetting plastics become cross-linked, due to which it forbids slippage when pressure and heat are reapplied. Some typical thermosets are (www.encyclopedia.com) as follows;

- Bakelite.
- Epoxy.
- Melamine.
- Polyesters.
- Polyurethane.

Starting Monomer: Based on the monomer units selected for polymerization, the following types of plastics are available commercially:

- Polyethylene terephthalate (PET).
- Polyvinyl chloride (PVC).
- Polypropylene (PP).
- Polystyrene (PS).
- High-density polyethylene (HDPE).
- Low-density polyethylene (LDPE).

Polyethylene: Polyethylene is formed through ethylene polymerization. The ethylene molecule, C_2H_4 is $CH_2 = CH_2$, two CH_2 groups connected by a double bond (Fig. 12.1).

Polyethylene is divided into various categories mostly based on its branching and density. Mechanical properties of PE significantly depend on different variables such as the type of branching and extent, the crystal structure, and the molecular weight:

- UHMWPE (ultrahigh molecular weight PE).
- HMWPE (high molecular weight PE).
- HDPE (high-density PE).
- HDXLPE (high-density cross-linked PE).
- PEX (cross-linked PE).
- MDPE (medium-density PE).
- LDPE (low-density PE).
- LLDPE (linear low-density PE).
- VLDPE (very low-density PE).

LDPE: It is thermoplastic and unreactive at room temperature, but some strong oxidizing agents, like solvents, may cause swelling of LDPE. It can withstand temperature of 80 °C continuously and 95 °C for a short time. Made in translucent or

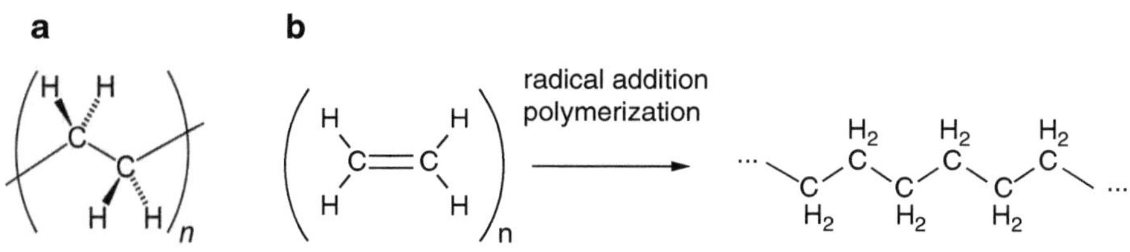

Fig. 12.1 (**a**) Stereochemistry of ethene: the repeating unit of polyethylene. (**b**) polymerization of ethene into polyethylene

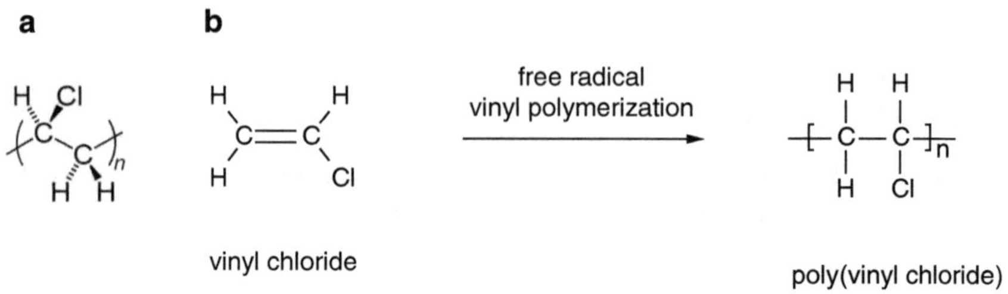

Fig. 12.2 (**a**) Stereochemistry of vinyl chloride: the repeating unit of PVC. (**b**) Polymerization of vinyl chloride into polyvinyl chloride

opaque variations, it is quite tough and flexible to the degree of being almost unbreakable (www.wikipedia.com). It is highly branched and has weak intermolecular forces of attraction between its molecules. It has low tensile strength and low density (Fig. 12.1).

Polyvinyl Chloride: It is a vinyl polymer. It is alike to polyethylene; however, on every second carbon in the backbone chain, one of the hydrogen atoms is replaced with a chlorine atom. It is produced by the free radical polymerization of vinyl chloride (Fig. 12.2).

The product which is obtained after polymerization process is unmodified PVC. By addition of some additives such as heat stabilizers, lubricants, UV stabilizers, plasticizer processing aids, impact modifiers, thermal modifiers, fillers, flame retardants, biocides, blowing agents, and smoke suppressors, and optionally pigments into a compound of PVC, we can made the PVC into a finished product (ChemSystems, November 2008, Polyvinyl Chloride (PVC) 07/08-7 Report).

Polystyrene (PS): Most plentiful plastics are shaped worldwide and largely used for the materials which are used in disposable dishware as well as packaging for food (Plastics Europe, 2017). Equally like other artificial plastics, PS is promptly used for its good mechanical properties and limited cost. PS is also used in food containers, packaging foam, cassette boxes, construction materials (insulation), compact disks, plates, disposable cups, and cutleries. About 21 million tons of polystyrene are manufactured worldwide as reported in 2013 by Yang et al. (2015a, 2015b). They are exceptionally stable polymers with strong hydrophobic character and high molecular weight that build these polymers recalcitrant to biodegradation (Ho et al., 2018).

Polypropylene (PP): Polypropylene, expressed as CnH_2n, is the most extensively used linear polymer among the synthetic polymers. A methyl group substituted in place of hydrogen is present in PE on every other carbon in the case of PP, which gives three stereoisomeric forms: isotactic, syndiotactic, and atactic (Baker & Mead, 2002). Properties of PP are almost similar to polyethylene while more resistant to heat and slightly harder. PP is considered as the second most commodity plastic that is produced widely (after PE) and mostly used for labeling and packaging. Various ranges of applications are reported like food packaging, lab equipment, textiles, and automotive components (Mohanan et al., 2020).

12.2.1 Polyethylene Terephthalate (PET)

Major synthetic petro-plastics are shaped in very large ranges globally. Worldwide production of PET accounted about 56 million tons as reported in 2013 by Neufeld et al. (2016). It consists of polyesters of aromatic compounds with a glass transition temperature (Tg), approximately 75–80 °C in air. However, Tg drop down to 60–65°C in solution which is aqueous in nature (Kawai et al., 2014). Temperatures above Tg (glass transition temperature), the amorphous portion in the PET turn into flexible and easy to microbial degradation or attack of microbial enzyme (Odusanya et al., 2013). Applications of PET are very wide, mostly in manufacturing industries, such as textile fibers, bottles, containers, and films (Mohanan et al., 2020). Diverse groups of microorganisms have the capacity to metabolize these polymers, which include the bacterium *Ideonella sakaiensis* 201-F6, for PET polymer depolymerization and utilize the subunits of terephthalate as an energy source and carbon for metabolism and growth (Yoshida et al., 2016).

12.3 Impact of Plastics on Environment

Plastic products are durable, have a minimal cost of production, and have tremendous applications in several sectors, making it highly demanding (Sharuddin et al., 2016). About 30–40% of plastics are manufactured over the world for packaging purposes, which ultimately contributes to the rise of plastic wastage of about 12% per year (Jaiswal et al., 2020). Due to different properties of plastics made them very strong like a functional chemical group that make the plastics' surface hydrophilicity and hydrophobicity, melting temperature, molecular weight, and chemical structure, which significantly reduced the deterioration rates of the plastic (Jaiswal et al., 2020). Plastic is also accountable for the loss of about 100,000 marine mammals and turtles per year, one million sea birds, largely due to suffocation as well as strangulation (Carr et al., 2020; Isangedighi et al., 2021).

Impacts that affect our environmental are broadly grouped into direct impact (ingestion, entanglement, and ecosystem alteration) and indirect impact (ecosystem alteration and invasive sp.).

Ingestion Plastic ingestion by seabirds, sea turtles, fish, and marine mammals often occurs through marine debris that they mistake for food. Marine debris ingestion can harm marine life seriously. For example, sea turtles and whales often ingest mistaken plastic bags for squid, and plastic pellets are ingested by birds often mistaken for fish eggs. Furthermore, in a study of about 38 green turtles, 61% is found to have ingested marine debris including cloth, plastic bags, and rope or string (Pavani & Rajeswari, 2014). Sometimes, accidentally, debris present in the marine can be eaten by animals in place of natural food, which sense them to filled stomach and leads to nutrient deficiency, when the marine debris collects in the animal's stomach causing the animal to sense filled. Similar starvation occurs when marine debris is ingested by animal's system preventing vital nutrients from being absorbed. Some internal injuries as well as infections may also result from the mistake of ingestion. Some marine debris, especially some plastics, contains toxic substances that can cause death or failure in the reproductive system of fish, shellfish, or any marine life (Moore et al., 2002).

Entanglement The life of marine ecosystem turns out to be entangled in the debris of marine environment creating dangerous injury or death, which leads to starvation, drowning, suffocation, increased vulnerability to predators, or other injuries. Debris related to marine environment can constrict and entangled the movement of animal, which leads to exhaustion or development of a serious infection from deep wounds caused by hardening of materials. For example, volunteers who participated in the International Coastal Cleanup event in 2008 revealed about 443 birds and animals were trapped by marine debris (2008 ICC Report, Ocean Conservancy).

Ecosystem Alteration Marine debris impacts are not only limited to mobile animals. While plants and other organisms which are immobile in nature are also sensitive, they are harmed by marine debris. Coral reefs can be damaged by derelict fishing gear that breaks or suffocates coral. Plants can be smothered by plastic bags and fishing nets. The ecosystems regarding ocean floor can be damaged and distorted by the movement of an abandoned yacht or other marine debris.

Invasive Species Floating marine debris can bring invasive species from different locations that consumed the debris in the marine environment as a "raft" to move from one body of water to another. British Antarctic Survey in 2002, estimated that debris produce through human being found in the oceans has about doubled the number of different species found in the subtropics.

Economic Impacts Marine debris can harm three important components of our economy: tourism, fishing, and navigation.

(i) **Tourism**: Marine debris is unsightly and unwelcoming to beachgoers, which can result in lost revenue from tourism. In the case of severity, marine debris can create beach closures. The costs to remove and dispose of the debris from the marine environment can be high, and the loss of tourism dollars can be even higher. In an attempt to stop the draining of trash to the ocean, the Los Angeles County's Department of Public Works and the Flood Control District spend $18 million each year on street sweeping, catch basin cleanouts, cleanup programs, and litter prevention and education efforts (L.A. County Boards of Supervisors Staff Report, 2007).

(ii) **Fishing**: Capturing fisheries experience significant impacts on economic from marine debris. Commercial fisheries are impacted when commercial fish and shellfish become bycatch in lost fishing nets or other fishing gears. This type of bycatch can result in both immediate losses in the standing stock of available seafood and decreases in the long-term sustainability of the stock due to negative impacts on its reproductive ability. For example, Marine Fisheries Commission in the Gulf States has expected blue crab ghost fishery causes a loss of crabs up to four to ten million in a year in Louisiana alone (Virginia Institute of Marine Science, 2006). Fisheries also can be financially affected when fishing gear and vessels are entangled or damaged by marine debris. The high cost of replacing fishing gear and vessels, as well as loss of days at sea for fishing, can cause small fisheries to go out of business.

(iii) **Navigation**: Floating marine debris is a navigational hazard that entangles propellers and clogs cooling water intake valves. Repairing boats damaged by marine debris is both time-consuming and expensive.

Harm to Marine Wildlife Countless marine animals and sea birds turn out to be entangled in the debris of marine environment or ingest it. This can cause them serious harm and often results in their death. Due to ingestion of these plastic debris as well as packaging band entanglement, drift nets, synthetic ropes, and lines, it is considered as the threat in marine life—mechanical primarily. Application of plastics continues to increase, which leads to promotion of pollution in the marine environment. It is also found that the gut content of thousands of birds in two separate studies revealed that the plastics ingested by seabirds had drastically increased during 10–15-year interval between studies. A study in the North Pacific found plastic particles in the stomachs of 8 of the 11 seabird species. These species point out that marine debris is affecting a major number of species. It affects worldwide about 267 species, including 44% of all seabird species, 86% of all sea turtle species, and 43% of all marine mammal species. A scientist reported that plastics made up 80–85% of the seabed debris in Tokyo Bay, an impressive figure considering that most plastic debris is buoyant (Pavani & Rajeswari, 2014). Microplastics are pieces of plastic that are about 5 millimeters long, which originate from a wide variety of sources including pre-production pellets spilled during transport, road runoff containing fragments of tires from vehicle and paint fragmentation of plastic debris, and outflow of water containing microbeads from cosmetics and fibers (Nelms et al., 2019). These microplastics are probably ingested through accidental consumption or through trophic transfer where predators eat preys that have microplastics in their body (Nelms et al., 2019). In the marine life environment, microplastics are also ingested by organisms (Markic et al., 2019). A wide research, the digestive tracts of different species of cetaceans and the pinnipeds of two different species have been found to have a presence of microplastics (Nelms et al., 2019). Various scientists were looking for ways to overcome the load of microplastics that is consumed by different individuals (Nelms et al., 2019). Assessment of entire digestive system of each individual which ingested microplastic, were reported that microplastics in all mammals examined is very limited, which recommended that the interest of presence of microplastic particles are transitory, and have capacity to passed by the animal through excretion (Nelms et al., 2019). This does not indicate that all particles of microplastic consumed are transitory; conversely, some researchers reported that the mammals' stomachs contained larger amount of microplastics compared to the intestines which depicted that few of them are retained in their systems (Nelms et al., 2019). So the comparative analysis of researcher establishes a potential relationship between the death group of the mammals and the presence of microplastic (Nelms et al., 2019). Consequently, the above information is very important in terms of ocean pollution as well as in to different animals like dolphins, whales, and seals that measured important health indicators of marine ecosystem because majority of them expands their natural life very long, high trophic level feeding status, and their susceptible nature of bioaccumulation and biomagnification by contaminants related to chemicals (Nelms et al., 2019). Microplastics not only are limited to marine and terrestrial ecosystems and organisms but also participate in human food chains. This involvement can lead to unpleasant health effects including congenital diabetes, immune disorders, and cancer in organism (Llorca & Farré, 2021). Carcinogen materials from plastic waste enter into the food chain through ingestion by different plankton and humans and either directly or indirectly from the soil, water, and their food where these plastics are "bioaccumulated" in the food web, finally escalating the volume inside the animals and humans (Jaiswal et al., 2020). In addition, different animals and humans are also exposed to microplastics and nanoplastics (MNPLs) from side-to-side ingestion accidentally and inhalation as well as through skin contact (Llorca & Farré, 2021). Nanoplastics are ranges from 1 to 1000 nanometers in size, which is 0.0001 millimeters or smaller—these have captured attentions because they are smaller enough that may pass through biological membranes, harming the functioning of the cells. MNLP comprises various additives such as plasticizers (that enhance softness as well as flexibility of the plastic), flame retardants, fillers, UV stabilizer metal colorants, and coating finishers (Llorca & Farré, 2021). It is basically the different additives that appreciably harm the environment and body when they are broken down and released. Some additives like polybrominated diphenyl ethers (PBDEs), bisphenol A, and some phthalates have been conscious in the few last years, that are suggested by researchers and scientist to be linked to carcinogenicity, obesity, neurotoxicity, and disruption of the endocrine system (Llorca & Farré, 2021).

12.4 Impact of Microbes on Plastics

Degradation of plastic requires keen attention of several different lenses like microbiology, economics, ecology, macrobiology, organic chemistry, and marine biology. All of these disciplines investigate how these plastics directly affect the environment. For example, biologists are fascinated toward how plastic or biodegradable products impact on the environment as well as organisms, while economics reveals how communities are monetarily impacted by plastics and recycling (Eriksen et al.,

2018). Plastic recycling is expensive for businesses and communities from an economic point of view.

Majority of the research believe that biodegradable products are better as they degrade "naturally," but it is not as easy as we judge. Majority of the "biodegradable" products still require some alteration and processing in a precise manner that unusually heats them to a towering temperature to degrade them, so it indirectly enhances the temperature in the environment (Thompson, 2018). Nowadays, various research are initiated, which uses heterotrophic composite of microbes and their enzymes as a valuable mechanism for the degradation of plastic. Initially, synthetic polymers like PET are considered as resistant to microbial attacks, but some researchers have evaluated that certain microbes can secrete different moieties that modify and break down the plastic (Carr et al., 2020). A well-studied plastic-eating microbe is the eubacteria *Ideonella sakaiensis*, which is reported to have utilized PET as a source of carbon as well as energy (Carr et al., 2020). This indicates that *I. sakaiensis* has the ability to degrade by catalyzing the oxidation-reduction reaction to break down the bonds in the plastic polymers. PET degradation depends on different physical factors of the plastic, like chemical structure, molecular weight, functional groups present, surface area, melting temperature, crystallinity, and hydrophilicity/hydrophobicity (Jaiswal et al., 2020). Polyesters have the ability to degrade slowly due to the presence of side chains compared to those that don't have side chains. Among various factors for degradation, the molecular weight is an important one, because it makes the plastic very slower for their degradation (Jaiswal et al., 2020). Different tools like genetic and protein engineering are employed to enhance the ability of microorganisms to degrade plastic. These tools help to modifying the amino acid sequence of a protein to increase their enzymatic activity and tolerate reaction conditions (Carr et al., 2020). This is a scientific advancement for the betterment of our environment; it may be fruitless if the degraded plastic and the organisms acting for degradation of plastics release toxic chemicals into the environment (Carr et al., 2020). The suggestion that free-living environmental microbes participate for plastic degradation was published approximately 30 years after the first commercial plastic production, in 1974, when Fields et al. (1974) showed that *Aureobasidium pullulans* a fungus was proficient for degradation of PCL. In a study by Gambarini et al. (2020), information over 400 publications recount degradation of different plastic types about 72 by 436 fungal species as well as bacterial species. Major presumptive microbes identified for plastic-degrading fit into five different bacterial and three fungal phyla. In the phyla of bacterial group, Firmicutes, Actinobacteria, and Proteobacteria have the maximum number of species, while Cyanobacteria and Bacteroidetes have lower. In the phyla of fungal group *Ascomycota* has maximum number while *Basidiomycota* and *Mucoromycota* have lower. Figure 12.3 presents various bacteria regarding the degradation of PVC plastics (Anwar, 2011). Plastics are mainly biodegraded aerobically in wild environment while anaerobically in the system of landfills as well as sediments; in composts and soil, it is degraded partly aerobically and partly anaerobically.

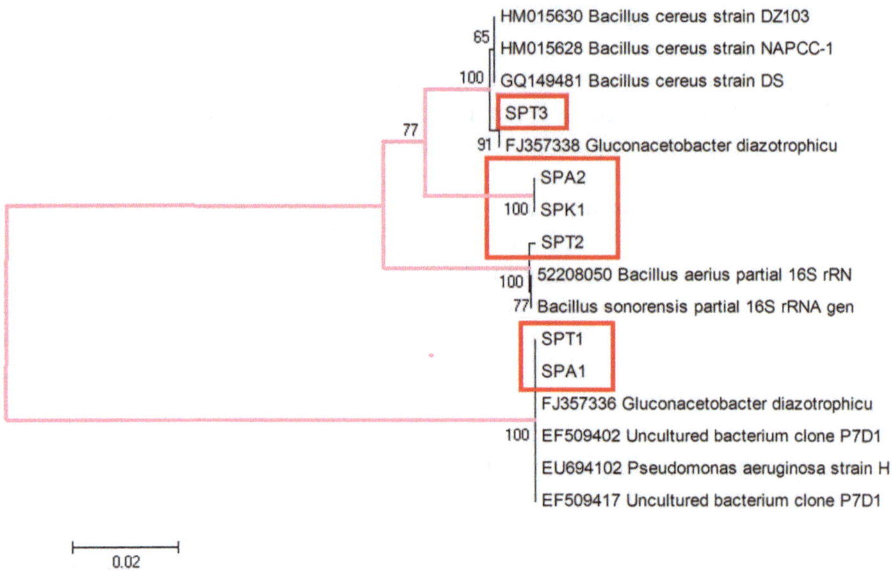

Fig. 12.3 Phylogenetic tree for PVC-degrading bacteria using 16S rDNA sequences by MEGA4 software and neighbor-joining method, reported by Anwar, 2011 and Anwar et al., 2013

Mechanism biodegradation of plastic under the condition aerobic environment has been explained diagrammatically (Fig. 12.4). Where the changes in polymer properties occurs structurally as well as functionally due to chemical, physical, or biological reactions in bond scission and subsequent chemical transformations have been categorized as polymer degradation. Degradation has been revealed by alteration in the properties of material like as optical or electrical and mechanical, characteristics, in crazing, discoloration, cracking, erosion, delamination. The changes also include addition of different new functional groups along with bond scission and subsequent chemical transformations, finally mixing to the environment or soil (Pospisil & Nespurek, 2011). Biodegradation of another thermoplastic polymer, i.e., polystyrene, has also been reported (Atiq et al., 2010). Further, there have been reports describing microbial degradation of other synthetic polymers, viz., polyurethane (Shah et al., 2008; Cosgrove et al., 2009), polyvinyl chloride (Anthony et al., 2004), and polycarbonates (Goel et al., 2008; Artham & Doble, 2010). Recently, in vitro biodegradation of epoxy and its silicone blend (ESB) by bacterial consortia has been reported by Negi et al. (2009). A list of different microorganisms degrading different groups of plastics has been summarized (Table 12.2).

Enzyme Responsible for Degradation of Plastic The amount of production of plastic and the rate of its production in the world have augmented since the last 5 years (2015–2020) which indicates that plastic waste management is very critical for the health of the environment (Wanga et al., 2020). Many microorganisms like fungi and bacteria had been reported that possess polyester-degrading mechanisms because of secretion of various enzymes from these organisms (Kaushal et al.,

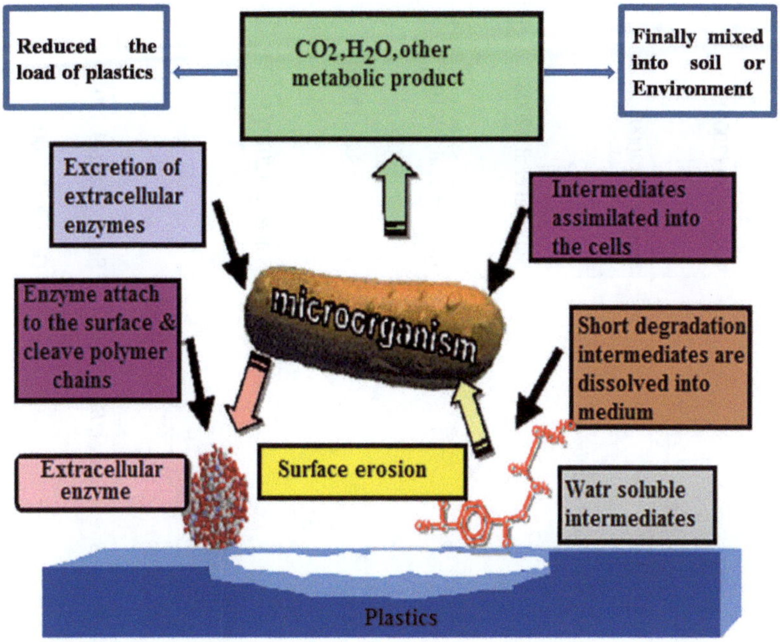

Fig. 12.4 Mechanism biodegradation of plastic under the condition aerobic environment (Tokiwa et al., 2009)

2021). Natural biodegradation of petro plastics is crucial. Still, some microorganisms can utilize certain polymers, under some special conditions, that lead to breakdown of polymer chains. These mechanisms are performed basically by the enzymes which are released by the microbes, which metabolized the plastic for the growth and development of microbes. Identification of these enzymes that are the main part of petro-polymer degradation, like PP and PE, and the mechanisms of these enzymes are mostly unknown. However, advancement to recognize these enzymes that are involved in the degradation of petro-polymer has been documented, on the basis of their activities that include oxidation, hydrolysis, and hydroxylation, resulting in breakdown of the polymer chains into smaller one like oligomers and monomers. On the basis of enzymatic degradation, petro-polymers can be categorized into hydrolyzable (PUR and PET) and non-hydrolyzable (PS, PP, PE, and PVC). The pathways involved for hydrolyzable and non-hydrolyzable biodegradation are appreciably different (Wei & Zimmermann, 2017). Some bonds in these polymers, like ester bond, are similar to the bonds present in other polyester polymers, so the enzymes that break down ester bonds are considered as enzyme esterases. Different groups of esterase are reported, which are varied in the structure of protein, biological functions, and substrate specificity. Hydrolyzable ester bonds containing polymers in their backbones, for instance, PUR and PET, are considered as more vulnerable to biodegradation compared to without hydrolysable ester bonds, like carbon chain backbones, that include PP, PS, PVC, and PE. Different enzymes for

Table 12.2 List of different microorganisms involved in utilizing a variety of plastics

S. no.	Plastic	Microorganism	References
Synthetic plastics			
1	Polyethylene	*Bacterium* Te68R *Microbacterium* sp. *Pseudomonas putida*	Kapri et al. (2009)
		Bacterium Te68R strain PN12 *Bacillus cereus* strain PN24 *Bacillus cereus* strain RGR10 *Proteobacterium* sp. strain CP1 a *Arthrobacter luteolus* strain LNR3	Soni et al. (2009)
		Bacillus cereus *Bacillus pumilus* *Arthrobacter* sp.	Satlewal et al. (2008)
		Aspergillus niger *Aspergillus flavus* *Aspergillus oryzae*	Konduri et al. (2010)
		Brevibacillus *Pseudomonas* *Rhodococcus*	Nanda and Sahu (2010)
		Microbacterium sp. strain MK3 *Pseudomonas putida* strain MK4 *Bacterium Te68R* strain PN12 *P. aeruginosa* strain PS1 *P. putida* strain PW1 *P. aeruginosa* strain C1	Sah et al. (2010)
		Comamonas, *Pseudomonas*, *Rhodococcus*, *Staphylococcus*, *Streptomyces*, *Bacillus*, *Acinetobacter*, *Micrococcus*	Sangeetha et al. (2019)
2	Polycarbonate	*Arthrobacter* sp. *Enterobacter* sp.	Goel et al. (2008)
		Engyodontium album MTP091SF1 *Penicillium* spp. MTP093 SF3	Artham and Doble (2010)
		Pseudoxanthomonas sp. strain NyZ600	Wenlong et al. (2021)

3	Polyurethane	*Nectria haematococca* *Penicillium viridicatum* *Penicillium ochrochloron* *Mucoromycotina* sp.	Cosgrove et al. (2009)
		Aspergillus sp. *FUN1* *Aspergillus* sp. *FUN5* *Chryseobacterium* BAC1	Cangemi et al. (2008)
		Bacillus sp. *AF8* *Pseudomonas* sp. *AF9* *Micrococcus* sp. *10* *Arthrobacter* sp. *AF11* *Corynebacterium* sp. *AF12*	Shah et al. (2008)
4	Polyvinyl chloride (PVC)	*Pseudomonas putida* AJ *Ochrobactrum* TD	Anthony et al. (2004)
		P. chrysosporium ME-446 *Polyporus versicolor*	Kirbas et al. (1999)
		Bacillus aerius SPT2 *Pseudomonas otitidis* SPT1 *Acanthopleuribacter pedis* strain SPT3 *Bacillus cereus* strain SPK1 *Acanthopleurobacter pedis* strain SPA1 *Bacillus cereus* strain SPA2 *Bacillus cereus* strain SPK1	Anwar et al. (2016)
5	Plasticized PVC	*Pseudomonas fluorescens* FS1	Zeng et al. (2002)
		Coryneform bacterium *Mycobacterium* sp.	Nakamiya et al. (2005)
		Brevibacterium *Pseudomonas* sp.	Booth and Robb (2007)

(continued)

Table 12.2 (continued)

S. no.	Plastic	Microorganism	References
6	Blend PVC	*Phanerochaete chrysosporium CCB 478* *Aspergillus fumigates*	Campos and Franchetti (2005)
		Trichoderma koningii *Fusarium oxysporum*	Kazmarek and Bajer (2007)
		Phanerochaete chrysosporium	Ali et al. (2009)
		Trogia buccinalis *Phanerochaete chrysosporium*	Campos and Franchetti (2005)
7	BTA copolyester	*Pseudomonas* sp.	Marten et al. (2005)
		Clostridium sp.	Abou-Zeid et al. (2004)
Natural plastics			
1	(Poly(3-hydroxybutyrate))	*Schlegelella thermodepolymerans* *Caenibacterium thermophilum*	Romen et al. (2004)
		Penicillium *Aspergillus*	Lee et al. (2005)
2	(Poly(3-hydroxybutyrate-co-3-mercaptopropionate))	*Pseudomonas indica* K2	Elbanna et al. (2004)
3	(Poly(3-hydroxybutyrate) poly(3-hydroxybutyrate-co-3-hydroxyvalerate)) (poly(3-hydroxybutyrate-co-3-hydroxypropionate))	*Streptomyces* sp. SNG9 *Ralstonia pickettii* T1	Mabrouk and Sabry (2001)
4	(Poly(3-hydroxybutyrate-co-3-hydroxypropionate))	*Acidovorax* sp. TP4	Wang and Lee (1997)

5	(Poly(3-hydroxypropionate-*co*-3-hydroxybutyrate))	Recombinant *Escherichia coli*	Valentin et al. (2000)
	(poly(3-hydroxybutyrate-co-4-hydroxybutyrate))	*Delftia acidovorans*	Hsieh et al. (2009)
6	(Poly(3-hydroxybutyrate))	*Streptoverticillium kashmirense* AF1	Shah et al. (2008)
7	(Poly(3-hydroxybutyrate))	*Streptoverticillium kashmirense* AF1	Shah et al. (2008)
8	(Poly(3-hydroxybutyrate))	*Desulfotomaculum* sp.	Cetin (2009)
9	(Poly(3-hydroxybutyrate-co-3-hydroxyvalerate))	Actinomycetes	Sang et al. (2001)
		Aspergillus flavus *Aspergillus oryzae* *Aspergillus parasiticus* *Asparagus racemosus*	Sanyal et al. (2006)
10	Polycaprolactone	*Clostridium botulinum* *Clostridium acetobutylicum*	Abou-Zeid et al. (2001)
		Bacillus sp. PG01	Wu et al. (2007)
Polymer blends			
1	Epoxy and epoxy-silicone blends (ESB)	*Bacterium* Te68R *Microbacterium* sp. *Pseudomonas aeruginosa* *Pseudomonas putida*	Negi et al. (2009)
2	Starch/polyethylene	*Fusarium* sp. AF4	Shah et al. (2008)
		Pseudomonas aeruginosa	Borghei et al. (2010)

high molecular weight polymer degradation like PET and ester-based PU are known, and their activity details have been characterized. But, yet, no specific enzymes that have the ability to degrade PP, PE, PVC, or PS have been well characterized, and their mechanisms are currently unknown. It is noticed that polymers that have large and high molecular weight are degraded by extracellular enzymes produced by microorganisms into smaller subunits which are incorporated into the microbial cells. After enteric to the cell, these degraded products are channeled by following the classical catabolic pathways to give energy as well as reducing power for the growth of cell. So, the microbial degradation of petro-polymers is considered as a very slow process (Mohanan et al., 2020). Resistance of these polymers is mainly due to their high molecular weight as well as their extremely hydrophobic surfaces, which avert these molecules from transiting the cell wall. The levels of degradability depend on the degree of amorphous as well as crystalline forms, along with the presence of strong C-C bonds, which are mainly resistant to enzyme attack. Cutinases (EC 3.1.1.74), a subgroup of esterase enzymes, have expanded its importance due to the ability to hydrolyze polyesters of a high molar mass (Chen et al., 2013). Cutinases are either carboxylic ester hydrolases or hydrolases that initially take out from the fungi famous for plant pathogenicity, e.g., *Fusarium solani pisi* (Kollattukudy, 1981; Purdy & Kolattukudy, 1975; Heredia, 2003). A polyesterase for hydrolyzing aromatic polyesters was first accounted from *Thermobifida fusca* by a group of German researcher (Müller et al., 2005). A diverse group of cutinases was also purified and characterized from a wide variety of bacterial strains of thermophilic actinomycetes (Wei et al., 2014a; Chen et al., 2013; Zimmermann & Billig, 2011), for instance, *Thermobifida fusca* (Kleeberg et al., 2005), *Thermomonospora fusca* (Fett et al., 1999), *Thermobifida cellulosilytica* (Herrero et al., 2011), *Thermobifida alba* (Hu et al., 2010), and *Thermomonospora curvata* (Wei et al., 2014b; Islam et al., 2019). Cutinases have also been reported for degradation of aliphatic polyesters, e.g., PCL (Baker et al., 2012; Kleeberg et al., 2005), along with aliphatic aromatic co-polyesters such as PET (Ribitsch et al., 2012a; Wei et al., 2014b; Herrero et al., 2011; Roth et al., 2014; Gamerith et al., 2017) and polytrimethylene terephthalate (PTT) (Eberl et al., 2009). Several literature on PET degradation suggest that majority of known PET hydrolases comes under cutinase group (Zimmermann & Billig, 2011; Tokiwa & Calabia, 2007; Baker et al., 2012). For PET hydrolysis, enzymes follow two mechanisms: (1) modification of surface of polyester fiber by enzymatic degradation or (2) hydrolysis of the polymer by enzymatic degradation. Surface modification of PET is carried out by different enzymes or hydrolases, such as carboxylesterases, lipases, cutinases, and proteases. These enzymes degrade only the surface not the polymer chains of the PET. However, PET waste management requires extensive degradation in the chains of the PET. Among cutinases, very few are recognized as PET hydrolases that participate to degrade the inner block of PET (about 10%), since this is the first finding of PET hydrolase by Müller et al. (2005). For surface hydrophilization of PET fibers, various hydrolases have been tested, such as lipases from *Thermomyces lanuginosus* (Eberl et al., 2009), *Burkholderia* spp. (Lee & Chung, 2009), *Triticum aestivum* (Nechwatal et al., 2006), and *Candida antarctica* (Vertommen et al., 2005); fungal cutinases from *Penicillium citrinum* (Liebminger et al., 2007), *Fusarium solani* (O'Neill et al., 2007; Alisch-Mark et al., 2006), *Aspergillus oryzae*, and *Humicola insolens* (Ronqvist et al., 2009); and those from Actinomycetes, *Saccharomonospora viridis* (Kawai et al., 2017), *Thermobifida cellulosilytica* (Herrero et al., 2011), *T. fusca* (Brueckner et al., 2008), carboxylesterases (Billig et al., 2010), *Thermobifida alba* (Ribitsch et al., 2012a), and esterases (*Thermobifida halotolerans*) (Ribitsch et al., 2012a). Besides crystallinity, surface topology, hydrophobicity, and molecular size of synthetic polymers (Webb et al., 2013; Tokiwa et al., 2009; Wei & Zimmermann, 2017), enzymatic PET degradation is also dependent on the reaction temperatures and active site accessibility to the polymer surface of the enzyme structure (Islam et al., 2019;Zumstein et al., 2017). Polyesterases favorably hydrolyze amorphous regions of PET (Brueckner et al., 2008; Welzel et al., 2002; Ronqvist et al., 2009; Gamerith et al., 2017; Donelli et al., 2010). Higher crystallinity restricts the movement of polymer chains that leads to decreases in the ease of use of polymer chains from enzymatic attack. Polyester degradation was found to encourage at elevated temperatures. Increased temperatures (about 55 °C) contribute more disinfection of digested plastic waste (Islam et al., 2019). Polyhydroxyalkanoate-binding modules (PBM) have also been greatly used to degrade polyurethane polyester copolymers (Gamerith et al., 2016). The fusion of PBM (PA-PBM) and polyamidase fourfold enhanced the hydrolysis of PU polymers relative to the native enzyme (Gamerith et al., 2016). Several researches were carrying out the esterase activities of microorganism against the degradation of polyester polyurethane (Rowe & Howard, 2002: Akutsu et al., 1998). Examples of extracellular esterase include PueA and PueB (Ruiz et al., 1999) and of membrane associated PudA (Akutsu et al., 1998). Polyurethanases (PUases) were characterized in *P. fluorescens*, *Pseudomonas chlororaphis*, and *Comamonas acidovorans* TB-35. Like other polymers, degradation of polyurethane occurs in two steps with PU esterase PudA (membrane associated esterase). Initially, hydrophobic adsorption of PudA to the PU surface is done by surface-binding domain promotes then hydrolysis of ester bonds forms the second step (Akutsu et al., 1998). Islam et al. (2019) account and superior binding and depolymerization of polyester-polyurethane nanoparticles by fusion of Tachystatin A2 (TA2- the anchor peptide) to the cutinase from *Thermomonospora curvata* (Tcur1278). TA2 are metal-binding peptides that sticking together provide a resourceful tool for a targeted degradation of microplastics and some other polymers such as PET, PU, and PP at normal temperature in a suspension of high dilution. TA2 was manufactured via a campaign of directed evolution (Rübsam et al., 2018b) and utilize a particular miscellany generation protocol for shorter peptides to generate variants with amplified binding to PP and PS in the presence of LAS and Triton X-100 surfactants (Rübsam et al., 2018a). Fusion of TA2 and cutinase (Tcur1278) speeds up the degradation of polyester polyurethane nanoparticles by a fold of 6.6 compared to the wild-type Tcur1278. Similarly, blending of the plastics with certain types of natural polymers, like starch, has also been shown to augment their biodegradation (Zheng and Yanful 2005; Ammala et al., 2011; Karimi & Biria, 2019). Mechanisms for improved biodegradation are directly linked to the rapid enzymatic hydrolysis of starch that makes the polymer porous susceptible to communally biotic and abiotic degradations (Bonhomme et al., 2003; Liu et al., 2013). Biodegradation of LDPE starch blend samples through an alpha-amylase in aqueous solution is accounted by Karimi and Biria (2019). Amylase has a co-metabolic behavior that hydrolyzed the starch as primary specific substrate as well as PE molecules. Moreover, the incorporation of hydrophilic starch into hydrophobic PE enhances the efficiency and biodegradability of both polymers (Hoque et al., 2013). The AHs isolated from *Pseudomonas* sp. strain E4 was reported to play an important role in the biodegradation of a non-oxidized LMWPE (Yoon et al., 2012). Jeon and Kim (2016) published the functional characteristics of alkane monooxygenases (AlkB1 and AlkB2) from *P. aeruginosa* that involved in the biodegradation of LMWPE and established that regulation mechanism of these enzymes was different and the AlkB2 enzyme was more effective in the degradation of LMWPE compared to AlkB1 enzyme. Danso et al. (2018) used Markov model-based search strategy to demonstrate that an outsized variety of potential polyesterases still desires to be discovered particularly in microorganism, which are at present not measured as a major source for cutinases (Danso et al., 2018). The marine *Pseudomonas* lineage, which includes psychrophilic, halophilic, hydrocarbonoclastic, and some species which are heavy metal-tolerant, is one best example; these were considered as a source of enzymes for PU hydrolysis (Wilkes & Aristilde, 2017), while the polyesterase activity in *Pseudomonads* is revealed by Guebitz group at the Austrian Centre of Industrial Biotechnology, Austria (Wallace et al., 2017; Haernvall et al., 2017). In *Pseudomonas pertucinogena*, polyesterase genes have been confirmed by using sequence homology searches (Bollinger et al., 2018). Further, various *Pseudomonas* have also been reported for a wide range of recalcitrant plastics such as PS, PP, PE, and PVC (Giacomucci et al., 2019). Kaoshal et al. in 2021 reported that enzymes are biocatalysts which participate in a reaction, act on a particular substrate, and accelerate the process of conversion of that substrate into a valuable product; the main degradation of plastics takes place with the help of enzymes that are present in microbial cells (extracellular/intracellular), which cause breakdown of polymer chains through assimilation within their cell and releasing metabolic products like CO_2, H_2O, CH_4, and N_2 (Amobonye et al., 2021). Numerous enzymes have been isolated from different organisms like algae, bacteria, actinomycetes, and fungi which are reported to cause degradation in different forms of plastic materials (Urbanek et al. 2018).

12.4.1 Methods Used to Evaluate the Degradability of Polymers

Biodegradation rate has been measured by using a wide range of variables, like substrate weight loss and changes in chemical structure of the polymer as well as in the mechanical properties and the percentage of carbon dioxide emission. Near the beginning, biodegradation experiments related to microorganism demonstrate that degradation could result in changes in the physical characteristics of plastics, such as crystallinity, tensile strength, and water uptake. The chief characteristics that are concerned in recognition of plastic degradation are reported (Albertsson et al., 1987). The evaluation of polymer degradability may be tested by visual observations to SEM (Fig. 12.5; Cangemi et al., 2006; Ring et al., 2008, Anwar et al., 2016, Tianyuan et al., 2023), and AFM analysis has become a significant tool for the measurement of the surface roughness on the nanometer scale (Kikkawa et al., 2002; Chen & Huang, 2004; Anwar et al., 2016). We can measure weight of plastic by measuring residual polymer after completion of experiment of polymer degradation (Nanda & Sahu, 2010), Gas chromatography or GC -MS (Witt et al., 2001; Jafari & Donaldson, 2008; Sasek et al., 2009), HPLC (Atiq et al., 2010; Kopecká et al., 2022; Tianyuan et al., 2023) changes in mechanical properties and molar mass (Erlandsson et al., 1997), CO_2 evolution/O_2 consumption (Orhan et al., 2004; Zheng & Yanful, 2005; Szaraz & Beczner, 2003; Mohee et al., 2006; Vijaya & Reddy, 2007),

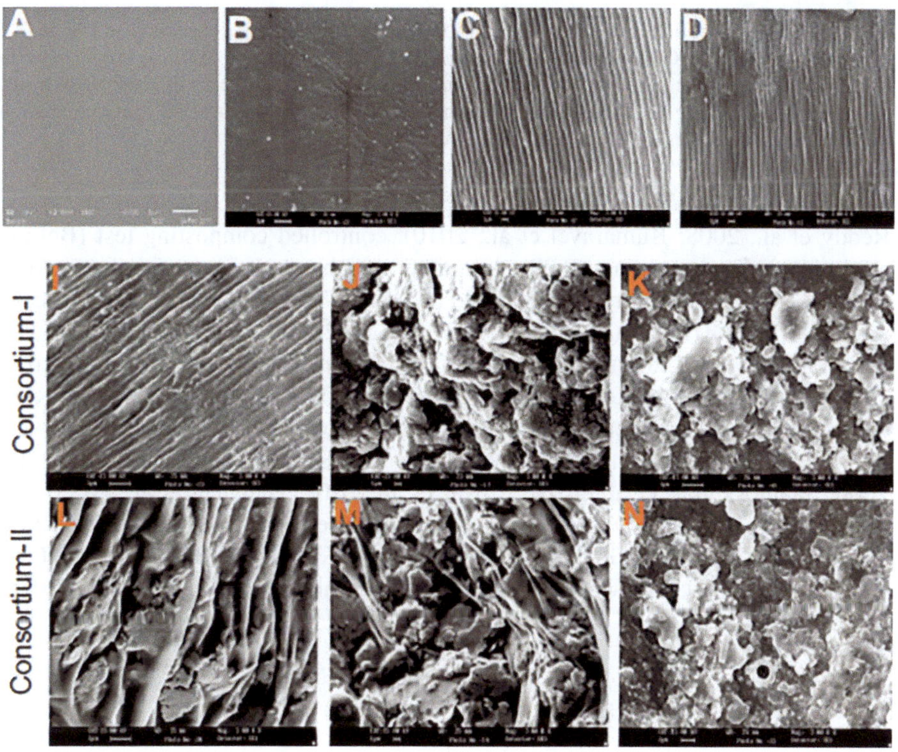

Fig. 12.5 Analysis of morphology of PVC film, against the absence/presence of microbial consortia with incubation period in the soil via SEM (**a–d**). In the absence of consortia, the morphology of the PVC film was changed slightly under SEM (**a–d**) after 0.0, 3.0, 6.0, and 9.0 months. However, in the presence of consortia, the degradation of PVC film was more obvious. The images representing in situ PVC biodegradation with consortium I (**i–k**) and consortium II (**l–n**) and treatments after incubation of 3.0, 6.0, and 9.0 months, reported by Anwar et al. (2016)

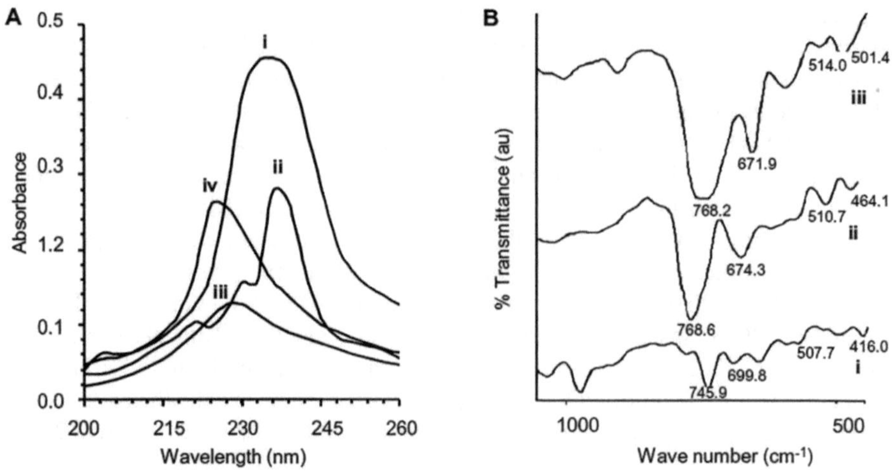

Fig. 12.6 Analysis of outcome of consortia in PVC degradation via FT-IR spectra and UV. In UV spectra (**a**) of fresh PVC film (without incubation and in the absence of consortia) (i), PVC film (9 months of incubation and in the absence of consortia) (ii), in the prescience of consortium I (iii), and consortium II (iv). FTIR spectra to examine the effect consortia on PVC film degradation under defined conditions (i–iii) (**b**) reported by Anwar et al. (2016)

radio-labeling (Vert et al., 2002), clear-zone formation (Abou-Zeid et al., 2001; Reddy et al., 2008; Kumaravel et al., 2010), controlled composting test (Bellina et al., 2000; Kunioka et al., 2009). A number of other techniques are also used to assess the biodegradability of polymeric material. These include differential scanning colorimetry (DSC) (Fig. 12.8; Ahmed et al., 2009; Anwar et al., 2016), UV spectroscopy (Fig. 12.6; Anwar et al., 2016), nuclear magnetic resonance spectroscopy (NMR), X-ray photoelectron spectroscopy (XPS), X-ray diffraction (XRD), contact angle measurements, and water uptake. Nowadays, two more techniques, viz., Fourier-transform infrared spectroscopy (FTIR) (Fig. 12.6; Cangemi et al., 2006; Santhoskumar et al., 2010; Anwar et al., 2016; Tianyuan et al., 2023) and simultaneous TG-DTG-DTA analysis, have been used to study the biodegradation of polymers (Fig. 12.7; Goel et al., 2008; Satlewal et al., 2008; Negi et al., 2009; Kapri et al., 2009; Tokiwa et al., 2009; Atiq et al., 2010, Anwar et al., 2016; Fig. 12.8).

12.5 Conclusions

Microbial as well as enzymatic degradation of polymers that are petroleum-derived has been an inventive area of research, which focused on resolving the plastic pollution in the environment. This chapter describes the microorganisms as well as their enzymes that are reported to degrade the wide range of synthetic polymers. Many strains of bacteria as well fungi with their enzymes were observed to degrade

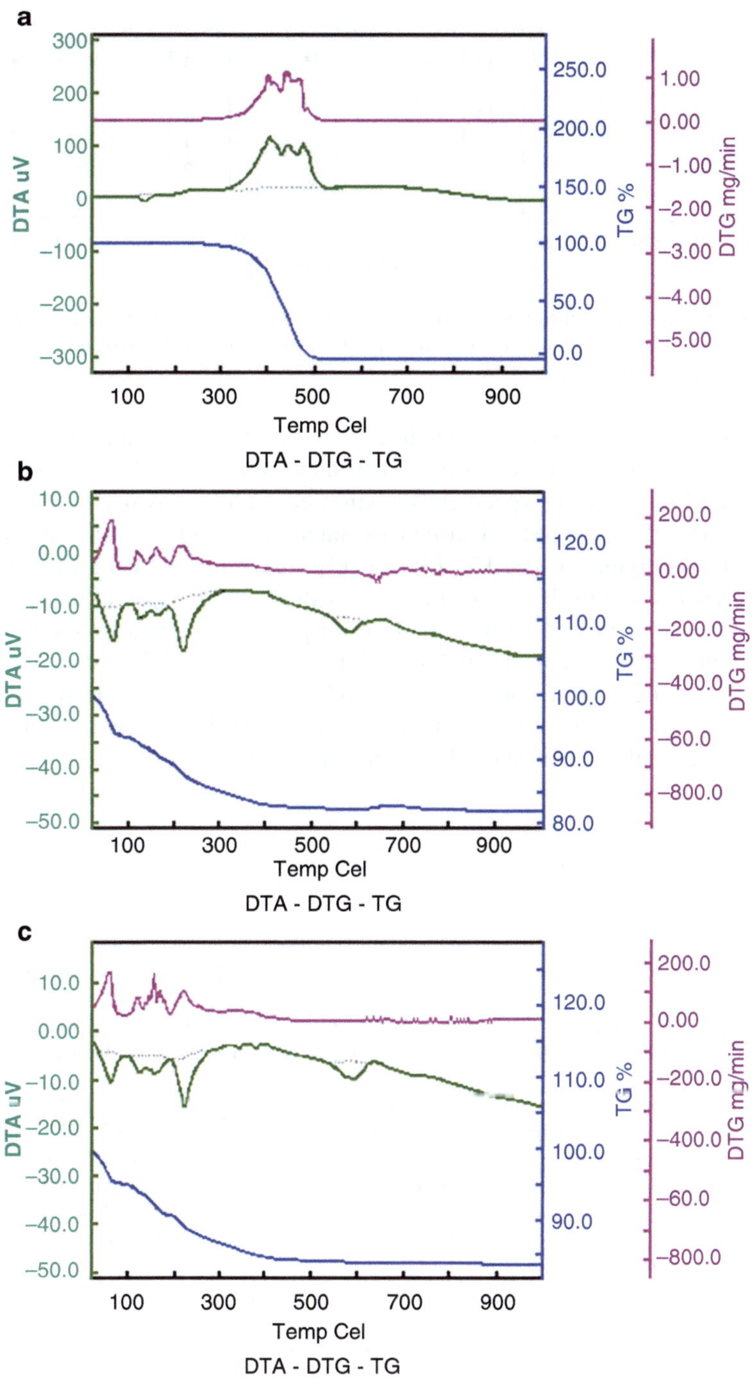

Fig. 12.7 Comparative TG-DTA-DTG of PVC (**a**) control (**b**) degraded by consortium CP I and (**c**) degraded by consortium CP I, reported by Anwar et al. (2016)

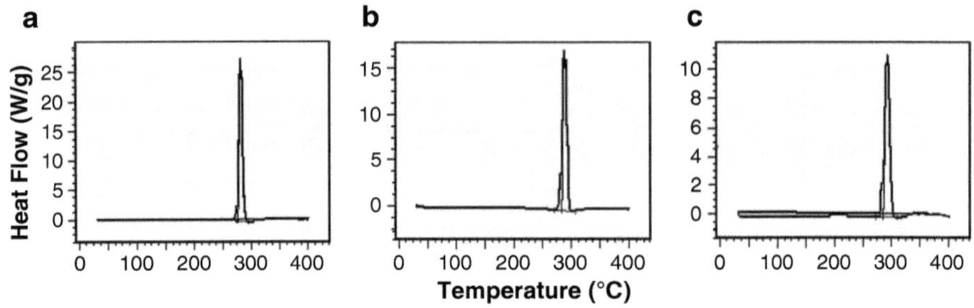

Fig. 12.8 DSC indicating the effect of consortia on the heat of fusion of PVC after incubation of 9 months under in situ conditions: control (**a**) degraded by consortium CP I (**b**) and consortium CP II (**c**) reported by Anwar et al. (2016)

recalcitrant, complex compounds and have been associated with the partial degradation of a wide range of different petro-plastics, including PP, PVC, PE, PS, PET, and ester-based PU. Enzymes specifically connected with depolymerization of ester-based PU and PET have been identified and intensively studied, while enzymes that effectively depolymerize PS, PE, PP, and PVC have not yet been identified and characterized, while for both categories of plastics various types of microbes were identified and categorized. Further analyses of plastic degradation were performed by different scientists using different techniques as mentioned. Based on this knowledge, genetic engineering approaches to produce genetically engineered microbial strains and their enzymes could be adopted as favored scheme to enhance biodegradation of the synthetic petroleum-based plastic waste.

References

Abou-Zeid, D. M., Müller, R. J., & Deckwer, W. D. (2001). *Anaerobic biodegradation of natural and synthetic polyesters, dissertation*. Technical University. http://opus.tu-bs.de/opus/volltexte/2001/246

Abou-Zeid, D. M., Müller, R. J., & Deckwer, W. D. (2004). Biodegradation of aliphatic homopolyesters and aliphatic-aromatic copolyesters by anaerobic microorganisms. *Biomacromolecules, 5*(5), 1687–1697.

Ahmed, J., Varshney, S. K., Zhang, J. X., & Ramaswamy, H. S. (2009). Effect of high pressure treatment on thermal properties of polylactides. *Journal of Food Engineering, 93*, 308–312.

Akutsu, Y., Nakajima-Kambe, T., Nomura, N., & Nakahara, T. (1998). Purification and properties of a polyester polyurethane-degrading enzyme from Comamonas acidovorans TB-35. *Applied and Environmental Microbiology, 64*, 62–67. https://doi.org/10.1128/AEM.64.1.62-67.1998

Albertsson, A. C., Andersson, S. O., & Karlsson, S. (1987). The mechanism of biodegradation of polyethylene. *Polymer Degradation and Stability, 18*, 73–87.

Ali, M. I., Perveen, Q., Ahmad, B., Javed, I., Razi-Ul-Hussnain, R., Andleeb, S., Atique, N., Ghumro, P. B., Ahmed, S., & Hameed, A. (2009). Biodegradation studies of cellulose blended polyvinyl chloride films. *International Journal of Agriculture and Biology, 11*, 577–580.

Alisch-Mark, M., Herrmann, A., & Zimmermann, W. (2006). Increase of the hydrophilicity of polyethylene terephthalate fibers by hydrolases from *Thermomonospora fusca* and *Fusarium solani* f. sp. pisi. *Biotechnology Letters, 28*, 681–685. https://doi.org/10.1007/s10529-006-9041-7

Ammala, A., Bateman, S., Deana, K., Petinakis, E., Sangwan, P., & Wong, S. (2011). An overview of degradable and biodegradable polyolefins. *Progress in Polymer Science, 36*, 1015–1049. https://doi.org/10.1016/j.progpolymsci.2010.12.002

Amobonye, A., Bhagwat, P., Singh, S., & Pillai, S. (2021). Plastic biodegradation: Frontline microbes and their enzymes. *Science of the Total Environment, 759*, 143536. https://doi.org/10.1016/j.scitotenv.2020.143536

Anthony, S. D., Meizhong, L., Christopher, E. B., Robin, L. B., & David, L. F. (2004). Involvement of linear plasmids in aerobic biodegradation of vinyl chloride. *Applied and Environmental Microbiology, 70*, 6092–6097.

Anwar, M. S. (2011). *Biodegradation studies of PVC and LDPE using indigenously developed microbial consortia.* PhD thesis. GBAU & T, Pantnagar, Uttarakhand.

Anwar, M. S., Negi, H., Zaidi, M. G. H., Gupta, S., & Goel, R. (2013). Biodeterioration studies of thermoplastics in nature using indigenous bacterial consortium. *Brazilian Archives of Biology and Technology, 56*(3), 475–484. https://doi.org/10.1590/S1516-89132013000300016

Anwar, M. S., Kapri, A., Chaudhry, V., Mishra, A., Ansari, M. W., Souche, Y., Nautiyal, C. S., Zaidi, M. G. H., & Goel, R. (2016). Response of indigenously developed bacterial consortia in progressive degradation of polyvinyl chloride. *Protoplasma, 253*, 1023–1032. https://doi.org/10.1007/s00709-015-0855-9

Artham, T., & Doble, M. (2010). Biodegradation of physicochemically treated polycarbonate by fungi. *Biomacromolecules, 11*(1), 20–28.

Atiq, N., Ahmed, S., Ali, M. I., Andleeb, S., Ahmad, B., & Robson, G. (2010). Isolation and identification of polystyrene biodegrading bacteria from soil. *African Journal of Microbiology Research, 4*(14), 1537–1541.

Audrézet, F., Zaiko, A., Lear, G., Wood, S. A., Tremblay, L. A., & Pochon, X. (2020). Biosecurity implications of drifting marine plastic debris: Current knowledge and future research. *Marine Pollution Bulletin, 111835*. https://doi.org/10.1016/j.marpolbul

Baker, A. M. M., & Mead, J. (2002). Thermoplastics. In C. A. Harper (Ed.), *Handbook of plastics, elastomers and composites* (4th ed., pp. 1–90). McGraw- Hill. ISBN 0-07-026714-6.

Baker, P. J., Poultney, C., Liu, Z., Gross, R. A., & Montclare, J. K. (2012). Identification and comparison of cutinases for synthetic polyester degradation. *Applied Microbiology and Biotechnology, 93*, 229–240. https://doi.org/10.1007/s00253-011-3402-4

Bellina, G., Tosin, M., & Degli-Innocenti, F. (2000). The test method of composting in vermiculite is unaffected by the priming effect. *Polymer Degradation and Stability, 69*, 113–120.

Billig, S., Oeser, T., Birkemeyer, C., & Zimmermann, W. (2010). Hydrolysis of cyclic poly(ethylene terephthalate) trimers by a carboxylesterase from Thermobifida fusca KW3. *Applied Microbiology and Biotechnology, 87*, 1753–1764. https://doi.org/10.1007/s00253-010-2635-y

Bollinger, A., Thies, S., Katzke, N., & Jaeger, K. E. (2018). The biotechnological potential of marine bacteria in the novel lineage of pseudomonas pertucinogena. *Microbial Biotechnology, 13*, 19–31. https://doi.org/10.1111/1751-7915.13288

Bonhomme, S., Cuer, A., Delort, A. M., Lemaire, J., Sancelme, M., & Scott, G. (2003). Environmental biodegradation of polyethylene. *Polymer Degradation and Stability, 81*, 441–452. https://doi.org/10.1016/S0141-3910(03)00129-0

Booth, G. H., & Robb, J. A. (2007). Bacterial degradation of plasticised PVC-effect on some physical properties. *Journal of Applied Chemistry, 18*(7), 194–197.

Borghei, M., Karbass, A., Khoramnejadia, S., Oromiehie, A., & Javid, A. H. (2010). Microbial biodegradable potato starch based low density polyethylene. *African Journal of Biotechnology, 9*(26), 4075–4080.

Brueckner, T., Eberl, A., Heumann, S., Rabe, M., & Guebitz, G. M. (2008). Enzymatic and chemical hydrolysis of poly(ethylene terephthalate) fabrics. *Journal of Polymer Science Part A: Polymer Chemistry, 46*, 6435–6443. https://doi.org/10.1002/pola.22952

Campos, A. D., & Franchetti, M. M. (2005). Biotreatment effects in films and blends of PVC/PCL previously treated with heat. *Brazilian Archives of Biology and Technology, 48*(2), 235–243.

Cangemi, J. M., Neto, S. C., Cheiric, G. O., & dosSantos, A. M. (2006). Study of biodegradation of a polymer derived from castor oil by scanning electron microscopy, thermogravimetry and infrared spectroscopy. *Polimeros, 16*(002), 129–135.

Cangemi, J. M., dosSantos, A. M., Neto, S. C., & Chierice, G. O. (2008). Biodegradation of polyurethane derived from castor oil. *Polímeros, 18*(3), 0104–1428.

Carr, C. M., Clarke, D. J., & Dobson, A. D. W. (2020). Microbial polyethylene terephthalate hydrolases: Current and future perspectives. *Frontiers in Microbiology, 11*(11), 571265. https://doi.org/10.3389/fmicb.2020.571265. PMID: 33262744; PMCID: PMC7686037.

Casabianca, S., Capellacci, S., Giacobbe, M. G., Dell'Aversano, C., Tartaglione, L., & Varriale, F. (2019). Plastic-associated harmful microalgal assemblages in marine environment. *Environmental Pollution, 244*, 617–626. https://doi.org/10.1016/j.envpol.2018.09.110

Cetin, D. (2009). Anaerobic biodegradation of poly-3-hydroxybutyrate (PHB) by sulfate reducing bacterium *Desulfotomaculum* sp. *Soil and Sediment Contamination, 18*(3), 345–353.

Chen, Y., & Huang, W. (2004). Numerical simulation of the geometrical factors affecting surface roughness measurements by AFM. *Measurement Science and Technology, 15*(10), 2005. https://doi.org/10.1088/0957-0233/15/10/010

Chen, S., Su, L., Chen, J., & Wu, J. (2013). Cutinase: Characteristics, preparation and application. *Biotechnology Advances, 31*, 1754–1767. https://doi.org/10.1016/j.biotechadv.2013.09.005

Cornwall, W. (2021). *Could plastic-eating microbes take a bite out of the recycling problem?* Waste Advantage Magazine.

Cosgrove, L., McGeechan, P. L., Handley, P. S., & Robson, G. D. (2009). Effect of biostimulation and bioaugmentation on degradation of polyurethane buried in soil. *Applied and Environmental Microbiology, 76*(3), 810–819. https://doi.org/10.1128/AEM.00534-09

Danso, D., Schmeisser, C., Chow, J., Zimmermann, W., Wei, R., & Leggewie, C. (2018). New insights into the function and global distribution of polyethylene terephthalate (PET)-degrading bacteria and enzymes in marine and terrestrial metagenomes. *Applied and Environmental Microbiology, 84*, e02773–e02717. https://doi.org/10.1128/AEM.02773-17

Donelli, I., Freddi, G., Nierstrasz, V. A., & Taddei, P. (2010). Surface structure and properties of poly(ethylene terephthalate) hydrolyzed by alkali and cutinase. *Polymer Degradation and Stability, 95*, 1542–1550. https://doi.org/10.1016/j.polymdegradstab.2010.06.011

Eberl, A., Heumann, S., Brueckner, T., Araujo, R., Cavaco-Paulo, A., & Kaufmann, F. (2009). Enzymatic surface hydrolysis of poly(ethylene terephthalate) and bis(benzoyloxyethyl) terephthalate by lipase and cutinase in the presence of surface active molecules. *Journal of Biotechnology, 143*, 207–212. https://doi.org/10.1016/j.jbiotec.2009.07.008

Elbanna, K., Lütke-Eversloh, T., Jendrossek, D., Luftmann, H., & Steinbüchel, A. (2004). Studies on the biodegradability of polythioester copolymers and homopolymers by polyhydroxyalkanoate (PHA)-degrading bacteria and PHA depolymerases. *Archives of Microbiology, 182*(2–3), 212–225.

Eriksen, M., Marcus, T., Prindiville, M., & Kiessling, T. (2018). *Freshwater microplastics: Microplastic: What are the solutions?, by Martin Wagner and Scott Lambert* (pp. 273–298., link.springer.com/chapter/). Springer. https://doi.org/10.1007/978-3-319-61615-5_13

Erlandsson, B., Karlsson, S., & Albertsson, A. C. (1997). The mode of action of corn starch and a pro-oxidant system in LDPE: Influence of thermooxidation and UV-irradiation on the molecular weight changes. *Polymer Degradation and Stability, 55*, 237–245.

Fett, W. F., Wijey, C., Moreau, R. A., & Osman, S. F. (1999). Production of cutinase by *Thermomonospora fusca* ATCC 27730. *Journal of Applied Microbiology, 86*, 561–568. https://doi.org/10.1046/j.1365-2672.1999.00690.x

Fields, R. D., Rodriguez, F., & Finn, R. K. (1974). Microbial degradation of polyesters: Polycaprolactone degraded by *P. pullulans*. *Journal of Applied Polymer Science, 18*, 3571–3579. https://doi.org/10.1002/APP.1974.070181207

Gambarini, V., Pantos, O., Kingsbury, J. M., Weaver, L., Handley, K. M., & Lear, G. (2020). Phylogenetic distribution for plastic-degrading organisms. *ASM Journals, 6*(1). https://doi.org/10.1128/msystems.01112-20

Gamerith, C., Herrero Acero, E., Pellis, A., Ortner, A., Vielnascher, R., & Luschnig, D. (2016). Improved enzymatic polyurethane hydrolysis by tuning enzymesorption. *Polymer Degradation and Stability, 132*, 69–77. https://doi.org/10.1016/j.polymdegradstab.2016.02.025

Gamerith, C., Zartl, B., Pellis, A., Guillamot, F., Marty, A., & Herrero Acero, E. (2017). Enzymatic recovery of polyester building blocks from polymer blends. *Process Biochemistry, 59*, 58–64. https://doi.org/10.1016/j.procbio.2017.01.004

Geyer, R. (2020). Chapter 2: Production, use, and fate of synthetic polymers. In T. M. Letcher (Ed.), *Plastic waste and recycling* (pp. 13–32). Academic. https://doi.org/10.1016/B978-0-12-817880-5.00002-5

Giacomucci, L., Raddadi, N., Soccio, M., Lotti, N., & Fava, F. (2019). Polyvinyl chloride biodegradation by pseudomonas citronellolis and Bacillus flexus. *New Biotechnology, 52*, 35–41. https://doi.org/10.1016/j.nbt.2019.04.005

Goel, R., Zaidi, M. G. H., Soni, R., Lata, K., & Shouche, Y. S. (2008). Implication of Arthrobacter and Enterobacter species for polycarbonate degradation. *International Biodeterioration and Biodegradation, 61*, 167–172.

Haernvall, K., Zitzenbacher, S., Wallig, K., Yamamoto, M., Schick, M. B., & Ribitsch, D. (2017). Hydrolysis of ionic phthalic acid based polyesters by wastewater microorganisms and their enzymes. *Environmental Science & Technology, 51*, 4596–4605. https://doi.org/10.1021/acs.est.7b00062

Harrison, J. P., Boardman, C., O'Callaghan, K., Delort, A. M., & Song, J. (2018). Biodegradability standards for carrier bags and plastic films in aquatic environments: A critical review. *Royal Society Open Science, 5*, 171792. https://doi.org/10.1098/rsos.171792

Heredia, A. (2003). Biochemical and biophysical characteristics of cutin, a plant barrier biopolymer. *Biochimica et Biophysica Acta, 1620*, 1–7. https://doi.org/10.1016/S0304-4165(02)00510-X

Herrero Acero, E., Ribitsch, D., Steinkelner, G. T., Gruber, K., Greimel, K., & Eiteljoerg, I. (2011). Enzymatic surface hydrolysis of PET: Effect of structural diversity on kinetic properties of cutinases from Thermobifida. *Macromolecules, 44*, 4632–4640. https://doi.org/10.1021/ma200949p

Ho, B. T., Roberts, T. K., & Lucas, S. (2018). An overview on biodegradation of polystyrene and modified polystyrene: The microbial approach. *Critical Reviews in Biotechnology, 38*, 1–13. https://doi.org/10.1080/07388551.2017.1355293

Hoque, M. E., Ye, T. J., Yong, L. C., & Mohd Dahlan, K. Z. (2013). Sago starch-mixed low-density polyethylene biodegradable polymer: Synthesis and characterization. *Journal of Materials, 365380*, 1. https://doi.org/10.1155/2013/365380

Hsieh, W. C., Wada, Y., & Chang, C. P. (2009). Fermentation, biodegradation and tensile strength of poly(3-hydroxybutyrateco-4-hydroxybutyrate) synthesized by Delftia acidovorans. *Journal of the Taiwan Institute of Chemical Engineers., 40*, 143–147.

Hu, X., Thumarat, U., Zhang, X., Tang, M., & Kawai, F. (2010). Diversity of polyester-degrading bacteria in compost and molecular analysis of a thermoactive esterase from *Thermobifida alba* AHK119. *Applied Microbiology and Biotechnology, 87*, 771–779. https://doi.org/10.1007/s00253-010-2555-x

Huang, Y., Liu, Q., Jia, W., Yan, C., & Wang, J. (2020). Agricultural plastic mulching as a source of microplastics in the terrestrial environment. *Environmental Pollution, 260*, 114096. https://doi.org/10.1016/j.envpol.2020.114096

Isangedighi, I. A., David, G. S., & Obot, O. I. (2021). Plastic waste in the aquatic environment. In *Analysis of nanoplastics and microplastics in food essay* (1st ed., pp. 55–84). CRC Press, Taylor & Francis Group. eBook ISBN9780429469596.

Islam, S., Apitius, L., Jakob, F., & Schwaneberg, U. (2019). Targeting microplastic particles in the void of diluted suspensions. *Environment International, 123*, 428–435. https://doi.org/10.1016/j.envint.2018.12.029

Jafari, A. J., & Donaldson, J. D. (2008). Determination of HCl and VOC emission from thermal degradation of PVC in the absence and presence of copper, copper(II) oxide and copper(II) chloride. *E-Journal of Chemistry, 6*(3), 685–692.

Jaiswal, S., Sharma, B., & Shukla, P. (2020). Integrated approaches in microbial degradation of plastics. *Environmental Technology and Innovations, 17*. https://doi.org/10.1016/j.eti.2019.100567

Jeon, H. J., & Kim, M. N. (2016). Comparison of the functional characterization between alkane monooxygenases for low-molecular-weight polyethylene biodegradation. *International Biodeterioration and Biodegradation, 114*, 202–208. https://doi.org/10.1016/j.ibiod.2016.06.012

Kapri, A., Zaidi, M. G. H., & Goel, R. (2009). Nanobarium titanate as supplement to accelerate plastic waste biodegradation by indigenous bacterial consortia. *AIP Conference Proceedings, 147*, 469–474.

Karimi, M., & Biria, D. (2019). The promiscuous activity of alpha-amylase in biodegradation of low density polyethylene in a polymer-starch blend. *Nature Scientific Reports, 9*, 2612. https://doi.org/10.1038/s41598-019-39366-0

Kaushal, J., Khatri, M., & Arya, S. K. (2021). Recent insight into enzymatic degradation of plastics prevalent in the environment: A mini – Review. *Cleaner Engineering and Technology, 2*, 100083.

Kawai, F. (2010). The biochemistry and molecular biology of xenobiotic polymer degradation by microbes. *Bioscience, Biotechnology, and Biochemistry, 74*(9), 1743–1759.

Kawai, F., Oda, M., Tamashiro, T., Waku, T., Tanaka, N., & Yamamoto, M. A. (2014). Novel Ca^{2+}-activated, *thermostabilized polyesterase* capable of hydrolyzing polyethylene terephthalate from *Saccharomonospora viridis* AHK190. *Applied Microbiology and Biotechnology, 98*(24), 10053–10064. https://doi.org/10.1007/s00253-014-5860-y

Kawai, F., Kawase, T., Shiono, T., Urakawa, H., Sukigara, S., & Tu, C. (2017). Enzymatic hydrophilization of polyester fabrics using a recombinant cutinase Cut190 and their surface characterization. *Journal of Fiber Science and Technology, 73*, 8–1. https://doi.org/10.2115/fiberst.fiberst.2017-0002

Kazmarek, H., & Bajer, K. (2007). Biodegradation of plasticized poly (vinyl chloride) containing cellulose. *Polymer Science, 45*, 903–919.

Kesti, S. S., & Shivasharana, C. T. (2018). The role of insects and microorganisms in plastic biodegradation. *International Journal of Scientific Research in Biological Sciences, 5*(6), 75–79.

Kikkawa, Y., Abe, H., Iwata, T., Inoue, Y., & Doi, Y. (2002). Crystal morphologies and enzymatic degradation of melt-crystallized thin films of random copolyesters of (R)-3-hydroxybutyric acid with (R)-3-hydroxyalkanoic acids. *Polymer Degradation and Stability, 76*, 467–478.

Kirbas, Z., Keskin, N., & Guner, A. (1999). Biodegradation of polyvinylchloride (PVC) by white rot fungi. *Bulletin of Environmental Contamination and Toxicology, 63*, 335–342.

Kleeberg, I., Welzel, K., Vandenheuvel, J., Müller, R. J., & Deckwer, W. D. (2005). Characterization of a new extracellular hydrolase from *Thermobifida fusca* degrading aliphatic-aromatic copolyesters. *Biomacromolecules, 6*, 262–270. https://doi.org/10.1021/bm049582t

Kollattukudy, P. E. (1981). Structure, biosynthesis, and biodegradation of cutin and suberin. *Annual Review of Plant Physiology, 32*(1), 539–567. https://doi.org/10.1146/annurev.pp.32.060181.002543

Konduri, M. K. R., Anupam, K. S., Vivek, J. S., Kumar, R. D. B., & Narasu, M. L. (2010). Synergistic effect of chemical and photo treatment on the rate of biodegradation of high density polyethylene by indigenous fungal isolates. *International Journal of Biochemistry and Biotechnology, 6*(2), 157–174.

Kopecká, R., Kubínová, I., Sovová, K., Mravcová, L., Vítěz, T., & Vítězová, M. (2022). Microbial degradation of virgin polyethylene by bacteria isolated from a landfill site. *SN Applied Sciences, 4*(11), 1–12.

Kumaravel, S., Hema, R., & Lakshmi, R. (2010). Production of Polyhydroxybutyrate (bioplastic) and its biodegradation by *pseudomonas lemoignei* and *aspergillus Niger*. *E-Journal of Chemistry, 7*(S1), S536–S542.

Kunioka, M., Ninomiya, F., & Funabashi, M. (2009). Biodegradation of poly (*butylene succinate*) powder in a controlled compost at 58 °C evaluated by naturally-occurring carbon 14 amounts in evolved CO_2 based on the ISO 14855-2 method. *International Journal of Molecular Sciences, 10*(10), 4267–4283.

Lacerda, A. L. D. F., Rodrigues, L. D. S., Van, S. E., Rodrigues, F. L., Ribeiro, L., & Secchi, E. R. (2019). Plastics in sea surface waters around the Antarctic peninsula. *Scientific Reports, 9*(1), 3977. https://doi.org/10.1038/s41598-019-40311-4

Lear, G., Kingsbury, J. M., Franchini, S., Gambarini, V., Maday, S. D. M., Wallbank, J. A., Weaver, L., & Pantos, O. (2021). Plastics and the microbiome: Impacts and solutions. *Environmental Microbiomes, 16*, 2. https://doi.org/10.1186/s40793-020-00371-w

Lee, C. W., & Chung, J. D. (2009). Synthesis and biodegradation behavior of poly(ethylene terephthalate) oligomers. *Polymer Korea, 33*, 198–202.

Lee, K. M., Gimore, D. F., & Huss, M. J. (2005). Fungal degradation of the bioplastic PHB (Poly-3-hydroxybutyric acid). *Journal of Polymers and the Environment, 13*(3), 213. https://doi.org/10.1007/s10924-005-4756-4

Li, R. L., Zhang, S. P., Zhang, L. L., Yu, K. F., Wang, S. P., & Wang, Y. H. (2020). Field study of the microplastic pollution in sea snails (Ellobium chinense) from mangrove forest and their relationships with microplastics in water/sediment located on the north of Beibu gulf. *Environmental Pollution, 263*(Pt B), 114368. https://doi.org/10.1016/j.envpol.2020.114368

Liebminger, S., Eberl, A., Sousa, F., Heumann, S., Fischer-Colbrie, G., & Cavaco-Paulo, A. (2007). Hydrolysis of PET and bis-(benzoyloxyethyl) terephthalate with a new polyesterase from *Penicillium citrinum*. *Biocatalysis and Biotransformation, 25*, 171–177. https://doi.org/10.1080/10242420701379734

Liu, X., Yu, L., Xie, F., Petinakis, E., Sangwan, P., & Shen, S. (2013). New evidences of accelerating degradation of polyethylene by starch. *Journal of Applied Polymer Science, 130*, 2282–2287. https://doi.org/10.1002/app.39421

Llorca, M., & Farré, M. (2021). Current insights into potential effects of micro-nanoplastics on human health by in-vitro tests. *Frontiers in Toxicology, 3*(29), 752140. https://doi.org/10.3389/ftox.2021.752140

Mabrouk, M. M., & Sabry, S. A. (2001). Degradation of poly(3-hydroxybutyrate) and its copolymer poly(3-hydroxybutyrate-co-3-hydroxyvalerate) by a marine *Streptomyces* sp. SNG9. *Microbiological Research, 156*, 323–335.

Markic, A., Gaertner, J. C., Gaertner-Mazouni, N., & Koelmans, A. A. (2019). Plastic ingestion by marine fish in the wild. *Critical Reviews in Environmental Science and Technology, 50*(7), 657–697. https://doi.org/10.1080/10643389.2019.1631990

Marten, E., Müller, R. J., & Deckwer, W. D. (2005). Studies on the enzymatic hydrolysis of polyesters. II. Aliphatic–aromatic copolyesters. *Polymer Degradation and Stability, 88*(3), 371–381.

Mohanan, N., Montazer, Z., Sharma, P. K., & Levin, D. B. (2020). Microbial and enzymatic degradation of synthetic plastics. *Frontiers in Microbiology, 11*, 580709. https://doi.org/10.3389/fmicb.2020.580709

Mohee, R., Unmar, G. D., Mudhoo, A., & Khadoo, P. (2006). Biodegradability of biodegradable/degradable plastic materials under aerobic and anaerobic conditions. *Waste Management, 28*, 1624–1629.

Montazer, Z., Habibi Najafi, M. B., & Levin, D. B. (2019). Microbial degradation of low density polyethylene and synthesis of polyhydroxyalkanoate polymers. *Canadian Journal of Microbiology, 65*(3), 224–234. https://doi.org/10.1139/cjm-2018-0335

Moore, S. L., Moore, S. B., Weisberg, G. L., Lattin, A. F., & Zellers. (2002). A comparison of neustonic plastic and zooplankton abundance in southern California's coastal waters. *Marine Pollution Bulletin, 44*, 1035–1038.

Moore, R. E., Millar, B. C., & Moore, J. E. (2020). Antimicrobial resistance (AMR) and marine plastics: Can food packaging litter act as a dispersal mechanism for AMR in oceanic environments? *Marine Pollution Bulletin, 150*, 7. https://doi.org/10.1016/j.marpolbul.2019.110702

Mueller, R. J. (2006). Biological degradation of synthetic polyesters—Enzymes as potential catalysts for polyester recycling. *Process Biochemistry, 41*, 2124–2128.

Müller, R. J., Schrader, H., Profe, J., Dresler, K., & Deckwer, W. D. (2005). Enzymatic degradation of poly(ethylene terephthalate): Rapid hydrolysis using a hydrolase from T. fusca. *Macromolecular Rapid Communications, 26*, 1400–1405. https://doi.org/10.1002/marc.200500410

Nakamiya, K., Hashimoto, S., Edmonds, J. S., Yasuhara, A., & Morita, M. (2005). Microbial treatment of bis (2-ethylhexyl) phthalate in polyvinyl chloride with isolated bacteria. *Journal of Bioscience and Bioengineering, 99*(2), 115–119.

Nanda, S., & Sahu, S. S. (2010). Biodegradability of polyethylene by *Brevibacillus, pseudomonas,* and *Rhodococcus* spp. *New York Science Journal, 3*(7), 95–98.

Nechwatal, A., Blokesch, A., Nicolai, M., Krieg, M., Kolbe, A., & Wolf, M. (2006). A contribution to the investigation of enzyme catalysed hydrolysis of poly(ethylene terephthalate) oligomers. *Macromolecular Materials and Engineering, 291*, 1486–1494. https://doi.org/10.1002/mame.200600204

Negi, H., Kapri, A., Zaidi, M. G. H., Satlewal, A., & Goel, R. (2009). Comparative in-vitro biodegradation studies of epoxy and its silicone blend by selected microbial consortia. *International Biodeterioration & Biodegradation, 63*(5), 553–558.

Nelms, S. E., Parry, H. E., Bennett, K. A., Galloway, T. S., Godley, B. J., & Santillo, D. (2019). What goes in, must come out: Combining scat-based molecular diet analysis and quantification of ingested microplastics in a marine top predator. *Methods in Ecology and Evolution, 10*(10), 1712–1722. https://doi.org/10.1111/2041-210X.13271

Neufeld, L., Stassen, F., Sheppard, R., & Gilman, T. (2016). *The new plastics economy rethinking the future of plastics* (Vol. 7, p. 58). World Economic Forum.

O'Neill, A., Araújo, R., Casal, M., Guebitz, G., & Cavaco-Paulo, A. (2007). Effect of the agitation on the adsorption and hydrolytic efficiency of cutinases on polyethylene terephthalate fibres. *Enzyme and Microbial Technology, 40*, 1801–1805. https://doi.org/10.1016/j.enzmictec.2007.02.012

Odusanya, S. A., Nkwogu, J. V., Alu, N., Etuk Udo, G. A., Ajao, J. A., & Osinkolu, G. A. (2013). Preliminary studies on microbial degradation of plastics used in packaging potable water in Nigeria. *Nigerian Food Journal, 31*, 63–72. https://doi.org/10.1016/S0189-7241(15)30078-3

Ojeda, T. (2013). Polymers and the environment. In F. Yılmaz (Ed.), *Polymer science* (pp. 1–34). InTech. https://doi.org/10.5772/51057

Orhan, Y., Hrenovic, J., & Buyukgungor, H. (2004). Biodegradation of plastic compost bags under controlled soil conditions. *Acta Chimica Slovenica, 51*, 579–588.

Pavani, P., & Rajeswari, T. R. (2014). Impact of plastics on environmental pollution. *Journal of Chemical and Pharmaceutical Sciences*. National Seminar on Impact of Toxic Metals, Minerals and Solvents leading to Environmental Pollution. 0974-2115. ISSN: 0974-2115.

Plastics Europe. (2017). *Plastics – The facts 2017: An analysis of European plastics production, demand and waste data.*

Pospisil, J., & Nespurek, S. (2011). Highlights in chemistry and physics of polymer stabilization. *Macromolecular Symposia, 115*(1), 143–163. https://doi.org/10.1002/masy.19971150110

Purdy, R. E., & Kolattukudy, P. E. (1975). Hydrolysis of plant cuticle by plant pathogens. Properties of cutinase I, cutinase II, and a nonspecific esterase isolated from *Fusarium solani pisi*. *Biochemistry, 14*, 2832–2840. https://doi.org/10.1021/bi00684a007

Reddy, S. V., Thirumala, M., & Mahmood, S. K. (2008). Biodegradation of polyhydroxyalkanoates. *International Journal of Microbiology, 4*(2).

Ribitsch, D., Herrero Acero, E., Greimel, K., Dellacher, A., Zitzenbacher, S., & Marold, A. (2012a). A new esterase from *Thermobifida halotolerans* hydrolyses polyethylene terephthalate (PET) and polylactic acid (PLA). *Polymers, 4*, 617–629. https://doi.org/10.3390/polym4010617

Ring, A., Goertz, O., Steinstraesser, L., Kuhnen, C., Schmitz, I., Muhr, G., Steinau, H. U., & Langer, S. (2008). Analysis of biodegradation of copolymer dermis substitute in the dorsal skinfold chamberbof balb/c mice. *European Journal of Medical Research, 11*, 471–478.

Romen, F., Reinhardt, S., & Jendrossek, D. (2004). Thermotolerant poly(3-hydroxybutyrate)-degrading bacteria from hot compost and characterization of the PHB depolymerase of *Schlegelella* sp. KB1a. *Archives of Microbiology, 182*, 157–164.

Ronqvist, Å. M., Xie, W., Lu, W., & Gross, R. A. (2009). Cutinase-catalyzed hydrolysis of poly(ethylene terephthalate). *Macromolecules, 42*, 5128–5513. https://doi.org/10.1021/ma9005318

Roth, C., Wei, R., Oeser, T., Then, J., Foellner, C., & Zimmermann, W. (2014). Structural and functional studies on a thermostable polyethylene terephthalate degrading hydrolase from *Thermobifida fusca*. *Applied Microbiology and Biotechnology, 98*, 7815–7823. https://doi.org/10.1007/s00253-014-5672-0

Rowe, L., & Howard, G. T. (2002). Growth of *Bacillus subtilis* on polyurethane and the purification and characterization of a polyurethanase-lipase enzyme. *International Biodeterioration and Biodegradation, 50*, 33–40. https://doi.org/10.1016/S0964-8305(02)00047-1

Rübsam, K., Davari, M. D., Jakob, F., & Schwaneberg, U. (2018a). KnowVolution of the polymer-binding peptide LCI for improved polypropylene binding. *Polymers, 10*, 423. https://doi.org/10.3390/polym10040423

Rübsam, K., Weber, L., Jakob, F., & Schwaneberg, U. (2018b). Directed evolution of polypropylene and polystyrene binding peptides. *Biotechnology and Bioengineering, 115*, 321–330. https://doi.org/10.1002/bit.26481

Ruiz, C., Main, T., Hilliard, N. P., & Howard, G. T. (1999). Purification and characterization of twopolyurethanase enzymes from *pseudomonas chlororaphis*. *International Biodeterioration and Biodegradation, 43*, 43–47. S0964-8305(98)00067-5. https://doi.org/10.1016/S0964-8305(98)00067-5

Sah, A., Kapri, A., Zaidi, M. G. H., Negi, H., & Goel, R. (2010). Implications of Fullerene-60 upon in-vitro LDPE biodegradation. *Journal of Microbiology and Biotechnology, 20*(5), 908–916.

Sandra, M., Franchetti, M., Egerton, T. A., & White, J. R. (2010). Morphological changes in poly (Caprolactone)/poly (vinyl chloride) blends caused by biodegradation. *Journal of Polymers and the Environment, 18*, 79–83.

Sang, B. I., Hori, K., Tanji, Y., & Unno, H. (2001). A kinetic analysis of the fungal degradation process of poly (3-hydroxybutyrate-co-3-hydroxyvalerate) in soil. *Biochemical Engineering Journal, 9*(3), 175–184.

Sangeetha, D. R., Ramya, R., Kannan, K., Robert, A. A., & Rajesh, K. V. (2019). Investigation of biodegradation potentials of high density polyethylene degrading marine bacteria isolated from the coastal regions of Tamil Nadu, India. *Marine Pollution Bulletin, 138*, 549–560.

Santhoskumar, A. U., Palanivelu, K., Sharma, S. K., & Nayak, S. K. (2010). A new synthesis of nickel 12-hydroxy oleate formulation to improve polyolefin's degradation. *Journal of Bioremediation & Biodegradation, 1*(2), 1000108.

Santos, R. G., Andrades, R., Demetrio, G. R., Kuwai, G. M., Sobral, M. F., & Vieira, J. D. S. (2020). Exploring plastic-induced satiety in foraging green turtles. *Environmental Pollution, 265*, 114918. https://doi.org/10.1016/j.envpol.2020.114918

Sanyal, P., Samaddar, P., & Paul, A. (2006). Degradation of poly(3-hydroxybutyrate) and poly(3-hydroxybutyrate-co-3-hydroxyvalerate) by some soil *aspergillus spp*. *Journal of Polymers and the Environment, 14*(3), 257-263(7).

Sasek, V., Vitasek, J., Chromcova, D., Prokopova, I., Brozek, J., & Nahlik, J. (2009). Biodegradation of synthetic polymers by composting and fungal treatment. *Folia Microbiologica, 51*(5), 425–430.

Satlewal, A., Soni, R., Zaidi, M. G. H., Shouche, Y., & Goel, R. (2008). Comparative biodegradation of HDPE and LDPE using an indigenously developed microbial consortium. *Journal of Microbiology and Biotechnology, 18*(3), 477–482.

Shah, A. A., Hasan, F., Hameed, A., & Ahmed, S. (2008). Biological degradation of plastics: A comprehensive review. *Biotechnology Advances, 26*, 246–265. https://doi.org/10.1016/j.biotechadv.2007.12.005

Shahnawaz, M., Sangale, M. K., & Ade, A. B. (2019). Chapter: In situ bioremediation technology for plastic degradation. In *Bioremediation Technology for Plastic Waste*. ISBN: 978-981-13-7491-3. https://doi.org/10.1007/978-981-13-7492-0

Sharuddin, S. D. A., Abnisa, F., Ashri, W. M., Daud, W., & Aroua, M. K. (2016). A review on pyrolysis of plastic wastes. *Energy Conversion and Management, 115*, 308–326. https://doi.org/10.1016/j.enconman.2016.02.037

Shimao, M. (2001). Biodegradation of plastics. *Current Opinion in Biotechnology, 12*, 242–247. https://doi.org/10.1016/s0958-1669(00)00206-8

Soni, R., Kapri, A., Zaidi, M. G. H., & Goel, R. (2009). Comparative biodegradation studies of non-poronized and poronized LDPE using indigenous microbial consortium. *Journal of Polymers and the Environment, 17*, 233. https://doi.org/10.1007/s10924-009-0143-x

Steinmetz, Z., Wollmann, C., Schaefer, M., Buchmann, C., David, J., & Troeger, J. (2016). Plastic mulching in agriculture. Trading short-term agronomic benefits for long-term soil degradation? *Science of the Total Environment, 15*(550), 690–705. https://doi.org/10.1016/j.scitotenv.2016.01.153

Szaraz, L., & Beczner, J. (2003). Optimization processes of a C02 measurement setup for assessing biodegradability of polymers. *International Biodeterioration & Biodegradation, 52*(2), 93–95.

Thompson A (2018)"Solving microplastic pollution means reducing, recycling-and fundamental rethinking." Scientific American., https://www.scientificamerican.com/article/solving-microplastic-pollution-means-reducing-recycling-mdash-and-fundamental-rethinking1/

Tianyuan Su, Zhang T, Liu Pan, Bian J, Zheng Y, YuanY, Li Q, Liang Q, · Qingsheng Q (2023). Biodegradation of polyurethane by the microbial consortia enriched from landfill. Applied Microbiology and Biotechnology 107:1983–1995.

Tokiwa, Y., & Calabia, B. P. (2007). Biodegradability and biodegradation of polyesters. *Journal of Polymers and the Environment, 15*, 259–267. https://doi.org/10.1007/s10924-007-0066-3

Tokiwa, Y., Calabia, B. P., Ugwu, C. U., & Aiba, S. (2009). Biodegradability of plastics. *International Journal of Molecular Science, 10*(9), 3722–3742. https://doi.org/10.3390/ijms10093722

Urbanek, A. K., Rymowicz, W., & Mirończuk, A. M. (2018). Degradation of plastics and plastic-degrading bacteria in cold marine habitats. *Applied Microbiology and Biotechnology, 102*, 7669–7678. https://doi.org/10.1007/s00253-018-9195-y

Valentin, H. E., Mitsky, T. A., Mahadeo, D. A., Tran, M., & Gruys, K. J. (2000). Application of a propionyl coenzyme a synthetase for poly(3-Hydroxypropionate-*co*-3-Hydroxybutyrate) accumulation in recombinant *Escherichia coli*. *Applied and Environmental Microbiology, 66*(12), 5253–5258.

Vert, M., Santos, I. D., Ponsart, S., Alauzet, N., Morgat, J. L., Coudane, J., & Garreau, H. (2002). Degradable polymers in a living environment: Where do you end up. *Polymer International, 51*, 840–844.

Vertommen, M. A. M. E., Nierstrasz, V. A., van der Veer, M., & Warmoeskerken, M. M. C. G. (2005). Enzymatic surface modification of poly(ethylene terephthalate). *Journal of Biotechnology, 120*, 376–386. https://doi.org/10.1016/j.jbiotec.2005.06.015

Vijaya, C., & Reddy, R. M. (2007). Impact of soil composting using municipal solid waste on biodegradation of plastics. *Indian Journal of Biotechnology, 7*, 235–239.

Wallace, P. W., Haernvall, K., Ribitsch, D., Zitzenbacher, S., Schittmayer, M., & Steinkellner, G. (2017). PpEst is a novel PBAT degrading polyesterase identified by proteomic screening of *pseudomonas pseudoalcaligenes*. *Applied Microbiology and Biotechnology, 101*, 2291–2303. https://doi.org/10.1007/s00253-016-7992-8

Wang, F., & Lee, S. J. (1997). Poly(3-hydroxybutyrate) production with high productivity and high polymer content by a fed-batch culture of *Alcaligenes latus* under nitrogen limitation. *Applied and Environmental Microbiology, 63*, 3703–3706.

Wanga, C., Zhaob, L., Lim, M. K., Wei-Qiang, C., & Sutherland, J. W. (2020). Structure of the global plastic waste trade network and the impact of China's import ban. *Resources, Conservation and Recycling, 153*, 104591.

Webb, H. K., Arnott, J., Crawford, R. J., & Ivanova, E. P. (2013). Plastic degradation and its environmental implications with special reference to poly(ethylene terephthalate). *Polymers, 5*, 1–18. https://doi.org/10.3390/polym5010001

Wei, R., & Zimmermann, W. (2017). Microbial enzymes for the recycling of recalcitrant petroleum-based plastics: How far are we? *Microbial Biotechnology, 10*, 1308–1322.

Wei, R., Oeser, T., Barth, M., Weigl, N., Lübs, A., & Schulz-Siegmund, M. (2014a). Turbidimetric analysis of the enzymatic hydrolysis of polyethylene terephthalate nanoparticles. *Journal of Molecular Catalysis B: Enzymatic, 103*, 72–78. https://doi.org/10.1016/j.molcatb.2013.08.010

Wei, R., Oeser, T., Then, J., Kühn, N., Barth, M., & Zimmermann, W. (2014b). Functional characterization and structural modeling of synthetic polyester-degrading hydrolases form *Thermomonospora curvata*. *AMB Express, 4*, 44. https://doi.org/10.1186/s13568-014-0044-9

Welzel, K., Müller, R. J., & Deckwer, W. D. (2002). Enzymatischer Abbau von polyester-nanopartikeln. *Chemie Ingenieur Technik, 74*, 1496–1500. https://doi.org/10.1002/1522-2640(20021015)74:10<1496::AID-CITE1496>3.0.CO;2-P

Wenlong, Y., Chao-Fan, Y., Limin, S., Jie, Z., Ying, X., & Ning, Y. Z. (2021). Biodegradation of bisphenol-A polycarbonate plastic by *Pseudoxanthomonas* sp. strain NyZ600. *Journal of Hazardous Materials, 416*(15), 125775.

Wilkes, R. A., & Aristilde, L. (2017). Degradation and metabolism of synthetic plastics and associated products by *pseudomonas* sp.: Capabilities and challenges. *Journal of Applied Microbiology, 123*, 582–593. https://doi.org/10.1111/jam.13472

Witt, U., Einig, T., Yamamoto, M., Kleeberg, I., Deckwer, W. D., & Muller, R. J. (2001). Biodegradation of aliphatic–aromatic copolyesters: Evaluation of the final biodegradability and ecotoxicological impact of degradation intermediates. *Chemosphere, 44*, 289–299.

Wu, K. J., Wu, C. S., & Chang, J. S. (2007). Biodegradability and mechanical properties of polycaprolactone composites encapsulating phosphate-solubilizing bacterium *bacillus* sp. PG01. *Process Biochemistry, 42*, 669–675.

Yang, Y., Yang, J., Wu, W. M., Zhao, J., Song, Y., & Gao, L. (2015a). Biodegradation and mineralization of polystyrene by plastic-eating mealworms. Chemical and physical characterization and isotopic tests. *Environmental Science & Technology, 49*, 12080. https://doi.org/10.1021/acs.est.5b02661

Yang, Y., Yang, J., Wu, W. M., Zhao, J., Song, Y., & Gao, L. (2015b). Biodegradation and mineralization of polystyrene by plastic-eating mealworms: Part 2. Role of gut microorganisms. *Environmental Science & Technology, 49*, 12087–12093. https://doi.org/10.1021/acs.est.5b02663

Yoon, M. G., Jeon, J. H., & Kim, M. N. (2012). Biodegradation of polyethylene by a soil bacterium and AlkB cloned recombinant cell. *Journal of Bioremediation & Biodegradation, 3*, 145.

Yoshida, S., Hiraga, K., Takehara, T., & Oda, K. (2016). A bacterium that degrades and assimilates poly (ethylene terephthalate). *Science, 351*, 1196–1199. https://doi.org/10.1126/science.aad6359

Zeng, F., Cui, K., Fu, J., Sheng, G., & Yang, H. (2002). Biodegradation of di(2-ethylhexyl) phthalate by Pseudomonas fluorescens FS1. *Water, Air, and Soil Pollution, 140*, 297–305.

Zheng, Y., & Yanful, E. K. (2005). A review of plastic waste degradation. *Critical Reviews in Biotechnology, 25*, 243–250.

Zimmermann, W., & Billig, S. (2011). Enzymes for the biofunctionalization of poly(ethylene terephthalate). *Advances in Biochemical Engineering/Biotechnology, 125*, 97–120. https://doi.org/10.1007/10_2010_87

Zumstein, M. T., Rechsteiner, D., Roduner, N., Perz, V., Ribitsch, D., & Guebitz, G. M. (2017). Enzymatic hydrolysis of polyester thin films at the nanoscale: Effects of polyester structure and enzyme active-site accessibility. *Environmental Science & Technology, 51*, 7476–7485. https://doi.org/10.1021/acs.est.7b01330

Chapter 13
Microbial Biological Degradation of Polymers: Recent Trends

Hemant Sharma, Saurav Gurung, Binu Gogoi, Rabina Gurung, Yadika Subba, and Arun Kumar Rai

Abstract Biodegradation greatly depends on microorganisms, such as bacteria, algae, and fungi, to degrade polymers through their metabolic activity under aerobic or anaerobic conditions. Synthetic plastics are greatly linked to our modern lifestyle; therefore, their accumulation in the environment is a major issue and of great concern to the environment and humans. Products such as polyvinylchloride, polystyrene, polyurethane, polyethylene, polypropylene, etc. are resistant to biodegradation pathways under natural conditions. Polymer-degrading microbes have been isolated and identified under in vitro conditions. Microbial degradation and depolymerization of waste plastics into higher-value bioproducts, such as biodegradable polymers, have been a great strategy. This chapter aims to provide valuable information pertaining to the microbial degradation of polymers through their enzymatic activity.

Keywords Biodegradation · Microbial degradation · Biodegradable polymers

13.1 Introduction

Polymers are macromolecules comprised of monomers that are simpler chemical molecules (Tauer, 2007). Based on the origin, polymers are classified as synthetic or natural. Synthetic polymers include plastics, rubber, fibers, etc. that are man-made or derivatives of chemical processes, whereas natural polymers include cellulose, starch, nucleic acids, etc. that occur naturally (Bhatia, 2016).

Further, depending on the degradability of the polymers by enzymes or microorganisms, they are classified into biodegradable or nonbiodegradable polymers (Adeleye et al., 2020). Biodegradable polymers include those polymers that can be

H. Sharma · S. Gurung · B. Gogoi · R. Gurung · Y. Subba · A. K. Rai (✉)
Department of Botany, Sikkim University, Gangtok, Sikkim, India
e-mail: akrai@cus.ac.in

R. Soni et al. (eds.), *Advanced Strategies for Biodegradation of Plastic Polymers*, https://doi.org/10.1007/978-3-031-55661-6_13

degraded to smaller monomers or simple products such as water, salts, carbon dioxide, etc. through the activity of microorganisms such as algae, fungi, bacteria, etc., whereas nonbiodegradable polymers include those polymers that remain undegraded in the environment for a very long time (Naeem et al., 2023).

The biological degradation of polymers by microorganisms represents the process of breaking down synthetic or natural polymers into simpler molecules by bacteria, fungi, or algae (Cai et al., 2023). The microorganisms can be used for various industrial, environmental, or medical applications, such as reducing plastic waste, biofuel production, and treating infections (Weng et al., 2008; Kalelkar et al., 2021; Skariyachan et al., 2022). However, possible limitations are associated with the degradation of polymers by microorganisms, such as the specificity of the enzymes involved, the toxicity of some of the residues of the polymers, etc.

In this article, we will review some of the recent trends and developments in the field of microbial degradation of polymers, focusing on the types of polymers that can be degraded, the mechanisms and factors that influence the degradation process, and the potential applications and implications of this technology.

13.2 Biodegradable Polymers

Biopolymers are natural polymers produced by the cells of living organisms, such as plants, animals, bacteria, and algae (Gobi et al., 2021). Unlike synthetic polymers, biopolymers have a well-defined structure, which determines their biological functions and applications. Biopolymers consist of monomeric units that are covalently bonded in chains to form larger molecules. There are three main classes of biopolymers: polynucleotides, polypeptides, and polysaccharides. Polynucleotides are long polymers of nucleotides, such as DNA and RNA, which store and transmit genetic information. Polypeptides are polymers of amino acids, such as proteins and peptides, which perform various structural and enzymatic roles in living systems. Polysaccharides are linear or branched chains of sugar carbohydrates, such as starch, cellulose, and alginate, which provide energy storage and structural support for cells (Runnels et al., 2018; Biswas et al., 2022).

Biodegradable polymers are materials or polymers that can be degraded into low-molecular-weight compounds by biological processes or spontaneous hydrolysis under natural or physiological conditions (Nair & Laurencin, 2007). Biodegradable polymers have many applications in various fields, such as packaging, agriculture, medicine, and biotechnology (Vroman & Tighzert, 2009).

Natural polymers: These are polymers of proteins, polysaccharides, nucleic acids, etc. derived from living organisms which are degraded enzymatically by certain specific enzymes, viz., collagen, gelatin, and fibrin (Shabbir Ali et al., 2023) (Table 13.1).

Synthetic biopolymers: These are polymers that are produced artificially and can be degraded under different physiological conditions which include polyanhydrides, polyamides, polycarbonates, polyesters, polyphosphoesters, etc. (Kubicek, 2016;

Table 13.1 Natural polymers

Biopolymer	Source	Example
Polynucleotide	Nucleus or cytoplasm of eukaryotic and prokaryotic cells, respectively	Ribonucleic acid, deoxyribonucleic acid
Polypeptide	Ribosomes in the cytoplasm or endoplasmic reticulum	Actin, fibrin, collagen, etc.
Polysaccharide	Chloroplasts of plant cells or extracellular matrix of bacterial cells	Starch, alginate, starch, etc.

Table 13.2 Modified biopolymers

Biopolymer	Modification or blend	Applications
Starch	Plasticized with glycerol or blended with polyethylene	Packaging films, foam, etc.
Collagen	Cross-linked with glutaraldehyde or blended with gelatin	Tissue scaffold, medical implants, etc.

Deb et al., 2019). Some of the examples of synthetic polymers are polylactic acid, polyglycolic acid, and their copolymers which can be used as implants, sutures, drug delivery systems, etc. (Jem & Tan, 2020).

Biopolymers have many advantages over synthetic polymers, such as biocompatibility, biodegradability, renewability, and diversity (Gowthaman et al., 2021). Biopolymers can be utilized for numerous applications in the food and medical sectors, viz., edible films, emulsions, materials for packing, systems made for delivery of drugs, medical implants, healing of wounds, tissue scaffolds, materials used for dressings, etc. (Udayakumar et al., 2021). However, biopolymers also have some limitations, such as low mechanical strength, high water sensitivity, high cost, and limited availability (Rhim & Ng, 2007). Therefore, biopolymers are often modified or blended with other materials to improve their properties and performance (Khademian et al., 2020) (Table 13.2). Some examples of biopolymer modifications and blends are given in the table below (Rosa et al., 2010; Goodarzi et al., 2019; Nashchekina et al., 2020).

13.3 Biopolymer Structure and Functions

Most of the biopolymers are extracted from algae, plants, animals, insects, etc. and are readily available in the nature. Some of the biopolymers such as agar and alginates are extracted from algae. Being one of the biopolymers, hyaluronic acid is commonly found in the umbilical cords, while some of the biopolymers are synthesized under in vitro conditions such as the synthesis of DNA molecules using polymerase chain reaction, secretion of biopolymers with the fermentation process, etc. (Rao et al., 2014). Starch, a carbohydrate polymer, is secreted in huge amounts by crop plants. The organic polymer basically contains cross-linked chains of glucose

Table 13.3 Classification of biopolymers

Sl. no.	Biopolymer	Properties
1	Polylactides	A polymer of lactic acids manufactured through vacuum forming, blowing, and injection molding. Exhibits water resistance
2	Dextrans	A polymer of glucose, obtained through hydrolysis of starch. It is enzymatically produced with the help of *Enterococcus faecalis*
3	Cellulose	A polymer of glucose, obtained naturally from plants, cotton, etc.
4	Polyglycolide acid	Synthetic biodegradable polymer, highly crystalline, high tensile strength, used in sutures
5	Polylactic acid	Slow degrading polymer, semicrystalline, less elongation but high tensile strength, and load-bearing properties make it useful in orthopedic fixation
6	Polycaprolactone	A semicrystalline polymer, useful in transferring drugs, enzymes, etc.
7	Polyanhydride	An excellent drug delivery vehicle due to high biocompatibility
8	Polyorthoesters	A degradable polymer

units which are α-1,4-linked and α-1,6-linked. Different enzymes are typically involved in the breakdown of this complex carbohydrate. Apart from being a source of energy for animals and humans, starch is being utilized continuously in the industrial sector for an impressive array of uses such as adhesives, extensive preparation of dyes, furniture, fuels based on alcohol, pharmaceutical sectors, etc. (Apriyanto et al., 2022).

Biopolymers can be classified under various categories based on the materials which is listed as under (Hassan et al., 2019) (Table 13.3).

13.4 Biological Degradation

The mechanism of degradation of polymers by microorganisms depends on the structure and type of polymer, the type of microorganisms involved in the degradation, and the environmental factors (Lenz, 1993). Depending on the place of degradation of the polymers, it can be classified into two broad categories:

(a) Extracellular: Degradation process involves enzyme secretion by the microorganisms that hydrolyze the polymer outside the cell. Some examples of extracellular enzymes are ligninases, cellulases, cutinases, polyhydroxyalkanoate depolymerases, etc. These enzymes degrade biodegradable plastics produced by some microorganisms such as lignin, cellulose, cutin, and polyhydroxyalkanoates (PHAs).

(b) Intracellular: Degradation process involves degradation inside the cell after the uptake of polymer fragments by the microorganisms. Some of the examples of enzymes are lipases, esterases, and glycosidases that are capable of degrading polymers such as polyesters, lipids, and some polysaccharides.

Generally, there are two types of mechanisms of enzymes involved in the degradation process of polymers (Murad & Azzaz, 2011):

(a) Exo-type: It acts on terminal end of the polymers and releases dimer or monomer units.
(b) Endo-type: It randomly acts on the chains of polymers and cleaves it, thus releasing varying lengths of oligomers.

The process of biodegradation of polymers includes two broad steps (Orlando et al., 2023):

1. Depolymerization of polymers into smaller units of oligomers, dimers, and monomers.
2. The smaller units are further metabolized by microorganisms.

Some of the enzymes involved in the degradation of polymers include:

- Proteases: This category of enzymes involves in the catalyzation of hydrolysis of peptide bonds in polypeptide molecules. These enzymes are secreted by plants, animals, and some of the microorganisms belonging to bacteria and fungi such as *Bacillus subtilis* (Keshapaga et al., 2023), *Aspergillus oryzae* (Niu et al., 2023), etc.
- Lipases: Most of the enzymes of this category are exo-type enzyme that is involved in hydrolysis of ester bonds in polyester chains of lipids into dimers and monomers, while endo-type lipases release oligomers. Some of the microorganisms that secrete lipase enzymes include *Rhizopus oryzae* (Yasvanthrajan et al., 2023), *Pseudomonas cepacia* (Martínez Gil et al., 2022), *Candida antarctica* (Shu et al., 2023), etc.
- Cellulases: Different enzymes act synergistically to degrade cellulose which are classified into cellulase enzymes (Sethi & Scharf, 2013). Cellulase enzymes can be further classified into three categories, viz.:

  (a) Endoglucanase: An endo-type enzyme that releases oligomers after random cleavage of β-1,4-glycosidic bonds of cellulose chains.
  (b) β-glucosidases: Exo-type enzymes that hydrolyze cellobiose into glucose,
  (c) Exoglucanase: An exo-type enzyme that cleaves β-1,4-glycosidic bonds at nonreducing or reducing ends of the cellulose chains and releases cellobiose.

Examples of cellulase-secreting microorganisms include *Aspergillus niger* (Nurdin et al., 2023), *Trichoderma reesei* (Li et al., 2023), *Clostridium thermocellum* (Nhim et al., 2022), etc.

- Esterases: Hydrolyze ester bonds in polyesters such as polyethylene terephthalate (PET), which is a common plastic used for bottles and fabrics. *Klebsiella oxytoca* has been known to secrete esterase enzyme (Zhang et al., 2023).

13.5 Mechanism of Biodegradation

The biological degradation of polymers represents a gradual process by which organisms break down the polymer chains into smaller fragments that can be assimilated or converted into simpler compounds. The complex mechanisms of biodegradation depend on the type, structure, and specific properties of the polymers, as well as the environmental conditions and microbial diversity. Some of the key factors that positively affect biodegradation are molecular weight, functional groups, percentage of crystallinity, hydrophobic nature, additives, and specific surface area of the polymer (Glaser, 2019). Biodegradation can be classified into two types: abiotic and biotic. Abiotic biodegradation typically includes physical and/or chemical agents, viz., temperature, light content, water, oxygen concentration, or radicals, which are responsible for initiating the breakdown of polymer chains. Biotic biodegradation deals with biological agents such as microorganisms including fungi, bacteria, algae, or animals that release enzymes for the efficient metabolism of the fragments of polymers (Lenz, 1993). Hydrolysis represents the possible reaction of polymer bonds and water molecules that results in the gradual formation of small molecules of ester, organic acid, or alcohol. Catalytic oxidation involves a reaction between oxidizing agents, such as oxygen, and chains of polymers resulting in the formation of peroxides and carboxylic or carbonyl groups. In the case of enzymatic cleavage, specific enzymes cleave the bonds of the polymer that results in the formation of smaller units of monomers or oligomers which can be degraded by microorganisms and assimilated or recycled (Ghatge et al., 2020). Some notable examples of biodegradable polymers include starch, polylactic acid, polyhydroxyalkanoates, cellulose, polyesters, polyamides, and polyurethanes. Such polymers are typically obtained from plants or synthesized from monomers of petroleum. Biodegradable polymers have a plethora of potential applications in packaging industries, pharmaceutical sectors, and modern agriculture and for managing waste (Brodhagen et al., 2015; Vohlídal, 2021).

13.6 Enzymatic Degradation of Polymer

Enzymes involved in polymer degradation are usually secreted by microorganisms that use the smaller units of polymer as a source of carbon and energy. The general mechanisms of biodegradation typically include hydrolysis, oxidation, and cleavage through enzymatic action. Degradation process of polymers undergoes different stages (Silva et al., 2023):

(a) Biodeterioration: Biodeterioration represents the process of change in the distinctive characteristics of a material due to the activity of plants, animals, and microorganisms such as bacteria, fungi, algae, etc. (Cozzolino et al., 2022; Di Carlo et al., 2022). The complex mechanism of biodeterioration involves the

Table 13.4 Polymer-degrading microorganisms

Sl. no.	Microorganisms	Enzyme secreted	Action	References
1	*Thermobifida fusca* and *Humicola insolens*	Cutinase	Degrades polyethylene terephthalate (PET)	Flores-Castañón et al. (2022)
2	*Pseudomonas chlororaphis*	Esterase	Degrades polyurethane (PU)	Kemona and Piotrowska (2020)
3	*Aspergillus Niger*	Lipases and proteases	Degrades polylactic acid (PLA)	Chai et al. (2022)
4	*Phanerochaete chrysosporium*	Lignin peroxidase (LiP) and manganese peroxidase (MnP)	Degrades lignin	Biko et al. (2020)
5	*Saccharomyces cerevisiae*	Cellulases	Degrades cellulose	Arnthong et al. (2022)

functional activity of organisms on the surface or within the structure of a material that results in the considerable alteration of properties of the material due to the secretion of certain acids, enzymes, pigments, etc. One such example includes the secretion of lipase enzyme by microorganisms that results in the degradation of organic materials such as paper (Table 13.4). Deterioration involving certain microorganisms has been reported from different habitats. Some of the species belonging to the genus of *Bacillus*, *Pseudomonas*, *Aspergillus*, etc. have been reported to utilize the plastic polymers as a source of carbon under in vitro conditions resulting in the reduction in the biomass of the polymers (Tareen et al., 2022).

(b) Bio-fragmentation: A process in which the enzymes or other catalytic agents that are released by certain microorganisms promote the reduction of molecular weight of polymers by cleaving action and production of small monomeric units (Kaabel et al., 2021). A study was conducted to evaluate the process of bio-fragmentation of polyethene microplastics in *Danio rerio*, without fatality of the organism. It was reported that the zebra fish was able to fragment the microplastics into nanoplastics within a short span of time (Khan & Ali, 2023).

(c) Mineralization: The process wherein the polymers are converted to simpler molecules through metabolization and excretion that can be assimilated or absorbed by microorganisms or the environment. The excreted metabolites are used either as substrates or as a source of energy by certain microorganisms. The mineralization process can be utilized to monitor the biodegradation process of a polymer by assessing the utilization or release of certain molecules such as the utilization of oxygen or release of carbon dioxide in aerobic or anaerobic fermentation processes, respectively (Kawai, 2015).

(d) Assimilation: This process involves the utilization of smaller products of polymers after the degradation process by the microorganisms as a source of nitrogen or carbon. The assimilation process helps the microorganisms to grow and further utilize the substrate present in the environment (Oliveira et al., 2020; Zeenat et al., 2021). Assimilation of polymers is an important step in recycling process of organic substances which helps in the reduction of polluting substances from the environment.

13.7 Factors Affecting Biodegradation

Degradation of polymers involves various complex steps and processes. Some of the factors responsible for the degradation of polymers include (Leja & Lewandowicz, 2010; Kanwal et al., 2022):

(a) Chemical structure of the polymers that is responsible for reactivity, stability of the polymer, hydrophilicity nature, etc.
(b) Physico-mechanical properties such as molecular weight of the substance, crystalline or amorphous nature, elasticity, porosity, etc.
(c) Degree of polymerization.
(d) Tacticity of the main polymeric chain.
(e) The type and nature of groups attached to the main chain and the number of substitutions.
(f) Type and nature of microorganisms involved in the degradation process and their population.
(g) Type and nature of the enzymes.
(h) Potential of biofilm growth of the surface of the materials.
(i) Abiotic factors such as sunlight, temperature, pH, etc.

13.8 Future Prospects and Constraints

Biopolymers are sustainable materials for food and medical applications because they can appreciably reduce the environmental impact and chief dependence on fossil fuels. Biopolymers can additionally provide novel functionalities and potential benefits for human health and economic well-being. However, biopolymers still face some formidable challenges in practical terms of successful production, processing, characterization, and regulation. The microbial degradation of polymers typically represents a promising strategy for adequately reducing plastic pollution and utilizing renewable resources. However, there are many challenges and functional limitations to overcome, such as slow degradation rate, low specificity and efficiency of key enzymes, adverse effects of additives and metabolites on microorganisms, and optimization of environmental conditions. Therefore, further research and development are typically needed to properly explore the potential and overcome the limitations of biopolymers in the foreseeable future as well as improve the microbial degradation of polymers and make it more feasible and sustainable.

References

Adeleye, A. T., et al. (2020). Sustainable synthesis and applications of polyhydroxyalkanoates (PHAs) from biomass. *Process Biochemistry, 96*, 174–193. https://doi.org/10.1016/j.procbio.2020.05.032

Apriyanto, A., Compart, J., & Fettke, J. (2022). A review of starch, a unique biopolymer – Structure, metabolism and in planta modifications. *Plant Science, 318*, 111223. https://doi.org/10.1016/j.plantsci.2022.111223

Arnthong, J., et al. (2022). Overexpression of LAS21 in cellulase-displaying saccharomyces cerevisiae for high-yield ethanol production from pretreated sugarcane bagasse. *Fermentation, 8*(11), 652. https://doi.org/10.3390/fermentation8110652. MDPI.

Bhatia, S. (2016). In S. Bhatia (Ed.), *Natural polymers vs synthetic polymer BT – Natural polymer drug delivery systems: Nanoparticles, plants, and algae* (pp. 95–118). Springer. https://doi.org/10.1007/978-3-319-41129-3_3

Biko, O. D. V., Viljoen-Bloom, M., & van Zyl, W. H. (2020). Microbial lignin peroxidases: Applications, production challenges and future perspectives. *Enzyme and Microbial Technology, 141*, 109669. https://doi.org/10.1016/j.enzmictec.2020.109669. Elsevier.

Biswas, M. C., et al. (2022). Recent advancement of Biopolymers and their potential biomedical applications. *Journal of Polymers and the Environment, 30*(1), 51–74. https://doi.org/10.1007/s10924-021-02199-y. Springer.

Brodhagen, M., et al. (2015). Biodegradable plastic agricultural mulches and key features of microbial degradation. *Applied Microbiology and Biotechnology, 99*(3), 1039–1056. https://doi.org/10.1007/s00253-014-6267-5

Cai, Z., et al. (2023). Biological degradation of plastics and microplastics: A recent perspective on associated mechanisms and influencing factors. *Microorganisms, 11*(7), 1661. https://doi.org/10.3390/microorganisms11071661. MDPI.

Chai, X., et al. (2022). Degradation of wheat straw/polylactic acid composites by aspergillus Niger. *Polymer Composites, 43*(3), 1823–1831. https://doi.org/10.1002/pc.26500. Wiley Online Library.

Cozzolino, A., et al. (2022). The role of lichens, mosses, and vascular plants in the biodeterioration of historic buildings: A review. *Plants, 11*(24), 3429. https://doi.org/10.3390/plants11243429

Deb, P. K., et al. (2019). Pharmaceutical and biomedical applications of polymers. In *Basic fundamentals of drug delivery* (pp. 203–267). https://doi.org/10.1016/B978-0-12-817909-3.00006-6. Elsevier.

Di Carlo, E., Barresi, G., & Palla, F. (2022). Biodeterioration. In F. Palla & G. Barresi (Eds.), *Biotechnology and conservation of cultural heritage* (pp. 1–30). Springer. https://doi.org/10.1007/978-3-030-97585-2_1

Flores-Castañ6n, N., Sarkar, S., & Banerjee, A. (2022). Structural, functional, and molecular docking analyses of microbial cutinase enzymes against polyurethane monomers. *Journal of Hazardous Materials Letters, 3*, 100063. https://doi.org/10.1016/j.hazl.2022.100063

Ghatge, S., et al. (2020). Biodegradation of polyethylene: A brief review. *Applied Biological Chemistry, 63*(1), 27. https://doi.org/10.1186/s13765-020-00511-3

Glaser, J. A. (2019). Biological degradation of polymers in the environment. In *Plastics in the environment*. IntechOpen. https://doi.org/10.5772/intechopen.85124

Gobi, R., et al. (2021). Biopolymer and synthetic polymer-based nanocomposites in wound dressing applications: A review. *Polymers, 13*(12), 1962. https://doi.org/10.3390/polym13121962. MDPI.

Goodarzi, H., et al. (2019). Preparation and in vitro characterization of cross-linked collagen–gelatin hydrogel using EDC/NHS for corneal tissue engineering applications. *International Journal of Biological Macromolecules, 126*, 620–632. https://doi.org/10.1016/j.ijbiomac.2018.12.125

Gowthaman, N. S. K. et al. (2021). Advantages of biopolymers over synthetic polymers. In S. Thomas, S. Gopi, & A. B. T. –B. Amalraj and their I. A. (Eds), *Biopolymers and their industrial applications* (pp. 351–372). Elsevier. https://doi.org/10.1016/B978-0-12-819240-5.00015-8

Hassan, M. E. S., Bai, J., & Dou, D.-Q. (2019). Biopolymers; definition, classification and applications. *Egyptian Journal of Chemistry, 62*(9), 1725–1737. National Information and Documentation Centre (NIDOC), Academy of Scientific ….

Jem, K. J., & Tan, B. (2020). The development and challenges of poly (lactic acid) and poly (glycolic acid). *Advanced Industrial and Engineering Polymer Research, 3*(2), 60–70. https://doi.org/10.1016/j.aiepr.2020.01.002. Elsevier.

Kaabel, S., et al. (2021). Enzymatic depolymerization of highly crystalline polyethylene terephthalate enabled in moist-solid reaction mixtures. *Proceedings of the National Academy of Sciences, 118*(29), 1–6. https://doi.org/10.1073/pnas.2026452118

Kalelkar, P. P., Riddick, M., & García, A. J. (2021). Biomaterial-based antimicrobial therapies for the treatment of bacterial infections. *Nature Reviews Materials, 7*(1), 39–54. https://doi.org/10.1038/s41578-021-00362-4. Nature Publishing Group UK London.

Kanwal, A., et al. (2022). Polymer pollution and its solutions with special emphasis on poly (butylene adipate terephthalate (PBAT)). *Polymer Bulletin, 79*(11), 9303–9330. https://doi.org/10.1007/s00289-021-04065-2. Springer.

Kawai, F. (2015). Biodegradation of polymers (bioassimilation, biomineralization, biodisintegration, compost), overview. In *Encyclopedia of polymeric nanomaterials*.

Kemona, A., & Piotrowska, M. (2020). Polyurethane recycling and disposal: Methods and prospects. *Polymers, 12*(8), 1752. https://doi.org/10.3390/polym12081752

Keshapaga, U. R., et al. (2023). Characterization of high-yield Bacillus subtilis cysteine protease for diverse industrial applications. *Brazilian Journal of Microbiology, 54*(2), 739–752. https://doi.org/10.1007/s42770-023-00992-6

Khademian, E., et al. (2020). A systematic review on carbohydrate biopolymers for adsorptive remediation of copper ions from aqueous environments-part A: Classification and modification strategies. *Science of the Total Environment, 738*, 139829. https://doi.org/10.1016/j.scitotenv.2020.139829

Khan, D., & Ali, S. A. (2023). On the novel process of pristine microplastic bio-fragmentation by zebrafish (Danio rerio). *Archives of Environmental Contamination and Toxicology, 84*(3), 299–306. https://doi.org/10.1007/s00244-023-00987-2

Kubicek, C. P. (2016). Synthetic Biopolymers. In A. Glieder et al. (Eds.), *Synthetic biology* (pp. 307–335). Springer. https://doi.org/10.1007/978-3-319-22708-5_9

Leja, K., & Lewandowicz, G. (2010). Polymer biodegradation and biodegradable polymers – A review. *Polish Journal of Environmental Studies, 19*(2), 255–266.

Lenz, R. W. (1993). Biodegradable polymers. In R. S. Langer & N. A. Peppas (Eds.), *Biopolymers I* (Advances in polymer science) (Vol. 107, pp. 1–40). Springer. https://doi.org/10.1007/BFb0027550

Li, J., et al. (2023). Simultaneous production of cellulase and β-carotene in the filamentous fungus trichoderma reesei. *Journal of Agricultural and Food Chemistry, 71*(16), 6358–6365. https://doi.org/10.1021/acs.jafc.3c00690. ACS Publications.

Martínez Gil, J. M., et al. (2022). Biodiesel production from transesterification with lipase from pseudomonas cepacia immobilized on modified structured metal organic materials. *ACS Omega, 7*(46), 41882–41904. https://doi.org/10.1021/acsomega.2c02873. ACS Publications.

Murad, H. A., & Azzaz, H. H. (2011). Microbial pectinases and ruminant nutrition. *Research Journal of Microbiology, 6*(3), 246–269. https://doi.org/10.3923/jm.2011.246.269. Academic Journals Inc.

Naeem, S., et al. (2023). Biodegradable polymer challenges 11. In *Handbook of biodegradable materials* (p. 293). Springer Nature.

Nair, L. S., & Laurencin, C. T. (2007). Biodegradable polymers as biomaterials. *Progress in Polymer Science, 32*(8–9), 762–798. https://doi.org/10.1016/j.progpolymsci.2007.05.017. Elsevier.

Nashchekina, Y. A., et al. (2020). Biological and rheological properties of collagen cross-linked with glutaraldehyde. *Technical Physics, 65*(9), 1535–1540. https://doi.org/10.1134/S1063784220090224. Springer.

Nhim, S., et al. (2022). Biological cellulose saccharification using a coculture of clostridium thermocellum and Thermobrachium celere strain A9. *Applied Microbiology and Biotechnology, 106*(5–6), 2133–2145. https://doi.org/10.1007/s00253-022-11818-0
Niu, C., et al. (2023). Isolation, identification and application of aspergillus oryzae BL18 with high protease activity as starter culture in doubanjiang (broad bean paste) fermentation. *Food Bioscience, 51*, 102225. https://doi.org/10.1016/j.fbio.2022.102225
Nurdin, M., et al. (2023). Productivity of Aspergillus niger InaCC F57 isolate as cellulase agent in OPEFB hydrolysis for glucose high yield. *BioNanoScience, 13*(1), 114–127. https://doi.org/10.1007/s12668-023-01066-5. Springer.
Oliveira, J., et al. (2020). Marine environmental plastic pollution: Mitigation by microorganism degradation and recycling valorization. *Frontiers in Marine Science, 7*, 567126. https://doi.org/10.3389/fmars.2020.567126. Frontiers Media SA.
Orlando, M., et al. (2023). Microbial enzyme biotechnology to reach plastic waste circularity: Current status, problems and perspectives. *International Journal of Molecular Sciences, 24*(4), 3877. MDPI.
Rao, M. G., Bharathi, P., & Akila, R. M. (2014). A comprehensive review on biopolymers. *Scientific Reviews and Chemical Communications, 4*(2), 61–68.
Rhim, J.-W., & Ng, P. K. W. (2007). Natural biopolymer-based nanocomposite films for packaging applications. *Critical reviews in food science and nutrition, 47*(4), 411–433. https://doi.org/10.1080/10408390600846366. Taylor & Francis.
Rosa, D. S., et al. (2010). Starch plasticized with glycerol from biodiesel and polypropylene blends. *Journal of Thermal Analysis and Calorimetry, 102*(1), 181–186. https://doi.org/10.1007/s10973-010-0828-3. Akadémiai Kiadó, co-published with Springer Science+ Business Media BV ….
Runnels, C. M., et al. (2018). Folding, assembly, and persistence: The essential nature and origins of biopolymers. *Journal of Molecular Evolution, 86*(9), 598–610. https://doi.org/10.1007/s00239-018-9876-2. Springer.
Sethi, A., & Scharf, M. E. (2013). Biofuels: Fungal, bacterial and insect degraders of lignocellulose. *eLS, 10*(1002/9780470015902), a0020374. Wiley.
Shabbir Ali, A., et al. (2023). Applications of natural polymers in regenerative medicine and tissue engineering. *Pharmaceutical Communications, 2*(01 SE-Review Articles), 47–61. https://doi.org/10.55627/pharma.002.01.0299
Shu, L., et al. (2023). Transesterification of phosphatidylcholine with DHA-rich algal oil using immobilized Candida Antarctica lipase B to produce DHA-phosphatidylcholine. *Enzyme and Microbial Technology, 169*, 110266. Elsevier.
Silva, R. R. A., et al. (2023). Biodegradation of polymers: Stages, measurement, standards and prospects. *Macromolecules, 3*(2), 371–399. https://doi.org/10.3390/macromol3020023
Skariyachan, S., et al. (2022). Recent advances in plastic degradation – From microbial consortia-based methods to data sciences and computational biology driven approaches. *Journal of Hazardous Materials, 426*, 128086. https://doi.org/10.1016/j.jhazmat.2021.128086. Elsevier.
Tareen, A., et al. (2022). Biodeterioration of microplastics: A promising step towards plastics waste management. *Polymers, 14*(11), 2275. https://doi.org/10.3390/polym14112275
Tauer, K. (2007). *Polymer-dispersionen*. Max-Planck-Institut.
Udayakumar, G. P., et al. (2021). Biopolymers and composites: Properties, characterization and their applications in food, medical and pharmaceutical industries. *Journal of Environmental Chemical Engineering, 9*(4), 105322. https://doi.org/10.1016/j.jece.2021.105322
Vohlídal, J. (2021). Polymer degradation: A short review. *Chemistry Teacher International, 3*(2), 213–220. https://doi.org/10.1515/cti-2020-0015. De Gruyter.
Vroman, I., & Tighzert, L. (2009). Biodegradable polymers. *Materials, 2*(2), 307–344. https://doi.org/10.3390/ma2020307. Molecular Diversity Preservation International (MDPI).
Weng, J.-K., et al. (2008). Emerging strategies of lignin engineering and degradation for cellulosic biofuel production. *Current Opinion in Biotechnology, 19*(2), 166–172. https://doi.org/10.1016/j.copbio.2008.02.014. Elsevier.

Yasvanthrajan, N., et al. (2023). Production of biodiesel using immobilised Rhizopus oryzae lipase in a microchannel reactor. *Journal of The Institution of Engineers (India): Series E, 104*(1), 165–170. https://doi.org/10.1007/s40034-022-00257-1. Springer.

Zeenat, A., et al. (2021). Plastics degradation by microbes: A sustainable approach. *Journal of King Saud University – Science, 33*(6), 101538. https://doi.org/10.1016/j.jksus.2021.101538

Zhang, Y., et al. (2023). Characterization of feruloyl esterase from Klebsiella oxytoca Z28 and its application in the release of ferulic acid from de-starching wheat bran. *Microorganisms, 11*(4), 989. https://doi.org/10.3390/microorganisms11040989

Chapter 14
Plastic Polymers and Their Impact on the Environment: An Assessment

Aruna Dhamija and Sanjay Kumar Joshi

Abstract In this chapter, environmental policy impact is discussed considering biodegradable plastic polymers. An attempt is made to study various elements, namely, plastic/bioplastic biodegradation; biodegradation of herbicides, pesticides, and insecticides; agricultural crop residue biodegradation; biodegradation of oil; air pollution due to biodegradation; and water pollution due to biodegradable plastics. The current research is analyzed through several initiatives adopted by governments and nonprofit organizations that increase the usage of biodegradable plastics. The effects of biodegradable plastics on the ecosystem are also explored in current research. The article documented the pros and downsides of using biodegradable polymers in an effort to lessen plastic waste in the environment. Finally, it is concluded that for a sustainable future, there is the need of well-designed regulations and a thorough knowledge of the effects of biodegradable plastics.

Keywords Environmental policy · Biodegradable plastic polymers · Biodegradation · Sustainable future

14.1 Introduction

The proliferation of plastic waste has emerged as a serious threat to the environment, necessitating the development of effective countermeasures. Biodegradable polymers, which are designed to decompose back into their natural components, are a possible alternative. The purpose of this article is to investigate the environmental

A. Dhamija (✉)
GLA University, Mathura, Uttar Pradesh, India
e-mail: aruna.dhamija@gla.ac.in

S. K. Joshi
Agri-Business, Indira Gandhi Krishi Vishwavidyalaya, Krishak Nagar Raipur, Chhattisgarh, India

Agri-Business and Rural Management, Indira Gandhi Krishi Vishwavidyalaya, Raipur, India

R. Soni et al. (eds.), *Advanced Strategies for Biodegradation of Plastic Polymers*, https://doi.org/10.1007/978-3-031-55661-6_14

policies and results of using biodegradable polymers. The success of policies enacted to encourage the use of these polymers in decreasing plastic waste and avoiding environmental impact may be evaluated by learning more about their composition and breakdown processes.

Starch, cellulose, and polylactic acid are just a few examples of the renewable raw resources used to create biodegradable polymers. Biodegradable plastics are manufactured to break down in the presence of certain environmental factors, such as heat, moisture, or microbes, in contrast to regular plastics, which may stay in the environment for decades. Reducing plastic waste and keeping nondegradable materials from piling up in landfills and ecosystems might both benefit from the controlled breakdown of these polymers into harmless chemicals.

Various policies have been enacted by governments and organizations to promote the use of biodegradable plastics. These initiatives often take the form of legislation, incentives, or voluntary programs designed to increase the use of biodegradable plastics in manufacturing and retail. Policymakers aim to increase the demand for environmentally friendly plastics by businesses and consumers by providing a hospitable environment for innovation and investment.

However, it is essential to conduct a thorough analysis of how biodegradable plastics affect the environment. They may have benefits in that they produce less garbage, but there might be downsides that you should know about. In the case of biodegradable plastics, for example, the manufacturing process still consumes energy and resources, which may increase greenhouse gas emissions and other negative impacts on the environment. Methane, a powerful greenhouse gas, may also be released during the degradation of biodegradable polymers. Therefore, it is important to conduct a comprehensive analysis of the pros and cons of using biodegradable plastics.

14.2 Plastic Waste and the Ecosystem

One of the major problems now affecting the globe is plastic garbage, which has far-reaching effects. It may impair human health, degrade ecosystems, and harm animals, especially marine species. It can be found fouling everything from the highest mountains to the deepest ocean trenches. As usage of this adaptable material has increased over the past few decades, the waste problem has gotten worse. Since the turn of the century, the amount of plastic produced worldwide has quadrupled, reaching over 400 million metric tons annually in 2021. While plastic items typically have a lifespan of 10 years, depending on their composition and method of disposal, plastics can take up to 500 years to degrade (Alves, 2023).

14.3 Global Plastic Trash Production

More than 350 million metric tons of plastic trash are presently produced annually by humans. Global plastic trash production is anticipated to quadruple by 2060, reaching a startling 1 billion metric tons if existing policies remain unchanged. The amount might fall to fewer than 700 million metric tons if new laws and regulations are passed, such as higher tariffs on the usage of plastics and a greatly enhanced worldwide recycling system. Though fewer than 10% of plastic garbage is presently recycled annually, this objective is still lofty. The great majority of produced plastic garbage is either landfilled or burned, which releases dangerous chemicals. Another quarter is mishandled or left in disarray (Alves, 2023).

14.4 Mishandled Plastic Waste: Impacts on the Ecosystem

Mishandled plastic garbage is primarily burnt in open pits or dumped illegally, but a sizeable quantity also escapes into waterways including rivers and seas. An estimated 30 million metric tons of plastics are projected to have collected in the ocean between 1970 and 2019, while more than 100 million tons are anticipated to have accumulated in rivers and lakes. Such significant plastic trash contamination in rivers can have catastrophic effects on ecosystems and marine life.

More than 80% of the plastic debris dumped into the ocean worldwide comes from Asia as depicted in Fig. 14.1. While wealthy places contribute around 5% of the ocean's plastic garbage, they frequently export massive volumes of plastic debris to underdeveloped countries for processing. The majority of these locations are unable to handle such large amounts of trash (Alves, 2023) (Fig. 14.1).

Abovementioned issues have led to the development of biodegradable counterparts of fossil-based plastics being used in most industries, offering another chance at regaining environmental balance and decreasing pollution levels.

14.5 Biodegradation

Biodegradation is the process by which microorganisms like bacteria and fungus break down organic substances. In contrast to composting, it is typically believed to occur naturally. Biodegradation happens in composting but only because of the human intervention.

The first step in biodegradation is biodeterioration, in which an object's mechanical structure weakens; the second is biofragmentation, in which materials are broken down by microorganisms; and the third is assimilation, in which the old material is incorporated into new cells.

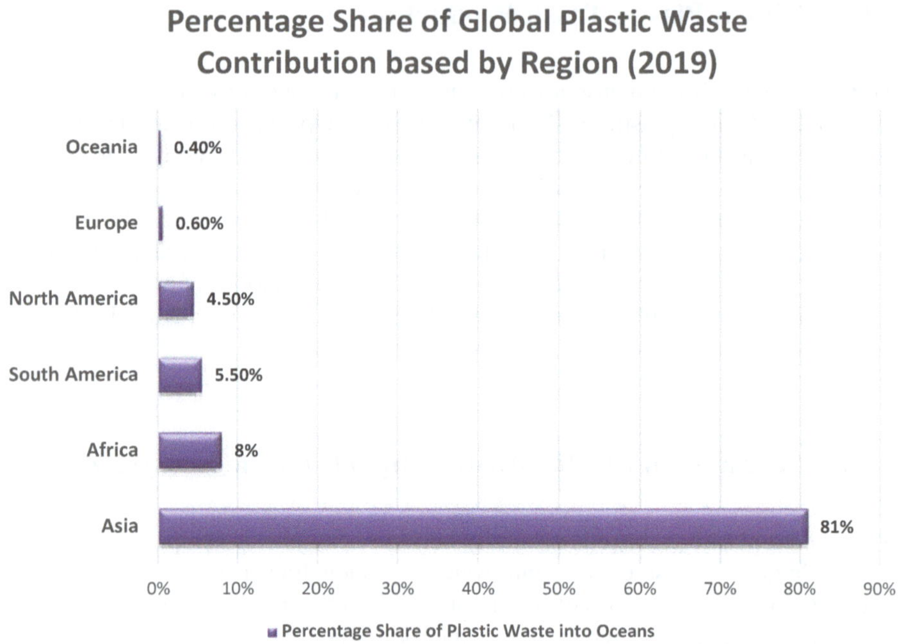

Fig. 14.1 Percentage share of global plastic waste contribution by different regions

In reality, biodegradation may occur with almost any chemical component or substance; time is the critical factor. Vegetables, for example, may disintegrate in a matter of days, whereas glass and other plastics might take thousands of years to break down. The European Union has adopted a biodegradability criterion requiring that more than 90% of the starting material be broken down into carbon dioxide (CO_2), water, and minerals within 6 months.

14.5.1 Mechanisms of Biodegradable Technology

Biodegradable technology is an established field that has found to be used in packaging, manufacturing, and even medicine. The compromise between biodegradability and performance is the main obstacle to broad adoption. As an example, lactide-based plastics lack the advantages of more conventional materials in terms of packaging.

The European Standards Organization, known as CEN, defines oxo-biodegradation as "degradation resulting from oxidative and cell-mediated phenomena, either simultaneously or successively." The precise name for a substance that degrades by the process of oxo-biodegradation as defined by CEN is "oxo-biodegradable," while it is also referred to as "oxo-fragmentable" or "oxo-degradable." Although oxo-biodegradable formulations speed up the biodegradation process, it needs a great deal of expertise to strike the right balance between the chemicals inside the formulations so that the product has a certain amount of time of use before degrading and biodegrading.

The biomedical industry makes extensive use of biodegradable technologies. In terms of use, biodegradable polymers may be broken down into the following subcategories, medicinal, ecological, and dual application, while in terms of origin they are split evenly between natural and synthetic varieties. In order to create polymer drug coatings out of biodegradable polymers, the Clean Technology Group is capitalizing on the usage of supercritical carbon dioxide, which is a solvent at high pressure and room temperature. The polymer (meaning a substance consisting of molecules with repeating structural units to create a lengthy chain) is based on lactic acid, a chemical commonly generated by the body, and is thus able to be expelled naturally after being used to encapsulate a medicine prior to injection in the body. With the coating's timed release properties, fewer injections are needed to provide the same therapeutic effect. After being injected into a patient, biodegradable polymers break down into soluble, nontoxic by-products, making them an intriguing candidate for use in medication administration, as stated by Professor Steve Howdle. It is possible to tune release rates by choosing polymers that undergo degradation at the correct rates once within the body.

Biodegradable, elastic shape-memory polymers have also found utility in biomedical settings. Degradable thermoplastic polymers have made biodegradable implant materials feasible for use in less invasive surgeries. The ability of these polymers to undergo shape change in response to an increase in temperature has led to the development of shape-memory capabilities and readily biodegradable sutures. This allows implants to be placed via narrow incisions, complicated deformations to be performed with ease, and the eventual biodegradation of sutures and other surgical aids.

14.6 Biodegradable Plastics: Market Size and Trends

The global market for biodegradable plastics is projected to reach USD 20.9 billion by 2028 at a CGR of 21.3% from USD 7.9 billion in 2023 (Biodegradable Plastics Market, 2023b). One of the main factors driving market expansion is governments banning the use of single-use plastic together with increased public awareness of the negative consequences of plastic waste.

Additionally, it is projected that increased usage of biodegradable plastics in agriculture and packaging would support the segment's growth. Plastics that don't break down are a worldwide problem. In order to combat this issue, governments all over the world are encouraging the use of biodegradable plastics and outlawing single-use plastics. Due to its eco-friendliness, people are also prepared to pay more for biodegradable plastics (Grand View Research, 2023).

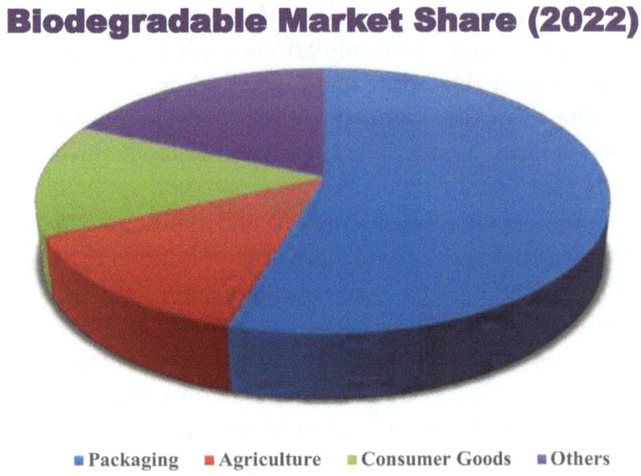

Fig. 14.2 Biodegradable market share by industry application

14.6.1 Market Trends by Industrial Application

With a revenue share of more than 60.3%, packaging was the biodegradable plastic market's largest application category in 2022 as depicted in Fig. 14.2. One of the most popular industries using biodegradable polymers is packaging. Both rigid and flexible packaging utilize it.

The biodegradable plastic industry is expanding as more and more conventional plastics are being replaced in food packaging materials such boxes, wraps, cups, and plates. A sizeable portion of the market was also occupied by the agriculture industry. The application of biodegradable mulch film promotes soil water conservation, control of weed growth, and preservation of a favorable soil temperature to promote quicker crop development (Fig. 14.2). These elements will probably serve as roadblocks in the market's ascent (Global Biodegradable Plastics Market research Report, 2023).

14.6.2 Market Trends by Type

With a sales share of more than 41.3%, starch-based was the leading product category for biodegradable plastics in 2022. The market is segmented into starch-based, PLA, PBAT, PBS, PHA, and other products depending on the kind of product. Throughout the forecast period, the starch-based segment is anticipated to rule the market in both value and volume terms. In application areas including packaging, autos, and agriculture, starch-based polymers are employed. Plastics made from the PLA product are also gaining ground in this market very quickly. One of the main factors influencing the industry is PLA's low cost in comparison to other products (Biodegradable Plastics Market, 2023a).

PHA is anticipated to expand significantly throughout the projection period. The PHA sector of the biodegradable plastic market is anticipated to develop as a result of the growing usage of PHA in medical and other applications. PHA is mostly employed in healthcare applications including medication encapsulation and implantable medical devices. Tissue engineering, bone marrow scaffolds, and bone plates are other medicinal applications. PHA is utilized in the manufacture of compost bags, agricultural foil and films, as well as packaging for food, drink, and consumer goods, in addition to its usage in medicine. These elements are anticipated to support PHA's expansion on a worldwide scale (Grand View Research, 2023).

14.6.3 Market Trend by Sector

With a sales share of more than 60.3%, packaging was the leading application area for biodegradable plastics in 2022. One of the most popular industries using biodegradable polymers is packaging. Both rigid and flexible packaging utilize it. The biodegradable plastic industry is expanding as more and more conventional plastics are being replaced in food packaging materials such boxes, wraps, cups, and plates (Grand View Research, 2023).

A sizeable portion of the market was also occupied by the agriculture industry. The application of biodegradable mulch film promotes soil water conservation, control of weed growth, and preservation of a favorable soil temperature to promote quicker crop development. Additionally, wealthy countries might see the growing usage of biodegradable plastics in consumer products (Grand View Research, 2023).

14.6.4 Market Trend by Region

With a revenue share of more than 43.15%, Europe was the largest geographical market for biodegradable plastics in 2022. The European Union's prohibition on single-use plastics and the general high awareness of plastic waste are expected to drive the regional market throughout the projected period. On the other hand, Pacific is anticipated to become the region with the quickest rate of growth (Global Biodegradable Plastics Market, 2023).

Pacific is anticipated to become the region with the quickest rate of growth worldwide. The region's market is expected to grow as a result of rising disposable income and rising living standards in developing nations like China and India.

In the market, Central and South America also showed promise as a travel destination. The availability of inexpensive raw materials should bode well for the local market. Throughout the projection period, the area is anticipated to see a CAGR of 9.6%. The demand for bio-based packaging in the food and consumer product industries is predicted to increase considerably over the course of the forecast period, which bodes well for the US market for biodegradable plastic.

In comparison to conventional plastics, biobags and bio-based plastics are easier to use, have a better tensile strength, and are environmentally beneficial. Because of this, the demand for biodegradable plastics is expected to increase throughout the course of the forecast period. The market for biodegradable plastic in Europe is expected to expand significantly (Grand View Research, 2023).

14.6.5 Companies Driving Biodegradable Market

Market competition varies from low to substantial from one region to another. Due to the vast number of players, there is a medium level of competition throughout Europe and North America. On the other hand, due to a lack of market participants, competitive rivalry is minimal in the Asia Pacific region. However, both developed and emerging countries are expected to see a growth in demand for biodegradable plastic. Important industry companies are investing a lot of money in research and development to bring new biodegradable plastic goods to market that are durable and easily degrade in the environment. Some of the prominent players in the global biodegradable plastic market include Cargill Incorporated, PTT MCC Biochem Co., Ltd., Biome Technologies plc, Plantic Technologies Limited, BASF SE, Total Corbion PLA, Synbra Technology BV, etc. (Grand View Research, 2023).

14.7 Government Policies for Handling of Plastic Polymers

Companies might be influenced to adopt new strategies by policies that are put in place on a regional or global scale. Since the 17 Sustainable Development Goals (SDGs) were established by the United Nations in 2015, several businesses have promoted sustainable policies. The objectives call for a shift toward the use of renewable sources in plastic manufacturing that does not negatively affect human health (SDG3), the environment (SDG13), marine life (SDG14), or terrestrial ecosystems (SDG15). Goal 11 (sustainable cities and communities) and Goal 12 (responsible production and consumption) may both be advanced by adopting a circular economy mindset. However, despite widespread support for the SDGs, progress toward more eco-friendly plastics is gradual and sometimes needs legislation that varies by nation.

The new EU Green Deal 2020 will focus on stopping Europe's "illegal" trash shipments to Asia. Simultaneously, a regulatory framework for bio-based and biodegradable polymers will be put into place. To further reduce the demand for biodegradable plastics, the ambitious strategy will also examine ways to enhance trash management on a local level. A new financial strategy for rural revitalization will provide incentives for both circular economy and bio-economies. Even though we covered how little of an influence biodegradable plastics have on land take, any increase in the quality and availability of soil will only help these plastics. Production of oxo-degradable plastics, which are believed to cause persistent microplastics in the environment, was banned by the European Union in 2019. However, the transfer of such expertise to bio-based plastic is not covered by the restriction. Increased biodegradability may be shown if oxidation is applied to a bioplastic biodegradability of plastics may be improved. The EU rule from 2018 makes sustainable investing easier by establishing transparent standards for rating green investment funds. In Europe, the extended producer responsibility (EPR) has led to the imposition of fees on plastic manufacturers. The EASAC research claims that these costs range from too low (E50–250 per ton) to too high. However, such charges are not levied elsewhere, and a higher price would do nothing except harm European businesses' ability to compete. However, it may be a considerable financial burden for businesses to manufacture polymers that are both recyclable and biodegradable. Europe's Horizon 2020 and Horizon Europe research funding programs provide the necessary resources. The latter has seen a shift in investment priorities toward digitalizing industry and protecting critical infrastructure at the expense of climate-related technology. Public-private partnerships (PPPs) are expected to account for half of the projected budget (initially E100Bn but still up for negotiation). Seventy percent of PPPs, or E35 billion, will go to creative SMEs for product creation at early stages of the technology adoption life cycle (TRL1-5, ideation/product development). Given the inexpensive cost of petroleum feedstocks and well-established industrial processes, bio-based and biodegradable plastics frequently lack the economic push toward commercialization (high TRLs).

Focus on sustainability is reserved for international conventions rather than national or regional legislation wherever in the world. There seems to be less of an emphasis on environmental protection in official US policy. The Clean Water Act from 1972 is still in effect and is used to punish those who litter. In contrast, the US government issued its National Bioeconomy Blueprint in early 2012 with regard to biomass. The emphasis of this legislation is on renewable energy sources, such as CO_2 fuels and biodegradable polymers manufactured from biomass. Only California is now working on legislation to phase out plastics that aren't composting or recyclable, and even that plan is meeting bureaucratic pushback. Even though China and other nations have weak regulations regarding biodegradable plastics, they nonetheless provide funds for related study. By restricting imports and eliminating single-use plastics, Rwanda is boosting its economy and serving as a model for other African nations. The study from the Organisation for Economic Co-operation and Development (OECD) provides further information on bioplastic policy on a national level.

Policymakers have been prompted to take action by the public outcry over microplastics. The use of microbeads in personal care products and household cleaning products was outlawed in the UK in 2018. Discharges of up to 500 million microplastics daily from UK wastewater treatment facilities prompted this prohibition. In order to make a well-informed decision on proposed new legislation, EU parliamentarians are reviewing papers. Intentionally added microplastics (such as microbeads in cosmetics) have lately been the subject of a socioeconomic review, such as that conducted by the European Chemicals Agency (ECHA). Using completely biodegradable materials as a replacement for these microbeads would be a huge step forward for environmental protection. However, it is important to carefully evaluate the need of using microbeads.

Any (bio)plastic with an annual output of more than 1 ton must register as a chemical with the appropriate authorities. All around the globe, governments have enacted laws to limit the spread of hazardous materials. Only 205 tiny molecules fall within the REACH regulation's definition of "substances of very high concern," which includes chemicals that are carcinogenic, mutagenic, or hazardous for reproduction (CMR), as well as those that are persistent and/or bioaccumulative. The cumulative and harmful impact of (micro)plastics suggests, however, that they belong on the list of SVHCs. All product and biodegradability testing must be completed by the producer in a GLP-certified laboratory before the review time (which is currently set at 45 days). The EU Green Deal may one day aid in the identification and control of potentially harmful materials in (emerging) waste streams.

To promote biodegradable plastics, the type of initiatives taken by authorities all over the world has been discussed below.

14.7.1 Single-Use Plastic Bans

One of the most prevalent initiatives is the ban of single-use plastics. For instance, in 2019, the European Union passed a directive banning certain single-use plastic items such as straws, cutlery, and plates by 2021. The "European Union Strategy for Plastics in a Circular Economy" was established by the European Commission as part of its CE Action Plan, and it addresses concerns such as recyclability, biodegradability, the presence of toxic compounds in some plastics, and marine litter (Calisto Friant et al., 2022).

The goal of EU regulations on single-use plastics is to avoid and lessen the harm that some plastic items may do to the environment, particularly the marine environment and human health. Additionally, they seek to advance the adoption of a circular economy via the use of cutting-edge and environmentally friendly business models, goods, and resources, therefore enhancing the smooth operation of the internal market (Single-use plastics, 2019).

Besides EU, many other countries have initiated bans related to the use of single-use plastics. Some of those countries are Canada, China, India, Japan, Kenya, Mexico, the Netherlands, etc.

14.7.2 Extended Producer Responsibility

Additionally, EU has mandated the development of extended producer responsibility (EPR) systems to manage waste plastic packaging. An EPR system is a mechanism for policy whereby the makers or importers of a product, as opposed to the government, are given the administrative, financial, and physical duty to manage the complete life cycle of a product, particularly the take-back, recycling, and final disposal (Calisto Friant et al., 2022).

In India, the Extended Producers Responsibility (EPR) regime is currently being implemented in the Plastic Waste Management Rules, 2016, and it is the producer, importer, and brand owner's responsibility to ensure processing of their plastic packaging waste through recycling, reusing, or end-of-life disposal (such as co-processing/waste-to-energy/plasticto-oil/roadmaking/industrial composting) (Centralized Extended Producers Responsibility Portal for Plastic Packaging, 2023).

14.7.3 Taxes

Taxes on plastic material, certain types of plastic polymer, or specific applications for plastic (such as single-use packaging) can aid in reducing unsustainable plastic material usage. Applying taxes to single-use plastic goods may assist to raise their price, which in turn may reduce demand for them and lead to substitution. Effective tax policies have the potential to encourage the use of stronger and/or more environmentally friendly substitutes. These alternatives might be non-plastic alternatives made of wood, metal, or glass or redesigned plastic choices that are more easily recyclable, biodegradable, or, in the opposite case, more durable (Taxes on single-use plastics, 2020).

Wales was the first state in the UK to impose a mandatory fee on all carrier bags in October 2011 in an effort to reduce their use beyond what had been accomplished through a voluntary arrangement with shops. The fee is applied to single-use bags that are not meant to be reused more than once and are manufactured entirely or primarily of plastic, paper, and plant-based starch. The fee has successfully reduced the use of carrier bags, which is projected to have decreased by 71% in a short interval between 2011 and 2014 (Taxes on single-use plastics, 2020).

14.7.4 Research and Development

In 2020, the BIM, which was supported by the Malaysia Implementation Council for Biotechnology and introduced in November 2011 by the Malaysian prime minister, was set to give the country a foundation for creating a high-income bioeconomy by creating a workable R&D and commercialization environment for industrial, agricultural, and medical biotechnology (Moshood et al., 2021).

China is reported to be receiving a substantial amount of governmental and scientific investment for biodegradable plastics. The National Development and Reform Commission created a specific equity fund. Institutions, like the University of Tsinghua and Sichuan University, the Institutes of Physics and Chemistry of the

Chinese Academy of Sciences, and others have been actively engaging in research. China has also boosted the development of polylactic acid (PLA) materials. A specific initiative has pushed the manufacturing and use of biodegradable polymers since 2005 (Moshood et al., 2021).

Similar strategies have been at work in countries like Japan, Thailand, etc. wherein investments are being made toward research and development into bioplastic development and making them more and more safe for the environment.

14.7.5 Incentivization

This includes an incentive for those engaged in the development or marketing of biodegradable plastics by way of providing subsidies or levies, hence encouraging the progress in bioplastic industry. An example of the Ministry of Japan suits in this category, where they had intended to provide financial support to businesses and academic institutions working to create technologies that would allow plastic packaging to be replaced with bioplastic packaging, as well as to expand and upgrade existing facilities to boost the production of bioplastic products (Renewable Carbon, 2018).

The Ministry also intended to research techniques for gathering and getting rid of discarded bioplastic goods. In its financial request for the next fiscal year, it has budgeted 5 billion (about $45 million) for these initiatives. By raising domestic bioplastic shipments to 1.97 million tons by fiscal 2030, according to the government's global warming prevention strategy, the nation could reduce 2.09 million tons of CO_2 emissions (Renewable Carbon, 2018).

14.7.6 Education and Awareness for Biodegradable Plastics

Moreover, governments and NGOs have been actively promoting the reduce-reuse-recycle approach to plastics. They have been advocating for the reduction of plastic consumption, encouraging the reuse of plastic products, and promoting recycling programs. For instance, the United Nations Environment Program launched the Clean Seas campaign in 2017, which aims to engage governments, businesses, and individuals in reducing marine plastic pollution through various actions, including promoting the use of biodegradable plastics.

Furthermore, governments have partnered with NGOs to raise awareness and educate the public about the problems associated with plastic waste. The Government of Malaysia has collaborated with environmental nongovernmental organizations to launch awareness campaigns, as part of their "Malaysia's Roadmap Towards Zero Single-Use Plastics: 2018–2030." These initiatives aim to transition toward a more sustainable future by encouraging the use of bio-based and biodegradable plastics (Terzopoulou et al., 2021).

Several countries, including Japan and South Korea, have established national programs and initiatives to support the development and adoption of bioplastics. These programs include funding for research and development, support for pilot projects, and the establishment of testing and certification standards. Furthermore, governments have recognized the importance of a circular bioeconomy and are implementing strategies to promote the recycling and upscaling of bioplastics. For instance, in the Netherlands, the government has launched a program called "Towards a Sustainable Plastics Industry" which aims to promote the recycling and reuse of bioplastics through the development of innovative technologies and infrastructure.

In addition, governments and NGOs are also working to increase public awareness and participation in the use of biodegradable plastics. For instance, in many countries, governments have implemented awareness campaigns to educate the public about the benefits of biodegradable plastics and how to properly use and dispose of them. This includes initiatives such as "No Plastic Bag Day," "3R campaign," and "Environmental Education" in Malaysia, organized by both the government and nonprofit organizations like World Wildlife Fund-Malaysia and the Malaysian Nature Society (Latip et al., 2022). These campaigns aim to educate citizens on the importance of reducing plastic consumption, promoting the reuse of plastics, and recycling programs.

14.8 Impacts of Biodegradable Plastic on the Environment

1. *Plastic/Bioplastic Biodegradation*

Only approximately 1% of the roughly 368 million tons of plastics generated annually are bioplastics. The bioplastic market is expanding and diversifying every day due to rising human requirements and the development of new biopolymers, cutting-edge applications, and innovative goods. European Bioplastics' most current market statistics shared with the nova-Institute predicts that by 2025 worldwide bioplastic production capabilities would have peaked at 2.87 million tons, up from 2.11 million tons in 2020. PLA, PP, and PHA are just a few examples of cutting-edge, newly developed biopolymers that show promising growth rates. Biopolymers, including PLAs, PHAs, thermoplastic starch, and plant-based components, are widely reported. Biomass-based products that are both biodegradable and renewable are preferable to nonbiodegradable petrochemicals. Figure 3 depicts the breakdown and regeneration of many polymers.

2. *Biodegradation of Herbicides, Pesticides, and Insecticides*

Pesticides and their degradable products can be degraded or transformed by microorganisms like bacteria, fungi, and algae in the presence of stimulating factors like the growth of some plant species, such as *Luffa cylindrica (L.) M. Roem. (Loofa)*, *Agave tequilana F.A.C. Weber*, and the beans of *Coffea arabica L.*

3. *Agricultural Crop Residue Biodegradation*

These by-products of crop production are significant natural capital, and there is an urgent need to manage them in order to make the most of the various inputs. Agricultural conservation relies heavily on the science-based practice of managing crop wastes. The quick shift from traditional farming to modern, input-intensive farming often results in increased levels of agricultural residue. Waste goods are produced at an alarming pace as a result of increased food production to feed a rising population. The by-products of agriculture help plants and soils thrive.

4. *Biodegradation of Oil*

Bacteria, yeasts, molds, fungus, algae, and protozoa are only some of the microorganisms that may break down oil components for energy. Bacterial biodegradation hydrocarbon is naturally dispersed in many regions of the world's seas and is notably abundant on chronically oil-polluted coastlines that are subjected to untreated sewage and industrial discharges from ships.

The biodegradation of oil in seawater is affected by a wide variety of interrelated biological, chemical, and physical factors, such as the presence of a specific microorganism, the availability of a suitable habitat, the presence of nutrients, the wind direction sea movement, the velocity of oil, the type of oil components, salinity, chemical bulk in the area, the amount of accumulated oil, and the natural degradation mechanism of fat over time.

5. *Air Pollution due to Biodegradation*

The biodegradation of acetone was conducted in a trickle-bed biofilter employing two columns of comparable size, one filled with coconut fibers and the other with plexiglass chips.

Acetone was biodegraded in a trickle-bed biofilter using two columns of similar size, one filled with plexiglass chips and the other with coconut fibers. Through botanical filtration, Epipremnum aureum, often known as golden pothos, has effectively eliminated indoor gas pollution in either its native soil or an activated root bed. The elimination efficiency of detectable levels of formaldehyde was crucial. Clean air delivery rate (CADR), which is typically used to detect air quality and has the cleaning ability of mobile air cleaners, was only ~5.1 $m^3/h/m^2$ bed when considering static plants in pots as in closed rooms but increased to ~233 $m^3/h/m^2$ bed when considering air flowing through the bed. Assuming microbial activity at 108 $m^3/h/m^2$ bed, we get a CADR value. Root-zone moisture was also critical for sustaining microbial life and absorbing water-soluble chemicals like formaldehyde. The unstable organic chemicals ingested by the plant were efficiently destroyed thanks to the plant's introduction and maintenance of a microbial population well suited for the task. The plant's presence improved the effectiveness of the filtering process.

There are a lot of useful chemical components in *Glycine max (L.) Merr.* (soybean), including about 18 distinct amino groups. At the molecular level, contaminants may be stopped by any chemical class. In order to expose the chemical groups in the soy protein to pollutants, acrylic acid was employed to untangle the protein. All chemical and particulate contaminants may be filtered out by the produced filter.

6. *Water Pollution due to Biodegradable Plastics*

By using microorganisms including bacteria, algae, fungi, protozoa, and green plants in the presence of enzymes, biodegradation is a biological treatment of sewage/wastewater that removes contaminants from the natural domain or modifies them into beneficial forms. Biodegradation is the process by which organic matter is decomposed into simpler, more readily excised molecules by the action of microbes. Because it is a highly sustainable method for cleaning polluted water, biodegradation has lately been applied in sewage purification. This natural phenomenon has various cost-effective and safety-conscious uses. In the end, it outperforms alternatives like chemical and physical eradication methods.

14.9 Limitations

In present study, biodegradation processes could have been considered as it helps keep soil ecosystems healthy, which is important for plant productivity. Soil microbes decompose organic matter, which benefits plant growth by increasing available nutrients, enriching soil structure, and boosting its ability to store water.

Biodegradable agriculture methods as having the potential to increase crop yields could be discussed to make soil healthy and fertile that can be enhanced by applying organic fertilizers made from natural materials like compost or manure. As a result, plant development and output can be improved.

Biodegradation can help boost productivity considering factors such as genetic traits, environmental conditions, and management practices, and the needs of individual species could be included in future study as its plays a vital role. Optimal productivity is difficult to attain without a systemic strategy that considers a wide range of elements and seeks to strike a balance between ecological sustainability and productivity targets.

14.10 Conclusion

In response to the environmental problems caused by conventional plastics, biodegradable polymers have developed as a viable alternative. The usage of biodegradable plastics has been actively encouraged by governments and organizations via the development of legislation and regulations, as can be justified from their increasing market size in different regions spread all over the world. A more sustainable future is the intended result of these regulations, which also seek to encourage innovation and cut down on plastic waste.

The environmental implications of biodegradable plastics should be carefully considered notwithstanding its useful features, such as decreased trash buildup and possible resource-saving. Energy usage, carbon emissions, and the possible release of methane are only some of the environmental problems unique to the manufacture and breakdown processes of biodegradable plastics. To make sure the use of biodegradable plastics doesn't have unanticipated negative effects, it's important to do a thorough analysis.

In conclusion, environmental implications of biodegradable plastics should be taken into account while formulating policy. Biodegradable plastics are a sustainable alternative to traditional plastics, and their widespread adoption would benefit from further study and development in the sector, as well as well-informed policy.

References

Alves, B. (2023, August 31). *Global plastic waste*. Retrieved from Statista.com: https://www.statista.com/topics/5401/global-plastic-waste/#topicOverview

Biodegradable Plastics Market. (2023a). Polaris Market Research. Retrieved from https://www.polarismarketresearch.com/industry-analysis/biodegradable-plastics-market

Biodegradable Plastics Market. (2023b). Markets and Markets. Retrieved from https://www.marketsandmarkets.com/Market-Reports/biodegradable-plastics-93.html

Calisto Friant, M., Lakerveld, D., Vermeulen, W. J. V., & Salomone, R. (2022). Transition to a sustainable circular plastics economy in The Netherlands: Discourse and policy analysis. *Sustainability, 14*(1), 190. https://doi.org/10.3390/su14010190

Centralized Extended Producers Responsibility Portal for Plastic Packaging. (2023). Retrieved from Ministry of Environment, Forest and Climate Change Government of India: https://eprplastic.cpcb.gov.in/#/plastic/home

Global Biodegradable Plastics Market. (2023). Data Bridge Market Research. Retrieved from https://www.databridgemarketresearch.com/reports/global-biodegradable-plastics-market

Global Biodegradable Plastics Market Research Report. (2023). Insight Ace Analytic. Retrieved from https://www.insightaceanalytic.com/report/global-biodegradable-plastics-market-/1187

Grand View Research. (2023). *Biodegradable plastic market size, share & trends analysis report by product (starch based, PLA, PHA, PBAT, PBS), by application (packaging, consumer goods, agriculture), by region, and segment forecasts, 2023–2030*. Grand View Research.

Latip, M., Sharkawi, I., Mohamed, Z., & Kasron, N. (2022). The impact of external stakeholders' pressures on the intention to adopt environmental management practices and the moderating effects of firm size. *Journal of Small Business Strategy, 32*(3), 45–66.

Moshood, T. D., Nawanir, G., Mahmud, F., Mohamad, F., Ahmad, M. H., & Abdul Ghani, A. (2021). Expanding policy for biodegradable plastic products and market dynamics of bio-based plastics: Challenges and opportunities. *Sustainability, 13*, 6170. https://doi.org/10.3390/su13116170

Renewable Carbon. (2018, September 10). *Japan to subsidize bioplastics development*. Retrieved from Renewable Carbon: https://renewable-carbon.eu/news/japan-to-subsidize-bioplastics-development/

Single-use plastics. (2019, July 2). Retrieved from European Commission: https://environment.ec.europa.eu/topics/plastics/single-use-plastics_en

Taxes on single-use plastics. (2020, June 4). Retrieved from OECD: https://www.oecd.org/stories/ocean/taxes-on-single-use-plastics-186a058b/

Terzopoulou, Z., Zamboulis, A., Bikiaris, D N., Valera, M Á., & Mangas, A. (2021, November 26). *Synthesis, properties, and enzymatic hydrolysis of poly(lactic acid)-co-poly(propylene adipate) block copolymers prepared by reactive extrusion*. https://scite.ai/reports/10.3390/polym13234121

Chapter 15
Current Status of Bioplastic Synthesis

Safina Ismail, Kalp Das, and Ravindra Soni

Abstract The chapter provides a comprehensive overview of the synthesis and significance of bioplastics as eco-friendly alternatives to conventional petro-based plastics. Petroleum-based plastics are linked to numerous environmental issues, such as pollution, persisting in both terrestrial and marine habitats, and greenhouse gas emissions. The escalating environmental concerns associated with traditional plastics have led to an urgent need for sustainable alternatives. Bioplastics, derived from renewable resources such as plants, microorganisms, or agricultural by-products, have emerged as a promising solution to mitigate the ecological impact of conventional plastics. Beginning with an introduction to the historical development and widespread use of plastics, the narrative delves into the environmental challenges posed by plastic pollution. The chapter emphasizes the urgent need for sustainable alternatives, leading to the exploration of bioplastics as a viable solution. The focus then shifts to the definition and classification of bioplastics based on their biobased or biodegradable properties.

Keywords Bioplastic · Starch · Sustainability · Polyhydroxyalkanoates

15.1 Introduction

The inception of plastics in the early twentieth century marked a revolutionary moment in materials science. Baekeland's creation of Bakelite in 1907 is often considered the starting point (Kumar et al., 2014). Plastics have since become an integral part of daily life, finding applications in packaging, construction, healthcare, and more. Plastics derived from petrochemicals, such as polyethylene terephthalate

S. Ismail · R. Soni (✉)
Department of Agricultural Microbiology, Indira Gandhi Krishi Vishwavidyalaya, Raipur, Chhattisgarh, India

K. Das
Department of Vegetable Science, Punjab Agricultural University, Ludhiana, Punjab, India

R. Soni et al. (eds.), *Advanced Strategies for Biodegradation of Plastic Polymers*, https://doi.org/10.1007/978-3-031-55661-6_15

(PET), polybutylene terephthalate (PBT), polypropylene (PP), polystyrene (PS), and polyvinyl chloride (PVC), are widely utilized in daily life because of their versatility, affordability, lightweight, and exceptional thermal qualities (Kumar et al., 2014; Chisti, 2014; Gadhave et al., 2018). Burning or incinerating plastic releases a number of harmful emissions, including methane and carbon dioxide, which are greenhouse gases that have an impact on global climate (Barker, 2007). Products based on petrochemicals lead to a number of issues, such as trash buildup on land and in natural environments like the sea and oceans (Thompson et al., 2009). The manufacturing of plastics requires the use of additives like brominated flame retardants and phthalate plasticizers (Barnes et al., 2009). Plasticizers are compounds that are added to plastics in order to improve its longevity, flexibility, transparency, and durability. As a result, scientists are currently searching for alternatives.

15.2 Bioplastic: A Sustainable Approach

A. *Definition*

Bioplastics, which are plastics that are either biodegradable or derived from renewable resources, offer a possible substitute (Guidotti et al., 2018; Garlotta, 2001; Smitthipong et al., 2015). While bio-derived bioplastics provide a significant reduction in the carbon footprint in the stage of resource extraction, biodegradable bioplastics have alternative methods of waste disposal, minimizing the quantity of plastic waste ending up in our environment (White et al., 2020).

B. *Desirable Properties*

In relation to bioplastics, attributes including mechanical, thermal, and physical qualities are taken into account. The physical characteristics taken into account are apparent viscosity, density, and mold shrinkage. Melting point, heat distortion temperature, and Vicat softening temperature are examples of thermal properties. Mechanical properties include hardness, bending module, moisture absorption, transparency, oxygen barrier, compressive yield strength, compressive modulus, flexural strength, Izod impact strength, shrinkage, and tensile elongation brake. Other characteristics including crystallinity, stackability, and puncture resistance are also taken into account (Pandey et al., 2010; Mathew, 2015).

C. *Classification*

Three categories can be applied to bioplastics:

1. Biodegradable plastics made from biobased materials including PLA (polylactic acid), PHAs (polyhydroxyalkanoates), TPS (thermoplastic starch), and PBS (polybutylene succinate).
2. Biodegradable plastics made from petrochemical resources, like PCL (polycaprolactone) and PBAT (polybutylene adipate terephthalate).

3. Nonbiodegradable or partially biodegradable plastics made of biobased monomers, such as bioPE (bio-polyethylene), bioPET (bio-polyethylene terephthalate), and bioPP (bio-propylene), as well as technical polymers produced from biotechnology, such as bioPTT (polytrimethylene terephthalate) and certain bioPAs (polyamides) (Fredi & Dorigato, 2021).

D. *Bioplastic Production*

Starch
It is a polysaccharide made up of glycosidic bonds binding amylose and amylopectin together. Starch-based polymers can be derived from any plant that has a greater starch content; however, they are primarily generated from sources of wheat, potatoes, rice, and corn. Complex mixtures of starch and biodegradable plastics, like PLA, PBAT, PBS, PCL, and PHAs, are known as starch-based plastics (Gadhave et al., 2018).
Cellulose
Being the most prevalent biopolymer, cellulose is a type of polysaccharide found in a wide range of organisms, including bacteria, plants, and mammals. Biodegradable cellulose-based bioplastics, such "cellophane film," are created by chemically treating cellulose xanthate. Numerous cellulose fillers, including hemp, eucalyptus pulp, flax, and tunicin, have been employed as agents in biocomposite stiffeners (Peterson et al., 1999).
Polyhydroxyalkanoates (PHAs)
It is commonly known that prokaryotic cells possess storage polymers of polyhydroxyalkanoic acids. A class of biobased polymers known as polyhydroxyalkanoates (PHAs) is members of the polyhydroxyester family, which includes 3-, 4-, 5-, and 6-hydroxyalkanoic acids (Pratt et al., 2019). Methane released from feedstock in landfills, composting facilities, farms, food processors, waste haulers, operators of biorefineries, and plastic compounders, in particular, can be used to produce PHA in a successful, low-cost commercial setting (Vigneswari et al., 2014; Semprini et al., 1991). In addition to the more costly biomass derived from edible crops, PHA can also be made from wood biomass, energy, grass, and crop leftovers (Renmatix, Pennsylvania, USA) (Curry & Pillay, 2012). *Methylobacterium rhodesianum*, *Cupriavidus necator*, *Bacillus megaterium*, and other bacteria create PHB, a commonly used PHA obtained from methane (Kuruppalil, 2011, Razza et al., 2009; Poirier et al., 1992).
Polylactic Acid (PLA)
From a packaging perspective, PLA is the most appealing material due to its superior processing ability, biodegradability, and biocompatibility. It is processed by extrusion, thermoforming, blow molding, and injection molding. It is made up of methyl group on alpha C atoms and lactic acid (2-hydroxy propionic acid) (Hakola, 1997). Packaging materials such as PET (polyethylene terephthalate), PS (polystyrene), LDPE (low-density polyethylene), and HDPE (high-density polyethylene) have been replaced by it.

Cyanobacteria-Based Bioplastic
Cyanobacteria blooms, which use sunlight to make chemicals through photosynthesis, have been used to produce bioplastics, according to recent studies. Cyanobacteria have the ability to convert glucose to acetyl-CoA, which is subsequently transformed to acetoacetyl-CoA, β-hydroxybutyryl-CoA, and PHB, as previously mentioned (Balaji et al., 2013). Additionally, it has been demonstrated that genetically modified cyanobacteria that consume sugars may also make polymers; this technique has the potential to replace processes that rely on fossil fuels (Juneja et al., 2013; Yim et al., 2011; Mohan et al., 2020).

15.3 Importance of Bioplastics

- *Renewable Resource*: Bioplastics are often derived from renewable resources like corn, sugarcane, or other plant-based materials. This reduces dependence on finite fossil fuels and helps in mitigating the environmental impact of plastic production.
- *Reduced Carbon Footprint*: The production of bioplastics typically results in lower greenhouse gas emissions compared to conventional plastics. This is because the raw materials for bioplastics are often derived from plants that absorb carbon dioxide during their growth.
- *Biodegradability*: Some bioplastics are designed to be biodegradable, meaning they can be broken down by natural processes over time. This contrasts with traditional plastics, which can persist in the environment for hundreds of years.
- *Reduced Dependency on Nonrenewable Resources*: Bioplastics help reduce the reliance on nonrenewable fossil fuels for plastic production. This contributes to long-term sustainability by using resources that can be replenished over time.
- *Diverse Feedstock Options*: Bioplastics can be produced from a variety of feedstocks, including agricultural residues and nonfood crops, providing flexibility in sourcing raw materials without affecting food supplies.
- *Energy Efficiency*: The production processes for some bioplastics require less energy compared to the production of traditional plastics. This can result in lower-energy consumption and a smaller environmental footprint.
- *Potential for Closed-Loop Systems*: Bioplastics can be integrated into circular economy models where products are designed for reuse, recycling, or composting. This can contribute to reducing overall waste and promoting a more sustainable approach to plastic use.
- *Consumer Preference*: With increasing awareness of environmental issues, there is a growing demand for sustainable products. Bioplastics align with consumer preferences for eco-friendly alternatives to traditional plastics (Niaounakis, 2016; European Bioplastics, 2017).

15.4 Global Production Scope

The global production scope of bioplastics has seen a significant expansion. Statistics from reputable sources illustrate the increasing adoption of bioplastics in various industries (European Bioplastics, 2022). Approximately 50% of the world's bioplastic is produced in Asia. In terms of bioplastic production capability, Europe comes in second. The economy of biobased and biodegradable materials is growing because of new biopolymers like PLA (polylactic acid) and PHAs (polyhydroxyalkanoates), which are 100% biobased and biodegradable. By 2020, bioplastics will account for 10% to 15% of the global plastic market, growing to 25%–30% of the market and surpassing 10 billion units (European Bioplastics, 2022). The European Bioplastics Association reported that the production of bioplastics worldwide in 2019 was 2.11 MT, which is less than 0.6% of the total amount of plastics produced (Niaounakis, 2019). By 2025, it is expected to reach 2.89 MT (Lamberti et al., 2020; European Bioplastic Association, 2019). The global bioplastic market, by product, includes polylactic acid, thermoplastic starch, biopolyamides (nylons), polyhydroxyalkanoates, biopolyols/polyurethane, cellulosics, biopolytrimethylene terephthalate, biopolyethylene, biopolyethylene terephthalate, and polybutylene succinate (Research and Markets, 2021).

15.5 Future Perspectives

For a number of reasons, the capacity and quality of bioplastics exhibit enormous promise in the future. Their reduced carbon footprint in comparison to other conventional forms of bioplastic is the primary justification. In addition, bioplastics have a higher-energy efficiency than regular plastics. The price of polymers manufactured from oil fluctuates due to oil scarcity and value fluctuations. Another rationale is "eco-safety," which refers to the reduction of greenhouse gas emissions through the use of bioplastic. Thus, it will contribute to a decrease in global warming (Garlotta, 2001). Concerning how bioplastics affect the environment, opinions are becoming more and more divided. Though they are frequently praised as superior substitutes for traditional plastics, bioplastics are also linked to drawbacks (Licciardello & Piergiovanni, 2020). Certain bioplastics that are either nondegradable or have limited degradation rates only decompose when exposed to high temperatures or in municipal composters or digesters (Hermann et al., 2011; Lancelot & Moriyama, 2010; Jouhara et al., 2017). Furthermore, under certain precise and tried-and-true circumstances, certain biodegradable polymers can only break down in particular active landfill sites (Mohanty & Misra Ma, 2000). Moreover, the production of bioplastics from plants like maize and grain involves the repurposing of land for the production of plastic rather than meeting food needs (Slater & Gerngross, 2000). Biofuels and bioplastics are produced on over 25% of the agricultural area used to grow grains, according to a recent statistical research. Food prices could rise significantly as more agricultural land is used to create biofuels and bioplastics, which would be detrimental to the less fortunate sections of society (Popp et al., 2014).

References

Balaji, S., Gopi, K., & Muthuvelan, B. (2013). A review on production of poly β hydroxybutyrates from cyanobacteria for the production of bio plastics. *Algal Research, 2*(3), 278–285.

Barker, T. (2007). *Climate change 2007: Mitigation* (Contribution of Working Group III to the fourth assessment, report of the Intergovernmental Panel on Climate Change, p. 104). Cambridge University Press/IPCC.

Barnes, D. K. A., Galgani, F., Thompson, R. C., & Morton. (2009). Accumulation and fragmentation of plastic debris in global environments. *Philosophical Transaction of the Royal Society B: Biological Sciences, 364*(1526), 1985–1998.

Chisti, Y. (2014). How renewable are the bioplastics? *Biotechnology Advances, 32*(7), 1361.

Curry, N., & Pillay, P. (2012). Biogas prediction and design of a food waste to energy system for the urban environment. *Renewable Energy, 41*, 200–209.

European Bioplastics. (2017). *Advantages of plastics from renewable resources*. European Bioplastics.

European Bioplastics. (2022). Market data. Retrieved from https://www.european-bioplastics.org/market/

European Bioplastics Association. (2019). *Bioplastics: Facts and figure*. European Bioplastics Association.

Fredi, G., & Dorigato, A. (2021). Recycling of bioplastic waste: A review. *Advanced Industrial and Engineering Polymer Research, 4*(3), 159–177.

Gadhave, R. V., Das, A., Mahanwar, P. A., & Gadekar, P. T. (2018). Starch based bioplastics: The future of sustainable packaging.

Garlotta, D. (2001). A literature review of poly(lactic acid). *Journal of Polymers and the Environment, 9*(2), 63–84.

Guidotti, G., Soccio, M., Lotti, N., Gazzano, M., Siracusa, V., & Munari, A. (2018). Poly(propylene 2,5-thiophenedicarboxylate) vs. poly(propylene 2,5-furandicarboxylate): Two examples of high gas barrier bio-based polyesters. *Polymers (Basel), 10*(7), 785.

Hakola, J. (1997). *Proceedings of renewable bioproducts: Industrial outlets for the 21st century*. EC Symposium held in Wageningen, Holland. Section 4.

Hermann, B. G., Debeer, L., De Wilde, B., Blok, K., & Patel, M. K. (2011). To compost or not to compost: Carbon and energy footprints of biodegradable materials' waste treatment. *Polymer Degradation and Stability, 96*(6), 1159–1171.

Jouhara, H., Czajczyńska, D., Ghazal, H., Krzyżyńska, R., Anguilano, L., Reynolds, A., et al. (2017). Municipal waste management systems for domestic use. *Energy, 139*, 485–506.

Juneja, A., Ceballos, R. M., & Murthy, G. S. (2013). Effects of environmental factors and nutrient availability on the biochemical composition of algae for biofuels production: A review. *Energies, 6*(9), 4607–4638.

Kumar, Y., Shukla, P., Singh, P., Prabhakaran, P. P., & Tanwar, V. K. (2014). Bioplastics: A perfect tool for eco-friendly food packaging: A review. *Journal of Food Product Development and Packaging, 1*(1), 01–06.

Kuruppalil, Z. (2011). Green plastics: An emerging alternative for petroleum-based plastics. *International Journal of Engineering Innovation, 3*(1), 59–64.

Lamberti, F. M., Román-Ramírez, L. A., & Wood, J. (2020). Recycling of bioplastics: Routes and benefits. *Journal of Polymers and the Environment, 28*(10), 2551–2571.

Lancelot, C. J., & Moriyama, C. (2010). *A review of biodegradation Technology in today's biologically active landfills*. Plastics Environmental Council.

Licciardello, F., & Piergiovanni, L. (2020). Packaging and food sustainability. In *The interaction of food industry and environment* (pp. 191–222). Academic Press.

Mathew, L. K., (2015). An Overview of Bioplastics dalam. *International journal of current research and academic review, 3*(9), 15–19.

Mohan, S. V., Hemalatha, M., Chakraborty, D., Chatterjee, S., Ranadheer, P., & Kona, R. (2020). Algal biorefinery models with self-sustainable closed loop approach: Trends and prospective for blue-bioeconomy. *Bioresource Technology, 295*, 122128.

Mohanty, A., & Misra Ma, G. (2000). Hinrichsen, biofibres, biodegradable polymers and biocomposites: An overview. *Macromolecular Materials and Engineering, 276*(1), 1–24.

Niaounakis, M. (2016). *Biopolymers: Processing and products*. William Andrew.

Niaounakis, M. (2019). Recycling of biopolymers – The patent perspective. *European Polymer Journal, 114*, 464–475.

Pandey, A., Kumar, P., & Singh, V. (2010). http://www.indiapackagingshow.com/ConfPDF2010/App-bioplastic.pdf

Petersen, K., Nielsen, P. V., Bertelsen, G., Lawther, M., Olsen, M. B., Nilsson, N. H., & Mortensen, G. (1999). Potential of biobased materials for food packaging. *Trends in food science and technology, 10*(2), 52–68.

Poirier, Y., Dennis, D. E., Klomparens, K., & Somerville, C. (1992). Polyhydroxybutyrate, a biodegradable thermoplastic, produced in transgenic plants. *Science, 256*(5056), 520–523.

Popp, J., Lakner, Z., Harangi-Rakos, M., & Fari, M. (2014). The effect of bioenergy expansion: Food, energy, and environment. *Renewable and Sustainable Energy Reviews, 32*, 559–578.

Pratt, S., Vandi, L. J., Gapes, D., Werker, A., Oehmen, A., & Laycock, B. (2019). Polyhydroxyalkanoate (PHA) bioplastics from organic waste. In *Biorefinery* (pp. 615–638). Springer.

Razza, F., Fieschi, M., Innocenti, D. F., & Bastioli, C. (2009). Compostable cutlery and waste management: An LCA approach. *Waste Management, 29*(4), 1424–1433.

Research and Markets. (2021). *The global market for bioplastics and biopolymers*. Research and Markets.

Semprini, L., Hopkins, G. D., Roberts, P. V., Grbic-Galic, D., & McCarty, P. L. (1991). A field evaluation of in-situ biodegradation of chlorinated ethenes: Part 3, studies of competitive inhibition. *Groundwater, 29*(2), 239–250.

Slater, S. C., & Gerngross, T. U. (2000 August). Gerngross, how green are green plastics? *Scientific American*.

Smitthipong, W., Chollakup, R., & Nardin, M. (2015). *Bio-based composites for high-performance materials: From strategy to industrial application*. CRC Press-Taylor & Francis Group.

Thompson, R. C., Moore, C. J., Vom, F. S., Saal, S. H., & Swan. (2009). Plastics, the environment and human health: Current consensus and future trends. *Philosophical Transactions of the Royal Society B. Biological Sciences, 364*(1523), 2153–2216.

Vigneswari, S., Bhubalan, K., & Amirul, A. (2014). Design and tailoring of polyhydroxyalkanoate-based biomaterials containing 4-hydroxybutyrate monomer. In *Biotechnology and bioinformatics: Advances and applications for bioenergy, bioremediation and biopharmaceutical research* (p. 281). Apple Academic Press.

White, E., Bassilakis, R., & Nogués, S. (2020). *From the plastics present to a sustainable future: The bioplastics innovation landscape, players and market opportunities*. Clarivate™.

Yim, H., Haselbeck, R., Niu, W., Pujol-Baxley, C., Burgard, A., Boldt, J., et al. (2011). Metabolic engineering of Escherichia coli for direct production of 1, 4-butanediol. *Nature Chemical Biology, 7*(7), 445–452.

Chapter 16
Role of Microbial Enzymes and Their Modification for Plastic Biodegradation

Anand Vaishnav, Jham Lal, N. Sureshchandra Singh, Bikash Kumar Pati, Naresh Kumar Mehta, and M. Bhargavi Priyadarshini

Abstract Plastic waste production is a significant global environmental issue that poses a threat to the health and survival of living organisms. The responsible management of these plastics is a significant challenge in the twenty-first century. Various plastic polymers have been developed over the past 150 years, replacing traditional materials like metal, glass, and wood in different applications. However, the nonreactive and nondegradable nature of plastic has resulted in a massive increase in plastic waste generation, causing serious environmental problems. Plastic waste accumulates in landfills, contaminates the soil, and contributes to increased greenhouse gas emissions. These are some of the harmful effects observed. It is essential to address this ecological problem due to the negative impact of plastic pollution on terrestrial ecosystems. Conventional methods of plastic waste disposal, such as incineration, landfilling, and recycling, have environmental drawbacks. Some synthetic plastics, such as polyethylene terephthalate, polyurethane, polystyrene, polypropylene, and polyvinyl chloride, have chemical structures, molecular weights, and degrees of crystallinity that make them highly resistant to natural biodegradation processes. Enzymes produced by microbes have long been considered a potential biological agent for the biodegradation of plastics; these enzymes can facilitate the hydrolysis or oxidation of polymer bonds, resulting in the formation of less complex and less harmful chemicals. Numerous microorganisms, particularly fungi and bacteria, have been isolated and characterized for their ability to degrade plastics under laboratory conditions. Some of the enzymes involved in plastic biodegradation have been identified and cloned, and their mechanisms of action have been elucidated. Furthermore, genetic engineering and protein engineering techniques have been employed to modify and enhance the activity, specificity, and

A. Vaishnav (✉) · J. Lal · N. S. Singh · B. K. Pati · N. K. Mehta · M. B. Priyadarshini
College of Fisheries, Central Agricultural University (Imphal), Lembucherra, Tripura, India

R. Soni et al. (eds.), *Advanced Strategies for Biodegradation of Plastic Polymers*, https://doi.org/10.1007/978-3-031-55661-6_16

stability of these enzymes. This chapter provides an overview of the current knowledge on the role of microbial enzymes and their modifications for plastic biodegradation and discusses the challenges and opportunities for the role of microbial enzymes and their modifications for plastic biodegradation.

Keywords Biodegradation · Protein · Microbial enzymes · Microorganisms · Polymer · Pollution

16.1 Introduction

Plastic pollution has emerged as a worldwide environmental crisis, with adverse effects on ecosystems, wildlife, and the health of humans. The extensive utilization of plastics across different industries has resulted in their accumulation in both terrestrial and marine settings, presenting a substantial challenge for waste management and ecological balance. Traditional plastic recycling methods have limitations, and thus, the search for sustainable and effective plastic biodegradation strategies has become imperative. Exploitation of natural resources without regulation has resulted in unprecedented environmental disruptions due to the rapid introduction of toxic chemicals. One of the major pollutants introduced into the environment is synthetic plastic, a man-made polymer. The year 1869 marked a significant milestone when John Hyatt produced the first plastic polymer, known as celluloid, intended as a substitute for ivory (Morris, 1986). Plastic is a synthetic polymer that has many advantages, such as durability, versatility, and low cost, but also poses a serious environmental problem due to its accumulation and persistence in nature. Plastic waste can cause various negative impacts, such as pollution, toxicity, greenhouse gas emissions, and loss of biodiversity (Fig. 16.1). A recent report reveals that since 1950 approximately 8.3 billion metric tons of plastic have been manufactured, with only 9% of it being recycled, while 79% has been disposed of in landfills or the natural environment (Mohanan et al., 2020). The report also predicts that by 2050 the amount of plastic in the oceans will exceed that of fish (Shilpa et al., 2022). In 2019, the global production of plastic reached an estimated 360 million tons. China emerged as the major contributor, producing 60 million tons, ahead of the United States with 38 million tons, Germany with 14.5 million tons, and Brazil with 12 million tons. The demand for plastics is primarily fueled by the packaging industry, with the building/construction, automobile, electrical/electronic, household/sports, and agriculture sectors also playing significant roles. Within landfills, the most prevalent type of plastic waste is low-density polyethylene (LDPE), predominantly in the form of plastic bags, accounting for 69.13% of the total waste (Mohanan et al., 2020).

Therefore, finding effective and sustainable ways to degrade and dispose of plastic waste is a major challenge for the twenty-first century. One of the possible solutions for plastic waste management is biodegradation, which involves the decomposition of materials made of organic or synthetic substances by microbes,

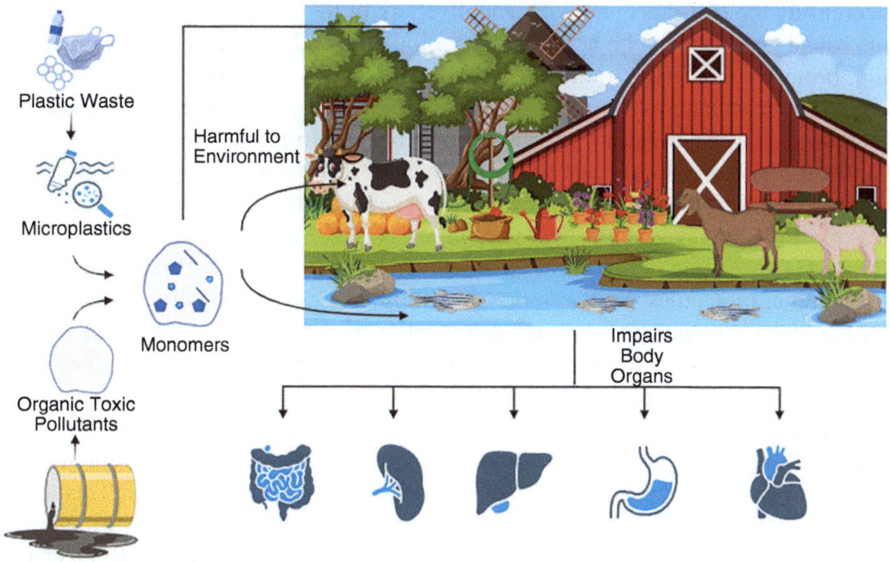

Fig. 16.1 Hazard effects of plastics in living organisms

plants, and algae. Biodegradation can reduce the amount and toxicity of plastic waste and also produce valuable products such as biogas, biofuels, and bioplastics. However, not all plastics are biodegradable, and the rate and extent of biodegradation depend on many factors, such as the type, structure, and size of the plastic, the condition of the environment, as well as the microbiome (De Jesus & Alkendi, 2023). One of the most important factors in plastic biodegradation is the presence of microbial enzymes. The degradation of plastics can be accelerated by the use of microbial enzymes, which are biological catalysts. Microbial enzymes can act on the surface or inside of the plastic and break down the polymer chains into smaller and simpler molecules, such as monomers, oligomers, or carbon dioxide and water. Microbial enzymes can be produced by various microorganisms, such as bacteria, fungi, and algae, that can use plastics as a source of carbon and energy. Some examples of microbial enzymes that can degrade plastics are cutinases, lipases, esterases, and peroxidases (Tania & Anand, 2023). However, the biodegradation of plastics by microbial enzymes is not efficient enough to cope with the increasing amount of plastic waste. There are several challenges and limitations that hinder the biodegradation of plastics by microbial enzymes, such as the low specificity, activity, and stability of the enzymes; the high crystallinity, hydrophobicity, and molecular weight of the plastics; the low availability of oxygen and nutrients; and the presence of additives and contaminants in the plastic waste (Bhardwaj et al., 2013). Therefore, there is a need to improve the biodegradation of plastics by microbial enzymes, and one of the possible ways to do so is to modify the enzymes or the plastics. The modification of microbial enzymes or plastics can enhance the biodegradation of plastics by increasing the affinity, efficiency, and diversity of the enzymes and by decreasing the resistance, toxicity, and diversity of the plastics. The modification of microbial enzymes can be achieved by various methods, such as genetic engineering, protein engineering, immobilization, covalent modification, and nanotechnology. The modification of plastics can be achieved by various methods, such as blending, copolymerization, grafting, oxidation, and irradiation.

Microbial enzymes are one of the key factors that determine the biodegradation of plastics. Microbial enzymes are biological catalysts that can accelerate the chemical reactions involved in the degradation of plastics. Microbial enzymes can act on the surface or inside of the plastic and break down the polymer chains into smaller and simpler molecules, such as monomers, oligomers, or carbon dioxide and water. Different microorganisms that can use plastics as a source of carbon and energy, such as bacteria, fungi, and algae, can produce microbial enzymes. The enzymatic pathways involved in plastic degradation vary depending on the type of plastic. As an example, polyethylene terephthalate (PET) is a common plastic that is utilized in many different products, such as bottles, containers, and textiles. The repetition of the ester bond between terephthalic acid and ethylene glycol forms PET. The degradation of PET into its monomers or other intermediates, such as bis(2-hydroxyethyl) terephthalate (BHET) and mono(2-hydroxyethyl) terephthalic acid (MHET), can be facilitated by microbes that possess the ability to hydrolyze ester linkages (Temporiti et al., 2022). *Ideonella sakaiensis* is just one of the several microbes that can fully breakdown PET with the help of two enzymes: PETase and MHETase. Microbes can absorb terephthalic acid and ethylene glycol, by-products of the hydrolysis of MHET from terephthalic acid, which occurs when PETase breaks down PET into MHET (Krakor et al., 2021). Another example of plastic degradation by microbial enzymes is polyethylene (PE), which is the most common plastic used for making bags, films, and pipes. PE is composed of repeating units of ethylene, linked by carbon-carbon bonds. Microbial enzymes that can oxidize carbon-carbon bonds, such as peroxidases and laccases, can degrade PE into smaller fragments, such as alcohols, aldehydes, and carboxylic acids. These enzymes require hydrogen peroxide or oxygen as co-substrates and can generate reactive oxygen species, such as hydroxyl radicals, that can attack the polymer chains (Amin et al., 2022). Some microorganisms, such as *Aspergillus tubingensis*, have been isolated and characterized for their ability to degrade PE using peroxidases and laccases. These enzymes can cause the oxidation, scission, and erosion of the PE surface, resulting in the formation of pits, cracks, and pores (Kim et al., 2022). Recent advancements in metagenomic studies have enabled the identification of microbial communities with specialized enzymatic capabilities for plastic degradation. Metagenomics is the study of the genetic material directly obtained from environmental samples without the need for culturing the microorganisms. With the use of genomics, we may learn more about the variety and roles played by microbes in various environments, including the plastisphere, water, and soil (biofilm that forms on the surface of plastic waste). Metagenomics can also help to discover novel genes and enzymes that are involved in plastic degradation (Gaur et al., 2022).

Moreover, scientists are exploring strategies to enhance the efficiency of microbial enzymes in plastic degradation. Genetic engineering techniques allow for the modification of enzymes to improve their catalytic activity, substrate specificity, and stability under different environmental conditions. By optimizing these enzymes, researchers aim to create more effective tools for breaking down specific plastic polymers. Some studies have used site-directed mutagenesis, directed evolution, or rational design to make PETase and MHETase work better and bind to PET and its intermediates better. These studies have identified several mutations that enhance the performance of these enzymes, such as increasing the flexibility of the active site, reducing the steric hindrance, and altering the charge distribution (interestingengineering.com). Another example is the use of fusion proteins, chimeric proteins, or protein complexes to combine the activities of different enzymes, such as cutinases, lipases, esterases, peroxidases, and laccases, for synergistic or sequential degradation of plastics. These studies have shown that the fusion or co-expression of different enzymes can increase the degradation rate and extent of various plastics, such as PET, PE, and polyurethane (Thakur et al., 2022). The role of microbial enzymes in plastic degradation represents a promising avenue for developing sustainable and eco-friendly waste management solutions. As research continues to unravel the complexities of enzymatic pathways and their interactions with various plastic types, there is a potential for the development of innovative technologies that can contribute to mitigating the environmental impact of plastic pollution. However, there are also some challenges and limitations that need to be addressed, such as the low availability and accessibility of plastics for microbial enzymes, the potential toxicity and accumulation of degradation products, the ethical and regulatory issues of using genetically modified microorganisms and enzymes, and the economic and social feasibility of implementing biodegradation processes at large scales (Mohanan et al., 2022).

16.2 Degradation of Plastic

Plastic degradation is the breakdown of complex polymer structures into simpler monomers, which depends on various factors such as the type of plastic, the surface characteristics, the shape and size, and the molecular weight of the polymers (Thakur et al., 2022; Harrison et al., 2018). This breakdown can happen through different natural or human-made mechanisms, each with different environmental impacts. Plastic degradation is important for tackling the global problem of plastic pollution and finding sustainable solutions. The degradation process involves different treatments that include physical, chemical, and biological variables. Physical variables include exposure to heat and radiation, chemical variables include the use of acids and bases, and biological variables include the use of microorganisms and insects. Polymer degradation can be classified based on the types of agents that interact with the plastic, such as photooxidative, thermal, ozone-induced, mechanochemical, and catalytic degradation and biodegradation (interestingengineering.com). Biodegradation is the process of plastic degradation by microorganisms such

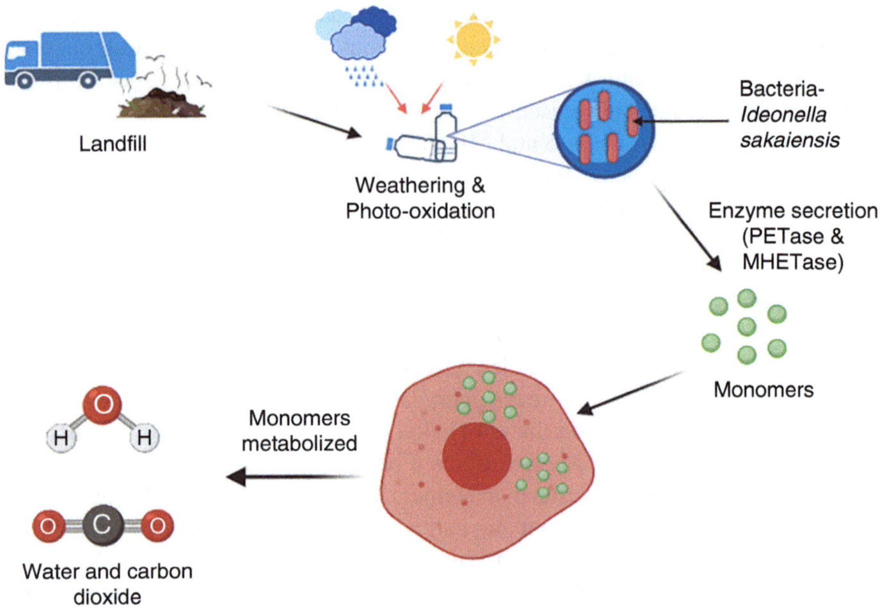

Fig. 16.2 Biodegradation of plastics through microbial enzymes

as bacteria, fungus, and actinomycetes, which convert the plastic into carbon and energy (Fig. 16.2).

16.3 Mechanism of Plastic Biodegradation

Microplastics are small pieces of plastic that can be degraded by microorganisms through depolymerization. The microorganisms produce free radicals and extracellular enzymes that can break the biodegradable polymers into smaller fragments. These fragments are then further depolymerized into monomers by the microorganisms, which can use them for growth. The monomers are transported into the cells, where they are mineralized. Mineralization is the process of converting the monomers into CO_2, H_2O, and CH_4 in anaerobic conditions or CO_2 and H_2O in aerobic conditions. Mineralization also produces biomass, which can be used as an energy source. The microbiological degradation of plastics involves four steps: colonization, hydrolysis, catalytic hydrolysis, and cracking. Colonization is the attachment or exposure of microorganisms to plastic surfaces, which is necessary for the subsequent enzymatic degradation. Hydrolysis is the combination of the polymer matrix and the enzymes, which leads to the cleavage of the polymer chain. Catalytic hydrolysis is the acceleration of hydrolysis by the enzyme hydrolase (Sangale et al.,

2019), which can be produced by the microorganisms inside or outside the cells. Cracking is the uptake and intracellular degradation of small oligomers or monomers by bacteria (Pramila & Ramesh, 2015).

Biodegradation by enzymes is easier to measure by weight loss and the introduction of functional groups than by oxidation, which is not effective for larger plastics or plastic pieces. One important change that happens in microplastics is the decrease in weight during biodegradation. This weight loss is calculated by comparing the initial weight of the microplastics with the weight after exposing them to microorganisms (Hung et al., 2016). The by-products that are produced by microbial degradation and metabolism are beneficial for the environment. A polymer's shape, especially its crystallinity, affects its biodegradability a lot. The enzymes attack the polymer's amorphous regions, which are less organized and more biodegradable, so this factor influences the biodegradation rate (Sangale et al., 2019). Therefore, microbial enzymes are very important for ecological health and environmental cleanup because they have efficiency and functionality, they take part in biodegradation and biotransformation, and they help biodegradation. Microorganisms such as bacteria, fungi, and actinomycetes can biodegrade various types of plastic. It is significant to note that various microbial species or genera target various groups of polymers in the case of plastic degradation. This specificity is due to the varying chemical compositions and structural characteristics of different plastic materials. The diversity of microbes capable of plastic degradation demonstrates the adaptability and versatility of these organisms in the face of environmental challenges. For example, polyethylene is a widely used synthetic plastic, and it can be categorized into different forms, such as low-density polyethylene (LDPE). Research has shown that some bacteria and fungi are particularly effective at degrading LDPE. Bacterial isolates of *Bacillus* and a fungal isolate of *Aspergillus* species have been identified as some of the most efficient degraders of LDPE. This indicates that specific microbial strains have evolved to target and metabolize certain types of plastics. They break down plastic into simpler compounds by acting as enzymes on the plastic polymer chains. This helps the biodegradation process in the long run (Fig. 16.3).

16.4 Microorganisms Involved in Plastic Degradation

16.4.1 Bacteria

Bacteria play a vital role in nutrient cycling on Earth, as they are involved in both the initial modification of nutrients and their subsequent cycling through the ecosystem. Their main function in the decomposition process is to ensure the release of carbon and nutrients from complex polymers. Research has shown that bacteria have the inherent ability to degrade long-chain fatty acids, which is the basis for

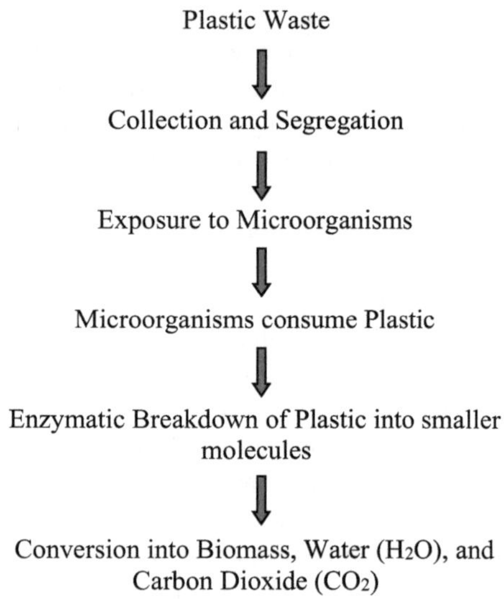

Fig. 16.3 Process of biodegradation of plastics

their capacity to degrade plastics. Pseudomonas is the most well-known and extensively studied bacterial genus for plastic polymer degradation (Wilkes & Aristilde, 2017). Biofilm formation is a key factor for bacterial plastic degradation, as it enhances the adhesion and persistence of bacterial colonies on plastic surfaces (Puglisi et al., 2019). One of the most remarkable discoveries about plastic degradation is that *Ideonella sakaiensis*, a new species that was found in a group of bacteria from a landfill, can efficiently degrade polyethylene terephthalate (PET), which is its main carbon and energy source (Yoshida et al., 2016). Several bacterial strains have also been associated with plastic degradation. A thermophilic bacterium, *Bacillus brevis*, was isolated from soil and has the ability to degrade polylactic acid; it is one of the many species in this group that possess this ability. Others species include *Azotobacter* sp., *Ralstonia eutropha*, *Halomonas* sp., *Acidovorax delafieldii*, *Paenibacillus amylolyticus*, *Bacillus pumilus*, *Bordetella petrii*, *Pseudomonas aeruginosa*, *Shewanella* sp., and *Bacillus megaterium*, among others. Studies on polyethylene-degrading bacteria have demonstrated the potential of various bacterial species such as *Acinetobacter baumannii*, *Arthrobacter* sp., *Viscosus* sp., *Pseudomonas* sp., *Arthrobacter viscosus*, *Bacillus amyloliquefaciens*, *Bacillus thuringiensis*, *Mycoides*, *Bacillus cereus*, *Bacillus pumilus*, *Staphylococcus cohnii*, *Xylosus* sp., *Pseudomonas fluorescens*, *Rahnella aquatilis*, *Micrococcus luteus*, *Micrococcus lylae*, *Paenibacillus macerans*, *Flavobacterium* sp., *Delftia acidovorans*, *Ralstonia* sp., *Rhodococcus erythropolis*, *Pseudomonas aeruginosa*, and *Bacillus brevis* in plastic biodegradation (Koutny et al., 2009). *Bacillus* sp., *Staphylococcus* sp., *Streptococcus* sp., *Diplococcus* sp., *Micrococcus* sp., *Pseudomonas* sp., and *Moraxella* sp. were some of the microbes that were detected in the polyethylene bags. Although many studies have focused on the biodegradability of pure bacterial strains, it is important to acknowledge that in natural environments bacteria often work together within consortia, a phenomenon that has been shown in several studies (Lwanga et al., 2018; Shah et al., 2008).

16.4.2 Fungi

Fungi, a dominant group among microorganisms, play a crucial role in maintaining biogeochemical cycles and the distribution of essential nutrients on Earth. Because they can use synthetic polymers as their main source of carbon, their ability to degrade various plastic polymers is becoming more and more obvious. Recent studies have highlighted different fungal species as potent agents for plastic biodegradation, with the *Aspergillus* genus emerging as a prominent fungal group with this capability. Some fungi have been found to have the ability to degrade plastics in various terrestrial habitats, especially *Aspergillus* species. Some examples of bacteria that can degrade plastics are *Acinetobacter clavatus*, *Aspergillus fumigatus*, and *Aspergillus niger* (Gajendiran et al., 2016; Osman et al., 2018; Usman et al., 2020). On the other hand, only *Penicillium roqueforti* and *Tritirachium album* can degrade polylactic acid (PLA) among all the fungal genera. Moreover, *Aspergillus niger* can degrade PVC (Mogil'nitskii et al., 1987). Many fungi can degrade polyhydroxybutyrate (PHB) and polyesters, such as *Acremonium*, *Cladosporium*, *Debaryomyces*, *Emericellopsis*, *Eupenicillium*, *Fusarium*, *Mucor*, *Paecilomyces*, *Penicillium*, *Pullularia*, *Rhodosporidium*, and *Verticillium*. Similarly, polycaprolactone (PCL) can be degraded by various fungi, such as *Aspergillus*, *Aureobasidium*, *Chaetomium*, *Cryptococcus*, *Fusarium*, *Rhizopus*, *Penicillium*, and *Thermoascus*. The polyesteracetals (PEA) can be degraded by the following microorganisms: *Aspergillus*, *Aureobasidium*, *Penicillium*, and *Pullularia*. Soil samples from areas where plastic has been disposed of have revealed the presence of fungi that can degrade plastic, such as *Alternaria solani*, *Spicaria* sp., *Aspergillus terreus*, *Aspergillus fumigatus*, and *Aspergillus flavus* (Raziyafathima et al., 2016). Other fungal strains that have been shown to have the potential for plastic degradation in recent years include *Aspergillus versicolor*, *Aspergillus flavus*, and *Chaetomium* sp. Polyethylene bags can be degraded by various fungi, such as *A. glaucus*, *A. flavus*, *A. candidus*, *A. nidulans*, *A. cremeus*, and *A. ornatus* (Kathiresan, 2003). It has also been shown that fungal consortia can degrade different types of plastic in a synergistic way, such as polyurethane (PU) and polyethylene (PE). Many other fungi, such as *Fusarium solani*, *Alternaria solani*, *Spicaria* sp., *Geomyces pannorum*, *Phoma* sp., and *Penicillium* sp., have strong plastic degradability. Some of the fungal species that have been studied for their ability to biodegrade bioplastics are *Paecilomyces lilacinus*, *Aspergillus flavus*, *Paecilomyces verrucosum*, *Paecilomyces album*, *Thermoascus aurantiacus*, and *Fusarium moniliforme* (Oda et al., 1995; Torres et al., 1996).

16.4.3 Actinomycetes

Actinomycetes are a large and varied family of filamentous bacteria found in many different habitats. Their significant biotechnological importance in areas like bioremediation, healthcare, and food business has made them famous. One category of actinomycetes that has shown remarkable plastic biodegradation capabilities is the *Streptomyces* group. Another group includes *Rhodococcus ruber*, *Actinomadura* sp., and *Thermoactinomyces* spp. (Auta et al., 2018; Jabloune et al., 2020). Some species of actinomycetes have demonstrated the ability to degrade specific kinds of plastic, such as polybutylene succinate (PBS), polylactic acid (PLA), and polycaprolactone (PCL). Species such as *Amycolatopsis thailandensis* (Chomchoei et al., 2011), *Streptomyces bangladeshensis* (Hsu et al., 2012), *Kibdelosporangium aridum*, *Saccharothrix* (Jarerat et al., 2003), *Actinomadura keratinilytica* (Sukkhum et al., 2011), and *Streptomyces thermoviolaceus* subsp. *Thermoviolaceus* (Chua et al., 2013) are have been particularly notable for their effectiveness in this regard. The remarkable plastic biodegradative potential of actinomycetes underscores their significance in addressing plastic pollution and offers a promising avenue for developing sustainable solutions in waste management. Furthermore, the versatility of actinomycetes and their adaptability to various environments make them valuable assets in a range of biotechnological applications beyond plastic biodegradation. Their roles in soil health, bioremediation of contaminants, and the production of bioactive compounds have expanded their significance in the scientific community.

16.5 Steps Involved in Plastic Biodegradation

Bacterial microbes control a cascade of metabolic reactions that break down plastic polymers. The degradation of plastic is characterized by changes in the physical properties of the polymers, such as a decrease in molecular weight, mechanical strength, and surface qualities (Ho et al., 2018). Through this process, various biochemical pathways are utilized, including biodeterioration, biofragmentation, assimilation, and mineralization. It is various enzymes and bond-cleaving processes that carry out each of these pathways (Gu, 2003; Pathak, 2017).

16.5.1 Biodeterioration

Biodeterioration is the initial process in the biodegradation of plastic polymers, serving as the initial step where microorganisms colonize the polymer's surface and initiate the breakdown of its chemical structure. This stage aims to diminish the resistance and durability of the plastic material, leading to modifications in its chemical, mechanical, and physical properties (Anjana et al., 2020). In order to improve the interaction between bacteria and polymeric surfaces, biofilm formation is essential for plastics with high surface hydrophobicity, like polyethylene (Schwibbert et al., 2019). Pseudomonas and other biofilm-forming bacteria are unique among microorganisms in their ability to adhere to and degrade low-density polyethylene (Tribedi et al., 2015). One interesting property of fungal cells is their hyphae, which allow them to connect to polymeric polymers and proliferate on various surfaces (Sánchez, 2020). Furthermore, when plastic polymers are attached to biofilms, exopolysaccharides are crucial in establishing strong adherence (Anjana et al., 2020). Biodeterioration sets the subsequent stages in plastic biodegradation. By initiating microbial colonization, surface modifications, and biofilm formation, biodeterioration enhances the accessibility of plastic polymers to enzymatic and chemical actions.

16.5.2 Biofragmentation

Biofragmentation is a key process in plastic degradation that involves the depolymerization of plastic polymers through the action of enzymes outside of cells and free radicals produced by microorganisms (Jenkins et al., 2019). This essential step contributes to the reduction of polymer size, ultimately breaking down the plastic into smaller units. The biofragmentation process is crucial for diminishing the molecular weight of plastics and creating lower-weight molecules that can be further utilized by microorganisms. Biofragmentation is a critical stage in plastic biodegradation as it initiates the cascade of events leading to the assimilation and mineralization of plastic polymers. By breaking down large, inert polymer chains into smaller, more reactive fragments, biofragmentation paves the way for subsequent steps in the biodegradation process. The reduced molecular weight and altered surface properties make the plastic more accessible to microbial attack, enabling further degradation and, ultimately, the return of plastic constituents to the environment in benign forms. Biofragmentation is also essential for developing effective strategies to harness microbial activities in plastic waste management and environmental remediation. Ongoing research in this field aims to unravel the intricacies of specific enzymes, microbial communities, and environmental conditions that optimize the biofragmentation process, offering promising avenues for addressing the global challenge of plastic pollution.

16.5.3 Assimilation

The assimilation stage involves the transfer of the biofragmentation-produced, lower-molecular-weight chemicals into the cytoplasm of the bacterium. The biofragmented products can be used as a carbon source by microbes through this process. Assimilation facilitates the growth of microorganisms and their active participation in the degradation process. The broken down components become a valuable resource for microbial metabolism.

16.5.4 Mineralization

Degradation of the polymer into oxidized metabolites occurs as a result of a cascade of enzyme processes occurring within the microbial cell. Standard examples of these metabolites are CO_2, N_2, CH_4, and H_2O (Ho et al., 2018). This mineralization phase is pivotal in returning the plastic's constituent elements to the environment in a benign form, thus completing the biodegradation cycle. All these steps in plastic biodegradation are essential for developing strategies to enhance the efficiency of this process and reduce the environmental impact of plastic pollution. Additionally, research in this field is ongoing, with a focus on optimizing biodegradation techniques and exploring the potential of microorganisms in plastic waste management.

16.5.5 Microbial Enzymes Involved in Degradation of Plastic

Microbial enzymes play a pivotal role in the degradation of plastics, and they can be broadly classified as either extracellular or intracellular enzymes (Gu, 2003). Extracellular enzymes, which are the more extensively studied group, exhibit a higher reactivity efficiency and can mediate oxidative and hydrolytic reactions with greater efficacy (Glaser, 2019). These enzymes are primarily responsible for depolymerizing the long carbon chains present in plastic polymers, leading to the formation of a mixture of oligomers, dimers, and monomers. Various microbes, including bacteria, fungi, and actinomycetes, produce these catalytic enzymes for plastic polymer degradation (Kaushal et al., 2021):

1. *Bacterial Enzymes*

Bacterial enzymes, such as alkane hydroxylase, PETase, cutinases, esterases, polyurethanases, amylases, alkane monooxygenases, and polyesterases, exhibit substrate-specific activities, targeting various plastic polymers (Kaushal et al., 2021). Bacteria, as prolific producers of these enzymes, exhibit remarkable reactivity and efficiency. For example, bacteria like *Pseudomonas* sp. E4 and *Pseudomonas aeruginosa* make alkane hydroxylases that are very important for breaking down polyethylene (PE). *Pseudomonas aeruginosa* uses a coordinated system of enzymes that includes alkane hydroxylase, rubredoxin, and rubredoxin reductase to do this. PETases, such as those found in *Ideonella sakaiensis*, are specialized enzymes capable of degrading polyethylene terephthalate (PET), a common plastic used in bottles (Yoon et al., 2012). Cutinases, produced by bacteria like *Thermobifida fusca* and *Thermobifida alba*, hydrolyze ester bonds and contribute to the degradation of various plastic polymers. Esterases, exemplified by enzymes from *Thermobifida halotolerans*, also play a role in breaking down plastics by hydrolyzing ester bonds. Polyurethanases are enzymes that break down polyurethane polymers. Bacteria such as *Pseudomonas chlororaphis*, *Pseudomonas fluorescens*, and *Comamonas acidovorans* TB-35 produce these enzymes (Jeon & Kim, 2015). The breakdown of starch-blended low-density polyethylene (LDPE) has been linked to amylases, which are recognized for their co-metabolic activities. The enzymes produced by different bacteria in the catalytic degradation of plastics are listed in Table 16.1.

2. *Fungal Enzymes*

Fungi represent another group of microorganisms that contribute to plastic degradation, producing a diverse array of enzymes, such as depolymerases, peroxidases, laccases, and cutinases (Kaushal et al., 2021). As an example, *Aspergillus fumigatus* releases a polyhydroxybutyrate (PHB) depolymerase that can break down polyhydroxybutyrate (PHB), polybutylene succinate (PBS), and polyethylene succinate (PES) (Jung et al., 2018; Kasuya et al., 1998). Manganese peroxidase, which is made by lignin-degrading fungi such as *Phanerochaete chrysosporium* and *Trametes versicolor*, has been linked to the breakdown of PE (Iiyoshi et al., 1998). The enzymes produced by fungi in catalytic degradation of plastics as listed in Table 16.2.

3. *Enzymes Produced by Actinomycetes*

Actinomycetes, a group of Gram-positive bacteria with filamentous growth characteristics, have demonstrated considerable potential in the biodegradation of plastics through the production of specialized enzymes (Gohain et al., 2020). One notable enzyme produced by actinomycetes is polyesterase. This enzyme is very good at breaking down ester bonds in many different types of plastic, such as polyethylene

Table 16.1 The enzymes produced by different bacteria in degradation of plastics

Enzyme	Bacteria involved	Functions	References
Alkane hydroxylase	*Pseudomonas* sp. E4, *Pseudomonas aeruginosa*	Breakdown of polyethylene (PE)	Mohan et al. (2016) and Ward et al. (2006)
PETase	*Ideonella sakaiensis*	Polyethylene terephthalate (PET) degradation	Shimpi et al. (2012) and Yoshida et al. (2016)
Cutinases	*Thermobifida fusca*, *Thermobifida alba*	Hydrolysis of ester bonds in various plastics	Motta et al. (2009) and Chen et al. (2013)
Esterases	*Thermobifida halotolerans*	Hydrolysis of ester bonds in plastic polymers	Ribitsch et al. (2012)
Polyurethanases	*Pseudomonas chlororaphis*, P. *fluorescens*, C. *acidovorans* TB-35	Degradation of polyurethane plastics	Akutsu et al. (1998)
Amylases	Various bacterial strains	Co-metabolic degradation of starch-blended LDPE	Karimi and Biria (2019)
Laccases	*Streptomyces scabiei*, *Nocardiopsis* sp., *Actinomyces* species	Hydrolytic enzymes for diverse plastic polymers	Giacomucci et al. (2019)

Table 16.2 The enzymes produced by fungi in degradation of plastics

Enzyme	Fungi producing enzyme	Functions	References
PHB depolymerase	*Aspergillus fumigatus*	Degradation of polyhydroxybutyrate (PHB), polybutylene succinate (PBS), and polyethylene succinate (PES)	Motta et al. (2009)
Manganese peroxidase	*Phanerochaete chrysosporium*, *Trametes versicolor*	Degradation of polyethylene (PE) by lignin-degrading fungi	Giacomucci et al. (2019)
Cutinase	*Fusarium solani*	Degradation of polybutylene succinate (PBS) plastic	Chen et al. (2013)
Peroxidases	*Aspergillus niger*	Involved in the degradation of various plastic polymers	Kawai et al. (2014)
Hydrophobins	Various fungal species	Enhances hyphal attachment to hydrophobic plastic surfaces	Quarantin et al. (2019)

terephthalate (PET) (Jabloune et al., 2020). Actinomycetes-derived polyesterases have shown promising enzyme activity, helping to break down PET into its monomers more efficiently. As a result of the enzyme action, ester linkages are broken, which makes it easier for PET to depolymerize and for actinomycetes to use carbon from plastic for growth and metabolism. This enzymatic strategy represents a sustainable and environmentally friendly approach to tackling plastic pollution. In addition to polyesterases, actinomycetes also produce other enzymes, such as lipases and proteases, which play crucial roles in the biodegradation of plastics. Lipases help break down lipid-based plastics like polyurethanes by speeding up the breakdown of ester bonds in these materials. Proteases, on the other hand, exhibit efficacy in the degradation of proteins associated with bioplastics. The enzymatic arsenal of actinomycetes underscores their versatility in addressing diverse plastic substrates. Harnessing the enzymatic potential of actinomycetes for plastic biodegradation presents an exciting avenue for developing sustainable solutions to the global challenge of plastic waste accumulation. Further exploration of the enzymatic capabilities of actinomycetes and optimization of their biodegradation pathways hold promise for advancing the field of environmentally conscious plastic waste management. An additional instance involves a type of endophytic actinomycete called *Nocardiopsis* sp., which was discovered in hibiscus and has the ability to break down PE and diesel fuel (Singh & Sedhuraman, 2015). The production of extracellular polymers by actinomycetes is particularly interesting, as these polymers help the bacteria attach to plastic surfaces (Gaytán et al., 2020). This attachment facilitates further microbial activity, which is crucial for the degradation process. Additionally, the formation of biofilms, a common feature among bacteria, has been identified as a significant factor in the colonization of plastics by actinomycetes. Biofilms are groups of microorganisms surrounded by a self-made matrix of extracellular polymeric substances. They help break down plastic more quickly by encouraging microbes to work together and share resources (Gilan & Sivan, 2013). The enzymes produced by actinomycetes in catalytic degradation of plastics are listed in Table 16.3.

Table 16.3 Enzymes produced by actinomycetes in degradation of plastics

Enzyme	Actinomycetes producing enzyme	Functions	References
Esterase	*Streptomyces scabiei*, *Nocardiopsis* sp.	Degradation of polyethylene terephthalate (PET)	Auta et al. (2017)
Laccases	Various actinomycetes species	Key role in the ability to grow on diverse plastic polymers	Ribitsch et al. (2012)
Cutinases	*Streptomyces scabiei*, *Nocardiopsis* sp., others	Hydrolysis of ester bonds in various plastic polymers	Chen et al. (2013) and Arkatkar et al. (2010)
Biofilm formation	Various Actinomycetes species	Facilitates attachment to plastic surfaces for subsequent microbial action	Cacciari et al. (1993)
Extracellular polymers	Various Actinomycetes species	Likely facilitate attachment to plastic surfaces	Sekhar et al. (2016)
Cutinases	*Streptomyces scabiei*, *Nocardiopsis* sp., others	Hydrolysis of ester bonds in various plastic polymers	Chen et al. (2013) and Arkatkar et al. (2010)

The diversity of microbial enzymes involved in plastic degradation is instrumental in advancing our knowledge of this process and exploring its potential applications in waste management and environmental remediation. The ongoing research in this field continues to unveil new enzymatic pathways and microbial strategies for breaking down plastics, offering hope for addressing the global issue of plastic pollution.

16.6 Factors Affecting Microbial Biodegradation of Plastics

Microbial biodegradation of plastics is influenced by a myriad of factors that determine the rate and efficiency of plastic breakdown in various environments. Understanding these factors is crucial for developing effective strategies to manage plastic waste and mitigate environmental impacts. This discussion will explore key scientific aspects of microbial plastic degradation, drawing insights from relevant studies and literature:

1. *Polymer Composition and Structure*

The chemical composition and structural complexity of plastic polymers significantly influence their susceptibility to microbial biodegradation. Microorganisms possess specific enzymes capable of breaking the chemical bonds within polymers. For instance, polyethylene (PE) and polypropylene (PP), being hydrophobic and chemically stable, are generally resistant to microbial attack. In contrast, polymers like polyethylene terephthalate (PET) have been shown to be susceptible to enzymatic degradation by specific microorganisms, such as *Ideonella sakaiensis* 201-F6, which can metabolize PET into its constituent monomers (Yoshida et al., 2016).

2. *Crystallinity and Morphology*

The crystalline structure and morphology of plastics also play a role in microbial degradation. Crystalline regions within polymers can act as barriers, hindering microbial enzymes' access to the polymer chains. Amorphous regions, on the other hand, may be more susceptible to enzymatic attack. The degree of crystallinity varies among plastics; for instance, high crystallinity is observed in polyethylene, while polyethylene used for packaging has lower crystallinity (Kawai et al., 2014). Understanding these structural features is essential for predicting and optimizing microbial degradation.

3. *Additives and Fillers*

Plastics often contain additives and fillers that can impact microbial biodegradation. These additives, including plasticizers, stabilizers, and flame retardants, may alter the chemical composition and properties of the polymer. Some studies suggest that the presence of additives can affect the microbial degradation process. For example, *Pseudomonas citronellolis* and *Bacillus flexus* have been found to biodegrade PVC film, with high depolymerizing activity toward PVC additives relative to the PVC polymer chains (Giacomucci et al., 2019).

4. *Environmental Conditions*

The environment in which plastics are exposed greatly influences microbial degradation. Factors such as temperature, humidity, and nutrient availability play crucial roles. Microbial activity is often higher in warmer and more humid conditions, as demonstrated by the increased degradation of plastics in composting environments. In addition, nutrient availability, especially carbon and nitrogen sources, can affect microbial growth and enzymatic activity. The use of plastic blends with carbohydrates, starch, or cellulose has been shown to improve biodegradation, as these blends provide additional carbon sources for microorganisms (Cacciari et al., 1993).

5. *pH and Oxygen Levels*

The pH of the environment can influence microbial enzyme activity. Different microorganisms have optimal pH ranges for their enzymatic activities, and variations outside these ranges can impact degradation efficiency. Similarly, oxygen availability is critical for microbial degradation processes. Aerobic microorganisms require oxygen for enzymatic activities, and their effectiveness may be compromised in anaerobic conditions. Understanding the interplay between pH, oxygen levels, and microbial activity is vital for predicting degradation rates in diverse environments.

6. *Microbial Diversity*

The diversity and composition of microbial communities in a particular environment determine the types of enzymes available for plastic degradation. Some microorganisms have evolved specific enzymes for breaking down certain plastic polymers. For instance, the discovery of PETase and MHETase enzymes in *Ideonella sakaiensis* highlights the role of microbial diversity in plastic degradation (Yoshida et al., 2016). The presence of a diverse microbial community with complementary enzymatic capabilities can enhance the overall efficiency of plastic biodegradation.

7. *Pretreatment Methods*

Pretreatment methods, such as irradiation, UV exposure, or thermal treatments, can modify the surface properties of plastics, making them more accessible to microbial attack. These treatments may induce chemical changes in the polymer, creating functional groups that microorganisms can recognize and enzymatically degrade. For example, UV-irradiated polypropylene has been shown to exhibit improved biodegradability, with *Bacillus flexus* demonstrating the ability to biodegrade UV-pretreated polypropylene (Arkatkar et al., 2010).

8. *Genetic and Metabolic Adaptations*

Microorganisms can undergo genetic and metabolic adaptations to enhance their ability to degrade plastics. Some studies have reported the isolation of bacterial strains with specific plastic-degrading capabilities. These strains may possess unique enzymatic pathways or express enzymes with higher efficiency in plastic degradation. Understanding the genetic and metabolic adaptations of microorganisms to plastic-rich environments is essential for harnessing their biodegradative potential.

16.7 Challenges and Future Directions

Despite significant progress in understanding microbial plastic degradation, challenges persist. The complex nature of plastic polymers, diverse microbial communities, and environmental variability pose challenges in predicting and controlling degradation rates. Additionally, the potential release of harmful by-products during degradation raises environmental and health concerns. Further research is needed to explore novel microbial strains, optimize degradation conditions, and develop sustainable strategies for plastic waste management:

1. *Characteristics of Polymers*

The rate of microbial degradation exhibits an inverse relationship with the molecular weight of the polymers, with higher-molecular-weight polymers undergoing slower degradation. Additionally, plastic polymers possessing a higher proportion of side chains are less conducive to microbial assimilation. Research has revealed that the noncrystalline regions of polymers are particularly susceptible to enzymatic degradation due to their more loosely packed and accessible nature. For instance, the enzyme lipase produced by *Rhizopus delemar* has the capability to degrade and mineralize low-molecular-weight polymer components, including monomers, dimers, and oligomers (Yoshida et al., 2016).

2. *Environmental Components*

Environmental factors, such as the prevailing level of humidity, dry or humid air conditions, the context of landfill sites, and composting environments, can influence the rate of microbial action. Parameters including light, temperature, moisture levels, and pH have been identified as enhancing factors that promote the breakdown of chemical bonds. These environmental factors also have the capacity to impact structural homogeneity and contribute to the formation of new functional groups within the plastic materials (Kawai et al., 2014).

3. *Chemical Factors*

The presence of chemical reagents and additives within the polymeric structures can exhibit a range of effects, including activation, inhibition, or catalysis of the biodegradation process. These chemical agents can influence functional groups and the hydrophilic or hydrophobic nature of the plastic. For example, the incorporation of surfactants like Tween 80 and sodium dodecyl sulfate has been observed to foster microbial degradation by increasing the hydrophilicity of the polymer surfaces (Giacomucci et al., 2019). Additionally, the polymer chains of the additives present in plastic formulations are not covalently bound to them, which makes it possible for them to leach from plastics and absorb into the environment. Comprehending the intricate interplay of these factors is vital for devising strategies to enhance microbial plastic biodegradation and for developing effective approaches to mitigate plastic pollution. Ongoing research continues to expand our knowledge of how these factors influence the biodegradation process and opens up possibilities for more sustainable waste management practices.

16.8 Ways to Improve Biodegradation of Plastics

To effectively manage plastic waste and mitigate its environmental impact, strategies involving recycling methods and the retrieval of microplastics from sewage are essential initial measures. These actions aim to address the adverse effects of additives, the attachment of persistent organic pollutants (POPs), and the breakdown of polymer chains (Li et al., 2020). Enhancing bacterial adhesion and promoting breakdown necessitate augmenting the hydrophilic properties of polymers through chemical or biological oxidation methods. Various pretreatment techniques, including high temperature, photooxidation catalysis, and catalysis by microbial enzymes, have demonstrated efficacy in improving biodegradation processes. Biodegradable plastics exhibit favorable physicochemical features, such as reduced molecular weight, enhanced flexibility, and increased functional groups, compared to conventional plastics, facilitating their processing and biodegradation (Matjašič et al., 2021). Optimizing PET through pretreatment, particularly temperature treatment, enhances the efficiency of PET-hydrolyzing enzymes, leading to improved degradation. PETase, the enzymatic catalyst, shows elevated activity at lower temperatures due to decreased molecular flexibility of PET molecules. Temperature treatment of plastics has been demonstrated to enhance their biodegradability, with specific bacteria playing a crucial role in influencing enzyme activity on long-chain polymer molecules. For instance, thermal pretreatment significantly decreased the tensile strength of thermally prepared high-density polyethylene (HDPE) (Narancic & O'Connor, 2019).

Recent progress in photocatalysis, an environmentally sustainable method, has utilized simple iron salts as efficient catalysts for oxidative bond cleavage when exposed to light. Aromatic hydrocarbons in the catalytic system can effectively break C–C bonds and convert them into benzoic acid. This approach has demonstrated effectiveness in decomposing commercial samples containing high-molecular-weight polystyrene (PS). Enzymatic modification of polymer fibers, utilizing enzymes like lipase and keratinase, enhances surface properties, including wettability, fastness, dyeing capabilities, and resistance to pilling. Whole-cell enzymes for biodegradation of polymers with C–C backbones offer a potentially improved approach. Utilizing whole-cell catalysis, particularly in PET biocatalysis, integrates enzyme synthesis and PET hydrolysis, resulting in effective weight reduction. *Clostridium thermocellum*, functioning as a whole-cell biocatalyst, demonstrated a 60% weight reduction in a PET film over a 14-day incubation at 60 °C (Yan et al., 2021). These findings suggest that whole-cell catalysis could be a valuable supplement for pretreatment catalysts.

16.9 Modification of Microbial Enzymes for Plastic Biodegradation

Modification of microbial enzymes for plastic biodegradation is an exciting area of research that holds promise for addressing the global issue of plastic pollution. Microbes naturally produce enzymes that can break down various components of plastics, but these enzymes are often not efficient enough for large-scale biodegradation. To improve the biodegradation capabilities of microbial enzymes, several strategies can be employed.

16.9.1 Protein Engineering

Protein engineering is a technique that involves the modification of the amino acid sequence or structure of a protein to alter its function or properties (Andler et al., 2022). Protein engineering can be used to modify the microbial enzymes for the biodegradation of plastic by either changing the amino acids at specific positions or by creating new combinations of protein domains or subunits. For example, the amino acid residues at the active site of the PETase enzyme from *I. sakaiensis* were mutated to increase its affinity and activity for PET (Neklesa et al., 2017). Similarly, the amino acid residues at the surface of the cutinase enzyme from *F. solani* were mutated to enhance its thermostability and hydrophobicity, which improved its ability to degrade PET and PU (Zhu et al., 2022). Protein engineering can also be used to create fusion proteins or chimeric proteins that combine different domains or subunits of different enzymes to create new functions or properties (Urbanek et al., 2021). For example, a fusion protein was made from the laccase domain from *T. versicolor* and the cellulose-binding domain from the cellulase enzyme from *Trichoderma reesei*. This protein made it easier for PS to bind and break down. The advantages of protein engineering are that it can fine-tune the activity, specificity, and stability of the enzymes and that it can create novel or synergistic functions or properties of the enzymes. However, the challenges of protein engineering are that it may require extensive screening and testing of the variants, that it may affect the folding or stability of the protein, and that it may encounter computational or experimental limitations such as protein design, synthesis, and expression. Protein engineering can be applied to these PET hydrolases to improve their catalytic efficiency, thermostability, substrate specificity, and product selectivity.

16.9.1.1 Some of the Methods That Can Be Used for Protein Engineering

Rational design: It involves the modification of specific amino acid residues based on the knowledge of the protein structure and function. For example, rational design was used to create a PETase variant with improved PET degradation rate by introducing two mutations that increased the flexibility of the active site.

Directed evolution: It involves the generation of a large library of protein variants by random mutagenesis, recombination, or insertion and then screening or selecting for the desired properties. For example, directed evolution was used to create a TfCut2 variant with enhanced thermostability and PET degradation activity by introducing four mutations that increased the hydrophobicity and rigidity of the protein.

Computational design: This involves the use of computational tools and algorithms to design novel proteins or optimize existing ones. For example, computational design was used to create a PETase variant with increased PET affinity and degradation activity by introducing three mutations that improved the electrostatic interactions and hydrogen bonding with the substrate.

Protein engineering of microbial enzymes for biodegradation of plastics is a promising strategy for solving the environmental problem of plastic pollution. By using protein engineering, it may be possible to create more efficient and versatile enzymes that can degrade various types of plastics under different conditions. However, there are still many challenges and limitations that need to be overcome, such as the low availability and stability of the enzymes, the high cost and complexity of the engineering process, the potential toxicity and accumulation of the degradation products, and the ethical and social implications of using genetically modified organisms. Therefore, more research and development are needed to optimize and evaluate the feasibility and sustainability of this approach.

16.9.2 Genetic Engineering

Genetic engineering is a technique that involves the manipulation of the genetic material of an organism to introduce or modify a desired trait (Urbanek et al., 2021). Genetic engineering can be used to modify the microbial enzymes for biodegradation of plastic by either transferring the genes encoding the enzymes from one organism to another or by altering the expression or regulation of the genes in the same organism (Anand et al., 2023). For example, the gene encoding the PET-degrading enzyme PETase from the bacterium *Ideonella sakaiensis* was cloned and expressed in *Escherichia coli*, resulting in a recombinant enzyme that showed higher activity and stability than the native enzyme (Maraveas et al., 2023). Similarly, the gene encoding the cutinase enzyme from the fungus *Fusarium solani* was inserted into the genome of the yeast *Pichia pastoris*, which enabled the production of large amounts of the enzyme that could degrade PET and PU (Mohanan et al., 2020). Genetic engineering can also be used to modify the promoter, terminator, or regulatory elements of the gene to enhance or suppress its expression. For instance, the expression of the laccase gene from the fungus *Trametes versicolor* was increased by replacing its native promoter with a strong constitutive promoter in *E. coli*, which resulted in a higher yield and activity of the enzyme that could degrade PS (Anand et al., 2023).

The advantages of genetic engineering are that it can introduce novel or improved enzymes into different host organisms and that it can control the expression level and timing of the enzymes. However, the challenges of genetic engineering are that it may face ethical and social issues, that it may cause unintended effects on the host organism or the environment, and that it may encounter technical difficulties such as gene transfer, integration, and expression.

16.9.3 Metabolic Engineering

Metabolic engineering is a biotechnological approach that can be used to modify the metabolic pathways and capabilities of microorganisms for biodegradation of plastics. Plastics are synthetic polymers that are widely used in various industries and applications, but they pose a serious environmental problem due to their persistence and accumulation in the natural ecosystems. One of the most common and problematic plastics is polyethylene terephthalate (PET), which is used for making bottles, containers, textiles, and packaging materials. PET is resistant to natural degradation due to its high degree of crystallinity and hydrophobicity, which prevent the access and attack of hydrolytic enzymes. However, some microorganisms have been discovered that can degrade PET by secreting enzymes that hydrolyze the ester bonds in the polymer chain. These enzymes are called PET hydrolases, and they belong to the family of cutinases, which are lipolytic enzymes that degrade plant cuticles (Mohanan et al., 2020).

Metabolic engineering can be applied to these microorganisms to improve their ability to degrade PET and to convert the degradation products into valuable bio-products, such as biodegradable polymers, biofuels, or biochemicals. Metabolic engineering involves the introduction, deletion, or modification of genes or enzymes that are involved in the degradation or bioconversion of PET. For example, metabolic engineering can be used to enhance the expression or activity of PET hydrolases, to expand the substrate range or specificity of the enzymes, to increase the stability or solubility of the enzymes, to optimize the cofactor or energy requirements of the enzymes, or to introduce novel enzymes that can degrade PET or its monomers (Wang et al., 2022). Metabolic engineering can also be used to modify the downstream pathways of PET degradation; to enable the utilization or assimilation of the degradation products, such as terephthalic acid (TPA), mono(2-hydroxyethyl) terephthalic acid (MHET), or bis(2-hydroxyethyl) terephthalate (BHET), by the microorganisms; and to direct the carbon flux toward the desired bioproducts, such as polyhydroxyalkanoates (PHAs), ethanol, or succinic acid (Benavides Fernández et al., 2022).

Metabolic engineering of microorganisms for biodegradation and bioconversion of plastics is a promising strategy for solving the environmental problem of plastic pollution, as well as for producing renewable and biodegradable bioproducts from plastic waste. However, there are still many challenges and limitations that need to be overcome, such as the low efficiency and specificity of the enzymes, the high cost and complexity of the engineering process, the potential toxicity and accumulation of the intermediates, and the ethical and social implications of using genetically modified organisms (Bollinger et al., 2020). Therefore, more research and development are needed to optimize and evaluate the feasibility and sustainability of this approach.

16.9.4 Immobilization Techniques

Immobilization is the process of attaching a biocatalyst (such as an enzyme or a cell) to an inert support (such as a solid matrix, a membrane, or a bead) to prevent its loss or diffusion from the reaction system. Immobilization has several advantages, such as increased enzyme stability, improved enzyme performance, reduced enzyme inhibition, facilitated enzyme recovery and reuse, and simplified downstream processing. Immobilization techniques can be classified into four main categories: adsorption, covalent bonding, entrapment, and cross-linking (Asaduzzaman & Salmon, 2022). Each immobilization technique has different examples of supports, biocatalysts, and applications for plastic biodegradation, as explained below:

(a) *Adsorption*: the physical attachment of the biocatalyst to the support surface by non-covalent interactions, such as electrostatic, hydrophobic, or van der Waals forces. Examples of supports are activated carbon, silica gel, alumina, and chitin. Examples of biocatalysts are PETase and MHETase from *Ideonella sakaiensis*, esterases from *Bacillus* and *Nocardia*, and lipases from *Thermomyces lanuginosus* and *Candida antarctica*. Examples of applications are the hydrolysis of PET into terephthalic acid and ethylene glycol and the transesterification of PET into cyclic acetals (Krakor et al., 2021).

(b) *Covalent bonding*: the chemical attachment of the biocatalyst to the support surface by forming covalent bonds, such as amide, ester, or thioester bonds. Examples of supports are agarose, cellulose, and polyacrylamide. Examples of biocatalysts are cutinases from leaf-branch compost and *Pseudomonas mendocina* and lipases from *Burkholderia* spp. and *Triticum aestivum*. Examples of applications are the hydrolysis of PET and PU into monomers and oligomers and the synthesis of biodegradable polyesters from PET monomers (Schwaminger et al., 2021).

(c) *Entrapment*: the inclusion of the biocatalyst within a porous or gel-like matrix, such as alginate, starch, or cellulose, without forming covalent bonds. Examples of biocatalysts are PETase and MHETase from *Ideonella sakaiensis* and PHB-accumulating bacteria, such as *Pseudomonas* and *Halomonas*. Examples of applications are the hydrolysis of PET into terephthalic acid and ethylene glycol and the bioconversion of PET monomers into polyhydroxyalkanoates (PHAs), which are biodegradable polymers (Krakor et al., 2021).

(d) *Cross-linking*: the formation of covalent bonds between the biocatalyst molecules or between the biocatalyst and a cross-linking agent, such as glutaraldehyde or polyethyleneimine, without using a support. Examples of biocatalysts are cutinases from leaf-branch compost and *Fusarium solani* and lipases from *Burkholderia* spp. and *Triticum aestivum*. Examples of applications are the hydrolysis of PET and PU into monomers and oligomers and the synthesis of biodegradable polyesters from PET monomers (Krakor et al., 2021).

16.9.5 Synthetic Biology Approaches

The main synthetic biology approaches for modifying and optimizing microbial enzymes for plastic biodegradation include the following:

1. *Nanotechnology*: The use of nanomaterials or nanoparticles to modify the properties or performance of microbial enzymes or their hosts. Examples of techniques are enzyme immobilization, enzyme encapsulation, enzyme conjugation, and enzyme biosynthesis. Examples of applications are the immobilization of PETase and MHETase from *Ideonella sakaiensis* on magnetic nanoparticles for improved stability and reusability, the encapsulation of cutinases from leaf-branch compost and *Pseudomonas mendocina* in silica nanoparticles for

enhanced activity and thermostability, and the conjugation of lipases from *Burkholderia* spp. and *Triticum aestivum* with gold nanoparticles for increased hydrolysis of PET and PU (Elegbede & Lateef, 2021).

2. *Posttranslational modification*: The modification of the amino acid residues or peptide chains of microbial enzymes after their translation to alter their structure, function, or interactions. Examples of techniques are glycosylation, phosphorylation, acetylation, and ubiquitination. Examples of applications are the glycosylation of PETase and MHETase from *Ideonella sakaiensis* to increase their solubility and stability, the phosphorylation of cutinases from leaf-branch compost and *Fusarium solani* to enhance their activity and thermostability, and the acetylation of lipases from *Burkholderia* spp. and *Triticum aestivum* to improve their hydrolysis of PET and PU (Shukla, 2019).
3. *Artificial intelligence*: The use of computational methods and algorithms to design, model, simulate, or optimize microbial enzymes or their hosts. Examples of techniques are machine learning, deep learning, neural networks, and genetic algorithms. Examples of applications are the design of novel PETase and MHETase variants from *Ideonella sakaiensis* with improved activity and thermostability using machine learning and deep learning, the modeling of the structure and function of cutinases from leaf-branch compost and *Pseudomonas mendocina* using neural networks, and the optimization of the reaction conditions and product yields of lipases from *Burkholderia* spp. and *Triticum aestivum* using genetic algorithms (Shukla, 2019).

16.10 Applications and Implications of Microbial Enzymes in Plastic Biodegradation

Plastic waste represents a dire environmental challenge, proliferating across ecosystems with detrimental impacts. Amidst this crisis, microbial enzymes emerge as a beacon of innovation and hope. These naturally occurring proteins, produced by microorganisms, catalyze chemical reactions necessary for the degradation of resilient plastic materials. They offer a promising avenue toward mitigating the ever-growing issue of plastic pollution. The strategic application of these enzymes in breaking down polymers into nontoxic and simpler compounds introduces a sustainable, efficient, and eco-friendly approach to managing and recycling plastic wastes.

Microbial Enzymes and Their Role in Biodegradation

Microbial enzymes, such as esterases, cutinases, and depolymerases, are adept at breaking down the long-chain polymers that constitute plastics into smaller, more manageable molecules. These enzymes target specific bonds within the plastic's molecular structure, resulting in simpler substances that are more amenable to further degradation and assimilation into natural cycles. For example, PETase, an enzyme recently discovered in certain bacteria strains, has shown promising results in degrading polyethylene terephthalate (PET), a widespread and persistent plastic. Understanding and harnessing these enzymes' mechanisms and specificities are pivotal in advancing biodegradation technologies.

Applications of Microbial Enzymes in Plastic Biodegradation
The application of microbial enzymes in mitigating plastic waste is multifaceted. Industrial processes have begun incorporating these enzymes to break down and recycle plastics such as PET bottles and packaging materials. This enzymatic recycling not only reduces plastic waste but also aids in recovering valuable materials that can be reused in manufacturing. Beyond industrial settings, there is a potential for these enzymes to be employed in environmental cleanup efforts, targeting polluted sites and water bodies with high concentrations of plastic waste. Research is also exploring the integration of these enzymes into waste treatment facilities to enhance the biodegradation of plastics in landfills and composting systems.

16.11 Implications of Microbial Enzymes in Plastic Biodegradation

Environmental Impact: The foremost implication is environmental, where enzymatic biodegradation represents a significant stride toward reducing plastic pollution. By converting plastics back into their constituent harmless compounds, the process can significantly reduce the volume of waste, lessen the impact on wildlife, and decrease the overall ecological footprint of plastic production and disposal.

Economic Considerations: Economically, the development and implementation of enzyme-based plastic degradation technologies promise new industrial pathways. These include the production of biodegradable plastics, the evolution of recycling industries, and the generation of valuable by-products. As these technologies mature and scale, they could stimulate economic growth and job creation within green technology sectors.

Social and Behavioral Change: On a societal level, the successful application of microbial enzymes in plastic waste management could foster a more sustainable and responsible approach to plastic use. As public awareness of these technologies grows, it can encourage more conscientious consumption patterns and greater support for sustainable products and practices.

16.12 Challenges and Future Directions

Despite its potential, the use of microbial enzymes in plastic biodegradation faces several challenges. The efficiency and speed of enzymatic degradation need enhancement to be viable on an industrial scale. Environmental factors such as temperature, pH, and presence of inhibitors can affect enzyme activity, necessitating controlled conditions for optimal performance. Additionally, the diversity of plastic types and additives used in their manufacture requires a broad range of enzymes to effectively tackle different polymers. Future research is thus directed toward enhancing the efficiency, stability, and range of these enzymes through genetic engineering, process optimization, and discovery of new biocatalysts. There's also an emphasis on developing integrated systems where microbial enzymes work in conjunction with mechanical and chemical recycling methods. Scaling up these technologies for widespread industrial and environmental application while ensuring they are economically viable and environmentally beneficial remains the ultimate goal.

16.13 Conclusion

One of the most important tools in the fight against plastic pollution is microbial enzymes. These enzymes, derived from various microorganisms, have shown remarkable efficiency in breaking down polymers that are otherwise resistant to degradation. Advancements in genetic engineering and biotechnology have further enhanced the effectiveness of these enzymes. By modifying microbial enzymes to optimize their stability, specificity, and activity under a range of environmental conditions, scientists have significantly improved the prospects for large-scale plastic degradation. These modifications not only increase the efficiency of plastic degradation but also expand the range of plastic types that can be processed, potentially offering a sustainable solution to the ever-growing issue of plastic waste.

Moreover, the integration of these modified microbial enzymes into waste management systems represents a promising step toward a more sustainable and environmentally friendly approach to plastic waste. The potential to convert waste plastics into valuable by-products through enzymatic degradation opens up new avenues for recycling and upcycling, which can contribute to a circular economy. However, there are still problems with making these processes bigger so they can be used in industry. These problems include improving enzyme production, making sure that enzyme-based degradation systems are cost-effective, and dealing with safety and regulatory issues. Continued research and collaboration between scientists, industry, and policymakers are crucial to overcome these challenges and fully realize the potential of microbial enzymes in plastic biodegradation. This endeavor not only addresses an urgent environmental issue but also sets a precedent for harnessing biotechnological innovations for sustainable development.

References

Akutsu, Y., Nakajima-Kambe, T., Nomura, N., & Nakahara, T. (1998). Purification and properties of a polyester polyurethane-degrading enzyme from Comamonas acidovorans TB-35. *Applied and Environmental Microbiology, 64*(1), 62–67. https://doi.org/10.1128/AEM.64.1.62-67.1998

Amin, M., Bhatti, H. N., & Bilal, M. (2022). Microbial lipases for polyester degradation. In *Enzymes for pollutant degradation* (pp. 71–92). Springer.

Anand, U., Dey, S., Bontempi, E., Ducoli, S., Vethaak, A. D., Dey, A., & Federici, S. (2023). Biotechnological methods to remove microplastics: A review. *Environmental Chemistry Letters, 21*, 1–24.

Andler, R., Tiso, T., Blank, L., Andreeßen, C., Zampolli, J., D'Afonseca, V., Guajardo, C., & Díaz-Barrera, A. (2022). Current progress on the biodegradation of synthetic plastics: From fundamentals to biotechnological applications. *Reviews in Environmental Science and Bio/ Technology, 21*(4), 829–850.

Anjana, K., Hinduja, M., Sujitha, K., & Dharani, G. (2020). Review on plastic wastes in marine environment – Biodegradation and biotechnological solutions. *Marine Pollution Bulletin, 150*, 110733.

Arkatkar, A., Juwarkar, A. A., Bhaduri, S., Uppara, P. V., & Doble, M. (2010). Growth of pseudomonas and bacillus biofilms on pretreated polypropylene surface. *Biodeterioration & Biodegradation, 64*, 530–536. https://doi.org/10.1016/j.ibiod.2010.06.002

Asaduzzaman, F., & Salmon, S. (2022). Enzyme immobilization: Polymer–solvent–enzyme compatibility. *Molecular Systems Design & Engineering, 7*(11), 1385–1414.

Auta, H. S., Emenike, C. U., & Fauziah, S. H. (2017). Screening of Bacillus strains isolated from mangrove ecosystems in Peninsular Malaysia for microplastic degradation. *Environmental Pollution, 231*, 1552–1559.

Auta, H. S., Emenike, C. U., Jayanthi, B., & Fauziah, S. H. (2018). Growth kinetics and biodeterioration of polypropylene microplastics by Bacillus sp. and Rhodococcus sp. isolated from mangrove sediment. *Marine Pollution Bulletin, 127*, 15–21.

Benavides Fernández, C. D., Guzmán Castillo, M. P., Quijano Pérez, S. A., & Carvajal Rodríguez, L. V. (2022). Microbial degradation of polyethylene terephthalate: A systematic review. *SN Applied Sciences, 4*(10), 263.

Bhardwaj, H., Gupta, R., & Tiwari, A. (2013). Communities of microbial enzymes associated with biodegradation of plastics. *Journal of Polymers and the Environment, 21*, 575–579.

Bollinger, A., Thies, S., Knieps-Grünhagen, E., Gertzen, C., Kobus, S., Höppner, A., Ferrer, M., Gohlke, H., Smits, S. H., & Jaeger, K. E. (2020). A novel polyester hydrolase from the marine bacterium pseudomonas aestusnigri–structural and functional insights. *Frontiers in Microbiology, 11*, 114.

Cacciari, I., Quatrini, P., Zirletta, G., Mincione, E., Vinciguerra, V., Lupattelli, P., & Giovannozzi Sermanni, G. (1993). Isotactic polypropylene biodegradation by a microbial community: physicochemical characterization of metabolites produced. *Applied and Environmental Microbiology, 59*(11), 3695–3700

Chen, S., Su, L., Chen, J., & Wu, J. (2013). Cutinase: characteristics, preparation, and application. *Biotechnology Advances, 31*(8), 1754–1767.

Chomchoei, A., Pathom-Aree, W., Yokota, A., Kanongnuch, C., & Lumyong, S. (2011). Amycolatopsis thailandensis sp. nov., a poly(L-lactic acid)-degrading actinomycete, isolated from soil. *International Journal of Systematic and Evolutionary Microbiology, 61*(4), 839–843.

Chua, T., Tseng, M., & Yang, M. (2013). Degradation of poly (ε-caprolactone) by thermophilic Streptomyces thermoviolaceus subsp. thermoviolaceus 76T-2. *AMB Express, 3*, 8–14.

De Jesus, R., & Alkendi, R. (2023). A minireview on the bioremediative potential of microbial enzymes as solution to emerging microplastic pollution. *Frontiers in Microbiology, 13*, 1066133.

Elegbede, J. A., & Lateef, A. (2021). Microbial enzymes in nanotechnology and fabrication of nanozymes: A perspective. In *Microbial nanobiotechnology: Principles and applications* (pp. 185–232). Springer.
Gajendiran, A., Krishnamoorthy, S., & Abraham, J. (2016). Microbial degradation of low-density polyethylene (LDPE) by Aspergillus clavatus strain JASK1 isolated from landfill soil. *3 Biotech, 6*, 1–6.
Gaur, V. K., Gupta, S., Sharma, P., Gupta, P., Varjani, S., Srivastava, J. K., Chang, J. S., & Bui, X. T. (2022). Metabolic cascade for remediation of plastic waste: A case study on microplastic degradation. *Current Pollution Reports, 8*(1), 30–50.
Gaytán, I., Sánchez-Reyes, A., Burelo, M., Vargas-Suárez, M., Liachko, I., Press, M., Sullivan, S., Cruz-Gómez, M. J., & Loza-Tavera, H. (2020). Degradation of recalcitrant polyurethane and xenobiotic additives by a selected landfill microbial community and its biodegradative potential revealed by proximity ligation-based metagenomic analysis. *Frontiers in Microbiology, 10*, 2986.
Giacomucci, L., Raddadi, N., Soccio, M., Lotti, N., & Fava, F. (2019). Polyvinyl chloride biodegradation by pseudomonas citronellolis and Bacillus flexus. *New Biotechnology, 52*, 35–41. https://doi.org/10.1016/j.nbt.2019.04.005
Gilan, I., & Sivan, A. (2013). Effect of proteases on biofilm formation of the plastic-degrading actinomycete Rhodococcus ruber C208. *FEMS Microbiology Letters, 342*(1), 18–23.
Glaser, J. A. (2019). *Biological degradation of polymers in the environment* (Vol. 1, p. 13). IntechOpen.
Gohain, A., Manpoong, C., Saikia, R., & De Mandal, S. (2020). Actinobacteria: Diversity and biotechnological applications. In *Recent advancements in microbial diversity* (pp. 217–231). Academic Press.
Gu, J. D. (2003). Microbiological deterioration and degradation of synthetic polymeric materials: Recent research advances. *International Biodeterioration & Biodegradation, 52*(2), 69–91.
Harrison, J. P., Boardman, C., O'Callaghan, K., Delort, A. M., & Song, J. (2018). Biodegradability standards for carrier bags and plastic films in aquatic environments: A critical review. *Royal Society Open Science, 5*(5), 171792.
Ho, B. T., Roberts, T. K., & Lucas, S. (2018). An overview on biodegradation of polystyrene and modified polystyrene: The microbial approach. *Critical Reviews in Biotechnology, 38*(2), 308–320.
Hsu, K. J., Tseng, M., Don, T. M., & Yang, M. K. (2012). Biodegradation of poly (β-hydroxybutyrate) by a novel isolate of Streptomyces bangladeshensis 77T-4. *Botanical Studies, 53*(3), 307–313.
https://interestingengineering.com/innovation/genetically-modified-bacteria-may-eat-up-ocean-plastic-waste. Accessed 27 Dec 2023.
Hung, C. S., Zingarelli, S., Nadeau, L. J., Biffinger, J. C., Drake, C. A., Crouch, A. L., Barlow, D. E., Russell, J. N., Jr., & Crookes-Goodson, W. J. (2016). Carbon catabolite repression and Impranil polyurethane degradation in Pseudomonas protegens strain Pf-5. *Applied and Environmental Microbiology, 82*(20), 6080–6090.
Iiyoshi, Y., Tsutsumi, Y., & Nishida, T. (1998). Polyethylene degradation by lignin-degrading fungi and manganese peroxidase. *Journal of Wood Science, 44*, 222–229.
Jabloune, R., Khalil, M., Moussa, I. E. B., Simao-Beaunoir, A. M., Lerat, S., Brzezinski, R., & Beaulieu, C. (2020). Enzymatic degradation of p-nitrophenyl esters, polyethylene terephthalate, cutin, and suberin by Sub1, a suberinase encoded by the plant pathogen Streptomyces scabies. *Microbes and Environments, 35*(1), ME19086.
Jarerat, A., Tokiwa, Y., & Tanaka, H. (2003). Poly (L-lactide) degradation by Kibdelosporangium aridum. *Biotechnology Letters, 25*, 2035–2038.
Jenkins, S., Quer, A. M. I., Fonseca, C., & Varrone, C. (2019). Microbial degradation of plastics: New plastic degraders, mixed cultures and engineering strategies. In *Soil microenvironment for bioremediation and polymer production* (pp. 213–238). Wiley.

Jeon, H. J., & Kim, M. N. (2015). Functional analysis of alkane hydroxylase system derived from Pseudomonas aeruginosa E7 for low molecular weight polyethylene biodegradation. *International Biodeterioration & Biodegradation, 103*, 141–146.

Jung, H. W., Yang, M. K., & Su, R. C. (2018). Purification, characterization, and gene cloning of an Aspergillus fumigatus polyhydroxybutyrate depolymerase used for degradation of polyhydroxybutyrate, polyethylene succinate, and polybutylene succinate. *Polymer Degradation and Stability, 154*, 186–194.

Karimi, M., & Biria, D. (2019). The promiscuous activity of alpha-amylase in biodegradation of low density polyethylene in a polymer-starch blend. *Scientific Reports, 9*, 2612. https://doi.org/10.1038/s41598-019-39366-0

Kasuya, K. I., Takagi, K. I., Ishiwatari, S. I., Yoshida, Y., & Doi, Y. (1998). Biodegradabilities of various aliphatic polyesters in natural waters. *Polymer Degradation and Stability, 59*(1–3), 327–332.

Kathiresan, K. (2003). Polythene and plastics-degrading microbes from the mangrove soil. *Revista de Biologia Tropical, 51*(3–4), 629–633.

Kaushal, J., Khatri, M., & Arya, S. K. (2021). Recent insight into enzymatic degradation of plastics prevalent in the environment: A mini-review. *Cleaner Engineering and Technology, 2*, 100083.

Kawai, F., Oda, M., Tamashiro, T., Waku, T., Tanaka, N., Yamamoto, M., et al. (2014). A novel Ca2+-activated, thermostabilized polyesterase capable of hydrolyzing polyethylene terephthalate from Saccharomonospora viridis AHK 190. *Microbiology and Biotechnology, 98*, 10053–10064. https://doi.org/10.1007/s00253-014-5860-y

Kim, D. W., Ahn, J. H., & Cha, C. J. (2022). Biodegradation of plastics: Mining of plastic-degrading microorganisms and enzymes using metagenomics approaches. *Journal of Microbiology, 60*(10), 969–976.

Koutny, M., Amato, P., Muchova, M., Ruzicka, J., & Delort, A. M. (2009). Soil bacterial strains able to grow on the surface of oxidized polyethylene film containing prooxidant additives. *International Biodeterioration & Biodegradation, 63*(3), 354–357.

Krakor, E., Gessner, I., Wilhelm, M., Brune, V., Hohnsen, J., Frenzen, L., & Mathur, S. (2021). Selective degradation of synthetic polymers through enzymes immobilized on nanocarriers. *MRS Communications, 11*(3), 363–371.

Li, X., Chen, L., Ji, Y., Li, M., Dong, B., Qian, G., ... & Dai, X. (2020). Effects of chemical pretreatments on microplastic extraction in sewage sludge and their physicochemical characteristics. *Water Research, 171*, 115379.

Lwanga, E. H., Thapa, B., Yang, X., Gertsen, H., Salánki, T., Geissen, V., & Garbeva, P. (2018). Decay of low-density polyethylene by bacteria extracted from earthworm's guts: A potential for soil restoration. *Science of the Total Environment, 624*, 753–757.

Maraveas, C., Kotzabasaki, M. I., & Bartzanas, T. (2023). Intelligent technologies, enzyme-embedded and microbial degradation of agricultural plastics. *AgriEngineering, 5*(1), 85–111.

Matjašič, T., Simčič, T., Medveščček, N., Bajt, O., Dreo, T., & Mori, N. (2021). Critical evaluation of biodegradation studies on synthetic plastics through a systematic literature review. *Science of the Total Environment, 752*, 141959.

Mogil'nitskii, G. M., Sagatelyan, R. T., Kutishcheva, T. N., Zhukova, S. V., Kerimov, S. I., & Parfenova, T. B. (1987). Disruption of the protective properties of the polyvinyl chloride coating under the effect of microorganisms. *Protection of Metals, 23*(1), 173–175.

Mohan, A. J., Sekhar, V. C., Bhaskar, T., & Nampoothiri, K. M. (2016). Microbial assisted high impact polystyrene (HIP) degradation. *Bioresource Technology, 213*, 204–207. https://doi.org/10.1016/j.biortech.2016.03.021

Mohanan, N., Montazer, Z., Sharma, P. K., & Levin, D. B. (2020). Microbial and enzymatic degradation of synthetic plastics. *Frontiers in Microbiology, 11*, 580709.

Mohanan, N., Wong, C.H., Budisa, N., & Levin, D.B. (2022). Characterization of polymer degrading lipases, LIP1 and LIP2 from Pseudomonas chlororaphis PA23. *Frontiers in Bioengineering and Biotechnology, 10*, 854298.

Morris, P. J. T. (1986). *Polymer pioneers: A popular history of the science and technology of large molecules* (Vol. 5). Center for History of Chemistry.

Motta, O., Proto, A., De Carlo, F., De Caro, F., Santoro, E., Brunetti, L., et al. (2009). Utilization of chemically oxidized polystyrene as co-substrate by filamentous fungi. *International Journal of Hygiene and Environmental Health, 212*, 61–66. https://doi.org/10.1016/j.ijheh.2007.09.014

Narancic, T., & O'Connor, K. E. (2019). Plastic waste as a global challenge: Are biodegradable plastics the answer to the plastic waste problem? *Microbiology, 165*, 129–137.

Neklesa, T. K., Winkler, J. D., & Crews, C. M. (2017). Targeted protein degradation by PROTACs. *Pharmacology & Therapeutics, 174*, 138–144.

Oda, Y., Asari, H., Urakami, T., & Tonomura, K. (1995). Microbial degradation of poly (3-hydroxybutyrate) and polycaprolactone by filamentous fungi. *Journal of Fermentation and Bioengineering, 80*(3), 265–269.

Osman, M., Satti, S. M., Luqman, A., Hasan, F., Shah, Z., & Shah, A. A. (2018). Degradation of polyester polyurethane by Aspergillus sp. strain S45 isolated from soil. *Journal of Polymers and the Environment, 26*, 301–310.

Pathak, V. M. (2017). Review on the current status of polymer degradation: A microbial approach. *Bioresources and Bioprocessing, 4*(1), 1–31.

Pramila, R., & Ramesh, K. V. (2015). Potential biodegradation of low density polyethylene (LDPE) by Acinetobacter baumannii. *African Journal of Bacteriology Research, 7*(3), 24–28.

Puglisi, E., Romaniello, F., Galletti, S., Boccaleri, E., Frache, A., & Cocconcelli, P. S. (2019). Selective bacterial colonization processes on polyethylene waste samples in an abandoned landfill site. *Scientific Reports, 9*, 1–13.

Quarantin, A., Hadeler, B., Kröger, C., Schäfer, W., Favaron, F., Sella, L., & Martínez-Rocha, A. L. (2019). Different hydrophobins of *Fusarium graminearum* are involved in hyphal growth, attachment, water-air interface penetration, and plant infection. *Frontiers in Microbiology, 10*, 751.

Raziyafathima, M. P. P. K., Praseetha, P. K., & Rimal, I. R. S. (2016). Microbial degradation of plastic waste: A review. *Chemical and Biological Sciences, 4*, 231–242.

Ribitsch, D., Herrero Acero, E., Greimel, K., Dellacher, A., Zitzenbacher, S., Marold, A., et al. (2012). A new esterase from Thermobifida halotolerans hydrolyses polyethylene terephthalate (PET) and polylactic acid (PLA). *Polymers, 4*, 617–629. https://doi.org/10.3390/polym4010617

Sánchez, C. (2020). Fungal potential for the degradation of petroleum-based polymers: An overview of macro-and microplastics biodegradation. *Biotechnology Advances, 40*, 107501.

Sangale, M. K., Shahnawaz, M., & Ade, A. B. (2019). Gas chromatography-mass spectra analysis and deleterious potential of fungal based polythene-degradation products. *Scientific Reports, 9*(1), 1599.

Schwaminger, S. P., Fehn, S., Steegmüller, T., Rauwolf, S., Löwe, H., Pflüger-Grau, K., & Berensmeier, S. (2021). Immobilization of PETase enzymes on magnetic iron oxide nanoparticles for the decomposition of microplastic PET. *Nanoscale Advances, 3*(15), 4395–4399.

Schwibbert, K., Menzel, F., Epperlein, N., Bonse, J., & Krüger, J. (2019). Bacterial adhesion on femtosecond laser-modified polyethylene. *Materials, 12*(19), 3107.

Sekhar, V. C., Nampoothiri, K. M., Mohan, A. J., Nair, N. R., Bhaskar, T., & Pandey, A. (2016). Microbial degradation of high-impact polystyrene (HIPS), an e-plastic with decabromodiphenyl oxide and antimony trioxide. *Journal of Hazardous Materials, 318*, 347–354.

Shah, A. A., Hasan, F., Akhter, J. I., Hameed, A., & Ahmed, S. (2008). Degradation of polyurethane by novel bacterial consortium isolated from soil. *Annals of Microbiology, 58*, 381–386.

Shilpa, Basak, N., & Meena, S. S. (2022). Microbial biodegradation of plastics: Challenges, opportunities, and a critical perspective. *Frontiers of Environmental Science & Engineering, 16*(12), 161.

Shimpi, N., Mishra, S., & Kadam, M. (2012). Biodegradation of polystyrene (PS)-poly(lactic acid) (PLA) nanocomposites using Pseudomonas aeruginosa. *Macromolecular Research, 20*, 181–187. https://doi.org/10.1007/s13233-012-0026-1

Shukla, P. (2019). Synthetic biology perspectives of microbial enzymes and their innovative applications. *Indian Journal of Microbiology, 59*(4), 401–409.

Singh, M. J., & Sedhuraman, P. (2015). Biosurfactant, polythene, plastic, and diesel biodegradation activity of endophytic Nocardiopsis sp. mrinalini9 isolated from Hibiscus rosa-sinensis leaves. *Bioresources and Bioprocessing, 2*, 1–7.

Sukkhum, S., Tokuyama, S., Kongsaeree, P., Tamura, T., Ishida, Y., & Kitpreechavanich, V. (2011). A novel poly (L-lactide) degrading thermophilic actinomycetes, Actinomadura keratinilytica strain T16-1 and pla sequencing. *African Journal of Microbiology Research, 5*(18), 2575–2582.

Tania, M., & Anand, V. (2023). The implementation of microbes in plastic biodegradation. *Journal of Umm Al-Qura University for Applied Sciences*, 1–11. https://doi.org/10.1007/s43994-023-00077-y

Temporiti, M. E. E., Nicola, L., Nielsen, E., & Tosi, S. (2022). Fungal enzymes involved in plastics biodegradation. *Microorganisms, 10*(6), 1180.

Thakur, S., Mathur, S., Patel, S., & Paital, B. (2022). Microplastic accumulation and degradation in environment via biotechnological approaches. *Water, 14*(24), 4053.

Torres, A., Li, S. M., Roussos, S., & Vert, M. (1996). Screening of microorganisms for biodegradation of poly (lactic-acid) and lactic acid-containing polymers. *Applied and Environmental Microbiology, 62*(7), 2393–2397.

Tribedi, P., Gupta, A. D., & Sil, A. K. (2015). Adaptation of pseudomonas sp. AKS2 in biofilm on low-density polyethylene surface: An effective strategy for efficient survival and polymer degradation. *Bioresources and Bioprocessing, 2*, 1–10.

Urbanek, A. K., Kosiorowska, K. E., & Mirończuk, A. M. (2021). Current knowledge on polyethylene terephthalate degradation by genetically modified microorganisms. *Frontiers in Bioengineering and Biotechnology, 9*, 771133.

Usman, M. A., Momohjimoh, I., & Usman, A. O. (2020). Mechanical, physical and biodegradability performances of treated and untreated groundnut shell powder recycled polypropylene composites. *Materials Research Express, 7*(3), 035302.

Wang, Y., Luo, L., Li, X., Wang, J., Wang, H., Chen, C., et al. (2022). Different plastics ingestion preferences and efficiencies of superworm (Zophobas atratus Fab.) and yellow mealworm (Tenebrio molitor Linn.) associated with distinct gut microbiome changes. *Science of the Total Environment, 837*, 155719.

Ward, P. G., Goff, M., Donner, M., Kaminsky, W., & O'Connor, K. E. (2006). A two-step chemo-biotechnological conversion of polystyrene to a biodegradable thermoplastic. *Environmental Science & Technology, 40*, 2433–2437. https://doi.org/10.1021/es0517668

Wilkes, R. A., & Aristilde, L. (2017). Degradation and metabolism of synthetic plastics and associated products by Pseudomonas sp.: Capabilities and challenges. *Journal of Applied Microbiology, 123*(3), 582–593.

Yan, F., Wei, R., Cui, Q., Bornscheuer, U. T., & Liu, Y. J. (2021). Thermophilic whole-cell degradation of polyethylene terephthalate using engineered clostridium thermocellum. *Microbial Biotechnology, 14*, 374–385.

Yoon, M. G., Jeon, H. J., & Kim, M. N. (2012). Biodegradation of polyethylene by a soil bacterium and AlkB cloned recombinant cell. *Journal of Bioremediation & Biodegradation, 3*(4), 1–8.

Yoshida, S., Hiraga, K., Takehana, T., Taniguchi, I., Yamaji, H., Maeda, Y., et al. (2016). A bacterium that degrades and assimilates poly (ethylene terephthalate). *Science, 351*(6278), 1196–1199. https://doi.org/10.1126/science.aad6359

Zhu, B., Wang, D., & Wei, N. (2022). Enzyme discovery and engineering for sustainable plastic recycling. *Trends in biotechnology, 40*(1), 22–37.

Chapter 17
Plastonomics: Impact of Plastic on Ecosystem and the World Economy

Sanjay Kumar Joshi and Akshita Vashishth

Abstract The worldwide annual consumption of plastic stands at an approximate of 450 million metric tons, which is almost 57 kg per person. Out of this yearly consumption, around 350 million metric tons are generated as waste, of which an estimate of 80 million metric tons falls under the category of mishandled plastic dump becoming the cause of plastic-generated pollution or plastic pollution. This plastic pollution costs both social and environmental aspects of economies – a minimum of USD 300 billion per annum. Hence, there is a need to develop eco-friendly technologies working toward recycling methods costing much cheaper than present alternatives, increase in secondary plastic usage, and managing single-use plastic pollutions.

Keywords Ecosystem · Plastic · Economics · Waste management

17.1 Introduction

This plastic pollution costs both social and environmental aspects of economies – a minimum of USD 300 billion per annum. Data suggests that with every dollar of GDP added to the world economy, the amount of required plastic also increased considerably. An eye-opening statement was put forward by an Ellen MacArthur Foundation paper in 2016, claiming that by 2050 there would be more plastic metric tonnage in the sea than fishes. This estimate is only expected to increase in light of the current policy mandates, economic activities, and lifestyle of both developed and developing countries (Plastic pollution: Pathways to net zero, 2023). As evident from Fig. 17.1, plastic use has considerably increased from 1960 to 2020 as compared to growth in population and GDP.

S. K. Joshi (✉) · A. Vashishth
Department of Agribusiness and Rural Management, Indira Gandhi Krishi Vishwavidyalaya, Raipur, Chhattisgarh, India
e-mail: skjoshi.igkv@gmail.com

R. Soni et al. (eds.), *Advanced Strategies for Biodegradation of Plastic Polymers*, https://doi.org/10.1007/978-3-031-55661-6_17

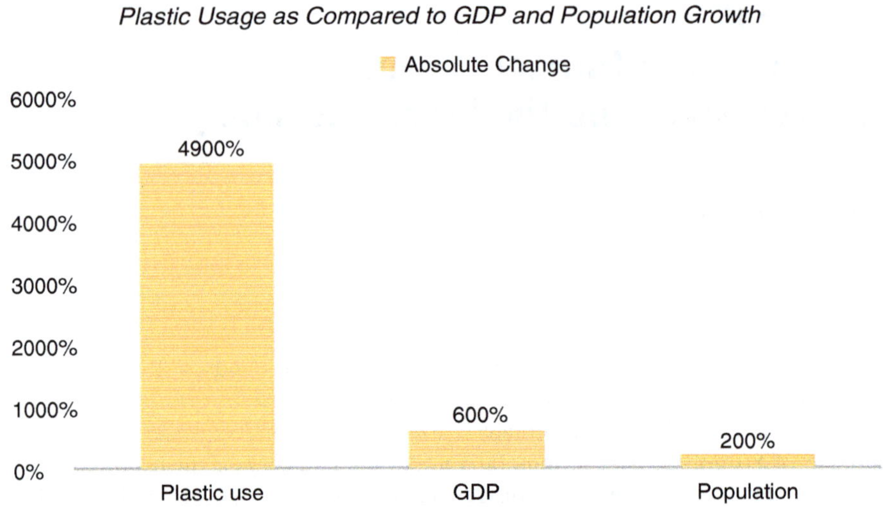

Fig. 17.1 Change in plastic usage, GDP, and population growth from 1960 to 2020

The dire situation can be understood from how developed nations have been paying developing nations like Ghana, Uganda, Tanzania, Ethiopia, Senegal, and Kenya to deal with their plastic waste; these developing nations see this waste as an added source of income for their citizens belonging to below poverty line categories. But problem arises when more often than not this waste ends up being burned or mismanaged than being recycled for further use. In 2020, around 40% of plastic waste generated in the UK was sent to Turkey, where this waste was illegally dumped and burned instead of being recycled (Wakunuma, 2021).

Considering the durability, flexibility, and application of plastic into the modern-world chase, banning the complete use of plastic is not a practical solution. Instead, plastic can be pulled out from current pit of world deprivation by way of two strategies: namely, mitigation and adaptation. Mitigation focuses on decreasing the volume and intensity of plastic waste generation, and adaptation focuses on methods to reduce mismanagement of plastic waste being generated day after day.

A noteworthy point in this direction is that although banning or charging taxes on plastic use can be an effective measure to curb public demand for plastic, this can also cause more damage than good. It has been highlighted that reverting only 1 kg of food waste to landfill has similar carbon footprint of landfilling 25,000,500 mL plastic bottles. This shows that avoiding plastic use for packaging of food can lead to greater food spoilage and hence a major damage to the environment. Hence, there is a need to develop technologies working toward recycling methods costing much cheaper than present alternatives and the increase in secondary plastic usage (Plastic pollution: Pathways to net zero, 2023).

17.2 Present Scenario for Plastic Consumption

Plastic is used in almost every aspect of modern life, including packaging, textiles, and consumer goods, which is a testament to its versatility, affordability, and durability. According to estimates (UNEP, 2023), the main plastic industry brings in between USD 600 and USD 700 billion annually.

The OECD's 2022 Global Plastics Outlook Database (OECD, 2022a) estimates that 450 million metric tons of plastic are used globally each year. More than 350 million metric tons of plastic are converted each year into plastic garbage, of which packaging accounts for 40% and consumer goods and textiles for the remaining 10–15%.

Fortunately, not all plastic garbage is improperly disposed of and turns into plastic pollution. Global plastic trash is disposed of in sanitary landfills in about 46% of cases, controlled incineration in 17% of cases, and recycling in 15% of cases. This leaves 22% of plastic garbage, or a total of about 80 million metric tons annually, which is improperly managed. Mismanaged plastic trash, often known as plastic pollution, has a variety of outcomes, some of which have a minor but nevertheless negative impact on the environment.

Figure 17.2 represents that plastic use in both OECD and non-OECD countries is at par with each other but mainly varies in their waste management practices.

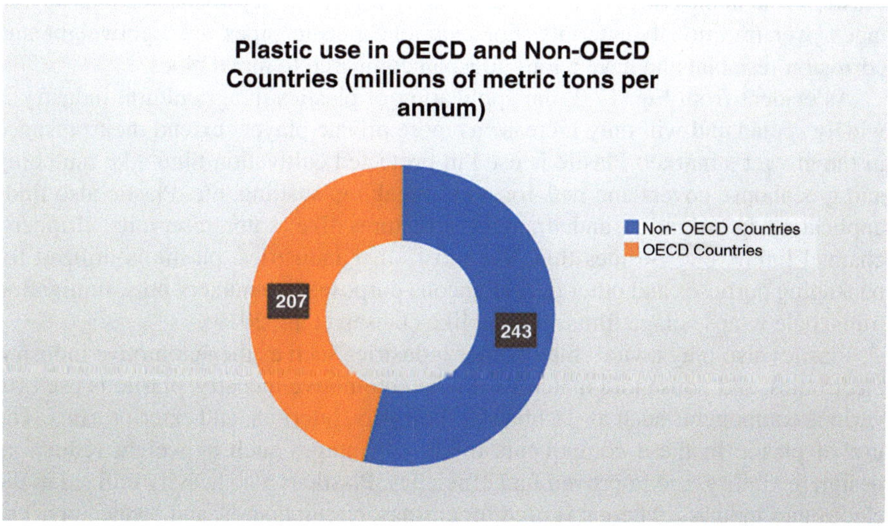

Fig. 17.2 Plastic use for OECD and non-OECD countries in millions of metric tons per annum, 2019

17.3 Varying Applications of Plastic

Plastic plays a crucial role in the global economy, making it an indispensable component in various industries and sectors. It is estimated that the packaging industry consumes approximately 36–45% of the total plastic produced worldwide, with other industries such as infrastructure and construction, textiles, transportation, electronic goods, and industrial machinery also contributing significantly to plastic production and usage (Barua et al., 2022).

The principal consumer of plastics is the packaging industry, which accounts for a significant portion of global plastic production. The use of plastic in packaging offers numerous advantages such as durability, impact resistance, flexibility, and low cost. These characteristics make plastics the material of choice for packaging, replacing traditional materials like metal, glass, and wood in many applications. Plastic's role in the packaging industry is driven by its cost-effectiveness and its beneficial physicochemical properties, such as gas barrier properties, malleability, and lightweight, incentivizing its extensive use in the packaging industry and other sectors (Microplastics in food commodities, 2022).

Moreover, plastics are also extensively used in the construction industry due to its versatility, durability, and cost-effectiveness. Plastics are commonly used in construction materials such as pipes, insulation, roofing materials, and electrical wiring. The use of plastics in construction offers numerous benefits, including improved energy efficiency, reduced maintenance costs, and enhanced design possibilities. Plastic's role in the construction industry is significant, as it provides various advantages over traditional materials. For example, plastic pipes are lightweight and corrosion-resistant and have a long lifespan compared to metal pipes.

As evident from Fig. 17.2, the application of plastics in agricultural industry is widely spread and will only increase as more private players extend their presence in the grower's market. Plastic is used in protected cultivation films like mulching and greenhouse covers and nets for wind-breaking, shading, etc. Plastic also finds application in irrigation and drainage structures like water reservoirs, drippers, channel lining, etc. Besides this, like most other industries, plastic is utilized for packaging purposes and other miscellaneous purposes like nursery pots, fumigation films, bale wraps, silage films, and the like (Jansen et al., 2019).

Plastics also play a vital role in other industries such as the automotive industry, electronics, and household materials. In the automotive industry, plastic is used for various components such as dashboards, bumpers, interiors, and exterior trims. The use of plastic in these components offers advantages such as weight reduction, design flexibility, and improved fuel efficiency. Plastic is also heavily utilized in the electronics industry, where it is used for casings, circuit boards, and connectors. The use of plastic in the electronics industry provides benefits such as insulation, heat resistance, and electrical conductivity. In addition to its use in packaging, construction, automotive, electronics, and household materials, plastic also plays a significant role in other sectors such as aerospace and medicine (Li et al., 2016). Also in agriculture, plastic is widely used for irrigation implements, mulch films, and other hardware components required to support plant growth. Beyond tools and implements, plastic-coated seeds and fertilizers are also available in the market, which pose a threat of being leached into soil and contaminate soil health as well as water bodies. The quality of food hence produced is also jeopardized due to varied applications of plastic in agriculture (Baker & Thygesen, 2022).

17.4 Poor Plastic Waste Disposal

Poor plastic waste management practices are contributing to a widespread pollution problem that affects the environment and human health. The increasing crisis of marine plastic waste management has become a global concern given its significant negative impact on humans and the environment (Ardiansyah et al., 2022).

Plastic material is one of the most serious solid waste pollutants, and its accumulation in the environment is highly resistant to biodegradation and unable to participate in substance recycling (Hashem et al., 2014). Accumulation of plastic waste in the environment causes wide-scale pollution with long-lasting effects making plastic waste management expensive and problematic.

In particular, the accumulation of plastic waste in aquatic environments is a major problem (Zimmermann, 2020). Around 69% of all plastic waste ever generated has ended up in landfills or the natural environment, making land-based plastic waste a significant contributor to marine plastic pollution (Tasseron et al., 2020). This pollution threatens various aspects of life, including the economy, ecosystem health, coastal sanitation and aesthetics, and human health.

An estimated eight million metric tons of plastic waste enters the oceans each year, leading to devastating impacts on marine life and ecosystems. Plastic waste in aquatic environments not only harms marine life but also negatively impacts human livelihood and ecosystem health (Tasseron et al., 2020). In addition, the accumulation of plastic waste has adverse effects on land environments as well (Bandara et al., 2017). This pollution not only affects wildlife and marine life but also poses a direct threat to human health.

Accumulated marine and terrestrial pollution problems, as well as the negative impacts of microplastic exposure on human health, have caused escalating public and governmental concerns. With the excessive use of plastics and increasing pressure on the capacity for plastic waste disposal, the demand for eco-friendly waste management has become of dramatic importance in the last few years (Cui et al., 2021).

Plastic waste generated per year by a person stands at an average of 221 kg in the United States, 69 kg in European OECD countries, and about 69 kg for Japan and Korea. A major part of the plastic pollution comes from improper collection and disposal of bigger plastic debris called macroplastics. But the leakage of microplastics, which are synthetic polymers less than 5 mm in size from sources like synthetic textiles, road construction, plastic pellets used in industries, and the like, poses serious issues to the ecosystem at large (Global Plastic Outlook, 2022).

Even though a number of bans and taxes exist for more than 120 countries on single-use plastics, their contribution in reducing plastic-generated pollution is not significant. Most of these regulations ban the usage and disposal of items like plastic bags which do not make a very significant part of the overall plastic waste generated in those countries. Taxes related to landfill and incineration which ultimately motivate recycling exist in only a few countries (Global Plastic Outlook, 2022).

Most plastics utilized today are virgin or, primarily, plastics derived from crude oil or gas. In order to address the pressing issue of plastic pollution and create a more sustainable future, it is crucial to enhance the market for recycled plastics. According to global data, the production of plastics from recycled sources has experienced significant growth, reaching 29.1 million tons in 2019, a fourfold increase from the year 2000. Despite this progress, recycled plastics still only account for 6% of total plastic production, indicating a significant gap that needs to be bridged. To effectively tackle the plastic crisis, it is necessary to shift the perception of recycled plastics as mere substitutes for virgin plastics and instead create a separate and well-functioning market for them. This can be achieved by setting clear targets for recycled content in products and investing in improved recycling technologies (Global Plastic Outlook, 2022).

17.5 Plastic-Generated Pollution

Plastic pollution is actually the mismanagement of plastic waste which then leaks into the environment and causes damage. An estimate of approximately 19 million metric tons of plastic waste is lost into aquatic and terrestrial environment through leakage. Plastic pollution has an estimated cost of USD 300 billion per annum, which has an upper limit of USD 1500 in some cases (Landrigan et al., 2023).

Impact on Climate

- According to estimates from the OECD (OECD, 2022b), more than 90% of plastics are produced using virgin feedstock derived from fossil fuels. It is difficult to calculate the overall greenhouse gas emissions attributable to the plastic sector. When our baseline plastic usage estimates are combined with the OECD's carbon intensity data, an annual plastic lifetime emission forecast of roughly 3.1 gigatons of CO2 equivalent is obtained for 2060.
- This level of emissions is equivalent to what would be produced annually by approximately 1500 coal-fired power plants with a capacity of 500 MW working at full capacity. India, in contrast, has more than 450 coal-fired power units (Global Energy Monitor, 2023). This demonstrates how large the carbon footprint of plastics is.

Impact on Biodiversity

- The toxicity of plastics may have an impact on biodiversity. These substances have the potential to leak into the environment, contaminating soils and streams and endangering both aquatic and terrestrial lives. Microplastics interact with and have an effect on the health of terrestrial fungus, plant pollinators, and invertebrates that live in the soil, upsetting vital ecosystems. Microplastics can build up in soil and water, affecting the availability of nutrients and microbial communities. This may have a negative effect on procedures like carbon sequestration and nutrient cycling.

- Plastics can suffocate marine life and obstruct streams. Plastic waste buildup in natural habitats can make it harder for animals to find food, shelter, and places to breed, ultimately causing habitat loss. Many creatures, including birds, marine mammals, and fish, may mistake plastics for food and absorb them, resulting in serious injury, choking, and even death. Plastics can cause species entanglement. According to research on scarring on whales from the Gulf of Maine, for instance, more than 80% of right whales and 50% of humpback whales had become tangled up in fishing gear (Knowlton et al., 2016; Robbins & Mattila, 2004).

Impact of Agriculture

- Agriculture suffers from the inappropriate disposal of plastic trash, including bags, bottles, and packaging materials. It might lead to soil pollution, which might have an effect on the productivity of agriculture. Plastics take a very long time to naturally decompose, which causes them to collect in the soil. Plant development and nutrient absorption might be hampered by plastic waste. Rivers, lakes, and oceans are commonly contaminated with plastic garbage. Runoffs, which transport polymers such as mulch films, can contaminate water. Plastic pollution in water bodies has the potential to endanger aquatic life and disturb the ecology. It might also contaminate irrigation water, which could harm crop quality and growth (Plastic pollution and agriculture, 2023).
- Plastics are widely employed in agriculture, from mulch film to seeds coated in plastic. They also end up in the biosolid fertilizer that is applied to fields. All of these items have contributed to higher agricultural yields, but there is mounting evidence that soil contamination from decomposed plastics is harming biodiversity and soil quality. Reduced production as a result may jeopardize long-term food security. Agricultural soil needs to be protected from further deterioration as a limited resource that is already under stress (Baker & Thygesen, 2022).
- Microplastics are the tiny fragments that are created over time as plastics degrade. Air, water, and even soil can be contaminated by these tiny particles. When microplastics are present in the soil, plants can absorb them and accumulate them in their tissues. The entry of microplastics into the food chain could come from this, which would have an impact on agricultural output and food safety. A buildup of plastic waste can also make it difficult for the soil to drain correctly, which can lead to waterlogging and lessen the oxygen available to plant roots. The growth of the crop's roots as well as the overall productivity of the farm may be negatively impacted by this (Plastic pollution and agriculture, 2023).
- According to research, depending on the size of the particles and the exposure level, microplastics can have varying effects (both positive and negative) on microbial communities, soil invertebrates, and soil physiochemical properties

(Yu et al., 2020; Zhu et al., 2019). Since most research is conducted in laboratories under varying settings and using different methodologies, it is challenging to compare the findings and determine the effects. Studies have shown that the presence of microplastics can affect the quantity, variety, movement, and rate of reproduction of soil biota, as well as the biomass of soil fauna and microbes (e.g., Zhu et al., 2019).

- Studies suggest that soil physiochemical properties (such as structure, water holding-capacity, density, etc.) can change when microplastics are present at levels that are significant to the environment (de Souza Machado et al., 2018). Reduced root growth and nutrient uptake are just a couple of the potential effects that modifications in soil physiochemical characteristics could have on plants (Rillig et al., 2019). Microplastics in soil undergo physical and chemical aging-related changes to their color, texture, chemical composition, surface qualities, and sorption ability (Ren et al., 2021). Some of these modifications increase the microplastics' capacity to absorb potential soil contaminants such as heavy metals and organic pollutants (Ren et al., 2021). This might reduce the availability of pollutants to soil biota (Baker & Thygesen, 2022).
- Furthermore, plastic garbage can act as a breeding ground for pests and diseases when it is left in agricultural fields. The risk of contracting a vector-borne disease may rise if water collects in used plastic bags or containers. Crop illnesses can also be caused by the presence of microorganisms that could infect plants in plastic garbage. Additionally, when the environment in agricultural areas is contaminated by plastic garbage, wildlife is at danger. Plastic products may be consumed by animals or get tangled in them, which could cause damage, choking, or even death. This upsets the ecosystem's natural equilibrium and might be harmful to agricultural biodiversity (Plastic pollution and agriculture, 2023).
- Decreasing the use of plastics in agriculture is a continuous effort, and much research is being conducted to increase the biodegradability of polymers used in agricultural goods. The majority of these are made from starch and used for biodegradable mulch, seed coatings, fertilizer coatings, etc. (Baker & Thygesen, 2022).

Impact on Society

- According to research, the levels of dangerous plastic pollution may have exceeded those that are acceptable for humans (Persson et al., 2022). When plastic pollution contaminates the food and water supplies, it poses a threat to human health. According to a study funded by the University of Newcastle, the average person could eat a credit card's worth of plastic per week or up to 5 grams (Dalberg Advisors et al., 2019).
- Human health is further harmed by air pollution brought on by open burning of plastics. Pregnant women and young children are more vulnerable to the effects of plastic exposure on reproductive, hormonal, metabolic, and brain activities. Infertility, obesity, cardiovascular disease, renal disease, preterm births, neurodevelopmental problems, and malignancies are all at higher risk due to plastic pollution (Azoulay et al., 2019).

- Plastic works as a serious danger for humans settled worldwide, especially those belonging to vulnerable communities. Plastic production requires raw material, the extraction of which leads to deforestation and human settlements being destroyed and misplaced. In line with that, plastic gives rise to water as well as air pollution leading to unhealthy surroundings snatching the rights of all living beings to a safe environment to thrive (UN Human Right, 2022).

17.6 Approaches to Curb Plastic Pollution: Adaptation and Mitigation Practices

(i) *Mitigation Practices* – It aims to control demand for plastic:

Taxation

- Taxation helps in increasing the overall cost of plastic, hence disincentivizing the manufacture as well as usage of plastic. This ultimately helps in decreasing the amount of waste generated by plastic-based products. The United Kingdom, Spain, and other such countries have been upfront with taxation-based approach to eventually control the production as well as use of plastic. The European Union has set a said amount for member countries which they would contribute based on the amount of nonrecycled plastic packaging waste produced by respective countries. The contribution is currently fixed at EUR 100 per metric ton.

Legislation

- This includes policy-based actions taken by countries to control plastic waste. For example, from October 2023, a ban was modeled to be placed in England on single-use plastic, hence curbing the amount of waste generated therefrom. Such bans on single-use cutlery, food packaging materials, shampoo bottles, etc. work well when accompanied by improvement measures aimed at developing environmentally friendly options for single-use purposes.
- In line with banning plastic use, what's usually ignored is the fact that plastic is a protective packaging material which cuts the contact of food with its environment, hence keeping it safe and healthy for delayed consumption. A study suggests that dumping 1 kg of food waste in landfill creates similar carbon footprint as that of landfilling 25,000,500 ml plastic bottles (Tylenda et al., 2022). Beside this, using options like cotton carry bags instead of their plastic counterparts can also prove to be more harmful than useful in light of the amount of water and pesticides required to grow cotton crop, covering only about 3% of the world's agricultural land.

Bioplastics

- Bioplastics are plastic materials manufactured from renewable sources of biomass, like vegetable fats and oils, recycled food waste or sawdust, etc. They work as an alternative to fossil-based plastics, hence having the capability to reduce plastic waste. Globally, bioplastics are being used at a rate of less than 0.5% at present (European Bioplastics, 2019). Although being a portable option as compared to fossil-based plastic, bioplastics are resistant to microbial degradation when placed in a natural environment. Therefore, it poses a threat to the environment if disposed of improperly, very much similar to fossil-based plastic. Bioplastics cost higher than fossil-based plastic and even require unreasonably high temperatures for composting. In line with that, bioplastics also have a competing carbon footprint similar to their fossil-based counterparts. Ethically speaking, many bioplastics require first-generation biomass, which is mostly edible, hence competing with food inflation as well as socioeconomic challenges like hunger and poverty.

Durability

- Increasing the durability of plastic can significantly decrease the need for their continuous manufacture. This can seem like a viable option, but again disposal mismanagement of durable plastic products can make it even more difficult to avoid pollution generation. This can be understood from the fact that usually the more durable is a product, the thicker will be its making, and a thicker surface will require more time to degrade (Chamas et al., 2020).

(ii) *Adaptation Practices* – In light of the advantage of plastic use and application in a number of industries, adaption practices aim at the uptake of improved waste management and disposal techniques for plastic:

Infrastructure for Waste Management

- An effective solution for getting rid of plastic waste with little to no damage to environment is technological advancement in infrastructure for waste management. This involves careful collection of waste materials, sanitary landfills, and proper incineration. It has been observed that about 90% of global improperly managed plastic waste comes from non-OECD countries (OECD, 2022a). It is further highlighted that out of all the plastic entering into oceans, 90% of that is sourced from just 10 rivers situated in Asia or Africa (Schmidt et al., 2017).
- In light of that, blocking leakage pathways can't be a much viable solution. Also, waste collection activities end up costing a lot to most municipalities, around 10–20% of their council budget for non-OECD countries (Kaza et al., 2018). The cost of mixed waste collection has been estimated to be between USD 40 and USD 86 per metric ton, and the costs are further increased when recycling is pursued for this collected waste. The cost of landfilling ranges between USD 90 and USD 150 per metric ton. Hence, it is easily demarcated that proper waste management requires high capital expenditure and can prove to be financially intrusive (Schmidt et al., 2017).

Recycling

- An estimated share of 15% annually generated plastic waste is sent for recycling, which is lower than the share of mismanaged waste (22%) (OECD, 2022a). Recycling helps in avoiding pollution caused due to leakage of plastic waste in water bodies and uncontrolled burning. Besides, recycling also results in the development of secondary plastics which reduces our dependence on primary plastics, hence decreasing the then generated carbon footprint created as a part of primary fossil-based plastic production.

- Given all of the above advantages, market for secondary plastic is highly underdeveloped and accounts for only about 6% of complete global plastic production. A major reason causing this significant shortfall in numbers is the difficulty faced in recycling of plastic products. The high inflammability of plastic, along with its toxicity and polymer diversity, causes recycling issues. Unlike metal and glass, plastics are not inert and hence can catch flame as well as release toxic substances in the environment. A study commissioned by the Canadian government highlighted that most of the plastic products and packaging materials are not well-suited for recycling in food-grade packaging (Stina Inc. and ECCC, 2021).
- New policy interventions are working toward increasing demands for recycled plastics. In the European Union, the amount of recycled plastic content in single-use plastic bottles needs to be at least 25% by 2025, which will go up to 30% by 2030 (EU, 2019). Such interventions by the government allow investments to come into recycling infrastructure and technology and will also motivate research toward the same. At present, about 40% of plastic waste collected for recycling is sent for incineration, hence unless an incentive is added toward recycling, either through legislative measures or creating an awareness for environmental safety, recycling will still be a theoretical concept being practiced only at rare occasions.

Figure 17.3 represents different categories of waste management, along with improperly managing or mismanaging plastic waste leading to pollution of both aquatic and terrestrial environment.

17.7 The Circular Economy

An urgent need is highlighted for transitioning away from single-use plastics toward a more sustainable and circular polymer economy (Nguyen et al., 2021). This transition is being led by the European Union, which is implementing legislative measures to eliminate excessive plastic waste.

The current linear economy model, which has been the basis for fossil-based plastic production, has led to significant environmental pollution (Visco et al., 2022). To address this issue, a shift toward a circular economy is necessary, where the end-of-life treatment of plastics is carefully planned. This alteration toward a circular economy involves promoting sustainable polymer technologies that decouple plastics from fossil feedstocks and increasing the quality and uptake of plastic recycling (Montava-Jorda et al., 2020). This brings forth the significance of eco-design and developing materials that are not only useful but are made with due consideration to their reuse and recyclability afterlife.

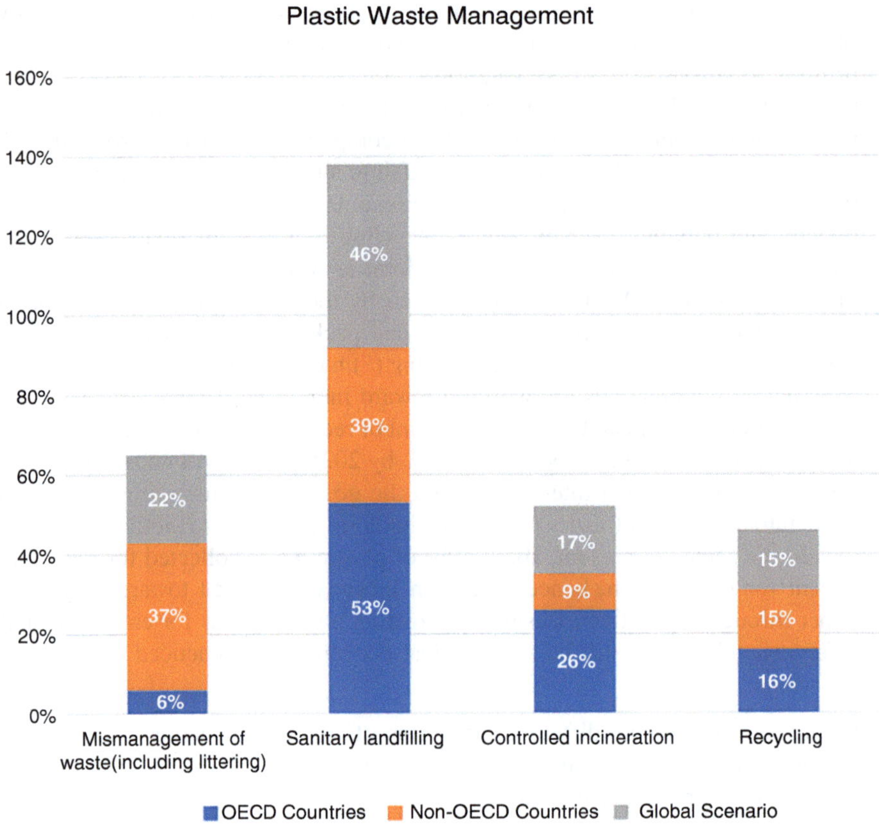

Fig. 17.3 Waste management by OECD countries, non-OECD countries and the world

The current plastic recycling methods are limiting for they primarily rely on mechanical reprocessing and often result in products of inferior quality. In order to achieve a truly sustainable circular plastics economy, concerted efforts from all stakeholders are of prime importance. This includes collaboration between policy-makers, industry players, and consumers to promote sustainable practices such as reducing plastic waste generation, improving recycling infrastructure, and developing innovative recycling technologies.

Furthermore, the concept of the "new plastics economy" is brought to international notice, which aims to reduce waste, maximize value, and use plastic efficiently (Wang et al., 2022). Great emphasis is made toward the urgent need for transitioning to a circular economy in the plastics industry and the role of sustainable polymer technologies and improved recycling methods in achieving this goal (Montava-Jorda et al., 2020). Such urgency is raised due to the heavy reliability of generations living today, as well as those arriving tomorrow on plastic, which is only expected to grow exponentially.

17.8 Future Outlooks for Plastic Consumption Worldwide

By 2060, the overall use of plastics is expected to almost triple globally in light of population increase all around the world, which would then propel economic output requiring increased plastic applications. The most affected areas are expected to be sub-Saharan Africa and Asia. With increased use of plastic, the waste would almost triple as well with only 20% or even less waste being recycled for repeated use and the rest ending up in landfills. The significant increase in plastic waste generation is depicted in Fig. 6 up till year 2060 (Fig. 17.4).

17.9 Conclusion

Primary plastics (or first-hand plastics) are foreseen to dominate the market in the coming years, but technology is expected to advance in order to introduce secondary plastics as viable use options, although covering only 12% of the total plastic share. Plastic pollution is going to become only more serious, as plastic leakage

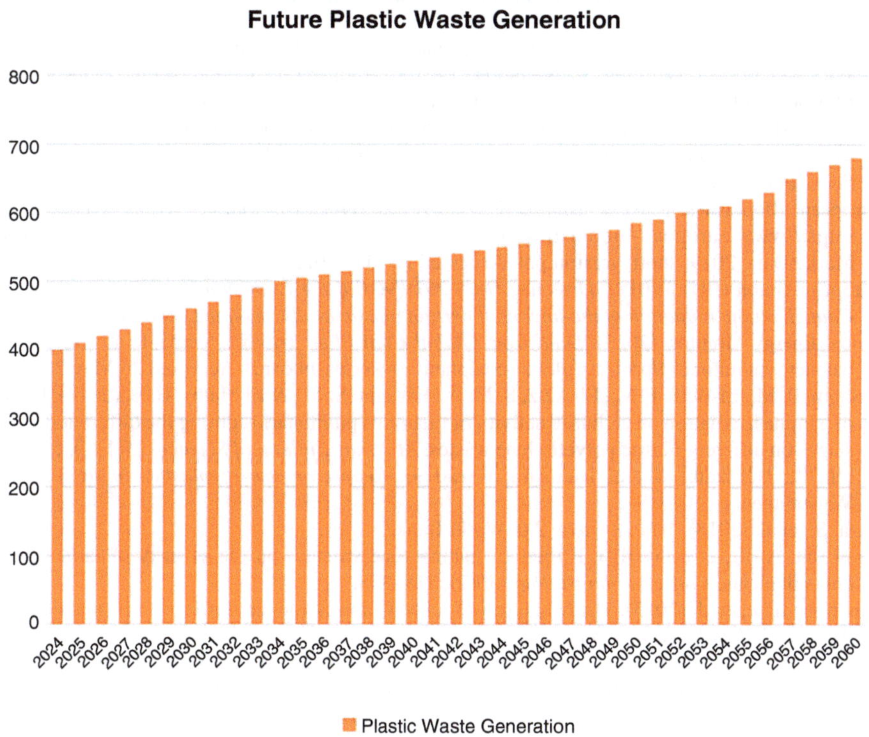

Fig. 17.4 Forecast for plastic waste generation in million metric tons per annum till the year 2060

doubles up to an estimate of 44 million tons per year giving rise to environmental damage. In line with that, the buildup of plastic residue in aquatic bodies, a severe issue for the world at present, is expected to triple its present value.

References

Advisors, D., De Wit, W., & Bigaud, N. (2019). *No plastic in nature: Assessing plastic ingestion from nature to people plastic*. World Wide Fund for Nature (WWF).

Ardiansyah, Damar, A., Hariyadi, S., & Machfud. (2022). Current source and distribution pattern of plastic waste leakage in the estuary of Jakarta Bay. *IOP Conference Series Earth and Environmental Science, 950*(1), 012057.

Azoulay, D., et al. (2019). *Plastic & health: The hidden costs of a plastic planet*. Center for International Environmental Law.

Baker, E., & Thygesen, K. (2022). *Plastics in agriculture – An environmental challenge*. UNEP Foresight. (2022). Global Plastic Outlook. OECD.

Bandara, W. R. L. N., Silva, R. M. D., Silva, K. N. D., Dahanayake, D., Gunasekara, S., & Thanabalasingam, K. (2017, January 1). *Is nano ZrO2 a better photocatalyst than nano TiO2 for degradation of plastics?*

Barua, P., Hasan, M. A., Tasnim, N., Islam, M. A., Debi, J., & Ahsan, R. (2022, December 17). *Eco-friendly entrepreneurship to promote plastic alternatives*. https://doi.org/10.32676/n.8.1.1

Chamas, A., et al. (2020). Degradation rates of plastics in the environment. *ACS Sustainable Chemistry & Engineering, 8*(9), 3494–3511.

Cui, Y., Chen, Y., Liu, X., Dong, S., Tian, Y., Qiao, Y., Mitra, R., Han, J., Li, C., Han, X., Liu, W., Chen, Q., Du, W., Tang, S., Xiang, H., Liu, H., et al. (2021). Computational redesign of a PETase for plastic biodegradation by the GRAPE strategy. *ACS Catalysis, 11*(3), 1340–1350.

de Souza Machado, A. A., Lau, C. W., Till, J., Kloas, W., Lehmann, A., Becker, R., & Rillig, M. C. (2018). Impacts of microplastics on the soil biophysical environment. *Environmental Science & Technology, 52*(17), 9656–9665.

European Bioplastics. (2019). *Bioplastics market development update 2019*. Available at: https://www.european-bioplastics.org/wp-content/uploads/2019/11/Report_Bioplastics-MarketData_2019_short_version.pdf. Accessed 26 May 2023.

European Union (EU). (2019). *Directive (EU) 2019/904 of the European Parliament and of the Council on the reduction of the impact of certain plastic products on the environment*.

Global Energy Monitor. (2023). *Data from Global Coal Plant Tracker*. Available at: https://globalenergymonitor.org/projects/global-coalplant-tracker/dashboard/. Accessed 26 May 2023.

Hashem, B. M., El-Beih, F. M., El-Aziz, S. M. A., Easanin, S. M., & El-Halby, H. E. (2014). Effect of gamma irradiation on cellulase activity of pleurotus pulmonarius during the different growth stages on different plastic polyethylene wastes. *Al-Azhar Bulletin of Science, 25*(1-C), 57–64.

Jansen, L., Henskens, M., & Hiemstra, F. (2019). *Report on use of plastics in agriculture*. Wageningen.

Kaza, S., et al. (2018). *What a waste 2.0*. World Bank.

Knowlton, A. R., et al. (2016). Effects of fishing rope strength on the severity of large whale entanglements. *Conservation Biology, 30*(2), 318–328.

Landrigan, P. J., et al. (2023). The Minderoo-Monaco Commission on Plastics and Human Health. *Annals of Global Health, 89*(1), 23.

Li, K., Lee, S. W., Yuan, G., Lei, J., Lin, S., Weerachanchai, P., Yang, Y., & Wang, J. (2016). Investigation into the catalytic activity of microporous and mesoporous catalysts in the pyrolysis of waste polyethylene and polypropylene mixture. *Energies, 9*(6), 431.

Microplastics in food commodities. (2022, October 17). https://scite.ai/reports/10.4060/cc2392en

Montava-Jorda, S., Lascano, D., Quiles-Carrillo, L., Muñoz, N. M., Boronat, T., Martinez-Sanz, A. V., Bou, S. F., & Torres-Giner, S. (2020). Mechanical recycling of partially bio-based and recycled polyethylene terephthalate blends by reactive extrusion with poly(styrene-co-glycidyl methacrylate). *Polymers, 12*(1), 174.

Nguyen, S. T., McLoughlin, E. A., Cox, J. H., Fors, B. P., & Knowles, R. R. (2021). Depolymerization of hydroxylated polymers via light-driven C–C bond cleavage. *Journal of the American Chemical Society, 143*(31), 12268–12277.

Organization for Economic Cooperation and Development (OECD). (2022a). *Global Plastics Outlook Database*. Available at: https://www.Oecdilibrary.org/environment/data/globalplasticoutlook_c0821f81-en. Accessed 26 May 2023.

Organization for Economic Cooperation and Development (OECD). (2022b). *Global Plastics Outlook: Economic drivers, environmental impacts and policy options.*

Persson, L., et al. (2022). Outside the safe operating space of the planetary boundary for novel entities. *Environmental Science & Technology, 56*(3), 1510–1521.

Plastic pollution and agriculture. (2023, June 05). *Plastic pollution: Pathways to net zero*. Credit Suisse Center for Sustainability (CfS). Retrieved from renature.co. https://www.renature.co/articles/plastic-pollution-and-agriculture/(2023)

Ren, Z., Gui, X., Xu, X., Zhao, L., Qiu, H., & Cao, X. (2021). Microplastics in the soil-groundwater environment: Aging, migration, and co-transport of contaminants – A critical review. *Journal of Hazardous Materials, 419*, 126455.

Rillig, M. C., Lehmann, A., de Souza Machado, A. A., & Yang, G. (2019). Microplastic effects on plants. *New Phytologist, 223*(3), 1066–1070.

Robbins, J., & Mattila, D. (2004). *Estimating humpback whale (Megaptera novaeangliae) entanglement rates on the basis of scar evidence: Report to the Northeast Fisheries Science Center*. National Marine Fisheries Service. Order number 43EANF030121, p. 21.

Schmidt, C., Krauth, T., & Wagner, S. (2017). Export of plastic debris by rivers into the sea. *Environmental Science & Technology, 51*(21), 12246–12253.

Stina Inc. & Environment and Climate Change Canada (ECCC). (2021). *Assessing the state of food grade recycled resin in Canada & the United States*. ECCC (Environment & Climate Change Canada). Retrieved from https://www.plasticsmarkets.org/jsfcontent/ECCC_Food_Grade_Report_Oct_2021_jsf_1.pdf

Tasseron, P., Zinsmeister, H., Rambonnet, L., Hiemstra, A., Siepman, D., & Emmerik, T. V. (2020). Plastic hotspot mapping in urban water systems. *Geosciences (Switzerland), 10*(9), 11.

Tylenda, E., et al. (2022). *The evolution towards a circular economy*. Goldman Sachs Group, Inc.

United Nations Environment Programme. (2023). *Beat plastic pollution: Global Report - World Environment Day 2023*. Available at: https://wedocs.unep.org/20.500.11822/43226.

UNEP Foresight. (2022). *Global Plastic Outlook*. OECD.

United Nations (UN) Human Rights. (2022). *UN human rights experts urge treaty to address 'plastic tide'*. Available at: https://www.ohchr.org/en/press-releases/2022/02/un-humanrights-experts-urge-treaty-address-plastic-tide. Accessed 26 May 2023.

Visco, A., Scolaro, C., Facchin, M., Brahimi, S., Belhamdi, H., Gatto, V., & Beghetto, V. (2022). Agri-food wastes for bioplastics: European prospective on possible applications in their second life for a circular economy. *Polymers, 14*(13), 2752.

Wakunuma, K. (2021, October 22). *Plastic waste is hurting women in developing countries – But there are ways to stop it*. Retrieved from The Conversation https://theconversation.com/plastic-waste-is-hurting-women-in-developing-countries-but-there-are-ways-to-stop-it-166596.

Wang, E., Chen, M., & Chen, X. (2022). Circular economy and the changing geography of international trade in plastic waste. *International Journal of Environmental Research and Public Health, 19*(22), 15020.

Yu, H., Fan, P., Hou, J., Dang, Q., Cui, D., Xi, B., & Tan, W. (2020). Inhibitory effect of microplastics on soil extracellular enzymatic activities by changing soil properties and direct adsorption: An investigation at the aggregate-fraction level. *Environmental Pollution, 267*, 115544.

Zhu, F., Zhu, C., Wang, C., & Gu, C. (2019). Occurrence and ecological impacts of microplastics in soil systems: A review. *Bulletin of Environmental Contamination and Toxicology, 102*(6), 741–749.
Zimmermann, W. (2020). Biocatalytic recycling of polyethylene terephthalate plastic. *Philosophical Transactions of the Royal Society A, 378*(2176), 20190273.

MIX
Papier aus verantwortungsvollen Quellen
Paper from responsible sources
FSC® C105338

If you have any concerns about our products,
you can contact us on
ProductSafety@springernature.com

In case Publisher is established outside the EU,
the EU authorized representative is:
Springer Nature Customer Service Center GmbH
Europaplatz 3, 69115 Heidelberg, Germany

Printed by Libri Plureos GmbH
in Hamburg, Germany